AF342386

STRUCTURAL APPLICATIONS OF MECHANICAL ALLOYING

Proceedings of an
ASM International® Conference
Myrtle Beach, South Carolina
27–29 March 1990

Edited by
F. H. Froes and J. J. deBarbadillo

Published by
ASM International®
Materials Park, Ohio 44073

Library of Congress Catalog Card Number: 90-062191
ISBN: 0-87170-390-4
SAN: 204-7586

ASM International®
Materials Park, Ohio 44073

Printed in the United States of America

ORGANIZING COMMITTEE

Dr. F. H. Froes, Co-Chairman
University of Idaho

Dr. J. J. deBarbadillo, Co-Chairman
Inco Alloys International, Inc.

Mr. R. Benn
Textron Lycoming

Mr. R. Schelling
Inco Alloys International, Inc.

Mr. A. Cox
Pratt & Whitney Aircraft

Dr. R. Schwarz
Los Alamos National Laboratories

Mr. B. Ewing
Allison Gas Turbine

Dr. C. Suryanarayana
Wright R&D Center

Mr. A. Johnson
General Electric

Mr. D. V. Varanese
ASM International®

Professor C. Koch
North Carolina State University

Dr. J. D. Whittenberger
NASA-Lewis Research Center

INTERNATIONAL LIAISON COMMITTEE

Dr. E. Arzt, FRG
Dr. P. Bridges, UK
Porf. D. Eliezer, Israel
Dr. M. Gutierrez, Spain
Dr. J-P. Hertemann, France
Dr. G. Jangg, Austria
Dr. S. Isobe, Japan
Dr. J. Marriott, CEC
Prof. D. Morris, Switzerland

Prof. P. Ramakrishnan, India
Dr. P. Rama Rao, India
Prof. O. V. Roman, USSR
Prof. P. Shingu, Japan
Dr. R. Singer, Switzerland
Dr. R. Sundaresan India
Dr. W. J. McG. Tegart, Australia
Prof. J. Wood, UK

FOREWORD

The mechanical alloying process was conceived in 1966 as a method for producing an oxide dispersion strengthened superalloy. In the ensuing years the development of the ODS alloys has evolved to the point where they are well established in aerospace and advanced industrial applications. Since the first technical paper in 1970 there have been literally hundreds of publications on the process and materials properties. However, only recently with the worldwide interest in advanced materials have the unique capabilities of the MA process become appreciated by the general metallurgical community. This renewed interest has inspired three international conferences in recent years starting with the DGM conference on "New Materials by Mechanical Alloying Techniques" in Calw-Hirsau West Germany in 1988, and the TMS conference on "Soilid State Powder Processing" in Indianapolis in 1989. The present conference which continued the theme was held as a topical conference under sponsorship of ASM International in Myrtle Beach, South Carolina on 26-29 March 1990.

The conference brought together 94 participants from 15 countries. Despite, or perhaps because of, the incomplete recovery of the region from Hurricane Hugo and the uncooperative weather, the conference facilitated an excellent exchange of information and views. In particular, the participation of the delegation from the Soviet Union afforded an initial insight into their activities in mechanical alloying.

The conference offered a unique blend of commercial users, systems and components fabricators, material suppliers, academics, and technology transfer specialists. The presentations ranged from preliminary observations on samples produced in gram lots to characterization of commercial materials with piece weight exceeding a ton. Perhaps most important, the conference fostered a dialogue between the alloy designers and application engineers.

The chairmen wish to thank ASM staff members Don Varanese and especially Lisa Hemeyer whose planning, patience and diligence enabled the conference to proceed smoothly. Also the help of the session chairmen and the international advisors is sincerely appreciated. In particular we wish to acknowledge the efforts of Dr. Tetsuo Kato of Daido Steel Co. whose advance promotional efforts produced a substantial participation from Japan. Finally, we would like to thank Inco Alloys International, Inc. and Elsevier Sequoia S.A. for financial support of conference functions.

Current plans are to hold conferences on mechanical alloying in Japan in 1991 (Dr. P.H. Shingu, Kyoto University, Chairman), in Europe in 1992 (Prof. E. Artz, Max Planck Institute, Stuttgart, Chairman), and back to the USA in 1993 (Drs. J.J. deBarbadillo, F.H. Froes and R. Schwarz, Chairmen).

<table>
<tr><td>Dr. F.H. Froes, Co-Chairman
University of Idaho
Moscow, ID</td><td>Dr. J.J. deBarbadillo, Co-Chairman
Inco Alloys International, Inc.
Huntington, WV</td></tr>
</table>

TABLE OF CONTENTS

HIGH TEMPERATURE ALUMINUM ALLOYS

LIGHT WEIGHT MATERIALS

OTHER MECHANICAL ALLOYING DEVELOPMENTS

DEVELOPMENT, TECHNOLOGY TRANSFER, AND APPLICATION OF ADVANCED AEROSPACE STRUCTURAL MATERIALS

F. H. Froes
Institute for Materials and Advanced Processes
University of Idaho
Moscow, Idaho 83843, USA

J. J. deBarbadillo
Inco Alloys International, Inc.
Huntington, West Virginia 25720, USA

C. Suryanarayana
WRDC/MLLS
WPAFB, Ohio 45433, USA

ABSTRACT

The increased performance requirements of structures and devices necessitates new and improved advanced materials and processes. New aerospace system requirements are discussed, and the necessary steps required to introduce new materials and processes into these systems presented. The state of maturity of one new process, mechanical alloying, the subject of the present conference, is discussed in detail.

RECENTLY IT WAS POINTED OUT that the three most important industries in driving technological change, national security considerations, and economic advances into the next century are information/communications systems (computers), biotechnology, and advanced materials and processes (1). Further, of these three, advanced materials and processes are the most critical and considered vital to advancements in the other two fields, hence the major emphasis being given to advanced materials and processes in many parts of the world.

Advanced materials may be defined as materials which have enhanced mechanical and physical characteristics compared to traditional materials, such as aluminum and steel, currently manufactured in large-volume assembly line type facilities. The characteristics either allow for very significant improvements in product or device performance or, of even greater significance, allow for new technologies that are not achievable using conventional materials (2). Examples of the latter are the high specific strength/specific stiffness of the polymeric composites which make possible the forward swept wing X-29, and the elevated temperature/low density behavior of the titanium aluminides slated for use in the National Aerospace Plane and advanced gas turbine concepts.

ADVANCED AEROSPACE MATERIALS

Enhanced structural materials are vital to aerospace system advances in the twenty-first century. However, the extreme demands of these aerospace systems will require new materials which are "stronger, stiffer, hotter and lighter" than traditional materials of construction. Additionally, the materials will be "tailored" to have the properties required for a given application by use of composite concepts (3-17). Cost can be a major concern, but an integrated design, manufacturing, and use approach can lead to cost effective application compared to conventional material use. These new materials must be considered as structures rather than in the same way as traditional materials such as metals.

The aerospace market is a "niche" market in which advanced structural materials tend to be technology driven, with the major emphasis on performance rather than cost, in contrast to other industries (Figure 1) (1), and cost strongly dictates sales volume (Figure 2). From the producer's perspective, volume sales require use in industries such as the automotive arena, but this will require low cost fabrication techniques to be developed, contrasting with the low-volume aerospace industry.

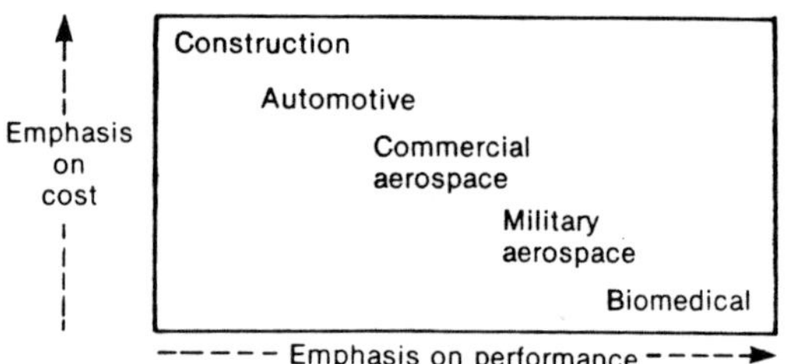

Barriers to the use of advanced materials decrease from upper left to lower right.

Figure 1: Relative Importance of Cost and Performance in Advanced Materials User Industries (1).

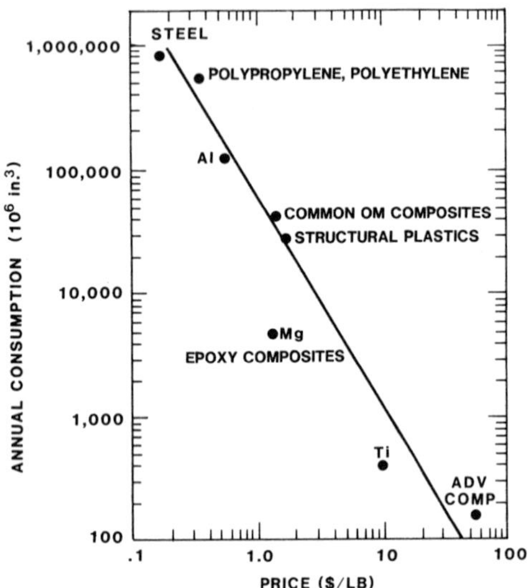

Figure 2: Influence of Price on Amount of Material Consumed.

The attraction of advanced materials is that they allow an enhanced materials behavior trend band to be achieved. That is to move from the trend band exhibited by the "basic" materials of today to an improved trend band of the "tailored" or "engineered" advanced materials of tomorrow by innovative chemistries, processes and microstructures (Figure 3) (10). However, in moving to this new trend band, it is necessary to balance the so-called "unconstrained" characteristics against the "constraining" characteristics. That is parameters such as strength, stiffness and temperature capability cannot be increased without

consideration for properties such as ductility and forgiveness (fracture toughness, fatigue crack growth rate, etc.). A feature of such advanced processes is that they involve large excursions from equilibrium in what has been termed "far from equilibrium" processing (Figure 4) (11). One such innovative process is that of Mechanical Alloying (MA), which leads to novel microstructures and improved mechanical properties in a variety of metal systems.

HISTORY OF AEROSPACE MATERIALS AND PERFORMANCE

The improvements which have been made in materials strength-to-density ratio over the years is shown in Figure 5 (18). The advanced materials of today exhibit a 50x enhancement compared to naturally occurring (primitive) materials.

In an attempt to look into the future of aerospace materials and performance a brief history of the materials of aerospace construction from the Wright brothers to the present day will first be considered, since it has been stated that "to look forward it is often helpful first to look back. A perspective requires a long view." (19) The Wright brothers of the turn of the century used the materials which had been previously used successfully to make gliders and kites: wood, wire and fabric (20). Wood was gradually replaced by steel tubes between 1910 and 1925. The first all-metal monoplane, which had wings made from pure aluminum, was designed by Hans Reissner in Germany and flew in 1912. In the early 1930's riveted stiffened stressed skin was first used for both military and commercial aircraft. Since this time aluminum alloys have been the predominant material of construction for the airframe of all types of aircraft.

Titanium emerged as a material of construction in the early mid-1950's, first for jet engines and later for airframes, especially for elevated temperature applications. Superalloys have played a major role in enhancing engine performance with improved melting practices (vacuum melting) and controlled solidification playing key roles.

The maximum speed capability of fighter aircraft is one indication of the performance enhancement to be gained by material performance

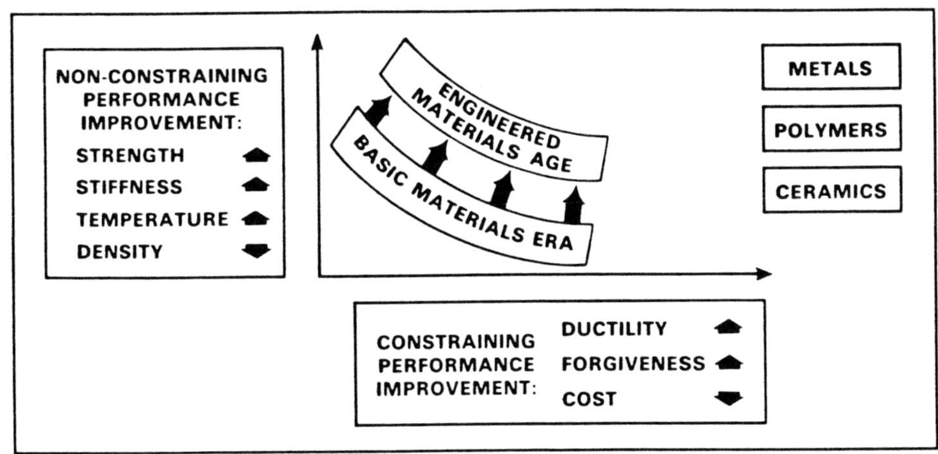

Figure 3: Trend Band Exhibited by Basic Materials, and Enhanced Trend Band Characteristic of Engineered Materials (10,11).

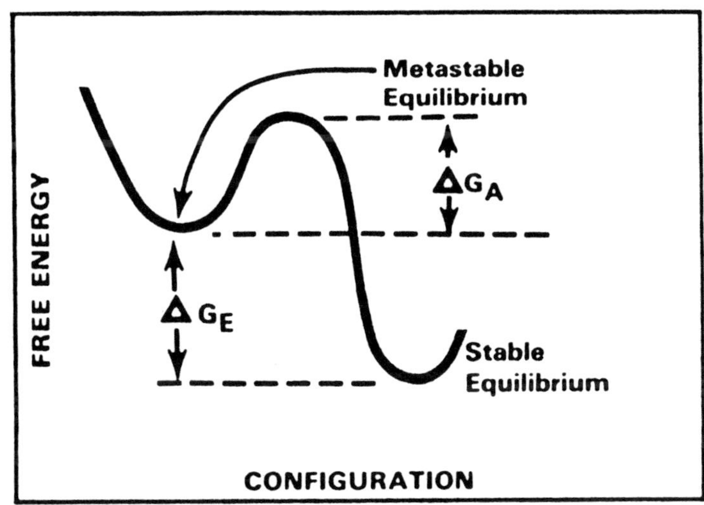

Figure 4: Schematic of "Far From Equilibrium" Processing. As the Free Energy Difference Between the Metastable and Equilibrium States (ΔG_E) Increases the Opportunity to Attain Novel Microstructures During Subsequent Relaxation Increases. These Novel Microstructures can Exhibit Enhanced Mechanical and Physical Properties (10,11).

improvement (20). Operational fighters almost doubled in maximum speed between 1934 and 1943; and by 1959, with the introduction of the F-106A which had a maximum speed of 1525 mph (Mach 2.3) at 40,000 ft. maximum speeds had increased by a factor of 6.5 times in a period of 25 years. A leveling off then occurred in large part because structural heating to 175°C meant that construction metals other than aluminum, such as titanium, were required for further advances; for example, the Mach 3+ SR71 surveillance aircraft.

Improved materials have also contributed significantly to improved engine performance; for example, the thrust-to-weight of the J79 turbojet which powered the F-4 in 1958 was only half that of the current F404 turbofan engine which powers the F18 (20). Efficiency depends directly on temperature, with present day engines limited to about 60% (18), whereas the maximum theoretical efficiency is about 80%. The dramatic increase in operating temperature of engines during this century made possible in large part by advances in materials is shown in Figure 6 (18).

Advanced organic matrix composites began to appear as viable structural materials by the late 1960's, first fiberglass/epoxy and later boron and graphite (carbon) and aramid (kevlar) reinforced composites (20). A density 45% less than aluminum combined with high strength and stiffness provides dramatic structural efficiency indices

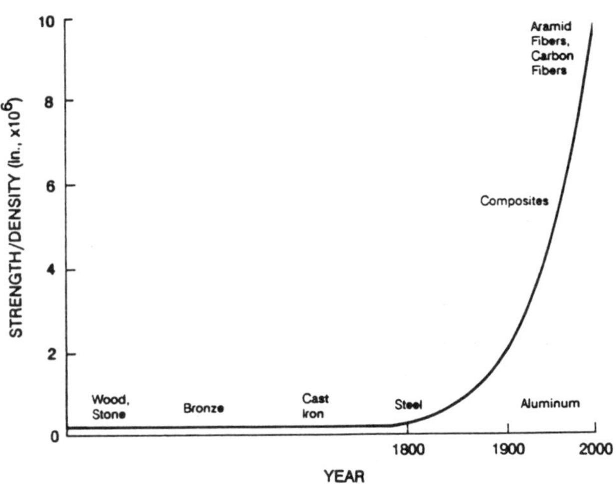

Figure 5: Progress in materials strength-to-density ratio as a function of time, showing a 50-fold increase in the strength of today's advanced materials compared to that of primitive materials (18).

and therefore, a great incentive to develop cost effective processing.

Since 1930, the specific tensile strengths of aerospace materials have increased by a factor of 3. The structural weight has changed little over the last 45 years (40% of the take-off gross weight of WWII fighters, 33% today) (20).

The materials menu for construction of both airframe and engine components in the 1980's expanded considerably (3-17). Now we have higher temperature composites, metals and metal matrix composites. There are also ordered intermetallic compounds, refractory metals, structural ceramics and carbon/carbon. An increased number of choices compared to prior years, but a menu which will become even more extensive as we move into the twenty-first century.

AEROSPACE SYSTEMS OF THE TWENTY-FIRST CENTURY

Twenty-first century systems will be more maneuverable, including very short take-off and landing (V/STOL) concepts to reduce dependence on large and fixed operating bases, and will also have increased speed capability. They will include aircraft with greatly increased range and improved fuel consumption over current systems. Payload will be increased while building-in varying degrees of stealth, instantly reactive knowledge-based defensive and offensive systems, and increased confidence (decreased inspection, increased reliability and

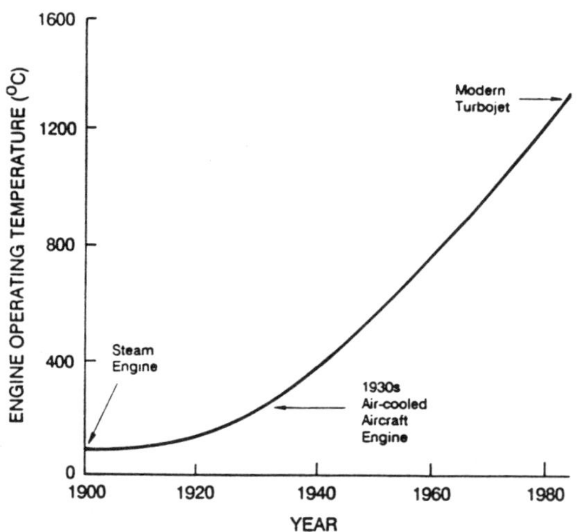

Figure 6: The steep climb in operating temperatures of engines during this century·made possible by modern materials (18).

decreased maintainability). Access to space in an affordable and predictable manner will be required, which greatly increases turn-around times and much larger payloads.

Among the systems under various stages of development are supersonic V/STOL concepts, air superiority fighters featuring agility with thrust vectoring, high Mach number interceptors and bombers, advanced helicopter, including tilt-rotor concepts, light-weight fighters and high speed civil transports (Mach 2-5) which combine low noise and emission with fuel efficiency. In the space arena major emphasis is on development of single-stage to orbit hypersonic flight vehicles (21). These include programs in the US, Russia, Britain, France, Germany, and Japan. A generic USAF transatmospheric vehicle of the next century is shown in Figure 7. There is likely to also be a trend towards unmanned autonomous systems.

The engines for the advanced systems will include low bypass turbofan power units at speeds up to Mach 2 and turbojet engines with afterburners to speeds above Mach 3 (Figure 8). Hypersonic aircraft will require multiple-mode transatmospheric cruise, and conventional landing. This could include take-off and landing with conventional turbine engines, acceleration to Mach 6-12 with a scramjet (supersonic combustion ramjet), and final attainment of orbital velocity with a rocket engine (22). Even more advanced concepts include use of antiproton, directed energy, and nuclear propulsion systems.

Figure 7: Generic USAF Transatmospheric Vehicle of the Next Century.

In the USA the Integrated High Performance Turbine Engine Technology (IHPTET) program is attempting to double the present thrust-to-weight ratio of the engine for fighter aircraft of the early next century while decreasing the fuel consumption to 50% (23). Basically, this means hotter running engines, more highly loaded components, and a decreased number of components in the engine. The material challenge involved is demonstrated by considering the limited capability of nickel-based superalloys to continue to meet increasing turbine inlet temperature (TIT), particularly avoiding internal cooling air; as much as 20% of the compressor delivery air is used for cooling turbine components in some current jet engines, elimination of which could require a $500^{\circ}C$ enhancement in material capabilities (24).

INTRODUCTION OF NEW MATERIALS

For the introduction of new materials or processes into production, four conditions must be satisfied: need, opportunity, maturity (availability) and cost (20). The need relates to the inability of current technology to achieve vehicle performance or cost goals. The opportunity for introduction can be negated if schedule and implementation cost are excessively high; and is generally much better for a new system than an existing system. However, severe problems such as excessive corrosion can even lead to introduction of new material/process in an existing system. Maturity relates to the risks in terms of "newness" of the new product. Cost can never be ignored, although in certain circumstances the design engineer may be willing to go to a very high figure if increased performance warrants the expense.

An interesting tabulation of the extended gestation period for a number of techniques is shown in Table I (25).

Generally, the fly-away price/pound of aircraft airframes increase in the order: light aircraft, medium singles and twins, large transports, business jets and turboprops, and the most costly fighter and attack aircraft (26). Thus it is much easier to justify use of a more expensive material in the latter categories, very difficult for light aircraft.

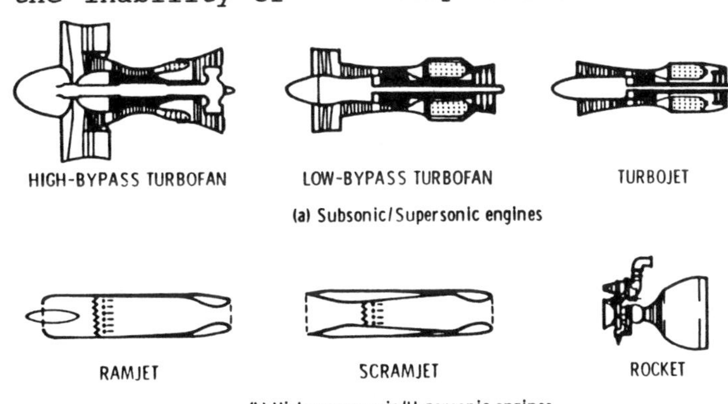

Figure 8: Types of Aircraft Engines Used for Different Flight Envelopes (22).

Table I: Gestation Period to go from Concept to Commercialization (Technology Transfer) (25).

ITEM	CONCEPT	COMMERCIALIZATION	TIME GAP
FLUORESCENT LIGHT	1852	1934	82 YEARS
BALL POINT PEN	1888	1938	50 YEARS
HELICOPTER	1904	1936	32 YEARS
TV	1907	1936	29 YEARS
TRANSISTOR	1940	1950	10 YEARS
ZIPPER	1891	1923	32 YEARS
RADAR	1887	1933	46 YEARS
DIESEL LOCOMOTIVE	1895	1934	39 YEARS

This means that while it is highly likely that large percentages of expensive organic composite materials will be used in the next generation fighter, it is equally unlikely that they will be seen in light aircraft in the near future (20,26).

The introduction of new materials has not been without problems (20). When the age-hardenable aluminum alloy, Duralumin, was first introduced in 1914, major quality problems occurred including exfoliation, embrittlement and disintegration in spots to a white powder. Humidity problems (with organic composites) are not a new problem. In 1917 humidity in association with poor quality control caused debonding of the wooden outboard tip rib on the upper wing resulting in loss of the aileron (20). Fatigue has been a concern over the years but this did not prevent the Comet crashes of the mid 1950's. More recently the F-111 was grounded in 1970 because of fatigue problems associated with the forged steel wing pivot. Currently fatigue testing accounts for major portion of the qualification cost of a new material; metal or organic composite (20).

For a new technology to reach full maturity it takes between 10 and 30 years; that is to a stage where it can be used with "zero" risk (20). Parts of this gestation period are involved with initial basic research, applied research, advanced development, adequate data gathering, scale-up and validation of structural design concepts. However, this is then followed by the go-ahead on a new system which can take 6-8 years to reach Initial Operating Capability (IOC). An interesting example of the maturation process is shown in Figure 9 for the case of polymeric composite (20,26). Even after all of the stages described above had been accomplished it still took 10 years to go from demonstration flying parts to "bill-of-material" components fabricated from this material. Just one example of the time and dollars required to bring a new material to the market place is shown in Figure 10 (27).

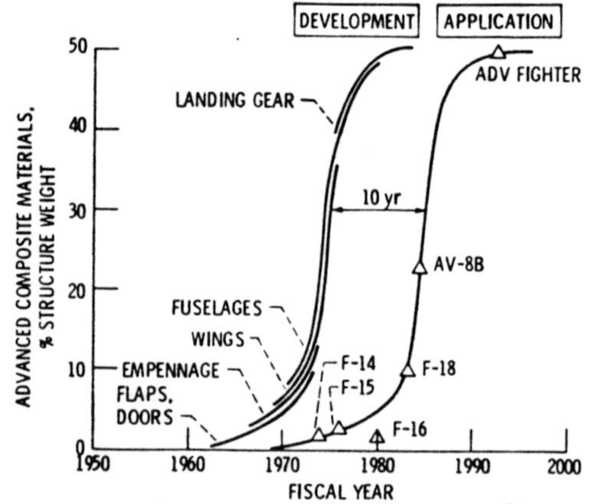

Figure 9: Polymer Matrix Composite Use in Airframes: Development and Application (20,26).

A comparison of the maturity of various types of structural materials has shown that commodity metals have passed their demand peak and usage is now declining, Figure 11 (28). In contrast the advanced materials are still in the mode of heavy investment in research and development.

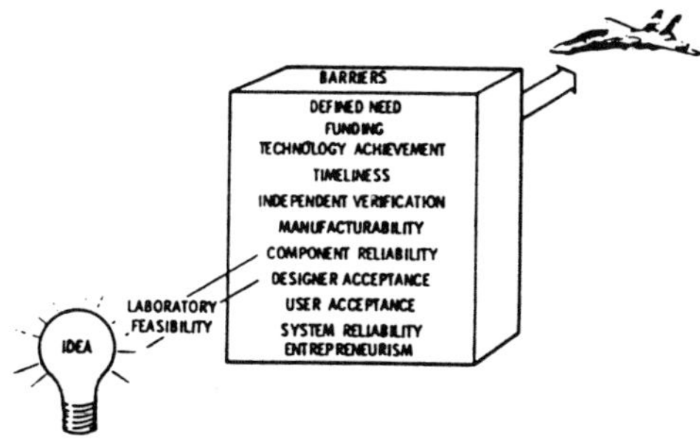

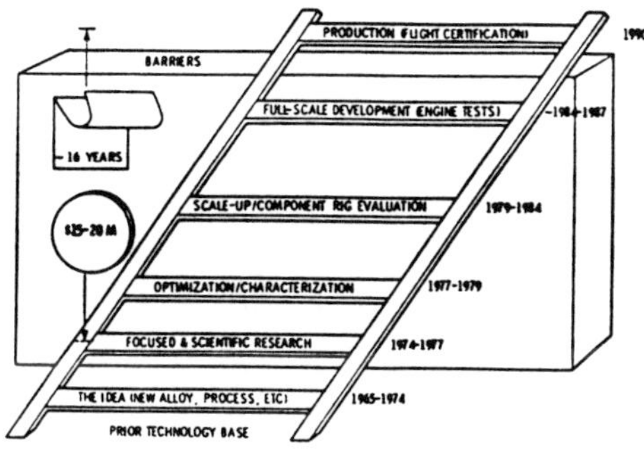

Figure 10: Technology Transfer Ladder to overcome Barriers to Introduction of a New Material into Aircraft Engines, using the Oxide Dispersion Strengthened Mechanically Alloyed MA6000 as an Example (27).

DEVELOPING METALS AND ALLOYS

In general to produce new and improved metals the trend will be away from traditional cast or cast and wrought (ingot) techniques toward nonequilibrium methods such as rapid solidification (RS) from the melt (either via a powder or directly by a spray method) or directly from the vapor phase, mechanical alloying, thermochemical processing (that is use of hydrogen as a temporary alloying element in systems such as titanium), nanostructures (structures with a dimension in the range of 10^{-9} - 10^{-7}m) all of which offer greatly increased flexibility in microstructure development and therefore, also in mechanical behavior (3-17).

Monolithic metals based on the light metals aluminum, titanium and magnesium produced by ingot, casting and powder metallurgy/rapid solidification are not standing still with the development of low density Al-Li alloys, elevated temperature, and high strength aluminum alloys (3-17). Other advances for the future include corrosion resistant magnesium alloys produced using ingot or RS methods, and even further reductions in the density of RS aluminum alloys utilizing the Al-Li-Be (3-17) system. Elevated temperature titanium alloys

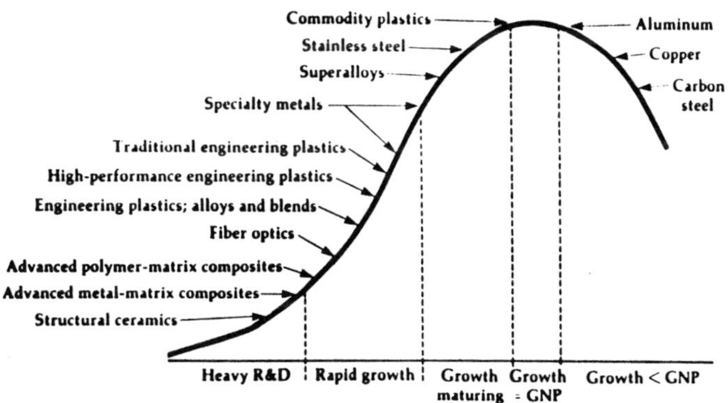

Figure 11: Materials Market Maturity Curve (28).

The maturation process has been formalized by the USAF in what are called Structural Integrity Programs (SIP), ENSIP for engine materials, ASIP for airframe materials (29). This requires new materials to achieve certain predictable developmental milestones in an orderly progression from the laboratory to use in full-scale production hardware. Thus when we are dealing with emerging materials today we are in essence dealing with the material of use for the year 2000 and beyond.

(to 600°C), higher strength (to 1700 MPa), and lower density (approaching the density of Al) titanium alloys could also be in use by the turn of the century. Perhaps mundane, but titanium castings are also likely to see increased use, particularly if thin sections can be successfully cast, and from advanced alloys. For future construction advanced aluminum alloys (possibly mechanically alloyed or intermetallics) and ordered intermetallics based on Ti_xAl (X=1 or 3) are also on the menu.

It has been suggested that superalloys are likely to be replaced in gas turbine engines by ceramic composites, intermetallic compounds, refractory metals or carbon/carbon composites in the near future. However, this euphoria must be tempered with some realism. Replacement of the superalloys in the hot section of the engine is only likely to be implemented in a few specialized applications by the year 2000. The advances which are occurring with superalloys include: single-crystal blades, intricate cooling passage concepts, directionally-solidified eutectics, improved coatings, and mechanically alloyed materials. At the same time, reliability is being enhanced by "building in quality" with reduced defect, fine-grained products (24). For example, a fine-grained PM René 95 billet can be sonic inspected at nine times greater sensitivity than a premium cast and wrought product.

In addition to the titanium aluminides, the nickel aluminide intermetallics (specifically Ni_3Al) offer lower density than conventional superalloys. Here potential applications include replacement of single crystal alloys in airfoil blades (23). The 33% density reduction translates to a blade made from the intermetallic weighing less than half that of the superalloy concept. In turn this leads to a 35 kg rotor system compared to a system more than 2 1/2 times heavier when fabricated from single crystal superalloy blades. At a very early stage of development are the beryllium intermetallics which offer the potential for high temperature strength combined with low density and oxidation resistance.

Some have summarily dismissed the refractory metals (melting points 2500-3000°C) based on what is considered to be an almost impossible situation in terms of enhancing their poor high temperature oxidation resistance (24). However, recent work indicates that judicious alloying using RS techniques may allow the necessary protective oxide to form (3-17). Coatings offer a much less desirable option. Further, the high density, up to twice that of superalloys (24), is a negative feature of the refractory metals.

Tailoring of surface regions using techniques such as powder metallurgy, chemical vapor deposition, ion implantation and reaction magnetron sputtering could lead to wear, fatigue and corrosion resistance enhancement. Another form of tailoring, which is already being evaluated, is the tailoring of the microstructure of disc hub and blade regions to meet fatigue and creep domination requirements, respectively.

MARKET ACTIVITY AND THE ADVANCED STRUCTURAL MATERIALS DILEMMA

Generally, the developers of the advanced metals are the traditional metal producers, such as Alcoa, Alcan and International Nickel. Thus the infrastructure to market these materials to the aerospace industry is in place. Currently there is an over-capacity of metals such as aluminum and steel both internally in the USA and world-wide. The threat of new material substitution has resulted in efforts to increase performance and decrease cost of the basic metals. Examples are the reduced density aluminum-lithium ingot alloys, and increased recycling of revert (scrap).

At the present time aluminum use in the aerospace industry represents slightly less than 3% of the total aluminum output of the USA; but almost 9% of the profit. A major factor in determining whether aluminum use will decrease is the cost of fuel. If fuel prices do not increase over the next 10 years, polymeric matrix composite replacement of aluminum could be relatively small (a factor of 2 x the 1986 price of $.55 per US gallon has been suggested as necessary for wide-spread replacement) (1). However, traditional aluminum use is likely to decrease in use by as much as 70% (1).

The aerospace industry is compared with seven other industries in Table II (18). In 1987 more than 800,000 workers were employed by the builders themselves, a number which doubled when supplier companies were also included.

Total aerospace sales exceeded $100 B, and extremely importantly the industry has an extremely healthy balance of trade, in excess of $15 B in 1987 (Table III).

The critical role played by advanced materials and processes in the aerospace industry (and for the other seven industries shown in Tables II and III) is shown in Table IV (18). At the present time the USA is in an overall leadership role, but extremely intensive foreign competition has developed. And many other nations have targeted advanced materials and

Table II: Economic Impact of Eight Industries (18).

Industry	1987 Employment[a] (thousands)	1987 Sales ($ billion)
Aerospace	835	105.6
Automotive	963	222.7
Biomaterials	—	>50
Chemical	1004	195.2
Electronics	1394	155.4
Energy	1229	375.8
Metals	629 (1230)[b]	98.9
Telecommunications	1007	146.0

[a]The statistics are taken from the *U.S. Industrial Outlook 1989*, published by the Department of Commerce, *International Trade Administration*, Washington, D.C.

[b]The 1980 to 1985 average based on a broader definition of the metals and mining industry used in *Employment Prospects for 1995*, Bulletin 2197 published by the Bureau of Labor Statistics, Washington, D.C. (1984).

Table III: International Trade Balances for Seven Industries (Billions of Dollars) (18).

Industries	1982	1984	1985	1986	1987
Aerospace	+ 11.1	+ 10.2	+ 12.3	+ 11.7	+ 15.1
Automotive	− 10.4	− 20.7	− 26.5	− 35.8	− 42.4
Chemical	+ 12.4	+ 10.7	+ 8.5	+ 8.5	+ 9.3
Electronics	+ 6.7	+ 2.5	+ 2.6	+ 0.6	− 0.1
Energy	− 53.3	− 52.7	− 44.2	− 30.8	− 38.3
Metals	− 9.5	− 12.9	− 11.6	− 9.6	− 10.8
Telecommunications	+ 0.2	− 1.0	− 1.2	− 1.3	− 1.7

SOURCE: Data are abstracted from *U.S. Industrial Outlook 1989*, published by the International Trade Administration, Department of Commerce, Washington, D.C.

Table IV: Material Needs of Eight Industries

Desired Characteristic	Aero.	Auto.	Bio.	Chem.	Elec.	Energy	Metals	Telecom.
Light/strong	✓	✓	✓					
High temperature resistance	✓			✓		✓	✓	
Corrosion resistance	✓	✓	✓	✓		✓	✓	
Rapid switching					✓	✓		✓
Efficient processing	✓	✓	✓	✓	✓	✓	✓	✓
Near-net-shape forming	✓	✓	✓	✓	✓	✓	✓	✓
Material recycling		✓		✓			✓	
Prediction of service life	✓	✓	✓	✓	✓	✓	✓	✓
Prediction of physical properties	✓	✓	✓	✓	✓	✓	✓	✓
Materials data bases	✓	✓	✓	✓	✓	✓	✓	✓

processes for emphasis in research and development leading to early application.

The dilemma is that aerospace is a small niche market in which advanced structural materials tend to be technology driven, with the major emphasis being on performance rather than cost. In other industries, such as construction and automobiles, cost strongly dictates sales volume (Figure 1) (1). In military aerospace and biomedical applications, performance is the overriding consideration. The commercial aerospace market falls in an intermediate position, with both cost and performance being of concern.

The producer must decide whether to invest in an initially small advanced materials market, where large volume production may be many years away. The planning horizon for most US-based companies covers a maximum of five years--a shorter time than is required to make the initially exotic tailored materials profitable. Adding to this problem are the export controls which restrict the exploitation of advanced military high-technology information in commercial applications. And the major potential market for advanced materials lies in commercial industries, rather than the low-volume military sector.

The dilemma, then, is whether to get into the initially small niche market, with the prospect of a slow growth into large volume markets, such as construction and automotive, or to stay with the tried and true basic materials industries where large volume already translates to low cost.

Hopefully, US industry will not make the wrong decision and give the future of advanced materials to other countries where near-term profits are given up in order to develop the production experience necessary to establish future markets. Without a forward-minded approach, US companies may find it very difficult to play catch-up in the future. The government can help develop the appropriate environment for this approach by providing tax incentives for long-term capital investments and by relaxing current product liability laws.

HISTORY OF MECHANICAL ALLOYING

A chronology of the development of mechanical alloying is shown in Figure 12.

Exploratory research which led to the invention of Mechanical Alloying was initiated by John Benjamin in 1966. His objective was to produce a material which combined the very high strength at low and intermediate temperatures ($600°C$ to $1000°C$) provided by γ' age hardening in nickel base alloys with high temperature ($1000°C$ to $1300°C$) strength conferred by a fine stable dispersion of oxide particles. A practical application was a gas turbine engine blade which encounters a wide range of temperatures during service.

As is true for most engineering advancements, there was a substantial legacy of structural and property knowledge to build upon. Age

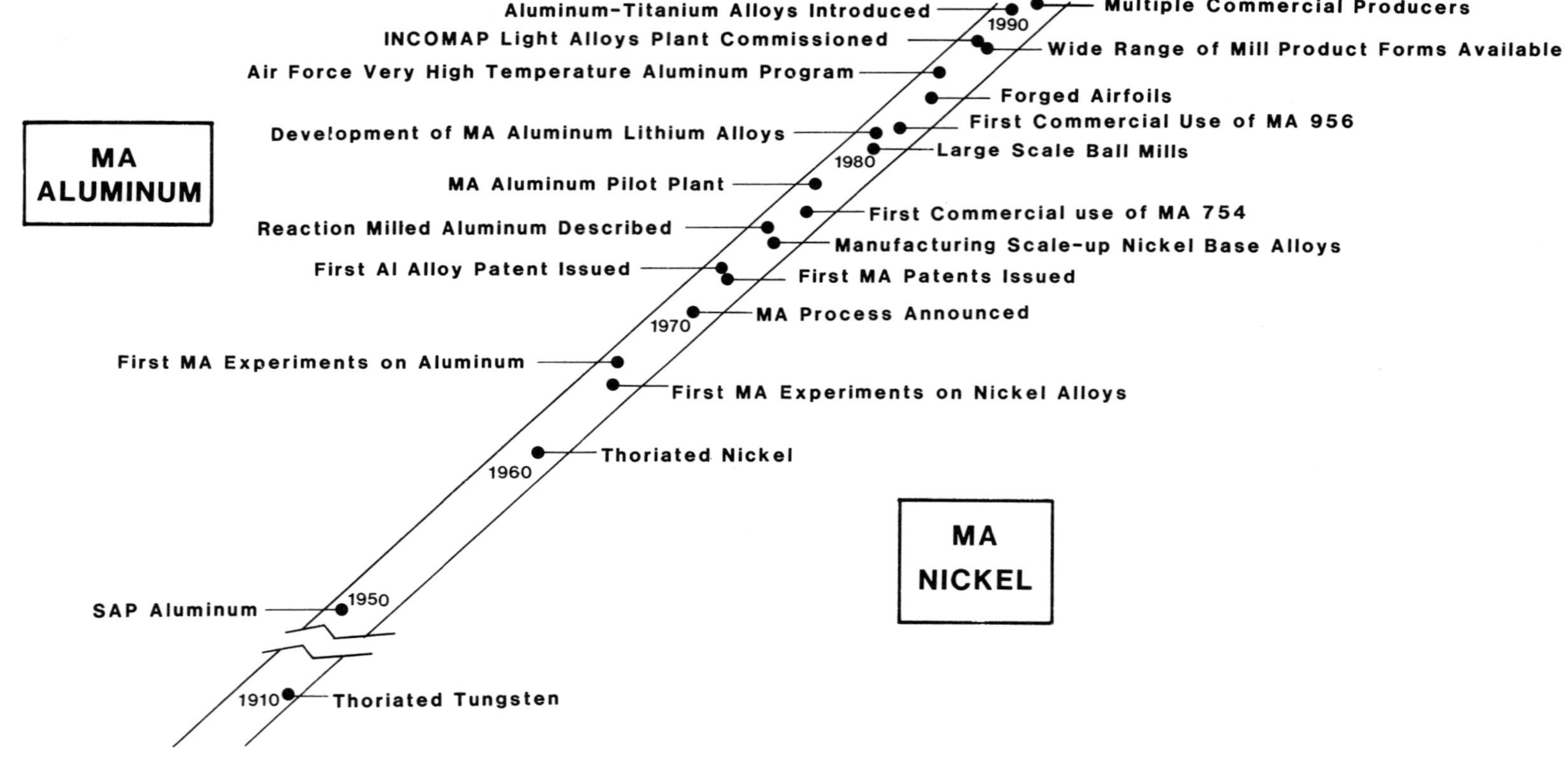

Figure 12: Chronology of Development of Mechanical Alloying.

hardening of nickel was first reported by Chevenard in 1929 (30). Rapid development of γ' age hardened alloys took place in the 1940's as an enabling materials technology for the aircraft gas turbine (31). By 1966 complex alloys consisting of ten or more elements and containing more than 50% of the coherent γ' strengthening phase were in common use.

Deliberate addition of a fine refractory oxide for strength and structural stability dates at least to the work of Coolidge (1910) on thoria dispersed tungsten (32). The principle was applied to aluminum by Irmann in 1949 (33) and nickel by Alexander in 1961 (34). Unfortunately, these pioneering methods for producing ODS metals could not be readily applied to alloys containing large amounts of the reactive or refractory elements needed for resistance to environmental degradation. By 1966 ODS had been demonstrated in a Ni-20% Cr alloy but the prospects appeared bleak for further progress using chemical processing techniques.

In his early work, Benjamin utilized conventional powder and melt injection approaches. When these proved unsuccessful, he turned to high energy mechanical milling in a laboratory scale shaker mill (35). The merits of the approach were immediately obvious as the desired dispersion of oxide in a complex alloy was readily achieved. It took some time, however, to produce an alloy with good high temperature strength. The answer came with the realization that a coarse grain structure (1-5 mm) could be generated through control of thermomechanical processing and secondary recrystallization. The first public announcement of the new process was made in 1970 (36).

Following some initial investigations of MA in other alloy systems to establish a broad patent position, Inco concentrated its efforts on the commercial development of ODS nickel-base alloys. The first use in a production engine occurred in 1977. This first product was a machining quality hot rolled and recrystallized bar of yttriated nickel - 20% chromium known today as INCONEL alloy MA 754. This alloy was used by General Electric Corporation for the nozzle guide vane assembly in the F404 engine for military fighter aircraft (37). In the years following, a number of other ODS nickel- and iron-base alloys have been introduced.

Continued process development work led to the introduction of large ball mills for powder processing in the early 1980's and subsequently to the availability of a wide range of standard mill forms such as plate, sheet, tube and wire. In addition a better understanding of the principles of thermomechanical processing resulted in the development of forging, spinning, ring rolling and drawing processes. The first forged component went into service in 1983. Today the wider range of alloys, product forms, fabrication methods and environmental properties ensures a much broader application of these materials in industrial as well as aerospace applications.

Exploratory work in the early 1970's at the Inco corporate laboratory in Suffern, N.Y. examined the applicability of MA to a very wide range of alloy systems. This work was undertaken when it was realized that MA could in principle be applied to almost any mixture of metals or metals and nonmetallic hard particles. For example the technique could be used to make an alloy from elements which were immiscible in the liquid state (copper-chromium) or which had widely differing boiling points (iron-magnesium). Systems examined at that time included copper, aluminum, magnesium, titanium, tool steels, sintered carbides, low temperature superconductors, catalysts and even ceramics. Although promising results were obtained in a number of systems, most were abandoned when it became clear that substantial financial investment would be needed for Inco to develop commercial products for markets with which it was unfamiliar. However, work was continued on aluminum alloys based on the extremely promising experimental results (38).

The subsequent commercial development for MA aluminum was slower than that for nickel alloys for several reasons. An obvious one was the need for Inco to develop expertise in a new alloy field. In addition, it quickly became apparent that the highly reactive nature of aluminum powder would necessitate the development of new processing procedures and equipment to prevent contamination and explosion. Finally, for economic reasons the acceptance of powder aluminum alloys by industry has been substantially slower than that of nickel alloys. Consequently, despite the early promise, it was not until 1988 that Inco (by that time IncoMAP

Light Alloys) commissioned its full scale production facility in Pittsboro, North Carolina.

Although oxide dispersions had previously been used exclusively to generate high temperature strength, the first commercial **MA** aluminum alloys were tailored for ambient temperature service. In fact, the first product objective was a high strength aluminum electrical conductor (39). It was found that the oxide-carbide dispersoids stabilized an extremely fine grain structure (0.1 - 1.0 μm) which provided strength in the absence of age hardening elements. This facilitated the development of strong, light weight alloys (such as AL-9052 and AL-905XL) which had exceptional corrosion resistance. These alloys will be used primarily for conventional aircraft and marine structural applications.

Laboratory work on aluminum alloys which could serve at high temperature began on an exploratory scale in 1978. Although modest advancement in alloy characterization was made in the early 1980's, the USAF Very High Temperature Aluminum Initiative provided a tremendous boost to the effort. Inco Alloys as a subcontractor to the University of Virginia on the AFWAL contract identified the aluminum-titanium alloy system as capable of meeting the goal of a 900°F (480°C) alloy. Subsequent work has identified several alloys within the nominal Al - 5/15% Ti range which meet specific objectives for ambient and high temperature strength, stiffness, density, corrosion resistance and formability (40). Commercial scale-up has started on the first alloy. These alloys are still very much developmental, but appear to be strong candidates for advanced engine and airframe applications.

As mechanical alloying entered its third decade, it underwent a significant rejuvenation. This may have been prompted by the impending expiration of the basic Inco patents, but more likely was simply coincident with the general explosion of interest in advanced materials and nontraditional alloy synthesis techniques. A significant milestone was the report by White of amorphous behavior in the Nb-Sn system (41). This observation attracted worldwide interest in academic circles because MA facilitated the study of the mechanisms of metal amorphization. Although a product with an amorphous structure generated by **MA** has yet to be commercialized, laboratories in several countries are actively exploring the opportunities. This upsurge in research is not limited to traditional ODS materials and amorphous metals. Today research groups are actively applying MA to intermetallics, metal matrix composites, ceramic precursors, sprayable powders and nanocrystalline materials. The role of MA as a viable process strategy for making unique materials now seems assured.

AN ADVANCED MATERIALS PERSPECTIVE

Those of us who call ourselves materials scientists and engineers are in an exciting period of time with much work to be done in developing new processes and materials to the level of maturity at which they become commercial entities. However as discussed in this paper, much creativity and hard work is required before our goal can be met; and the rest of the world is not standing still.

One good example of an advanced process/material has been discussed in this paper and is the subject of the present conference--mechanical alloying. Much excellent research, development and marketing has put nickel and aluminum based MA materials on the verge of full commercialization. Other MA materials such as titanium are at a much earlier stage of maturity. Thus the following papers exhibit an emphasis which varies with the alloy system under consideration. In total a very interesting example of an advanced process/material.

ACKNOWLEDGEMENTS

The authors would like to express their appreciation to Ms. T. Dahmen and Mrs. P. Bauer for formulating and typing this manuscript.

REFERENCES

1. Congress of the US, Office of Technology Assessment, "Advanced Materials by Design", June 1988.

2. A.R.C. Westwood, _Met. Trans. B_, vol. 19B, April 1988, 155.

3. Froes, F.H., "Aerospace and Defense Structural Materials for the Twenty-First Century," presented at the MPIF

Conference on <u>Powder Metallurgy in Aerospace and Defense</u>, Seattle, WA, Nov. 2-3, 1989, and to be published in the proceedings, ed. by F.H. Froes, 1990.

4. Froes, F.H., "Emerging Powder Metallurgy Technologies," to be presented at the MPIF sponsored meeting on <u>Powder Metallurgy - Key to Advanced Materials Technology</u>, Vancouver, B.C., 30 July - 1 Aug., 1990, and to be published in the proceedings.

5. <u>Scientific American</u>, 255, No. 4, Oct. 1986.

6. Froes, F.H., "Aerospace Materials for the Twenty-First Century", <u>Fourth Israel Materials Conf.</u>, Beer Sheba, Israel, Dec. 1988 and to be published in the proceedings.

7. Froes, F.H., <u>Material and Design</u>, May/June 1989, vol. X, No. 3, 110.

8. Froes, F.H., "Aerospace Materials for the Twenty-First Century", <u>Swiss Materials</u>, to be published 1990.

9. Froes, F.H., <u>Materials Edge</u>, No. 5, May 1988, 19.

10. Froes, F.H., "Emerging Technological Developments in Metals, Alloys, and Metal Matrix Composites", <u>US BOM Conference on Advanced Materials-Outlook and Information Requirements</u>, Washington, D.C., November 7-8, 1989, and to be published in the proceedings.

11. Froes, F.H., C. Suryanarayana and P.H. Shingu, work in progress 1989-90.

12. Froes, F.H., <u>Materials Edge</u>, May/June 1989, 17.

13. Froes, F.H. and C. Suryanarayana, <u>Workshop on Advanced Materials</u>, Minsk, USSR, May 29-June 2, 1989, "Rapid Solidification and Mechanical Alloying: State-of-the-Art and Technological Implications", and to be published in the proceedings.

14. Froes, F.H., Proceedings of <u>Second Int. SAMPE Metals and Metals Proc. Conf.</u>, "Space Age Metals Technology", eds. F.H. Froes and R.A. Cull, Dayton, OH, (1988), 1.

15. Froes, F.H. and Y.W. Kim, "Design of Engineered Structural Materials for the Twenty-First Century", presented at <u>Int. Conf. on Advanced Materials</u>, Milan, Italy, May 1989 and to be published in the proceedings.

16. Suryanarayana, C. and F.H. Froes, "Non-Equilibrium Processing of Light Metals for Aerospace Applications - Rapid Solidification and Mechanical Alloying", <u>IBC Conference on Rapid Solidification</u>, London, England, June 1-2, 1989, and to be published in the proceedings. Also Light Metals Age, vol. 47, June 1989, 18.

17. Froes, F.H. and J. Wadsworth, "Aerospace Structural Applications for Powder Metallurgy and Metal Matrix Composites", <u>BNF 7th Int. Conf.</u>, Oxford, England, June 3-5, 1989, and to be published in the proceedings.

18. <u>Materials Science and Engineering for the 1990's</u>, NRC, National Academy Press, 2101 Constitution Ave., NW, Washington, D.C., 1989.

19. Ashby, M.F., Phil. Trans. Roy. Soc., London, A322, (1987), 393.

20. Hadcock, R.N., J. of Aircraft, 17, No. 9, (Sept. 1980), 609.

21. Aviation Week, Oct. 10, 1988, 38.

22. McDaniels, D.L., T.T. Serafini and J.A. DiCarlo, J. of Energy Sys., 8, No. 1, (1986), 80.

23. Sprague, Robert A., Ad. Mats. and Proc., Jan. 1988, 67.

24. G.W. Meetham, "High Temperature Alloys for Gas Turbines and Other Applications", Reidel Publishing Co., (1986), 1.

25.   Burrus, D., MPIF, San Diego, CA., June 1989.

26.   Hadcock, Richard N., AIAA Aerospace Conf. and Tech. Show, Los Angeles, CA., Feb. 1985, 1.

27.   Stephens, Joseph R. and Michael V. Nathal, NASA Lewis Tech. Memo. #100903, and presented at Sixth Int. Sym. on Superalloys, Champion, PA., Sept. 18-22, 1988.

28.   Langer, Edward L., ASM News, Dec. 1989, 4.

29.   Lincoln, J., 1987, Structural Technology Transition to New Aircraft, ASD/ENFS, WPAFB, OH.

30.   Chevenard, P., <u>Comptes Rendus</u>, 1929, vol. 189, 846.

31.   Betteridge, W., <u>The NIMONIC Alloys</u>, Edward Arnold Ltd., London, 1959, 5.

32.   Coolidge, W.D., <u>Proc. Am. Inst. Elect. Eng.</u>, 1910, 961.

33.   Irmann, R., <u>Techn. Rundschau</u>, 1949, vol. 41 (36), 19.

34.   Alexander, G.B., R.K. Iler and S.F. West, U.S. Patent No. 2,972,529, Feb. 21, 1961.

35.   Benjamin, J.S., <u>New Materials by Mechanical Alloying Techniques</u>, E. Arzt, and L. Schultz, eds., DGM, Oberursel, West Germany, 3.

36.   Benjamin, J.S., <u>Met. Trans.</u>, 1970, vol. 1, 2943.

37.   Bailey, P.G., <u>Frontiers of High Temperature Materials</u>, J.S. Benjamin, ed., Inco Alloys Products Co. (1981), 57.

38.   Benjamin, J.S., and M.J. Bomford, U.S. Patent No. 3,740,210, June 19, 1973.

39.   Benjamin, J.S. and M.J. Bomford, <u>Met. Trans.</u>, vol. 8A, 1977, 1301.

40.   Mirchandani, P.K. and R.C. Benn, SAMPE, <u>Space Age Metals</u>, F.H. Froes and R.A. Cull, eds., August 2-4, 1988, vol. 2, 188.

41.   White, R.L.; Ph.D. Dissertation, Stanford University, 1979.

KEYNOTE ADDRESS
MA ALLOYS FOR AEROSPACE APPLICATIONS

I. C. Elliott, G. A. J. Hack
Inco Alloys International, Inc.
Holmer Road, Hereford HR4 9SL
United Kingdom

ABSTRACT

Of all the metallurgical process developments
which have taken place in recent decades,
mechanical alloying (MA) ranks among the most
novel in its ability to alloy powder in the
solid state and avoid the many problems of
melting and solidification in complex alloys.
A series of MA oxide dispersion strengthened
Fe-Cr, Ni-Cr and Ni-Cr-gamma prime alloys has
been developed which exhibit very high elevated
temperature strength, and in some cases also
remarkable oxidation resistance. Their
applications in aeroengine blade, vane and
sheet components are described. Whereas the
current alloys are aimed largely at direct
replacements of existing superalloys, the
future exploitation of MA will be in alloys and
alloy systems which are very difficult or
impossible to produce by any other means.

THE COURSE OF METALLURGICAL DEVELOPMENT has
been affected by numerous external factors,
one of the most profound of which was the
development of the aeroengine gas turbine.
This heralded a long period of extensive
alloy development from which the wide range of
well known alloys, termed the superalloys,
largely resulted. However, the development of
alloy chemistry to improve properties is
subject to the law of diminishing returns and
the approach of the practical limits in terms
of microstructural instability resulting in
hard phase formation, reduced corrosion
resistance and decreasing ductility caused the
emphasis to be switched to process development
(1). Thus the challenge thrown down from the
aeroengine manufacturers for stronger, stiffer,
lighter and more corrosion resistant materials
to deal with the ever increasing turbine
operating temperatures has been met, with clear
results on the extended range of many alloy
systems (Fig. 1).

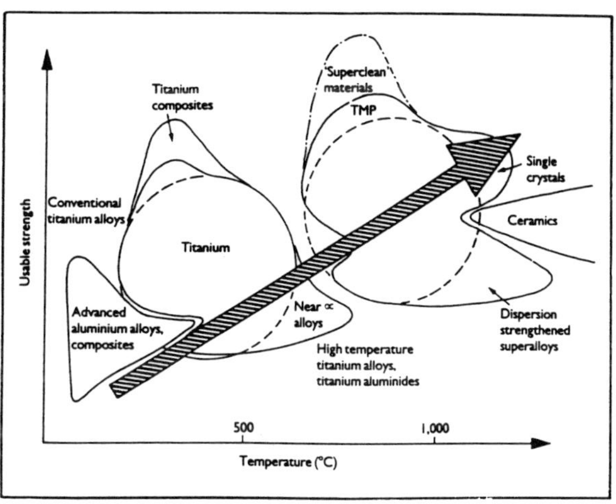

Fig. 1 - The spheres of influence
of major aeroengine materials
(after G. Meetham)

With the additional challenge of ceramics
and composites as serious engineering
materials, the future role of metals and alloys
will depend on the effectiveness and novelty of
process development. One such novel technology
which has again extended the available
latitudes for alloy development is that of
mechanical alloying, which exhibits a
remarkable ability to deal with normally
incompatible components. The process
flexibility extends beyond conventional alloys
to include metastable phases, amorphous
materials, intermetallics, and even cermets and
organic-ceramic-metallic material systems (2).
The manufacture of oxide dispersion
strengthened (ODS) alloys is an example of
this, their great advantage being the
retention of significant strength relatively

close to the absolute melting point. Although simple ODS alloys based on the $Ni-ThO_2$ system have been available for a number of years, the various chemically based processes used in their manufacture are not appropriate to more complex superalloys (3). Milling methods have been used for other simple ODS systems such as SAP (sintered aluminium powder) but to achieve the fine dispersion required for effective strengthening in superalloys calls for the higher energy, more complex process of mechanical alloying.

The attributes of mechanical alloying place it in a rather unique position and it is potentially one of the most powerful tools for material improvement that has emerged in this era of process development. However, to assess the extent of these potential new horizons, it is necessary first to consider some details of the MA process itself and the current state-of-the-art with respect to aeroengine applications.

THE MECHANICAL ALLOYING PROCESS

The fundamental process in mechanical alloying to produce metal powders with controlled microstructures is the repeated welding, fracturing and rewelding of a mixture of powder particles in a dry, high energy ball charge (4-7). Early development work utilised small attritors with the capability of producing about 1 kg of powder, and indeed such equipment is still used for R and D purposes. In an attritor the central shaft carries paddles which when rotated at high speed result in high energy collisions between the steel balls of the charge and allow mechanical alloy to take place (Fig. 2). However, as the size

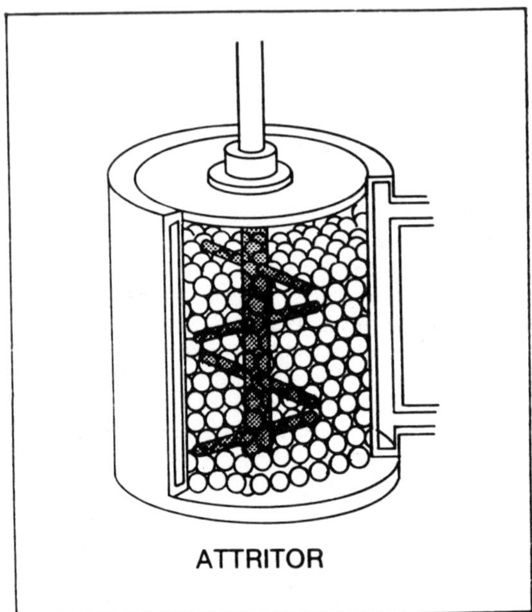

Fig. 2 - Schematic diagram of high energy attritor mill

of such vessels increases the problem of heat transfer increases and the practical limit is approximately 40 kg powder capacity. On a commercial scale mechanical alloying is carried out in ball mills of up to 1250 kg capacity which rotate on the horizontal axis and impart kinetic energy to the balls by virtue of the mill diameter (Fig. 3).

Fig. 3 - Ball mill facilities for mechanical alloying

Raw materials for mechanical alloying may include elemental powders, crushed master alloys and pre-alloyed powders, as well as oxide powder (usually yttria) to produce the fine dispersion of oxides which imparts many of the remarkable properties of the MA alloys. The raw material powders are pre-blended and processed in ball mills dedicated to a particular composition. The repetitive welding and fracturing process takes place at a rate determined by the mill speed, and hence the kinetic energy of the ball collisions, the temperature and atmosphere. After a short time the raw material powders (Fig. 4a) show evidence of welding and fracturing (Fig. 4b) although many of the original constituents are still identifiable. Full alloying is achieved when the layered structure is no longer optically visible (Fig. 4c), and although a lammellar structure is observable using electron optics (Fig. 4d) there is no detectable chemical inhomogenity. Furthermore, by this stage the yttria assumes a highly uniform dispersion with particles of approximately 30 nm diameter spaced at an optimum interparticle spacing of 500 nm.

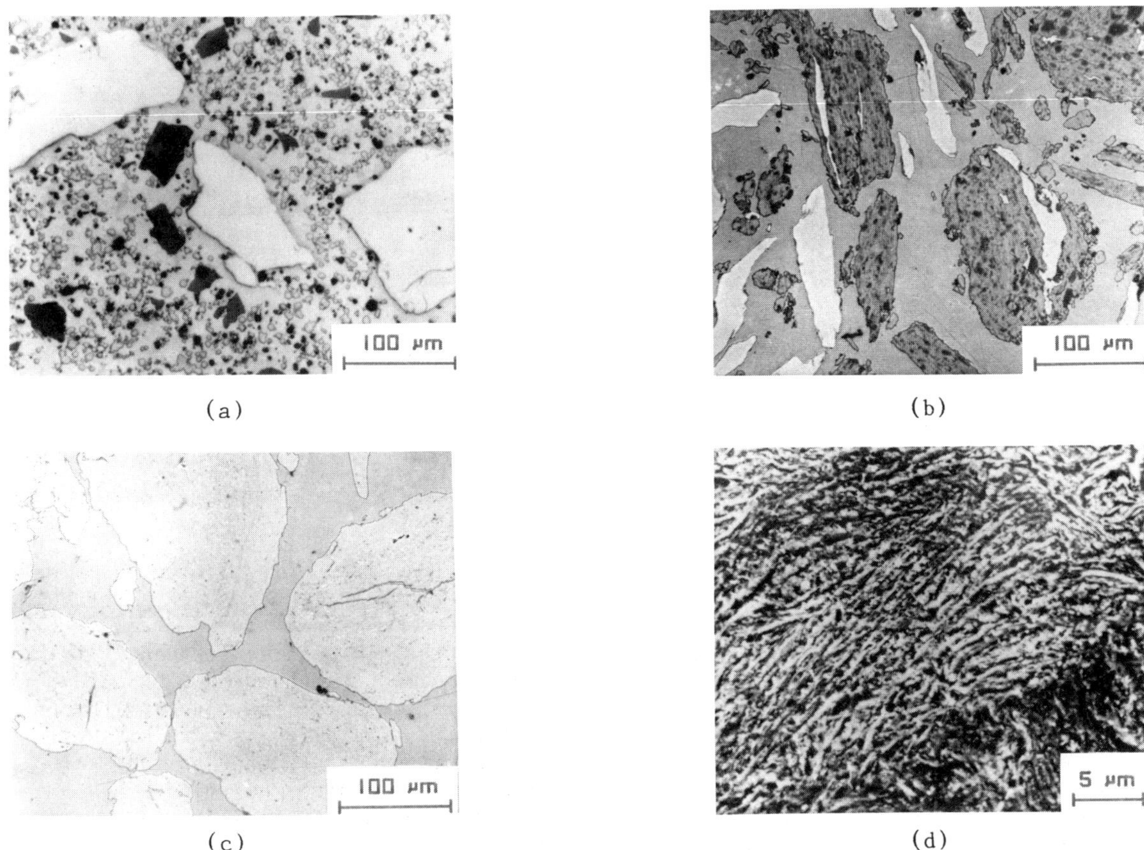

Fig. 4(a) - Raw material powders showing Cr (large light particles), master alloy (dark particles) and Ni powder (small light particles), (b) partly processed powder, (c) fully processed powder and (d) finely layered structure of processed powder (SEM)

In order to take advantage of this solid state alloying, a large number of process factors have required careful study and optimisation. The nature of the raw materials has a profound influence on the MA process, not too soft and not too hard but a balanced mixture of the two. Ball mills must be conditioned so that internal mill and steel ball surfaces are coated with a layer of the alloy being produced. As it is on this layer that the MA event takes place, its conformation is of critical importance to the balance of welding and fracturing, as well as preventing contamination of the powder charge. Mill atmosphere and temperature must be carefully controlled in a sequence which is often unique to a particular alloy, and the optimum milling time must be used to avoid underprocessing or at the other extreme, overhardening of the charge which results in grinding to very fine powder and radically affects the response during subsequent thermomechanical treatment. Powder handling, either of raw materials or fully processed powder must be carried out in scrupulously clean, dedicated containers and equipment to prevent cross contamination.

Under laboratory conditions it is a relatively simple matter to produce mechanically alloyed powder in very small quantities. However, what is not so easy is to do it without contamination by underprocessed and raw material powders when operating on a commerical scale involving individual powder batch sizes in excess of 1 tonne. Complex and sophisticated manufacturing and quality assurance procedures are thus required in order to produce high quality powder, which is essential for successful conversion to mill product.

Table 1 - Nominal compositions (wt %) of mechanically
alloyed oxide dispersion strengthened superalloys

	Ni	Fe	Cr	Al	Ti	C	Y_2O_3	Mo	W	Ta	B	Zr
INCOLOY* alloy MA 956	-	Bal	20	4.5	0.5	0.05	0.5	-	-	-	-	-
INCOLOY alloy MA 957	-	Bal	14	-	1.0	0.05	0.25	0.3	-	-	-	-
INCONEL* alloy MA 754	Bal	1.0	20	0.3	0.5	0.05	0.6	-	-	-	-	-
INCONEL alloy MA 758	Bal	1.0	30	0.3	0.5	0.05	0.6	-	-	-	-	-
INCONEL alloy MA 6000	Bal	-	15	4.5	2.5	0.05	1.1	2.0	4.0	2.0	0.01	0.15
INCONEL alloy MA 760	Bal	-	20	6.0	-	0.05	0.95	2.0	3.5	-	0.01	0.15
INCONEL alloy MA 757	Bal	-	16	4.0	0.5	0.05	0.60	-	-	-	-	-
Alloy 3002	Bal	-	20	4.0	0.5	0.05	0.60	-	-	-	-	-

THE ODS SUPERALLOYS AND THEIR APPLICATIONS

Early ODS alloys for aerospace applications largely consisted of TD nickel and TD nichrome, both made by chemically dependant processes. However, the subsequent range of yttria dispersion strengthened alloys resulting from the development of mechanical alloying covers a wider range of compositions and fall into three main groups (Fig. 5), Fe-Cr, Ni-Cr and Ni-Cr-gamma prime strengthened alloys. The nominal compositions of the main alloys are given in Table 1.

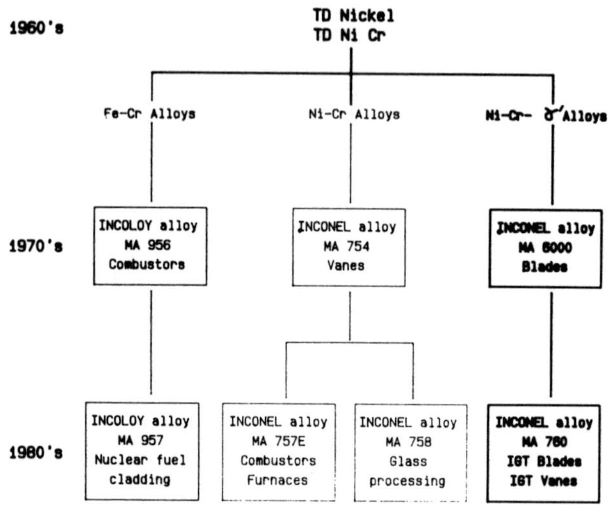

Fig. 5 - The development of ODS alloys

INCOLOY ALLOY MA 956 - This is a ferritic Fe-Cr-Al alloy strengthened with about one volume percent of yttria. The high Al content

(* Trademark of INCO family of companies)

results in the alloy being protected by an Al_2O_3 scale at high temperatures. Furthermore, a particular sequence of hot and cold rolling sheet products yield large "pancake" grains on heat treatment which ensures excellent isotopic properties in the plane of the sheet. Being ferritic the alloy is very amenable to cold forming and is produced as rod, bar, section, tube and wire, as well as hot and cold rolled sheet. It does, however, also have the characteristic ductile-brittle transition at between 0°C and 60°C (8) which must be recognised during processing and can be avoided by relatively simple expedients such as prewarming or steam heating before "cold" processing.

The oxide dispersion and large grain structure confers a degree of high temperature strength well beyond other common superalloy sheet materials (Fig. 6). It also has remarkable oxidation resistance. For instance, under cyclic oxidation conditions the highly adherent surface alumina provides a very

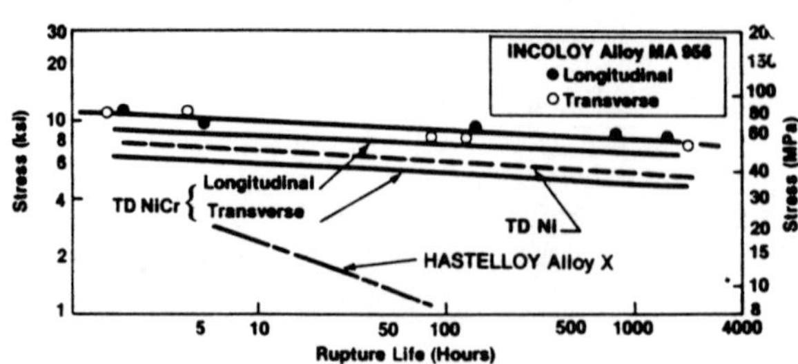

Fig. 6 - Stress rupture properties of INCOLOY alloy MA 956 at 1093°C compared to other sheet alloys
(HASTELLOY is a trademark of Cabot Corp)

stable protective scale which continues to
undergo parabolic oxidation to times at which
other alloys with less protective Cr_2O_3 scales
suffer severe weight loss (Fig. 7). The
alloy is also highly resistant to carburisation
and sulphidation (9).

Sheet, plate, spinnings, rings and
forgings in INCOLOY alloy MA 956 have
applications in combustion chambers,
afterburner and turbine casing sections where
the resistance of the alloy to creep, oxidation
and sulphidation allow higher metal temperatures
and longer component life. A prototype
combustion chamber lined with INCOLOY alloy MA
956 is illustrated in Fig. 8 and large,
seamless rings in Fig. 9.

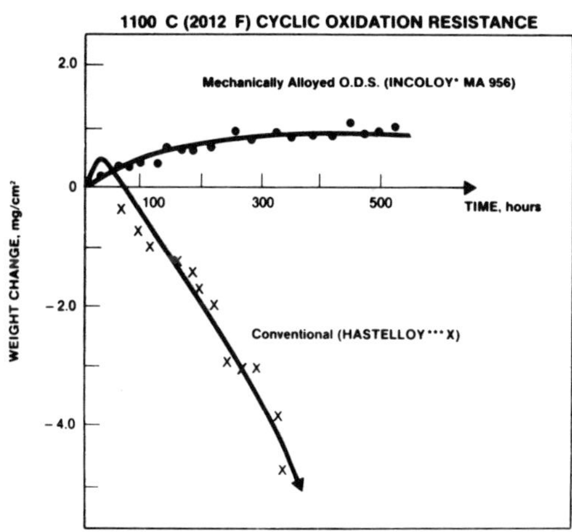

Fig. 7 - Cyclic oxidation resistance
of INCOLOY alloy MA 956 at 1100°C
compared with a conventional
superalloy

Fig. 8 - Prototype advanced high
temperature combustion chamber
using INCOLOY alloy MA 956.

Fig. 9 - Seamless rolled section
of INCOLOY alloy MA 956 for gas
turbine rings

Fig. 10 - Coarse elongated grain
structure formed in INCONEL alloy
MA 754 after high temperature
heat treatment

INCONEL ALLOY MA 754 - INCONEL alloy MA 754
is basically Ni-20%Cr strengthened by about
1 volume percent of yttria (Table 1). The
powder consolidation route including extrusion
and subsequent hot rolling results in very
coarse, elongated grain structures on high
temperature heat treatment (Fig. 10) with a
(100) texture in the longitudinal direction.
This imparts a low modulus of elasticity which
is associated with superior thermal fatigue
resistance, which combined with high creep
strength and resistance to oxidation make
the alloy highly suitable for aeroengine
vane applications (10). A brazed HPT nozzle
segment containing two INCONEL alloy MA 754
vanes for the GE F404-400 engine is illustrated
in Fig. 11.

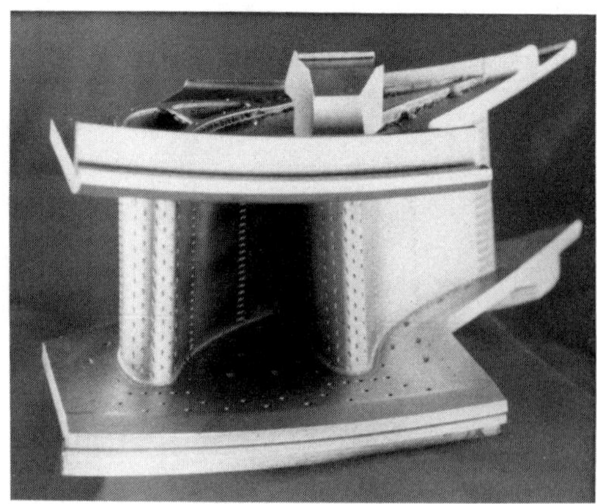

Fig. 11 - Brazed turbine vane
assembly fabricated from
INCONEL alloy MA 754

INCONEL ALLOY MA 6000 - INCONEL alloy MA
6000 is a Ni-Cr based alloy with gamma prime
precipitation and considerable matrix
strengthening through the presence of
refractory elements such as W and Mo. About
two volume percent of yttria provides strength
at very high temperatures and the general
balance of the composition being similar to
earlier cast alloys such as IN-792 and IN-738
is reflected in oxidation and corrosion
resistance. The alloy is produced in
extruded bar form with a very coarse, highly
elongated grain structure (Fig. 12) generated
by zone annealing.

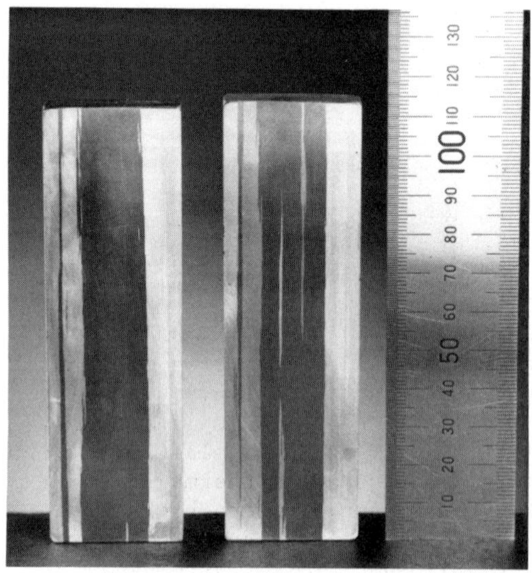

Fig. 12 - Zone annealed directionally
recrystallised INCONEL alloy MA 6000
bar illustrating the high grain
aspect ratio

The elevated temperature stress rupture
properties of INCONEL alloy MA 6000 are
compared to several other alloys in Fig. 13.
At intermediate temperatures (850°C) the
properties approach that of DS Mar-M200 + Hf,
a complex turbine blading alloy. At high
temperatures (1093°C) it is clear that INCONEL
alloy MA 6000 still retains useable strength
whereas the DS and single crystal alloys loose
most of their strength as the gamma prime phase
dissolves. Furthermore, long term stress and
creep rupture data show a unique upward
inflection of the rupture stress-life plot that
predicts higher design stresses for long term
service than would be predicted by short term
data (11). Such behaviour is a direct result
of the complimentary strengthening effects of
both gamma prime and the oxide dispersion.

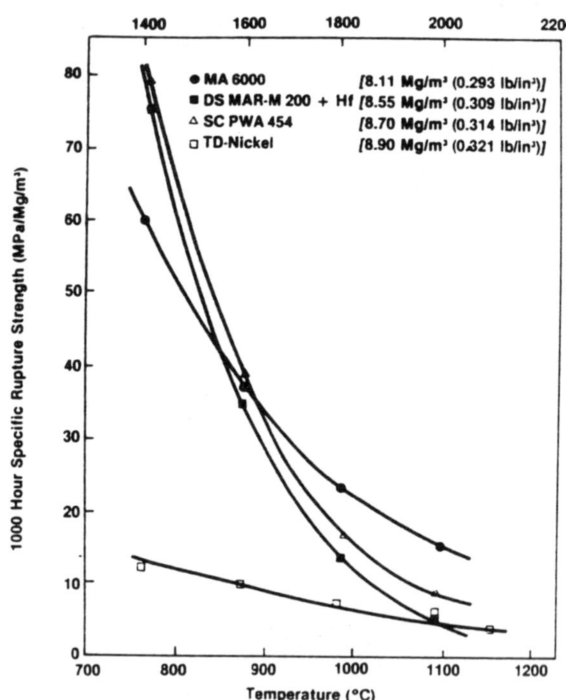

Fig. 13 - Comparison of 1000 hour
specific rupture strength of INCONEL
alloy MA 6000 with DS Mar-M200 + Hf,
TD Nickel and single crystal PWA
alloy 454

The major application for INCONEL alloy MA
6000 is that of small, highly stressed first
and second stage turbine blades (Fig. 14)
which can be machined from fully recrystallised
bar, although forging routes are also under
development with a consequent saving in
material (12, 13).

Fig. 14 - Prototype turbine blade
machined in INCONEL alloy MA 6000

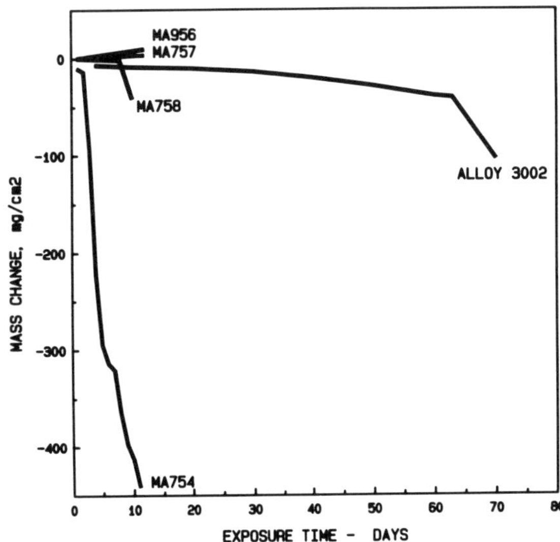

Fig. 15 - Comparison of the oxidation
resistance of several mechanically
alloyed ODS superalloys at 1300°C
in an air + 5% water vapour
atmosphere

FUTURE DEVELOPMENTS IN MA AEROSPACE ALLOYS

The alloys so far described have their
roots in the detailed laboratory work of the
early MA development and represent only a very
small example of the possibilities afforded by
mechanical alloying. In order to explore the
long term developments it is necessary first to
consider the immediate direction of work.

CURRENT DEVELOPMENTS - The ODS superalloy
family tree (Fig. 5) illustrates the type of
current developments in Fe-Cr, Ni-Cr and
Ni-Cr-gamma prime alloys, and reflects the
influence during the 1980's of market demands.
Hence a new swelling resistant nuclear canning
alloy (INCOLOY alloy MA 957) extended the Fe-Cr
group, and for aerospace applications ODS
Ni-Cr-Al alloys (Alloy 3002 and INCONEL alloy
MA 757) which exhibit excellent high
temperature strength and with the oxidation
resistance of INCOLOY alloy MA 956 (Fig. 15).
Typical applications could be combuster parts
manufactured from thin sheet. However, the
production of sheet in such alloys may require
considerable development, possibly using hot
isothermal rolling, pack rolling or complex
cold rolling techniques.

A recurrent theme in the evolution of the
ODS superalloys is enhanced corrosion
resistance. Thus a 30% chromium version of
INCONEL alloy MA 754 (designated MA 758) has
found applications in glass processing due to
its considerable resistance to attack by molten
glass. Similarly, the demands of industrial
gas turbine blades and vanes have resulted in
the development of INCONEL alloy MA 760, which
with its higher Cr and Al has better corrosion
resistance than INCONEL alloy MA 6000, albeit
at the expense of a little of the high
temperature strength (14). Nevertheless, in
aeroengine blading applications where it is
common practice to coat very high strength
alloys such as single crystals in order to

provide the required corrosion resistance, the
search for higher strength turbine blade ODS
alloys continues. In particular, lower Cr ODS
alloys with high levels of strengthening
elements like Mo, W and Al can exhibit gamma
prime volume fractions of approximately 70%
compared to INCONEL alloy MA 6000's 50% (11)
and consequent enhancement of intermediate
temperature strength. This is illustrated by
the comparison of experimental alloy 51 and
INCONEL alloy MA 6000 in Fig. 16.

Thus far the development of ODS
superalloys has allowed their use in a wide
range of applications, mainly solving existing

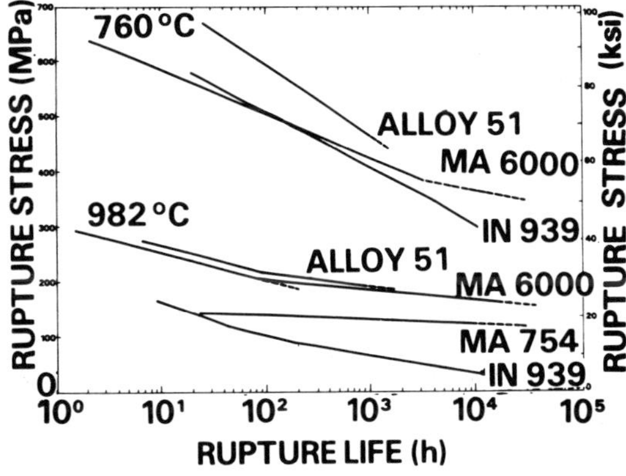

Fig. 16 - Effect of alloy design
on stress rupture characteristics

Table 2 - Comparison of Crystal Structures, Melting Points and
Densities of Selected Nickel, Iron, Titanium, Niobium
and Aluminium Intermetallics

Compound	Crystal Structure	Melting Point (°C)	Density (g/cm³)
Ni_3Al	$L1_2$	1390	7.5
$NiAl$	$B2$	1640	5.9
Fe_3Al	DO_3	1540	6.7
$FeAl$	$B2$	1330	5.6
Ti_3Al	DO_{19}	1600	4.2
$TiAl$	$L1_0$	1460	3.9
Nb_3Al	$A15$	1960	7.3
Al_3Ti	DO_{22}	1350	3.3

material problems by direct substitution of conventional Ni and Co alloys. However, new modifications and design of high temperature plant are now envisaged which could take full advantage of the opportunities which future MA alloys may offer.

FUTURE POSSIBILITIES - Although it is clear from the above that considerable scope remains for development of the current range of sheet, vane and blade ODS superalloys, their longer terms future is limited by the relatively low melting points of the conventional Ni and Co base alloys systems. To extend the operating temperatures and reduce weight of commercial and military engines, intermetallics and particularly aluminides are being extensively studied as they offer the opportunity of light, stiff and relatively strong alloys and composites with in many cases considerably higher melting points, and therefore higher ultimate operating capability. Some of the possible intermetallics are illustrated in Table 2. The main problem with such compounds is their generally low ambient temperature ductility (<2%). However, improvements in ductility have been achieved by alloying, particularly B additions and grain refinement by thermo-mechanical working and rapid solidification techniques (15, 16, 17). Much of the emphasis on processing has therefore been on powder metallurgy, but the constituents and melting points of many of the intermetallics makes preparation by melting techniques difficult. Mechanical alloying however has the ability to deal with these problems and allows dispersions to be introduced. The following is a brief review of some of the observations on MA intermetallics.

Nickel Aluminides - As with other alloys, MA processed intermetallics exhibit very fine grain sizes (1-10μ) depending on the details of thermal treatment, and grain refinement is a potent method for improving the strength of, for instance, Ni_3Al. Moreover, there are indications that not only strength improvements but also useful ductilities are obtained with fine grained structures (<10μ) in the otherwise brittle intermetallics (17). Not only does strength improve with grain refinement in MA Ni_3Al but compared to other processes, at the same grain size, MA intermetallics are inherently stronger due to the ability to introduce dispersions of oxides such as Y_2O_3 (Fig. 17). It has also been demonstrated that the more attractive NiAl intermetallic can be produced by laboratory scale MA (17).

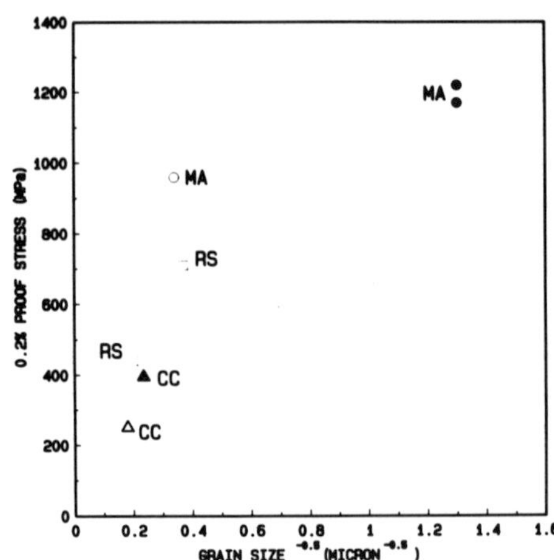

Fig. 17 - Relationship between yield strength and grain size for Ni_3Al intermetallics manufactured by conventional casting (CC), rapid solidification (RS) and mechanical alloying (MA).

Titanium Aluminides - Dispersion strengthened Ti_3Al has been produced by RS using precipitated dispersions of rare earth compounds e.g. E_2O_3, Ce_2O_3 or $Ce_4O_4S_3$ (18, 19, 20). However, after thermomechanical processing at 850-950°C, coarsening of the precipitates occurs leading to a weakened structure. Dispersions of Y_2O_3 have been introduced in Ti-6Al-4V alloy by MA (21) and Ti intermetallics, including Ti_3Al have been prepared by MA (22, 23). There would therefore appear to be considerable scope with MA Ti intermetallics prepared with dispersions and alloying additions e.g. Nb for enhanced ductility particularly for composite applications, although protection from oxidation would be necessary.

<u>Aluminium Intermetallics</u> - Although aluminium and its alloys is not generally considered in high temperature applications, Al-Ti alloys containing large volume fractions (up to 35 v/o) of submicron Al_3Ti have been produced by MA which have excellent long term stability up to 510°C (23). They exhibit low density and a superior combination of elevated temperature strength (Fig. 18) and elastic modulus when compared to conventional and RS Al alloys, such that they offer potential for applications in aeroengines where Ti alloys would normally be used.

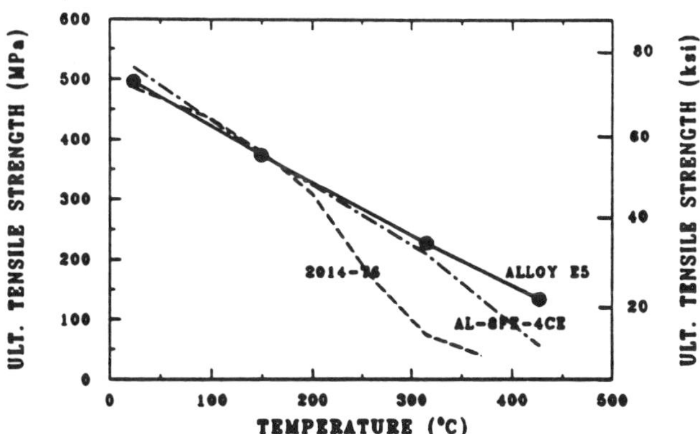

Fig. 18 - Comparison between the tensile strengths of alloy E5 (MA Al-Ti), cast and wrought 2014-T6 and Al-8Fe-4Ce.

<u>Niobium Intermetallics</u> - Niobium aluminides are potentially of interest for high temperature applications due to their very high melting points (>1850°C) and the possibility of improving the otherwise poor oxidation resistance of Nb. Normal melting routes with such compounds are difficult not only because of the high temperatures involved but also segregation. It has been demonstrated that Nb_3Al can be formed by MA (24).

It is clear that in the field of intermetallics there is enormous scope for the development of high temperature materials exhibiting low density, strength, stability, ductility and environmental resistance by mechanical alloying. Furthermore, such compounds may well be the basis of a new generation of metal matrix composites, which in themselves are an interesting challenge for mechanical alloying. Indeed, there is evidence to suggest that the current experience on relatively low volume fraction oxide dispersion alloys may be applied to much higher levels with interesting consequences. Mechanical alloying is a key that may open many doors and lead to the materials to fit the future needs of the aerospace industry in terms of reduced weight, operating temperatures beyond the capabilities of current alloys and consequential enhanced performance.

SUMMARY

A range of Fe-Cr, Ni-Cr and Ni-Cr-gamma prime alloys strengthened by a dispersion of yttrium oxide have been produced by mechanical alloying. They have found application in many areas, including aeroengines, where a unique combination of high temperature strength and environmental resistance make them suitable for blade, vane and sheet components. While these ODS superalloys substitute for conventional alloys used currently, a major use of MA for the future may be for advanced aeroengine alloys which are difficult or impossible to make by other means, such as aluminides, intermetallics and metal matrix composites based on Ni, Ti, Al and Nb.

ACKNOWLEDGEMENTS

The authors would like to thank Gaylord Smith for the data on oxidation resistance and Inco Alloys International for permission to publish these results.

REFERENCES

1. SIMS C.T., Proc. Fifth International Conference on Superalloys, Eds. M. Gell, C.S. Kortovich, R.H. Bricknell, W.B. Kent and J.F. Radovich, Pub. by TMS-AIME, 401-419, (1984).

2. BENN R.C. and McCOLVIN G.M., Proc. Sixth International Conference on Superalloys, Eds. D.N. Duhl, G. Maurer, S. Antolovich, C. Lund and S. Reichman, Pub. by TMS-AIME, 73-80, (1988).

3. HACK G.A.J., Met and Materials, <u>3</u>, Aug., 457, (1987).

4. BENJAMIN J.S., Met. Trans., <u>1</u>, Oct., 1-9, (1970),

5. BENJAMIN J.S. and VOLIN T.E., Met. Trans.,
    $\underline{5}$, Aug., 1929-1934, (1974).

6. BENJAMIN J.S., Proc. Conf. Gas Turbines
    and Fluids Engineering, Pub. ASME, New
    Orleans L.A., 1-9, (1976).

7. GILMAN P.S. and BENJAMIN J.S., Ann. Rev.
    Mater Sci., $\underline{13}$

8. DAVIDSON J.H., Proc. Second International
    Conference on Oxide Dispersion
    Strengthened Superalloys by Mechanical
    Alloying, Eds. J.S. Benjamin and R.C.
    Benn, 163-189, (1983).

9. CURWICK L.R., Proc. First International
    Conference on Oxide Dispersion
    Strengthened Superalloys by Mechanical
    Alloying, Ed. J.S. Benjamin, 3-10, (1981).

10. CRAWFORD W., Proc. Second International
    Conference on Oxide Dispersion
    Strengthened Superalloys by Mechanical
    Alloying, Eds. J.S. Benjamin and R.C.
    Benn, 272-286, (1983).

11. BENN R.C. and KANG S.K., Proc. Fifth
    International Conference on Superalloys,
    Eds. M. Gell, C.S. Kortovich, R.H.
    Bricknell, W.B. Kent and J.F. Radovich,
    Pub. by TMS-AIME, 319-326, (1984).

12. EWING B.A. and JAIN S.K., Proc. Sixth
    International Conference on Superalloys,
    Eds. D.N. Duhl, G. Maurer, S. Antolovich,
    C. Lund and S. Reichman, Pub. by TMS-AIME,
    131-140, (1988).

13. GRUNDY E., PATTON W.H., PRECIOUS C.J. and
    PINDER D., Proc. Second International
    Conference on Oxide Dispersion
    Strengthened Superalloys by Mechanical
    Alloying, Eds. J.S. Benjamin and R.C.
    Benn, 100-113, (1983).

14. BENN R.C. and McCOLVIN G., Proc. Sixth
    International Conference on Superalloys,
    Eds. D.N. Duhl, G. Maurer, S. Antolovich,
    C. Lund and S. Reichman, Pub. by TMS-AIME,
    73-80, (1988).

15. STOLOFF N.S., Int. Metall. Rev., $\underline{29}$, 123,
    (1984).

16. STEPHENS J.S. and NATAL M.V., Proc. Sixth
    International Conference on Superalloys,
    Eds. D.N. Duhl, G. Maurer, S. Antolovich,
    C. Lund and S. Reichman, Pub. by TMS-AIME,
    183-192, (1988).

17. BENN R.C., MIRCHANDANI P.K. and WATWE A.S.,
    Proc. Conf. on Solid State Powder
    Processing, Eds. A. Clauer and J.J. de
    Barbadillo, Pub. by TMS-AIME, 157-172,
    (1990).

18. KONITZER D.G., MUDDLE B.C. and FRASER H.L.,
    Scripta Met., $\underline{17}$, 963, (1983).

19. ROWE R.G., SUTLIFF J.A. and KOCH E.F.,
    Proc. Conf. On Rapid Solidification
    Technology of Titanium Alloys, Eds.
    F.H. Froes, D. Eylou and S.M.L. Sastry,
    Pub. by TMS-AIME, (1986).

20. ROWE, R.G., Proc. Elevated Temperature
    Titanium Alloys, TMS-AIME Meeting,
    Cincinnati OH, (1987).

21. CAIRNS R.L. and BENJAMIN J.S., U.S. Pat.
    No. 3,737,300, (1973).

22. SUNDARESAN R., JACKSON A.G., KRISHNAMURTHY
    S. and FROES F.H., Proc. Conf. Rapidly
    Quenched Metals VI, Montreal, (1987).

23. MIRCHANDANI P.K. and BENN R.C., Proc.
    Second International SAMPE Metals and
    Metals Processing Conference, Dayton OH,
    (1988).

24. LARSON J.M., LUHMAN T.S. and MERRICK H.F.,
    Proc. Conf. Manufacturing of
    Superconducting Matierals, Ed. R.W.
    Meyerhoff, 155, (1976).

STRUCTURAL DEVELOPMENT
IN FORGED MA 6000 BLADES

E. Grundy
Doncasters Monk Bridge Ltd.
Leeds, England

<u>ABSTRACT</u>

The enhanced elevated temperature
properties of Inconel Alloy MA 6000*
provide the opportunity to use uncooled
rotor blades in the hot stages of
advanced gas turbine engines. Whilst
the forging of such blades creates a
cost effective processing route, it is
essential to develop the same
unidirectional crystal structures and
mechanical properties as achieved in
the bar product.
Conventional forging routes, using a
protective thermal barrier layer
principle, have successfully been used
to fabricate blades from attritted
powder bar and various grades of ball
milled bar feedstsock. The development
of the correct post-forging structures
in blades has been achieved using a
commercially acceptable, low gradient
zone annealing technique. Interrupted
recrystallisation trials, producing
interfaces, has shown that the annealed
structure is mainly a function of
strain input and that the ease of
recrystallisation is dependent upon the
orientation of primary crystals. In
general, resultant crystal morphologies
are a product of powder condition,
forging parameters and zone annealing
variables.
Mechanical property levels of forged
blades are equivalent to the bar
product.

THE ADDITION OF oxide dispersion
strengthening to a superalloy matrix,
as in Inconel MA 6000, provides the

* Trademark Inco Family of Companies.

opportunity to increase the turbine
operating temperature or improve engine
efficiency through the use of uncooled
blades. Figure 1 illustrates these
advantages by comparing the creep
response of two nickel-base MA alloys
with single crystals.

The programme under consideration
here was progressed as a joint
development with Allison Gas Turbine to
investigate the potential of uncooled
MA 60000 second stage rotor blades in
their advanced T406 engine. Part of
this programme, including further
aspects of the exposure of MA 6000 to a
turbine environment, has already been
reported [1]. MA 6000 blades were to be
produced by a forging route, as opposed
to machining from recrystallised bar,
due to the inherently greater
efficiency of material utilisation with
this relatively expensive alloy. Many
varied aspects of investigation were
encompassed within the ultimate
objective of producing finished blades
for eventual engine testing.

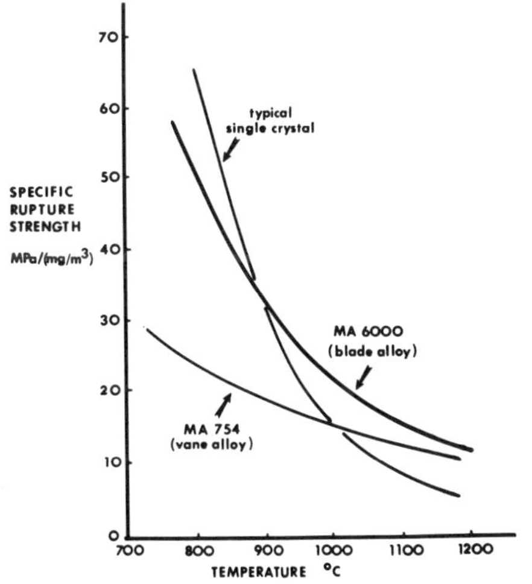

Fig. 1 - Specific Rupture Strengths

Sequential thermomechanical processing (TMP) stages had to create defect-free blades which would respond to post-forging, secondary recrystallisation and result in the same mechanical property levels as the recrystallised bar product. The dynamic recrystallisation process itself had to be capable of creating high aspect ratio (length to diameter) grain structures in pieces of variable section on a production scale basis. During the later stages of this present work, the small scale attritted route for producing MA 6000 powder was superceded by the more commercial, larger scale ball milling process. A comparative evaluation of bar feedstock from both of these techniques therefore had to be undertaken. Finally, since little was understood about the development of crystal structure in the alloy, a recrystallisation study was incorporated into the programme. It was anticipated that an increased understanding of grain growth and texture development may allow greater control to be exerted through TMP to optimise properties. A further possible consequence could be to change the texture away from the <110> type developed during secondary recrystallisation which adversely impacts mechanical properties at intermediate temperatures.

PRACTICAL CONSIDERATIONS

THERMOMECHANICAL PROCESSING - A comparison of MA 6000 with the simpler MA 754 alloy, compositions Table I, proves the former to have a more complex forging response [2] [3].

Table I.

MA ALLOY COMPOSITIONS

ALLOY	COMPOSITION WT%
INCONEL MA 754	Ni-20Cr-0.6 Y_2O_3
INCONEL MA 6000	Ni-15Cr-4W-2Mo 4.4 Al-2.5Ti-2.0Ta .05C-.01B-.15Zr- 1.1 Y_2O_3

Fig. 2 - MA 754 Forgings   x 0.3

Figure 2 illustrates some examples of MA 754 gas turbine components forged using conventional screw press techniques. The unidirectional structures are achieved by controlling metal flow during deformation and then subjecting the pieces to static recrystallisation for 1 hour at 1315oC. The more highly alloyed MA 6000 suffers from die chilling and resultant cracking if forged as MA 754. The surface therefore has to be protected either by the use of thermal barrier layers during high energy conventional forging or by the alternative technique of isothermal forging in heated dies at slow strain rates. Although the latter method is preferable, the present programme necessitated a conventional forging approach, again using controlled metal flow during deformation.

Unlike MA 754, a dynamic zone annealing treatment is necessary for MA 6000 to develop high aspect ratio, unidirectional structures on secondary recrystalliation. For blade forgings, this introduces further problems. Previous studies have mainly been confined to the high gradient i.e. induction-assisted, zone annealing of axisymmetric MA 6000 bar which has been worked in a highly directional fashion. Forged blades experience much more lateral material flow and also contain sections of widely varying thicknesses. As will be seen later, secondary recrystallisation is strongly influenced by strain input. An induction heating method for complex blade shapes is not economically realistic and hence a lower gradient, indirect heating technique had to be developed to zone anneal forged MA 6000 blades.

PRACTICAL APPROACH - Although most forging work was progressed on attritted powder bar an evaluation of ball milled powder bar of three different batches was included. These batches were produced at Inco Alloys International, Hereford [4] as part of a proprietory development exercise. Effects of the different ball milling conditions on the subsequent response to forging and structural development were to be evaluated.

Input stock whether attritted or ball milled, for all forgings was as-extruded, non-heat treated (un-recrystallised') bar supplied by Inco Alloys International, Hereford. The main forging stages for the subsequent production of the double- ended blades were:-
(1)   Prepare slug
(2)   Extrude using 2.75:1 ratio
(3)   Coin root end
(4)   Coin shroud end
(5)   Finish Forge

A thermal barrier to maintain the MA 6000 at temperature was applied to the slug initially and again at the pre-coining and pre-finish forging stages.

Following forging, the thermal barrier layer was removed and the blades coated in an oxidation-protective medium prior to zone annealing in air. The annealing cycle was completed in a low-gradient, resistance heated furnace depicted in Figure 3.

Fig. 3 - Zone Annealing Furnace

Here up to four blades could be zone annealed in one cycle by placing them end to end in a fixed Incoloy* tube. The resistance furnace was pre-heated to stabilise the zone annealing temperature and then driven over the tube at a preselected rate. A thermocouple probe introduced through

* Trademark Inco Family of Companies.

one end of the tube was used to monitor temperature distribution near or on the blades. The furnace produced temperature gradients of around 50oC/cm in the neighbourhood of the recrystal-lisation interfaces. During an examination of the effects of zone annealing variables on blade structures, a range of furnace translation rates varying from 2 cm/hr. to 10 cm/hour were used in combination with furnace temperature values between 1200oC and 1260oC.

For the mechanical property evaluation part of the programme, zone annealed blades were subjected to a 4-stage heat treatment process to simulate the application of a blade coating cycle and subsequent precipitate development [1]. This cycle was:-
1/2 hr. at 1232oC, air quench + 4 hr. at 1080oC, air quench.
+ 4 hr. at 843oC, air cool + 16 hr. at 760oC, air cool.
Mechanical test pieces were mainly sectioned longitudinally from the aerofoil, and had a gauge diameter of 2.3 mm, gauge length of 9 mm.

At one stage, blades were prepared for possible engine testing. Here pieces in the zone annealed condition had aerofoils and platforms processed to finished dimensions by EDM and polishing prior to any further heat treatment.

Work on interrupted interfaces to observe recrystallisation behaviour was pursued on both blade forgings and segments of extrusions produced at Stage 2 of the blade production route listed above. Such structural fronts were attained by placing a thermocouple at the desired location of the interface and terminating the zone annealing cycle when a temperature of 1160oC was reached at this position. Whilst metallographic analysis was done within the present programme, subsequent interface orientation work and further structural analysis was undertaken at Oxford University [5]. Crystal orientations in the relatively coarse grained recrystallised material were derived by selected area channelling patterns (SACPs) using a rocking beam. For the much smaller unrecrystallised grains, electron diffraction techniques were used on individual sub-micron sized grains [5] [6].

RESULTS AND DISCUSSION

FORGED BLADES - The application of correct TMP parameters and thermal barrier layer characteristics resulted in the production of crack-free blades. Figure 4 shows a forged blade, with the barrier layer removed, and compares it with one which has finish-machined aerofoil and platforms.

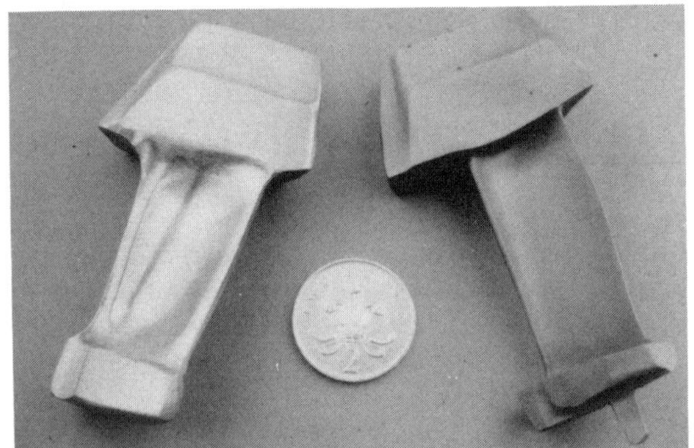

Fig. 4 - Forged and partially machined
         MA 6000 blades x 0.52

Close control of the barrier layer was essential to give a material distribution from which a blade of the correct engineering dimensions could be achieved through subsequent machining.

The low gradient zone annealing technique proved extremely effective in producing high aspect ratio structures, through all sections of the blades, necessary for optimum mechanical properties, i.e. crystal length : diameter greater than 20:1 [7]. This is clearly shown in Figure 5 which compares a statically recrystallised blade with a zone annealed one.

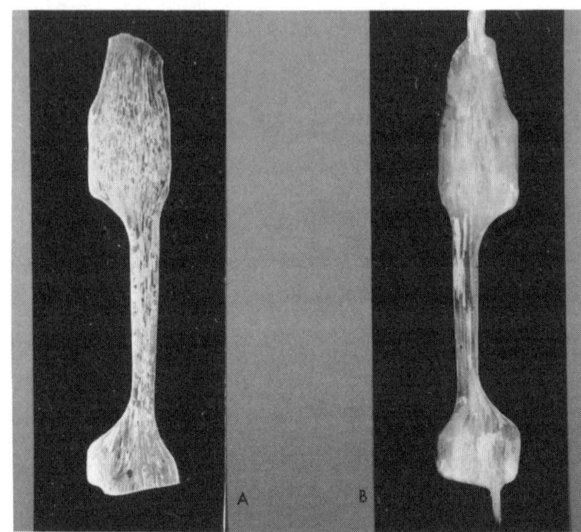

Fig.5 - Statically recrystallised (a)
        & zone annealed (b) structures
        x 0.67.

A high degree of structural control could also be exerted by independent variation of TMP parameters or zone annealing variables. Figure 6 depicts unidirectional structures with crystal diameters varying from 0.2mm to 4mm produced by varying the zone annealing speed and temperature for fixed TMP parameters. Figure 7 gives a transverse aerofoil view.

A similar situation is experienced with variable TMP and fixed zone annealing conditions.

Small crystal diameters are realised when the alloy is more heavily worked i.e. greater strain input. Similar structures are created by zone annealing at higher speeds and higher temperatures. For the smaller grain diameters, aspect ratios greater than 40:1 are easily achieved. With larger crystals, the aspect ratios are limited by the aerofoil length. The processing conditions standardised for blade production resulted in aspect ratios of around 30:1 in the aerofoil sections.

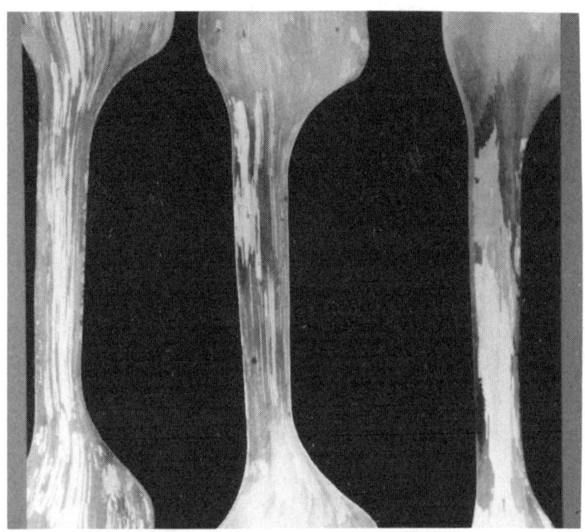

Fig.6 - Changes in grain size produced
        by varying zone annealing
        conditions   x 0.62

Fig.7 - Transverse aerofoil view
        x 0.75

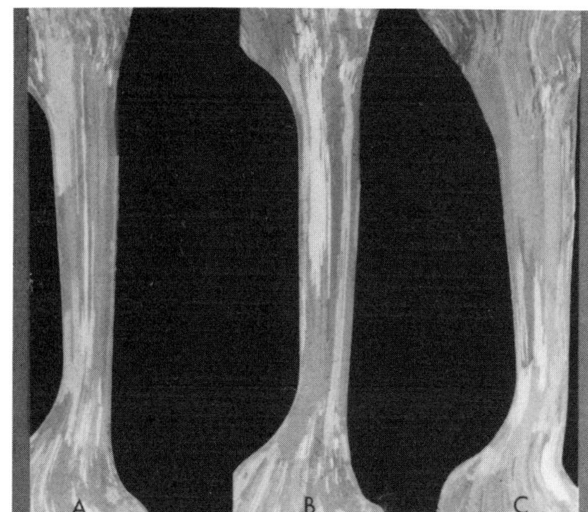

Fig. 8 - Blades forged from different ball milled batches. x 0.56

In the case of the ball milled powder bar feedstock, the three batches responded differently to the standardised forging process. Figure 8 depicts the slight structural variation and shows condition A to have produced some irregularly shaped crystals. Blades from condition C contained a large number of small crystals interspersed with more elongated grains. Condition B produced the most uniform, high aspect ratio structures, essentially the same as the best structures produced with the attritted powder bar. The ball milled type of material generally had a cleaner microstructure than the attritted type. Obviously, powder production conditions influence final structural development even after several stages of thermomechanical processing.

MECHANICAL PROPERTY RESPONSE - Forged and heat treated blades responded similarly to the recrystallised bar product [6]. Testing at 982oC, 816oC and room temperature for creep, stress rupture and tensile, both longitudinal aerofoil and transverse root block, proved this point for intermediate temperatures. The results from elevated temperature tests, which were most relevant to this programme, are listed below. All tests were completed in the longitudinal direction in the aerofoil section on fully heat treated specimens. A comparison is made with results from CMSX-2 single crystal specimens of similar dimensions [1] [8].

Table II

STRESS RUPTURE TESTING OF MA 6000 BLADES

STRESS	TEMP	LIFE	% El	(CMSX-2) [1] [7]
MPa	oC	Hr.		Hr.
103	1177	93	10	
87	1177	195	9	[17]
110	1121	406	6	[47]
138	1093	20	9	
124	1093	511	-	[74]

Rupture lives of the MA 6000 blades are obviously superior to the single crystal alloy. The forged blade rupture properties were around 7 MPa less than those published for recrystallised bar [9]. Previous work has, however, shown this to be mainly a consequence of the small specimen size and that results were the same when a common, minature specimen was used for both [1]. This was attributed to exaggerated oxidation attack promoting earlier failure. A further factor could be the specimen to grain size ratio effect. In small specimens, one or two predominant crystals can adversely influence rupture behaviour.

Further comparative stress rupture tests on aerofoil sections indicated that blades forged from ball milled powder bar had improved lives over those from attritted bar, Table 3. Respectable results were achieved in all aerofoil locations, reflecting the respective presence of high aspect ratio structures.

TABLE III

Forged Blade Stress Rupture Results

	Powder Type	Aerofoil Location	Life hr.
a)	96.5 MPa / 1177oC		
	Attritted	Leading Edge	7.1
	Ball Milled A	" "	6
	Ball Milled B	" "	10.25
b)	117.2 MPa / 1121oC		
	Attritted	Mid	28.5
	Ball milled A	"	35.5
	Ball milled B	"	32
c)	131 MPa / 1093oC		
	Attritted	Trailing Edge	29.5
	Ball milled B	" "	43.75

ANALYSIS OF INTERRUPTED INTERFACES - Partial zone annealing in the low gradient furnace revealed that secondary recrystallisation temperature, and hence interface shape, was non-linear and was determined by the distribution of work in the piece. Figure 9 compares interface shapes in blade aerofoil and root sections. The steeply angled front in the heavily worked aerofoil becomes flatter in the root area. This effect was more pronounced in pieces subjected to the extrusion cycle only. Figure 10 illustrates two interfaces, a and b, produced from a common extrusion :

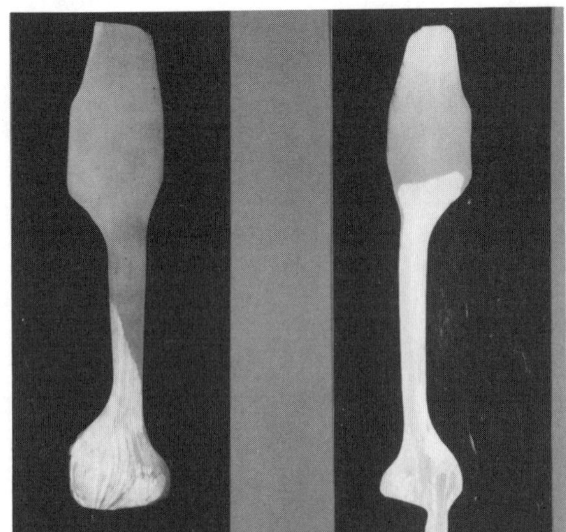

Fig. 9 - Interface shape in forged blades x 0.74

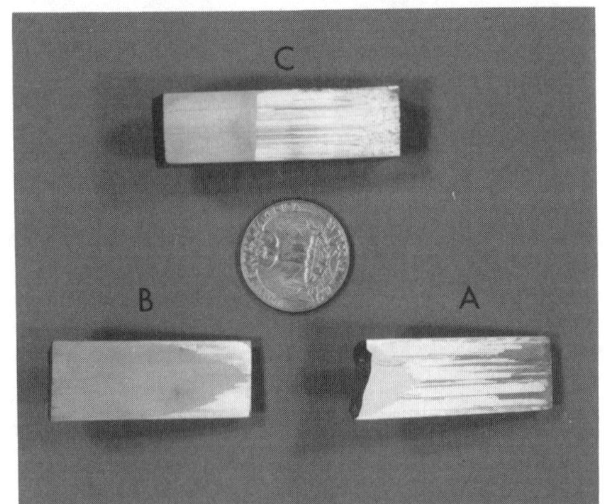

Fig 10 - Interface shapes in one extruded piece, a) and b) and from hot rolled bar c). x 0.6

section a) was from the leading end of the extrusion and section b) from the later stages of deformation. Clearly, the convex shape of the interface at the initial deformation stage becomes much more pronounced further up the extrusion.

For comparative purposes, an interrupted interface from hot rolled bar c) is shown. This obviously has a flat, linear recrystallisation front.

The leading portions of the interfaces in the extruded sections had a secondary recrystallisation temperature of about 1160oC (i.e. the γ' solvus temperature) and the trailing cusp area just above 1200oC. Knowledge of extrusion behaviour revealed that the lower secondary recrystallisation temperature was in the more heavily worked regions. Obviously, the progression of the interface is not solely dependent upon γ' dissolution [10] or by dispersoid coarsening [11], but contains a strain factor.

The analysis at Oxford University of the fine-grained and recrystallised regions around such interfaces [5] [6] led to a proposed model of grain coarsening. Here, it was assumed that since MA 6000 forms a preferred <110> orientation after zone annealing, the grain boundaries could be considered as {110} tilt boundaries, which are energetically favoured for growth.Due to the high atomic density of these tilt boundaries, they hold less solute atoms and hence experience less solute drag. They are therefore more mobile. It was also postulated that the driving force for secondary recrystallisation, Fd, is the reduction in grain boundary area which must be greater than the pinning forces. Fd was given by:

$$F_d = 2\left(\frac{1}{L_1} - \frac{1}{L_2}\right)\cdot\gamma \qquad -1$$

L_1 -unrecrystallised grain size
L_2 -recrystallised grain size
γ -grain boundary energy

Above the γ' solvus, Fd is less than the combined pinning forces of the yttria particles, Fp and the gamma prime precipitate, $F_{\gamma'}$, due to the large pinning force of the latter.
i.e. Fd << Fp + $F_{\gamma'}$ - below γ' solvus.
Recrystallisation does not occur. At the gamme prime solvus, $F_{\gamma'}$ becomes negligible, and Fd was shown to be larger than Fp and grain growth can occur.

Fd > Fp - above γ' solvus.

The lag in secondary recrystallisation with temperature across the extrusions was proposed to be an effect of crystal orientation in the fine grained region ahead of the interface. It was shown that there was a greater number of small crystals with a <110> orientation ahead of the interface in the more heavily worked

regions than in the centre. However,
in this latter location, the number of
<110> grains increased with increasing
temperature (due to easier diffusion)
and the interface lag was suggested to
be caused by a reorientation effect to
<110>. This would thus enhance the
mobility of the boundaries allowing
grain coarsening to occur.

Micrographs in Figures 11 and 12
show that the interface progresses in a
'finger-like' fashion which corresponds
to a longitudinally banded type of
structure. The latter is present in
the fine-grained region ahead of the
interface and continues through into
the coarse recrystallised material. A
single recrystallised grain also
contains several 'fingers'. Such
features, combined with the
orientation-influenced model of growth
[5] suggests a certain mode of
recrystallisation.

The banded structure in the
unrecrystallised material could be
alternating clusters of high and low
concentrations of <110> type grains
caused by differential strain
characteristics during deformation.
Secondary recrystallisation

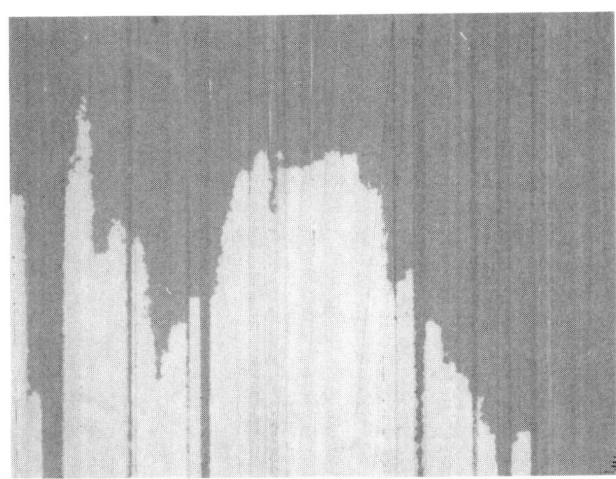

Fig. 12 - Interface micrograph   x 42.5

progresses in advance along the higher
density <110> bands. Fine grained
material in the inter-finger areas
could rotate at the higher temperatures
and recrystallise with the same
orientation as the surrounding
material, possibly aided by an
epitaxial type of growth, to create a
large recrystallised grain. The
appearance of the banded outline within
the recrystallised grain suggests that
small differences in lateral
orientation may occur in these final
stages. Grain boundaries in the
recrystallised material would
resultfrom large differences in
transverse orientation between clusters
of 'fingers'.

A solid state recrystallisation
process as described above has some
similarities with the liquid to solid
phase change in an unidirectional
solidification system. The finger-like
nature of a recrystallisation interface
is comparable to a cellular or
dendritic mode of freezing in the face
of supercooling. Here the last parts
to solidfy i.e. interdendritic sites
have a small orientation difference.
One major contrast is that there
appears to be little competitive growth
in the MA 6000 recrystallisation
process, even in the presence of a
highly curved interface, since growth
occurs in the direction of strain and
not at right angles to the interface.
This is a positive effect for forged
blades since high aspect ratio
structures can still be achieved in
complex shapes containing sections of
variable thickness, and hence different
work input, provided the direction of
strain is controlled. The lack of
competitive growth further suggests

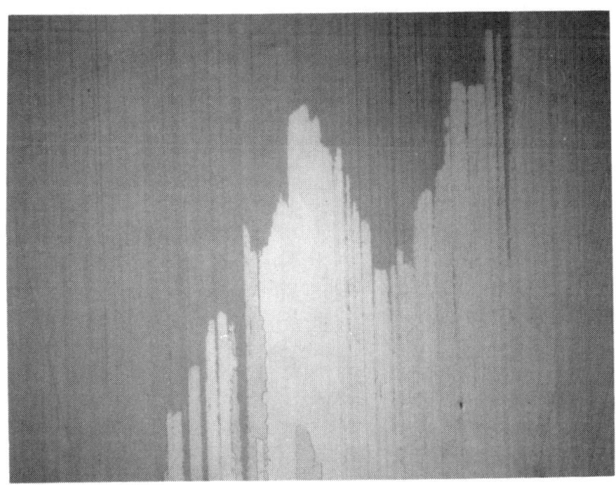

Fig. 11 - Interface microgrgaph   x 42.5

that the ultimate goal of single crystal MA 6000 blades will have to be achieved using other means. The influence of the {110} tilt boundaries on recrystallisation also indicates that improvements to intermediate temperature properties through a texture change would be difficult and a solution through alloy development may be more appropriate.

SUMMARY

The objective of forging crack-free MA 6000 blades by conventional, high strain rate press techniques has been met by the use of a `thermal barrier layer' principle to protect the alloy from die chilling and fracture. Control over shape within the layer was sufficient to produce blades of the correct engineering dimensions for subsequent machining to final tolerances by an EDM plus polishing method.

From a structural point of view, the desired crystals of aspect ratio greater than 30:1 were achieved in blades by control of TMP and zone annealing parameters. This allowed the stress rupture target of recrystallised bar properties to be met. The low gradient zone annealing technique used here is considered to be acceptable to complex shapes from both technical and commercial considerations.

A comparison of blades forged using input bar stock of attritted and ball-milled powder proved both to respond structurally in a similar fashion although the condition of the ball milled powder bar does influence development of final structures. The ball milled powder has a cleaner micro-structure which may be expected to improve mechanical properties. The limited amount of stress rupture testing done here suggests this may be the case.

Examination of recrystal-lisation fronts produced by the interrupted zone annealing of forged pieces revealed that the interface progressed in a curved fashion. Recrystallisation occurred first in the more heavily worked areas and suggested interface shape to be a function of strain and not thermal gradient. Further observations on the `finger-like' nature of localised interface areas lends more credibility to the `orientation' theory of recrystallisation [5] where concentrations of <110> crystals in the fine-grained region ahead of the interface are considered necessary for recrystallisation to occur. Since recrystallisation is dependent upon the direction of strain during deformation, completely longitudinal structures are produced in the forgings in spite of the curved interface.

The information generated during this work has allowed blades to be processed on a production basis.

REFERENCES

1. Ewing, B.A. and S.K. Jain, Proc. of Sixth International Symposium on Superalloys, Seven Springs, Pennsylvania, Sept. 18-22, 1988, 131.
2. Grundy, E. Proc. of International Conf. on PM Aerospace Materials, Luzern. November 2-4, 1987, 12.1.
3. Grundy, E. Proc. of Second Conf. on `Oxide Dispersion Strengthened Superalloys by Mechanical Alloying' London, May 22-25, 1983, 100.
4. McColvin, G. Inco Alloys Int. Hereford, England.
5. Marsh, J.M. and J.W. Martin. Conf. on `Microstructure and Mechanical Processing' Cambridge, March 1990.
6. Marsh, J.M. and J.W. Martin, Univ. of Oxford, Private Communication.
7. Singer, R.F. and E. Arzt. Proc. of Conf. `High Temperature Alloys for Gas Turbine and Other Applications', Deidel, Dordrecht, 1986 Vol. 1., 97.
8. Jain, S.K. Advanced Materials, Allison Gas Turbine, Indianapolis, Private Communication.
9. Incomap Mechanically Alloyed Products Data Sheet, Inconel Alloy MA 6000.
10. Hotzler, R.K. and T.K. Glasgow. Met Trans. 13A, 1665
11. Singer, R.F. and G.H. Gessinger, Met. Trans. 13A, 1463.

DESIGN AND DEVELOPMENT OF
Ni-BASE ODS SUPERALLOYS

Michio Yamazaki, Yozo Kawasaki, Katsuyuki Kusunoki
National Research Institute for Metals
Tokyo, Japan

ABSTRACT

The present authors were in charge of proposing new high γ' Ni-base ODS superalloys prepared by mechanical alloying. This was done in a national project performed from 1982 to 1989, in which four types of advanced alloys were treated. This paper summarizes the alloy design aspect of ODS alloys in this project. We had developed an alloy design program for γ plus γ'type Ni-base alloys and this was utilized for ODS alloys. One of the strongest conventionally cast alloys previously developed by this alloy design program was modified also by using this program, and this lead to an ODS alloy, TMO-2. This alloy is stronger than MA6000. To improve the intermediate temperature strength of TMO-2, the effect of γ' phase volume fractions was next investigated. The effects of the phase volume fraction on recrystallization properties, which are important in obtaining optimal grain structures, as well as on creep rupture strength are described.

INTRODUCTION

Oxide dispersion strengthened (ODS) Ni-base superalloys are considered to have potential applications in many high temperature applications. Especially, ODS alloys having high γ' contents are candidate materials which can be used for gas turbine blades instead of Ni-base single crystal alloys. The mechanical alloying technique developed by J.Benjamin[1] was applied to high γ' type Ni-base alloys to develop alloy MA6000[2]. This alloy, though it has not been used extensively for gas turbine or jet engine blades, many studies have been done for its processes, properties and applications.

A group in National Research Institute for Metals(NRIM) including two of the present authors and others developed a computer-aided alloy design program for γ + γ' type Ni-base superalloys and, by utilizing and improving this alloy design procedure, have been working to develop cast Ni-base superalloys including single crystal alloys[3,4,5),6].

The above mentioned alloy design technique was applied to develop Ni-base ODS superalloys of high volume fractions of γ' phase. The alloys are for gas turbine or jet engine blades. This alloy development was performed as a part of "Advanced Alloys with Controlled Crystalline Structures" in "JISEDAI" project by AIST of MITI. As in the other alloy developments in this project, the authors' group in NRIM was in charge of proposing new alloy compositions. Three companies as members of this project did ODS process studies by using the proposed composition. The present paper describes the alloy composition developmental aspect performed by the authors mainly in the above project. For this were the present authors responsible, the three companies, however, helped the authors make their ODS specimens.

BASIS OF ALLOY DESIGN

The above mentioned alloy design method for γ and γ' type Ni-base superalloys, originally developed for cast alloys, will be briefly described below.

The essential part of the alloy design program is made up of giving pairs of γ and γ' phase compositions in multi-component alloy systems, such as Ni-Al-Ti-Ta-W-Cr. The γ and γ' phases in a pair are materially equilibrium each other.

To select such pairs of γ and γ' phases, we do PHACOMP on γ phase compositions to avoid σ phase formation. We also evaluate solid solutioning degree of γ' phase by introducing a simple parameter named SI. SI stands for "solubility index" and if this is too large, some other phases are considered to appear.Lattice parameters of the two phases can be calculated and the absolute values of the parameters and the differences of them (mismatches) are used to select appropriate pairs.

A large numbers of such pairs selected as appropriate can be given. From any of them, one can obtain series of alloys with various volume fractions of γ' phase; in this series of alloys the compositions of the two phases are kept constant within the errors of calculations. By the program we can get some alloy properties as functions of γ' phase composition(we do not include γ phase composition for this purpose because it is determined as a function of γ' composition and is not independent).

In the first version of our alloy design program[3], we utilized published data of chemically analyzed compositions of some pairs of γ and γ' phases. Those data came from alloys with different heat treatment conditions and hence, together with inaccuracies of phase extraction and chemical analyses, are not ideal. The first version, however, works practically well. In the second version[6], we utilized EPMA analyzed data of the phases obtained by our group. For ODS alloy development we use the first version at the moment.

OUR TARGETS FOR ODS ALLOYS

Our "official" target in the above mentioned JISEDAI Project was as follows.

Rupture life at 1100 C and $14 kgf/mm^2$:
    More than 1000h.
Rupture elongation at that condition:
    More than 5%.

This target was determined by considering the properties of MA6000.

The target above is rather for higher temperatures and as it is known that Ni-base ODS alloy are not so strong at intermediate temperatures[7], we are trying to develop ODS alloys which are stronger at intermediate temperatures and satisfy the above target as well.

EXPERIMENTAL PROCEDURES

The experimental materials were prepared using the procedures as described by Benjamin.[1] Both elemental and master alloy powders were mixed with yttria powders in a ball mill for about three hours in an argon atmosphere and subsequently mechanically alloyed in a high-energy ball mill also with an argon atmosphere for 50 hours at about 50 C. The powders were put in evacuated steel cans at 500 C, cooled to room temperature and sealed..After preheating at 1050 C for 2 hours, they were consolidated by hot extrusion at the temperature with a reduction ratio of 16:1.

For most of creep test specimens, the extruded bars, after the removal of the cans by machining, were zone annealed at 1300 C or 1280 C(only for the alloy designed to have 75 volume % of γ' phase) at a rate of 100 mm/h. Isothermal annealing was also applied to creep test specimens of the alloy designed to have 55 volume % of γ' phase(alloy TMO-2) for the purpose of comparison.

Secondary recrystallization tests were performed for all the alloys by isothermal annealing.

Creep specimens, 4 mm in diameter in the gage portion and non-threaded, were tested in air after preheating at the test temperatures, 800 to 1050 C, for 20 hours.

RESULTS AND DISCUSSION

DEVELOPMENT OF ALLOY TMO-2[8] - We started from one of the strongest Ni-base conventionally cast superalloys developed previously by the authors' group by using the above mentioned alloy design method. It was alloy TM-220(see Table 1). This alloy was designed to have 65 vol. % of γ' phase. We considered that this γ' volume fraction was a little too high to begin with, and we set the fraction at 55 %. We also considered that W content of alloy TM-220 is high and Cr content a little too low, from density and hot corrosion points of view, respectively. From those considerations, we designed alloys TM-303 and TMO-2, the former being a conventionally cast alloy and the latter an ODS one(see Table 1). Alloy TM-303 was tested as a reference alloy in cast form; this has no yttria and contains a higher C content than alloy TMO-2.

Compositions of $\gamma'$and γ phases and vol. fraction of γ' phase of alloy MA6000 was calculated by our program and are given in Table 1.

Creep rupture tests were conducted for those alloys and the results are shown in Fig.1. Alloy TMO-2 shows superb creep rupture strengths at higher temperatures and it satisfies the official target of the project. The creep rupture elongation values were, however, 2 to 4 % and they were lower than the target value(5%).

Photo 1 shows a microstructure of alloy TMO-2 (isothermally annealed resulting in elongated grains) creep-tested and interrupted at about 50,000 hours. This specimen, after tested for 50,000 hours at 1000 C and a stress of 12 kgf/mm^2, showed virtually no deformation. This is considered to be the evidence for the existing of threshold stress in ODS alloys[9]. Photo 1 also shows that structure of alloy TMO-2 is quite stable even after heating for 50,000 hours at 1000 C, except some thin γ'-denuded

Table 1 - Compositions of alloys(mass%) and compositions of phases(atomic%) calculated by the alloy design program.

Alloy	γ'(%)	Ni	Al	Co	Cr	Ti	Ta	W	Mo	Y_2O_3	Note
TM-220	65	58.9	4.9	9.0	5.2	1.0	3.7	14.6	1.8		Hf:0.8,C:0.11
TM-303	55	59.1	4.2	9.9	6.0	0.8	4.8	12.5	2.3		Hf:0.6,C:0.1
TMO-2	55	58.4	4.2	9.7	5.9	0.8	4.7	12.4	2.0	1.1	
γ' phase (at%)		65.0	16.4	7.8	2.7	1.8	2.4	3.4	0.5		
γ phase (at%)		62.9	2.7	13.9	12.6	0.2	0.6	5.0	2.2		
TMO-19	0	55.8	1.1	12.4	9.9	0.2	1.9	14.3	3.2	1.1	
TMO-9	35	57.7	3.1	10.7	7.4	0.6	3.7	13.1	2.4	1.1	
TMO-8	45	58.3	3.7	10.2	6.6	0.7	4.3	12.7	2.2	1.1	
TMO-7	65	59.5	4.9	9.2	5.1	1.0	5.4	12.0	1.7	1.1	
TMO-20	75	60.1	5.5	8.7	4.3	1.1	6.0	11.6	1.5	1.1	
TMO-21	55	59.5	4.4	9.8	5.9	0.9	4.9	12.5	2.0	---	
MA 6000	52	68.4	4.6	---	15.2	2.5	2.0	4.0	2.0	1.1	Zr:0.15
γ' phase (at%)		72.8	15.5		3.9	5.2	0.8	0.8	0.5		
γ phase (at%)		62.7	3.5		29.6	0.5	0.2	1.4	1.9		

C:0.05, B:0.01, Zr:0.05.

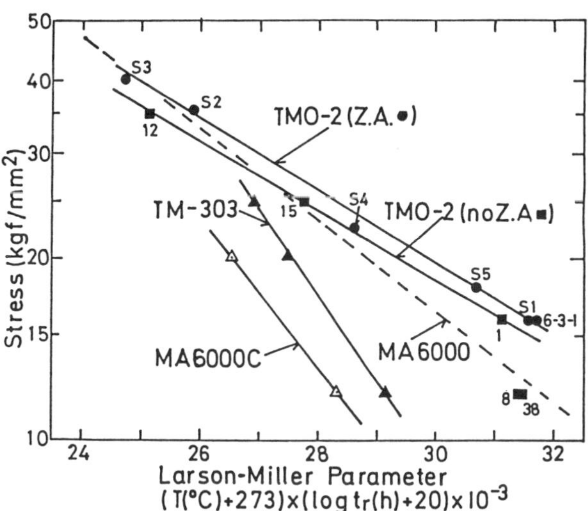

Fig.1 - Creep rupture lives of alloy TMO-2 and other alloys to be compared expressed by Larson Miller Parameters vs. stress. Alloys TM-303 and MA6000C are conventionally cast alloys of the compositions same as alloys TMO-2 and MA6000, respectively, except that they are free from yttria and contain higher C contents. Z.A. and no Z.A. stand for zone-annealed and isothermally-annealed, respectively.

Photo 1 - Microstructure of alloy TMO-2 (isothermally annealed to have elongated grains) after exposure at 1000 C for 50,000 hours at a stress of 12 kgf/mm² in air. Except thin γ'-denuded zone at surface, the microstructre is stable and little creep deformation is observed. The test condition is such that a strongest conventionally cast Ni-base superalloy would rupture in around 500 hours.

zones at the surface. The diameter of this specimen showed no reduction after the test; this also shows that the alloy is sufficiently oxidation-resistant at 1000 C.

EFFECT OF γ' VOLUME FRACTION ON CREEP[10] - To improve intermediate temperature strengths, the increases in γ' volume fractions were intended. For this purpose we rather studied the effect of γ' volume fractions on creep properties.

The γ and γ' phase compositions designed to exist in alloy TMO-2 were used to make alloys with various amounts of γ' phase, and thus we obtained a series of alloys with different amounts of γ' phase with the fixed(on the design basis, of course) compositions of γ and γ' phases.

Fig.2 shows the creep rupture test results for those alloys; see Table 1 for their compositions. It is seen that the intermediate temperature strengths of alloys TMO-7 and TMO-20 are improved; they are higher than those of a single crystal Ni-base superalloy PWA 1480.

The same data are plotted in Fig.3. This shows that the increase of γ' phase improves intermediate temperature strengths as is shown in literature.[7] Fig.4 shows that the increase of γ' phase is also effective for lowering intermediate temperature steady state creep rates.

Fig.5 shows the result of texture measurements by X ray. A {111} texture develops by increasing γ' phase volume fraction. It is known that this orientation is favorable for creep next to {100} orientation in single crystal experiments.

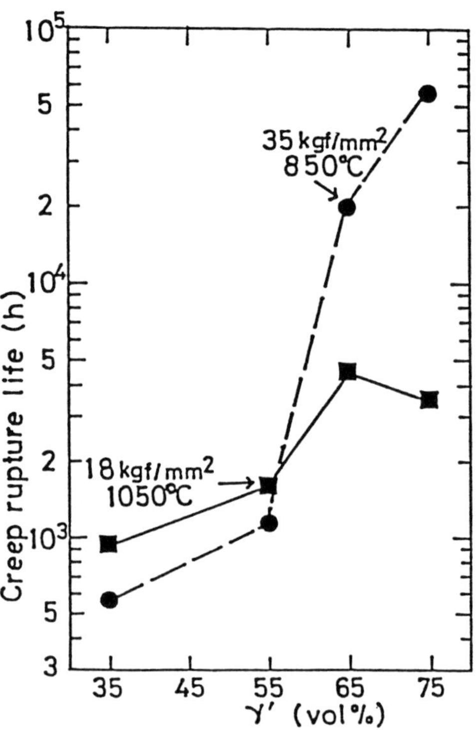

Fig.3 - Effect of γ' phase volume fractions of developed ODS alloys on creep rupture lives.

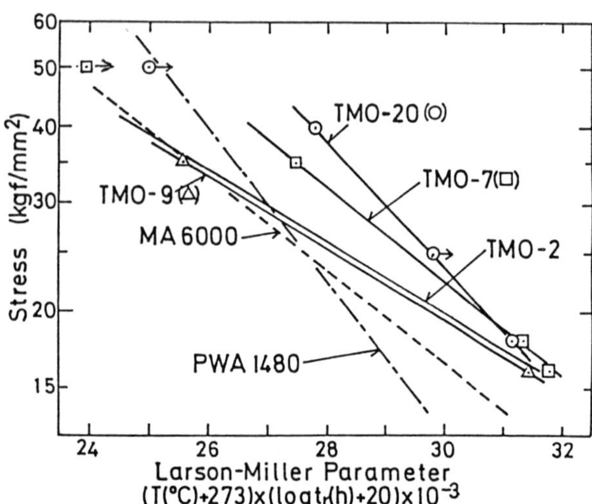

Fig.2 - Creep rupture lives (expressed by Larson Miller Parameters) of developed ODS alloys with various γ' phase volume fractions(see Table 1), MA6000 and PWA 1480(a single crystal Ni-base alloy). ODS alloys are all zone annealed.

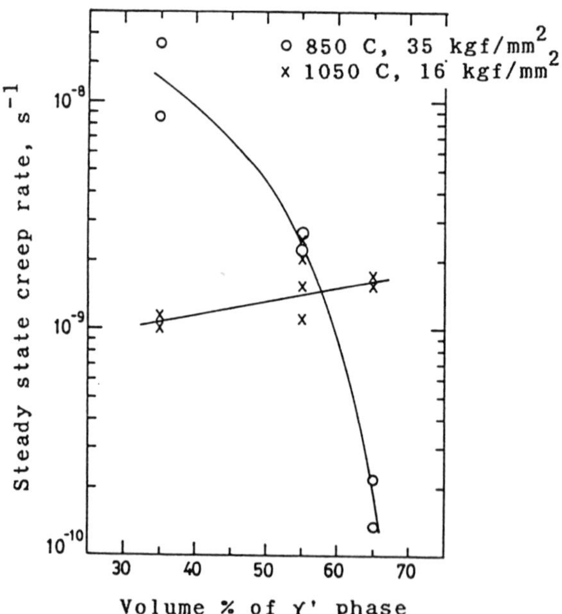

Fig.4 - Effect of γ' phase volume fraction on steady state creep rate of ODS alloys.

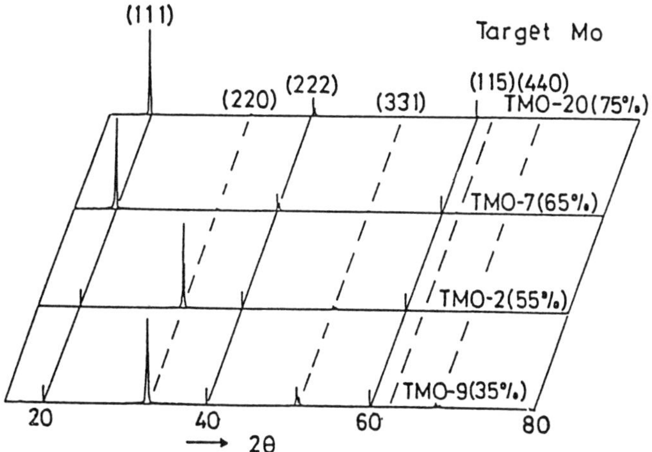

Fig.5 - Effect of γ' phase volume fractions on X ray diffraction patterns indicating the development of a favorable {111} texture by increasing γ' phase volume fraction up to 75 %.

EFFECT OF γ' VOLUME FRACTION ON SECONDARY RECRYSTALLIZATION RESPONSES[11]

- Secondary recrystallization responses of extruded bars (primarily recrystallized) are important for ODS materials. Elongated grains of aspect ratios greater than about 7 are required for good creep rupture properties[12]. When alloys show good secondary recrystallization by isothermal annealing, they can give good elongated grains by zone annealing. Consequently we examined secondary recrystallization responses of alloys by isothermal annealing.

Fig.6 shows Vickers hardness changes of extruded bars of alloys with various amounts of γ' phases. From this figure and SEM observations of specimens, secondary recrystallization temperatures were obtained and are plotted in Fig.7 as a function of γ' phase volume fractions.

Fig.8 shows primarily recrystallized grain size of as extruded bars as a function of γ' phase volume fractions.

By comparing the results of the above two figures, we can say that secondary recrystallization temperatures decrease as primary recrystallization grain sizes decrease by increasing γ' phase amounts. At around 60 % of γ' phase, the secondary recrystallization temperature coincides with the γ' phase solvus temperature. Beyond this γ' phase volume fraction, the secondary recrystallization is triggered by γ' dissolution and hence the recrystallization temperature goes up with increasing γ' phase amounts.

Fig.9 shows the effect of yttria contents on secondary recrystallization temperature (SRx) for the case of 55 volume % of γ'. The addition of yttria reduces the primary recrystallization grain size(Fig.8). In spite of this effect, the addition of yttria retards the secondary recrystallization.

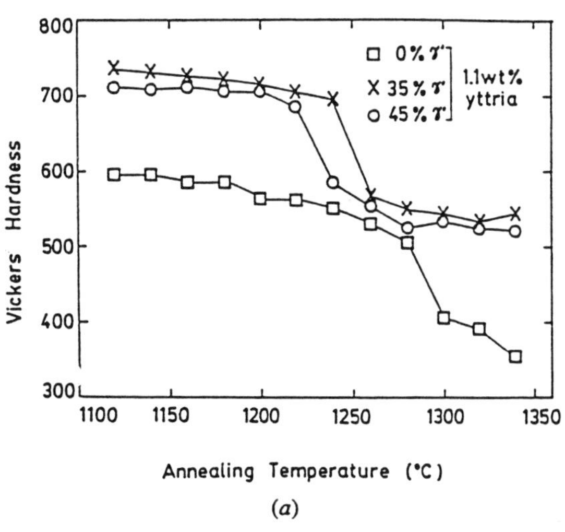

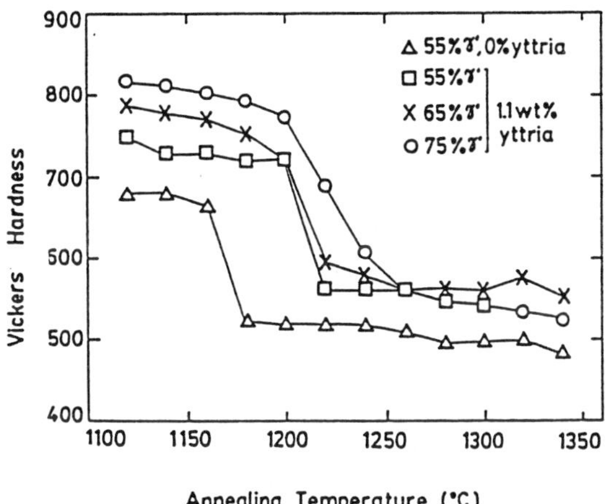

Fig.6 - Vickers hardness of the materials isothermally annealed at various temperatures for 1 hour; (a):alloys with γ' phase volume fractions of 0 to 45 % and (b):alloys with the fractions of 55 to 75 %.

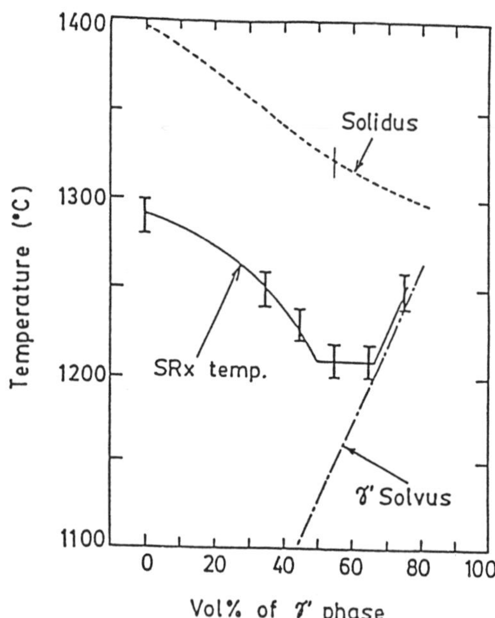

Fig.7 - Effect of Y' phase volume fractions on secondary recrystallization (SRc) temperature of alloys containing 1.1 mass % of yttria.

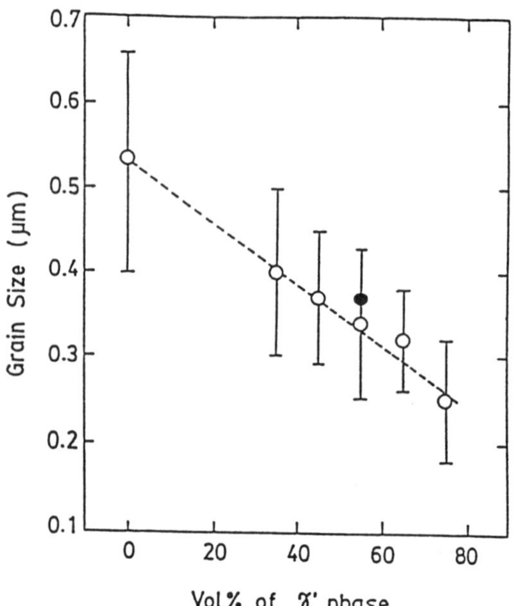

Fig.8 - Effect of Y' phase volume fractions on primary grain size (grain size of as extruded material); a closed data point is for alloy without yttria and others for alloys with 1.1 mass % of yttria.

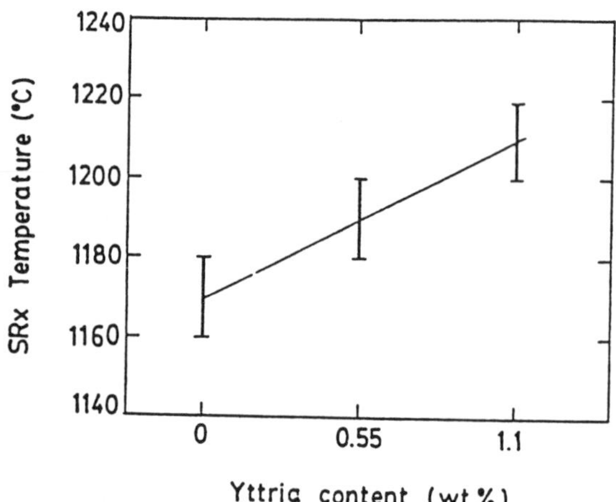

Fig.9 - Effect of yttria content on secondary recrystallization (SRx) temperature for alloys with 55 % of Y' phase.

CONCLUSIONS

1. A new strong ODS alloy, TMO-2 was developed by utilizing a computer-aided alloy design program.
2. To improve the intermediate temperature strength, alloys with higher Y' contents were developed to get alloys TMO-7 and TMO-20.
3. By increasing Y' phase content, the texture after zone annealing changed from {100} to {111}.
4. A specimen of alloy TMO-2 tested at 1000 C and 12 kgf/mm^2 in air for 50,000 hours and interrupted showed little deformation and oxidation.
5. The increase in Y' phase content lowered the secondary recrystallization temperature until about 55 % of the phase and beyond this, the secondary recrystallization. temperature coincided with the Y' solvus thus increased with increasing the phase content.

Acknowledgments

The authors are grateful to Mr S. Ochi and Mr N. Uenishi of Sumitomo Electric Industries Co. Ltd. and Dr K. Mino of IHI for their assistance in preparing ODS materials. Most of the present work was performed in "Advanced Alloys with Controlled Crystalline Structures", one of the themes in "Basic Technology for Future Industry(Jisedai Project)" sponsored by the Agency of Industrial Science and Technology, MITI.

REFERENCES

1) Benjamin, J. S., Met.Trans. $\underline{1}$, 2943(1970)
2) U.S.Patent, No.3,926,568, Dec.16, 1975, INCO.
3) Harada, H. and M. Yamazaki, Tetsu-to-Hagane $\underline{65}$, 1059(1977)
4) Harada, H., M. Yamazaki, Y. Koizumi, N. Furuya and H. Kamiya, "Proceedings of International Conference on High Temperature Alloys for Gas Turbines, held in Liege, Belgium, Oct., 1981", pp.721-735 D. Reidel Publishing Co.(1981)
5) Yamagata, T., H. Harada, S. Nakazawa, M. Yamazaki and Y. G. Nakagawa, "Superalloys 1984", p.147, The Metallurgical Society of AIME(1984)
6) Harada, H., K. Ohono, T. Yamagata, T. Yokokawa and M. Yamazaki, "Superalloys 1988", p.733, The Metallurgical Society(1988)
7) Gessinger, G.H., "Powder Metallurgy of Superalloys", p. 287, Butterworths, London (1984)
8) Kawasaki, Y., K. Kusunoki, S. Nakazawa and M. Yamazaki, Tetsu-to-Hagane $\underline{75}$, 529(1989)
9) Singer, R.F. and E. Arzt, "High Temperature Alloys for Gas Turbines and Other Applications(Proceedings for a conference held in Liege, Oct. 6-9, 1988)", p.97, D.Reidel Publishing Co.
10) Kusunoki, K., Y. Kawasaki, S. Nakazawa and M. Yamazaki, Tetsu-to-Hagane $\underline{75}$, 1588(1989)
11) Kusunoki. K., K. Sumino, Y. Kawasaki and M. Yamazaki, Met. Trans. $\underline{21A}$, 547(1990)
12) Benjamin, J.S. and M.J. Bomford, Met.Trans. $\underline{5}$, 615(1974)

GRAIN GROWTH OF γ' STRENGTHENED Ni-BASE ODS ALLOY

Noboru Uenishi, Yoshinobu Takeda
Sumitomo Electric Industries, Ltd.
Itami, Hyogo, Japan

ABSTRACT

Grain growth of γ' strengthened Ni base ODS alloy, TMO-2, were characterized experimentally. Mechanical alloying and hot powder extrusion were performed to make billets and grain growth properties during annealing of the billets were evaluated by using optical and transmission electron microscopy together with microhardness measurements, X-ray diffraction and ECP (electron channeling pattern). It was proved that grain growth was affected by some factors of manufacturing process conditions. The first factor is stored energy which was introduced into extruded billets during hot powder extrusions. Since it is the driving force for grain growth, the more stored energy there was, the larger grains grew. Another factor is contamination of MAed (mechanically alloyed) powder by oxygen, nitrogen and iron which acts as a factor to restrict grain growth. It was also proved that texture development has some co-relationship with the grain growth. As extrusion temperature was lowered, [111] and [100] fiber textures were more strongly developed and grains grew large during annealing to be over 1000 μm. Moreover, hot rolling of a extruded billet developed [110] fiber texture and grains in the rolled billet grew to over 5000 μm in rolling direction after annealing.

ODS ALLOYS are produced by using MAing and hot powder extrusion techniques(1-3). By MAing, Y_2O_3 particles are dispersed in matrix uniformly. The high strength of the alloy at elevated temperature originates from the dispersed Y_2O_3 particles. But the ODS alloys for high temperature use should have not only the dispersed particles but also large grains (more than 1000 μm in length) in order to avoid grain boundary sliding(4). To make ODS with these large grains, extruded bars which have very fine grains (less than 1 μm) are annealed. In the course of annealing, recrystallization takes place and then fine grains grow large. In the case of γ' strengthened Ni base ODS alloy(5), zone annealing method is used to get larger grains than normal annealing (isothermal annealing) could get. Therefore, it is important to make billets which are easy to recrystallize and to have large grains on annealing.

In this work MAing and hot powder extrusion were performed and grain growth property of γ' strengthened Ni base ODS alloy, TMO-2, was

Table 1 The nominal composition of TMO-2 alloy

ALLOY	γ' Vol%	Ni	Cr	Co	W	Mo	Ta	Ti	Aℓ	Zr	C	B	Y_2O_3
TMO-2	55	58.4	5.9	9.7	12.4	2.0	4.7	0.8	4.2	0.05	0.05	0.01	1.1

(wt%)

investigated. In the investigation isothermal annealing method was employed for the convenience of experiments and to avoid possible effects of zone annealing condition. The purpose of this paper is to point out several factors that intervene or promote grain growth.

EXPERIMENTAL PROCEDURE

The nominal composition of TMO-2 is listed in Table 1. Raw powders are elemental powders of Ni, Cr, Mo, Co, W, Ta and prealloyed master alloy powders of Ni-14wt%Al-28wt%Ti, Ni-46wt%Al, Ni-14wt%B, Ni-30wt%Zr and then yttria powder. All these powders were blended in such proportions as to give an overall composition indicated in Table 1 and then charged into a water cooled high energy ball mill, called an attritor. The chamber of the attritor was filled with 3 kg of blended powder and 50 kg of steel balls. The powder was milled for 180 ks at 200 rpm under an argon gas atmosphere. The resultant MAed powder was packed into vented mild steel cans and evacuated to less than 10^{-2} Pa for 7.2 ks at 733 K and then sealed. The cans were extruded to consolidate the MAed powder. The conditions employed to produce extruded bars (from 40 to 50 mm in diam.) were temperatures from 1343 K to 1373 K and extrusion ratios from 10 to 16. Following the extrusion, extruded bars were annealed isothermally for 3.6 ks at temperatures from 1513 K to 1593 K.

Oxygen, nitrogen and iron contents in MAed powders were analyzed. Microhardness measurement was carried out on extruded billets and annealed billets. Microstructures of both type were examined by optical microscope, and for one extruded billet that was extruded at 1343 K and extrusion ratio of 16, microstructure was examined by transmission electron microscopy. Texture analyses of extruded billets were performed by X-ray reflection method. Using filtered Cu Kα radiation, the intensities of (111), (200), (220) and (113) reflections were plotted to obtain the appropriate pole figures. ECP analyses were carried out on the large grains in annealed billets in order to examine the orientation of them.

Tensile specimens were taken from an extruded billet that was made at an extrusion temperature of 1343 K and extrusion ratio of 16. They were deformed uniaxially under conditions of 1323 K and strain rate from 10^{-4} to 10^{0} sec^{-1}. The same billet was hot rolled at 1373 K. By rolling, the billet, which had had a round cross section of 40 mm diam., had two flat edges with a space of 36 mm between them. Both the elongated specimen and the rolled billet were annealed isothermally at 1573 K. A part of these specimens was investigated by X-ray and ECP diffraction methods.

RESULTS AND DISCUSSION

EFFECT OF STORED ENERGY - Transmission electron micrograph of a extruded billet in the section parallel to the extrusion direction is shown in Figure 1. It is composed of very fine grains which contain particles the size of about a few hundred Å. An energy dispersive X-ray analysis revealed that these small particles contain yttrium and aluminum which means they are supposedly yttrium alumite. Yttrium alumite could be formed by reaction of yttria particles

Fig. 1 Transmission electron micrograph of a TMO-2 extruded billet in the section parallel to the extrusion direction.

0.5 μm

with aluminum in solution and oxygen from atmosphere or decomposition of fragmented oxides of master alloy. There is a wide variation in grain size and shape. Grain size is mostly between 0.1 and 0.6 μm in diameter. Most of the grains were equiaxed and had lattice defects, but there were a small amount of elongated grains and there were also some grains which had few lattice defects. These observations might indicate that dynamic recovery and dynamic recrystallization had taken place during hot extrusion and a small amount of static recrystallization had also taken place after that.

Three kinds of extruded billets were made at 1373 K. They had different extrusion ratios of 10, 12 and 16. The microhardness of the billets before and after annealing is shown in Figure 2. As annealing temperature was raised, the microhardness of them went down. In these three billets one that was extruded at an extrusion ratio of 10 showed a slow declination of microhardness and its grains didn't enlarge (less than 10 μm) during the annealing at any annealing temperature. On the other hand the other two billets showed steep declinations and large grains (more than 100 μm in their longitudinal directions) could be obtained at temperatures over 1533 K although the annealed billets with an extrusion ratio of 12 had regions without large grain. Over 1533 K the grain sizes and shapes were almost equal in each series of annealed billets. So it is apparent that when the extrusion temperature is 1373 K,

an extrusion ratio over 12 is necessary for billets to have large grains.

The effect of extrusion temperature was also investigated. Figure 3 shows two kinds optical macrostructures of annealed billets. Both billets were extruded at the same extrusion ratio of 16 and annealed at the same temperature of 1573 K for 3.6 ks. The difference in their manufacturing condition is extrusion temperature. As can be seen from the picture, one that was processed under lower extrusion temperature has bigger grains than the other. So in the examined temperature and extrusion ratio range, as the temperature was lowered and the extrusion ratio was raised, the grains obtained after annealing became large.

By the two observations about the effects of extrusion temperature and extrusion ratio, it is considered that stored energy introduced in the course of extrusion has the effect of promoting grain growth. Firstly, grain size was larger under condition of high extrusion ratio than low extrusion ratio, and high extrusion ratio gives billets a large amount of deformation which results in high stored energy in billets. Furthermore grain size was larger under condition of low extrusion temperature than in high extrusion temperature, and larger energy could be stored at low extrusion temperature than at high extrusion temperature because of low frequency of lattice defect annihilation at low temperature.

EFFECT OF CONTAMINATION - In the course of MAing MAed powder is more or less contaminated with oxygen and nitrogen from atmosphere and with iron from balls and chamber made of iron alloy. In executing mechanical alloying, argon gas atmosphere is employed and the surface of balls and chamber is coated with MAed powder. However, very little contamination might take place. Tow series of MAed powders were made to examine the effect of contamination on grain growth. Table 2 shows the compositions of

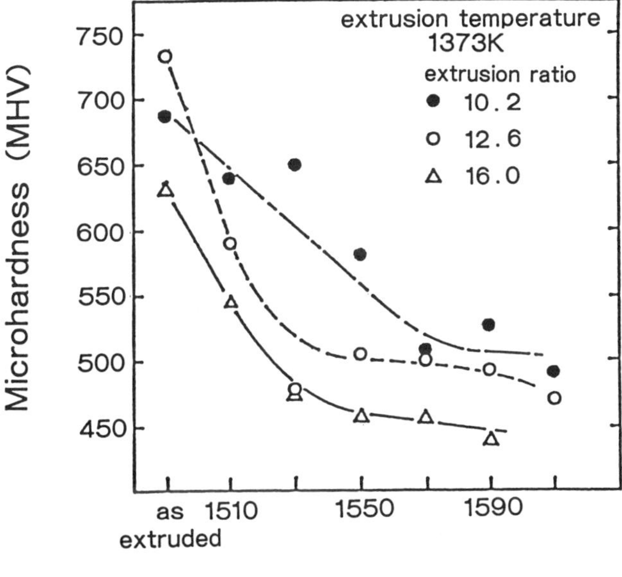

Fig. 2 Effect of annealing temperature and extrusion ratio on microhardness of annealed TMO-2 billets. (annealing time is 3.6 ks)

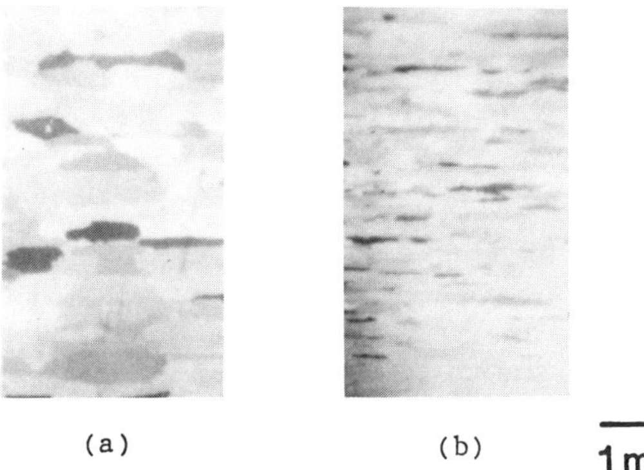

Fig. 3 Macrostructures of extruded and annealed TMO-2 billets

extrusion temperature: (a) 1343 K (b) 1373 K
extrusion ratio: 16
annealing condition: 1573 K x 3.6 ks

Table 2 Oxygen, nitrogen and iron contents in 2 kinds of 6 batches of MAed powders

	high contaminated powders	low contaminated powders
Oxygen	0.84 ~ 1.18	0.72 ~ 0.82
Nitrogen	0.07 ~ 0.11	0.02 ~ 0.03
Iron	0.49 ~ 1.1	0.15 ~ 0.55

wt%

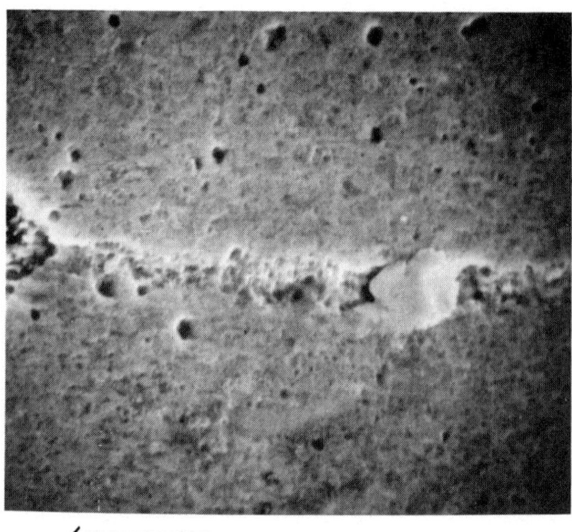

Fig. 5 Void formation seen in the annealed billet that contains high level contamination (on the plane parallel to extrusion direction)

oxygen, nitrogen and iron of the MAed powders. Each series contains 6 batches of MAed powders. Each series of the MAed powders were mixed and extruded at an extrusion temperature of 1353 K and extrusion ratio of 12 and then annealed. Microhardnesses of produced billets before and after annealing is shown in Figure 4. At all annealing temperatures microhardness of the low contaminated billet is lower than the high contaminated billet. The low contaminated billet took a sudden drop in microhardness at 1533 K and its grains grew large (more than 100

μm in their longitudinal direction) at a temperature over 1533 K, while the high contaminated billet didn't show either the sudden drop or large grains (they were less than 10 μm). As is seen from Figure 2 and Figure 4, a sudden drop in microhardness can be interpreted to reflect grain growth. The fact grains didn't grow large in the high contaminated billet indicate that these contaminants are obstacles for grain boundary to migrate. Figure 5 shows scanning electron micrograph in the section parallel to extrusion direction of the high contaminated billet after annealing. There were a lot of small voids which ran parallel to extrusion direction. There weren't any voids in as-extruded state, which means formation of voids during annealing and existence of some constituents that become gas and make voids. There weren't any voids visible in the low contaminated billet, therefore its grains became large.

EFFECT OF TEXTURE - So far there has been some discussion about two factors that have effects on grain growth. Whereas the contamination of MAed powder by oxygen, nitrogen and iron was proven to create obstacles to grain growth, it was also proved that stored energy given to billets by extrusion had the effect of promoting grain growth of TMO-2 alloy. In this study another possible factor was investigated. According to our explanation, stored energy is the driving force for grain growth. So in order to accumulate as much energy as possible, extrusion conditions of low extrusion temperature and high extrusion ratio should be employed to make extrusion billets with favorite grain growth response. However there must be a limit to the

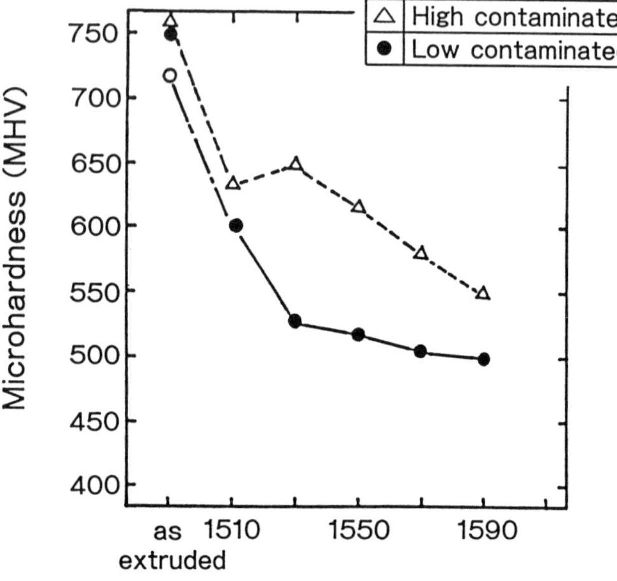

Fig. 4 Effect of annealing temperature and contamination level on microhardness of annealed TMO-2 billets
(annealing time is 3.6 ks, extrusion temp. and ratio are 1353 K and 16 respectively)

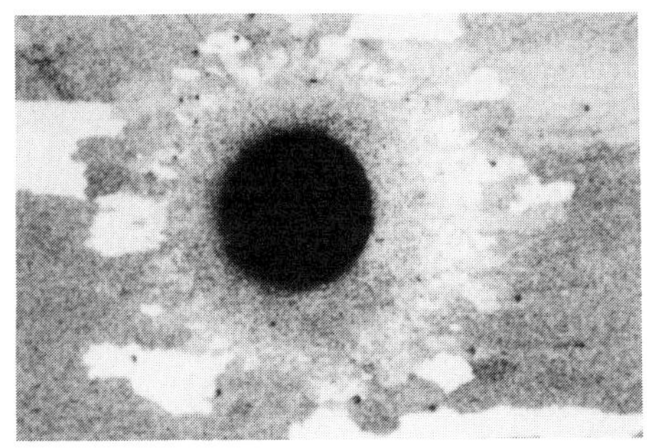

200μm

Fig. 6   Microstructure around indentation
after annealing at 1573 K for 3.6 ks   (indentation was made by Rockwell C hardness tester)
   extrusion temperature = 1353 K
   extrusion ratio = 12

condition we can choose. Since under too high
ratio and too low temperature conditions, very
high pressure will be required of extrusion
press equipment for a billet to break through.
So in any extrusion equipment it is natural that
there is a limit to the condition, which results
in the limit of improving grain growth property.
Finding out another factor that has an effect on
grain growth might lead to improved grain growth
property. The purpose for seeking another
factor is to know more about grain growth and to
improve the grain growth response of extruded
billets.

   A test piece with a plane parallel to
extrusion direction was taken from an extruded
billet and an indentation was made on the surface of the plane by an indenter and the test
piece was annealed at 1573 K. The microstruc-

ture around the indentation is shown in Figure
6. As seen from Figure 6, grains distant over
200 μm from the indentation were large, while
grains within about 200 μm from it were small
and their grain sizes were within 100 μm.
According to our explanation, stored energy is
the driving force of grain growth, and there
must have been a large amount of stored energy
around the indentation before annealing, but the
grains didn't become large. So the question
arises whether or not the more stored energy
there is the larger grains become after annealing. It might be considered that there is the
most suitable range of stored energy to make
grains large. Another conclusion that could be
derived from Figure 6 is about the shapes of
grains.   Grains near the indentation are
equiaxed. They aren't elongated in the extrusion
direction, whereas grains at a distance are
elongated parallel to the extrusion direction.
The reason why equiaxed grains were obtained
might be because anisotropy near the indentation
had disappeared before annealing was executed.
Because extrusion makes each MAed powder elongated, there is anisotropy in the extruded
billets. Compared with extrusion, the indentation gives material isotropic strain. So we
might have to take it into consideration that
direction has some relationship with grain
growth.

   Texture analysis of extruded billets were
performed by X-ray diffraction method.   The
intensities of {111}, {200}, {220} and {113}
reflections were plotted to make pole figures.
Figure 7 shows an example of the drown pole
figures. In this figure, extrusion direction is
parallel to ND (normal direction) and the
intensities are presented in units where average

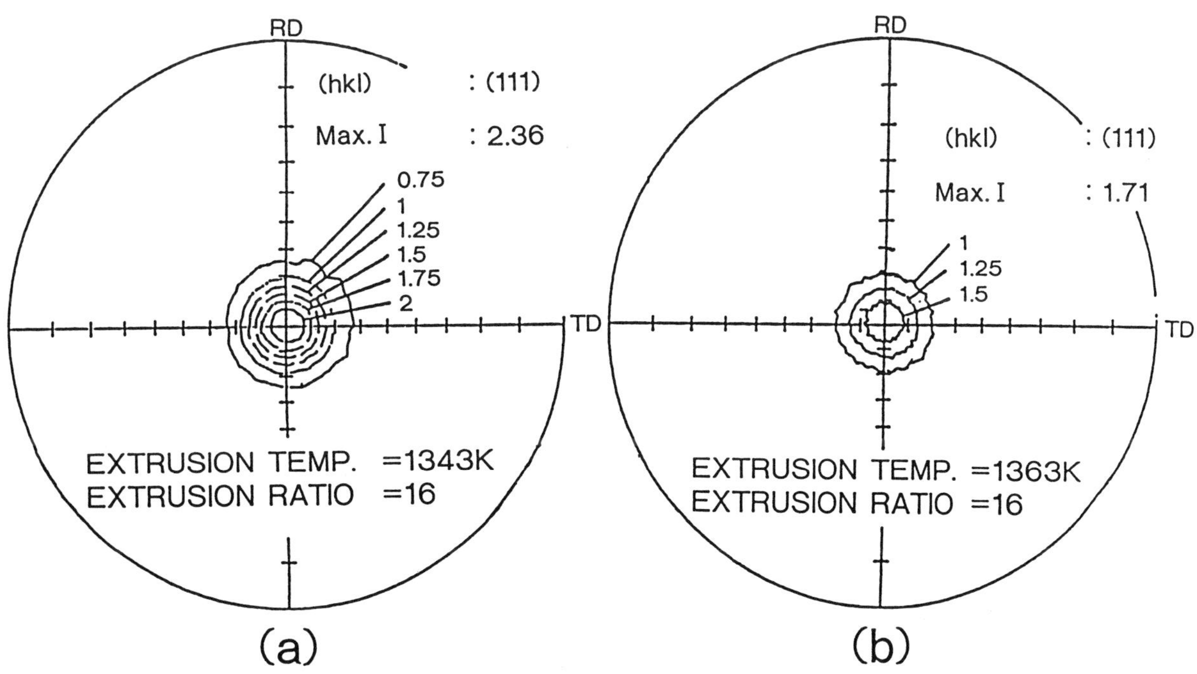

Fig. 7   {111} pole figures from the transverse
sections of extruded billets

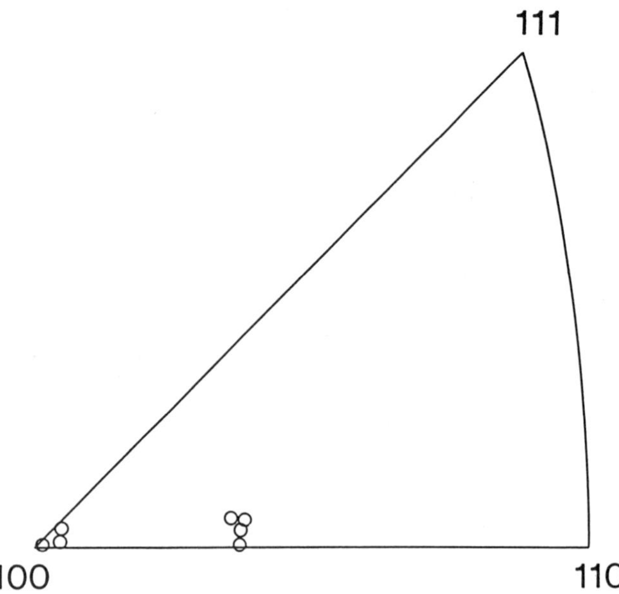

Fig. 8   Orientations of grains in an extruded
(extrusion temperature = 1363 K) and annealed
billet (each point shows the extrusion direction in a crystal)
            extrusion ratio: 16
            annealing condition: 1573 K x 3.6 ks

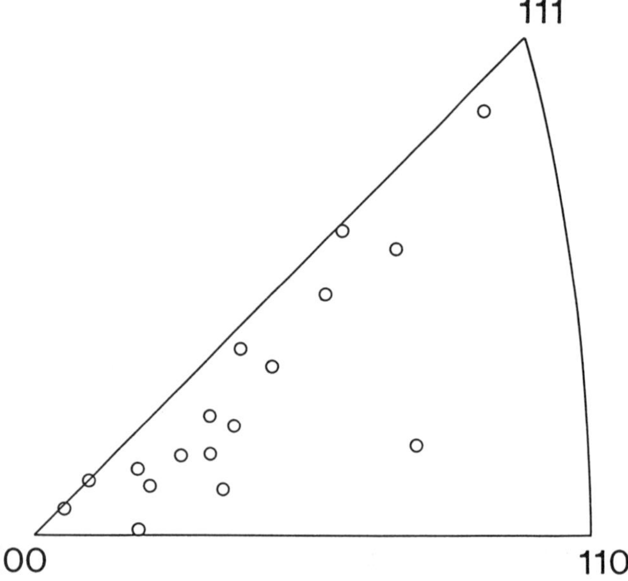

Fig. 9   Orientations of grains in an extruded
(extrusion temperature = 1343 K) and annealed
billet (each point shows the extrusion direction in a crystal)
            extrusion ratio: 16
            annealing condition: 1573 K x 3.6 ks

intensity value is unity. Although the range of
α between 60° and 90° is shown in the figure, the
range between 30° and 90° was also examined. As
seen from the figure, [111] fiber texture is
developed in the extruded billets examined.
From other pole figures, it was also proved that
[100] fiber texture was developed in the billets. The intensity of {220} was strong at a
little over 30° from the extrusion direction.
The intensity of {113} was weak, and it was the
strongest at the direction parallel to the
extrusion direction. In Figure 7 the texture of
a billet that was extruded at 1343 K is compared
with one that was extruded at 1363 K. Both
billets were made at the same extrusion ratio of
16 but, after annealing, grains grew larger in
the former than in the latter. The reason for
this might be due to the difference of stored
energy according to the former discussion. But
Figure 7 shows there was also a difference of
developed textures. The [111] fiber texture are
more strongly developed in the former than in
the latter. So there is a possibility that the
stronger the fiber texture is developed the
larger grains grow during annealing.

The orientation of each grain that had
grown large by annealing was examined. Specimens were taken from two kinds of extruded
billets of which the textures were examined
respectively as seen in Figure 7. The specimens
were annealed for 3.6 ks at 1573 K. Figure 8
and 9 show the orientations of grains in the
annealed specimens. In these figures the extrusion direction in each crystal is plotted in the
stereographic triangle. In Figure 8 the extrusion directions are concentrated in the vicinity
of [100] and [310], while they are lying along
the [100]-[111] line in the stereographic triangle in Figure 9. As there are only 7 points
plotted in Figure 8, a result similar to Figure
9 might have been obtained if many measurements
had been executed in Figure 8. By comparing the
results about texture with Figure 8 and 9, it
can be seen that grains after annealing have
almost the same orientations with those before
annealing. The orientations near [310] may
correspond to the strong {220} intensity that
was seen at a little over 30 from the extrusion
direction.

Tensile specimens were taken from an extruded billet which was made at an extrusion
temperature of 1343 K and extrusion ratio of 16.
They were tensile tested at 1323 K and at the
strain rate of 10^{-4}, 10^{-3} and 10^{0} sec^{-1} . The
tensile direction is identical to extrusion
direction. The elongations measured were 85%,
150% and 10% respectively and the maximum
stresses were 45, 60 and 108 MPa. The tensile
tested specimens were then annealed. Figure 10
shows macrostructure of the tensile tested and
annealed specimen. The left side of the figure
is the region where the test piece was deformed
severely. There were not any large grains
visible in this region and X-ray diffraction
revealed that no strong texture exists in this
field. However on the right side of the figure,

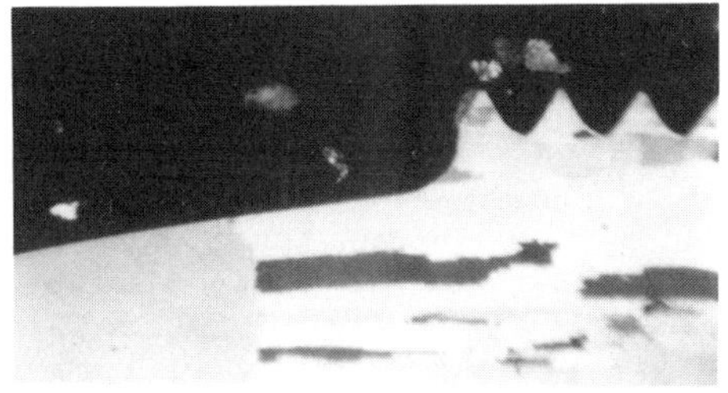

Fig.10   Macrostructure of a tensile tested specimen after annealing
        tensile strain rate: $10^{-3} sec^{-1}$
    tensile test temperature: 1323 K
        annealing condition: 1573 K x 3.6 ks

which is the chucking portion of the tensile test piece, there were very large grains of over 5 mm in the tensile direction. The grain size in this region is larger than that of the extruded and annealed billets which were shown in Figure 3. The textures of the regions in the three kinds of tensile specimens were examined. Pole figures were drawn and the angles between the tensile direction and peak intensity directions of {111}, {200} and {220} were measured. In the specimen of 10^{-4} sec^{-1} strain rate, the tensile direction was possibly [111] and [100]. In the specimen of 10^{-3} sec^{-1} strain rate, it was [211] and [310]. For that of 10^{0} sec^{-1}, it

was [310] and [110]. So it seems that tensile direction rotate from the [100]-[111] line toward [110] as tensile strain rate becomes higher. The rotation of crystal orientation and changes in the specimens' shapes reveal that this right side of the figure was deformed. ECP analysis revealed that the axis of tensile test was lying in the vicinity of [110] in the annealed state as shown in Figure 11.

Based on the examination of the tensile specimens, hot rolling was performed to obtain large grains that are similar to those obtained in the tensile tested specimens. Hot rolling was done with the same billet that was used for the tensile tests. In the hot rolling, rolling direction was identical to the extrusion direction of the billet. Figure 12 shows the macrostructure of the rolled and annealed billet. The grains in Figure 12 are almost as large as those of annealed tensile specimens and they are much larger than those of the annealed extrusion billets. The orientations of grains in Figure 12 are shown in Figure 13. In the figure the rolling direction in each crystal is plotted in the stereographic triangle. Although there is some scattering of the orientation, the rolling directions in many grains are lying near [110] and there are only few of them near the [100]-[111] line where the rolling direction (extrusion direction) existed before the rolling.

So far there has been discussion on some factors that have effects on grain growth. Of course, there may be some correlations between the factors. For example, even if much contamination is introduced in MAed powder, large grains could be obtained by employing a high extrusion ratio because such impurities as oxides can be broken into pieces by severely deforming the MAed powder severely. By employing a high extrusion ratio, stored energy is enhanced, which also recoups grain growth ability.

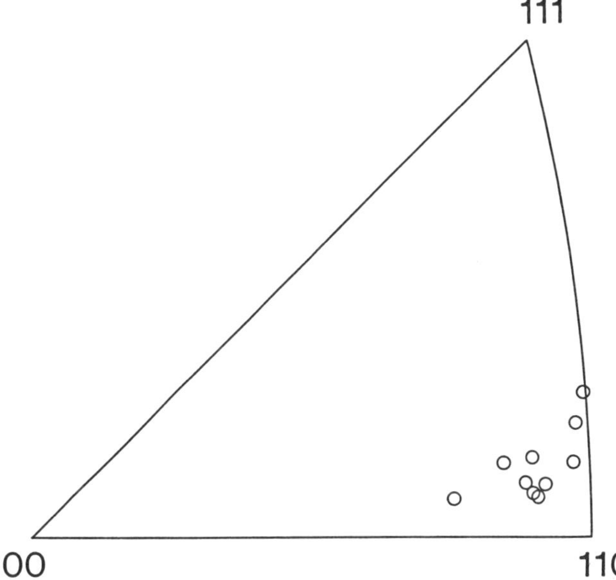

Fig.11   Orientations of grains in a tensile tested specimen after annealing (each point shows the tensile axis in a crystal)
        tensile strain rate: $10^{0} sec^{-1}$
    tensile test temperature: 1323 K
        annealing condition: 1573 K x 3.6 ks

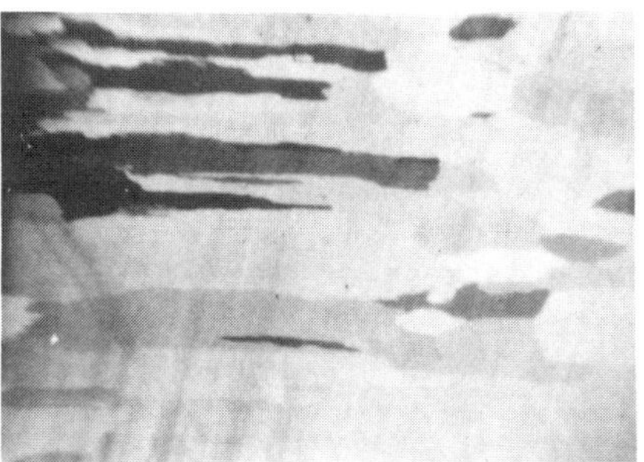

Fig.12   Macrostructure of hot rolled billet after annealing
        rolling temperature: 1373 K
        annealing condition: 1573 K x 3.6 ks

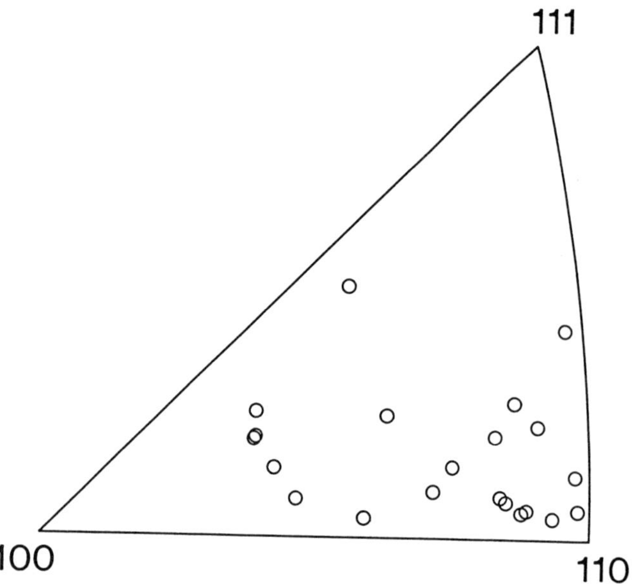

Fig.13   Orientations of grains in a hot rolled
billet after annealing
(each point shows the rolling direction in a
    crystal)

Although all of the effects of texture were
not clarified, it was proven that texture has
some relationship with grain growth. In Figure
10, the microhardness before annealing in the
severely deformed region (the left side of
Figure 10) was about 720 which was almost the
same with that in the chucking region, 725, (
the right side of the figure). The microhardness of the rolled billet was about 740 and that
of the before-rolling billet was 730, so there
wasn't much difference between them. If microhardness is a rough gage of the average stored
energy density, we could suppose the stored
energy in these regions were almost the same,
nevertheless the sizes of their grains after
annealing were different. The significant
difference between them was in texture as seen
in Figure 7, 8, 9, 11 and 13, so possibly the
effect of texture on grain growth exists. The
observation of recrystallization around indentation and the fact that decomposition of texture
at the severely deformed regions in the tensile
tested pieces arrested grain growth could also
support the existence of the effect of texture.

The reason of the texture decomposition in
the tensile tested pieces could be attributed to
the occurrence of superplastic deformation.
They showed total elongations of 85%, 150%, and
250% at strain rates of 10^{-4} , 10^{-3} and 10^{-2}
sec^{-1} respectively and showed m value over 0.3
at the strain rate range between 10^{-3} and 10^{-1}
sec^{-1} at 1323 K. In superplastic deformation
grain boundary sliding takes place, which results in the decomposition of texture. In the
test piece which was tensile tested at strain
rate of 10^{0} at 1323 K, total elongation was 10 %
and superplastic deformation didn't taken place

at any portion of the test piece. So very large
grains just like those in Figure 10 were observed all over the section examined.

The measured texture of extruded billets,
double fiber texture of [100] and [111], could
be considered to be natural for the FCC alloy
examined. However the reason of the orientation
rotation observed under tensile tests and hot
rolling is not clear. γ' phase deformation might
have something to do with the rotation. In γ'
phase, slip plain at low temperature is {111},
but {100} plane is also available at high temperature over 1073 K. So dislocation gliding on
{100} might be the reason for the rotation.

SUMMARY AND CONCLUSIONS

1 Stored energy introduced by hot powder
extrusion promoted the grain growth of TMO-2
alloy during annealing. As extrusion ratio was
raised and extrusion temperature was lowered,
grains grew larger to be about 1 mm in extrusion
direction.
2 Contamination of MAed powder by oxygen,
nitrogen and iron was shown to create obstacles
to grain growth. Grains didn't grow large
during annealing in billets which have a large
amount of these contaminants while those in
billets that contain a small amount of the
contamination grew large during annealing.
3 Pole figures revealed that extruded
billets have the double fiber texture of [111]
and [100]. It was proven that the texture was
strongly developed in the extruded billet of
which the grains grew large during annealing.
ECP showed that, after annealing, grown-up
grains have the orientations of extrusion direction lying along the [111]-[100] line in the
stereographic triangle. In the course of tensile tests or hot rolling of extruded billets,
orientation of grains rotated so that tensile
axis or rolling direction, which was identical
to the extrusion direction of the billets before
the deformations, came near [110] in each crystal though there was some scattering in the case
of the hot rolling, and recrystallization response of the rolled was much improved.

ACKNOWLEDGMENT

A part of this work was performed under
management of Research and Development Institute
of Metals and Composites for Future Industries
as a part of R & D of Basic Technology for
Future Industries sponsored by Agency of Industrial Science and Technology, MITI.

REFERENCES

1) J.S.Benjamin, Met.Trans.1,2943 (1970)
2) J.S.Benjamin and T.E.Volein, Met.Trans. 5,
   1929 (1974)
3) R.C.Benn, L.R.Curmick, G.A.J.Hack, Powder
   Metallurgy 4,191 (1981)
4) G.H.Gessinger, Met.Trans.7,1203 (1976)
5) G.H.Gessinger, Powder Metallurgy of
   Superalloys, Butterworths (1984)

ISOTHERMAL FORGING OF OXIDE DISPERSION STRENGTHENED NICKEL BASE SUPERALLOY BLADES

**Osamu Tsuda, Tomiharu Matsushita,
Nobuo Kanamaru, Kunihiko Nishioka**
Technical Development Group
KOBE STEEL, LTD.
Kobe, Japan

Abstract

For the production of oxide dispersion strengthened
nickel base superalloy turbine blades with cavities
for cooling, a new process has been postulated in
which a couple of blade parts can be forged precisely and bonded to form a blade.

Mechanical properties and post forging recrystallization behaviors were investigated to find out the
optimum forging conditions on designing of dies and
preforms.

Numerical simulations by rigid-plastic finite element method were also performed. The deformation
process shows that a plate shaped preform is desirable to achieve uniformity of strain and to assure
the grain growth recrystallization.

Blade parts were forged isothermally. The forged
specimens through the 'machining and forging' method
show good post forging recrystallization behaviors,
while the forged specimens with platforms through
the 'extruding and forging' method prove that it is
a near net shape forging process of the blade parts.

OXIDE DISPERSION STRENGTHENED (ODS) NICKEL BASE
SUPERALLOYS produced in the mechanical alloying process are thought to be applied in the next decade to
some engine parts like blades and vanes for a high
temperature gas turbine, because they have excellent
high temperature creep characteristics.

ODS nickel base superalloy turbine blades have usually been forged in both conventional forging and
isothemal fornging[1][2][3]. The net shape forging
technique for ODS alloys as well as the bonding
technique, however, has not been sufficiently established, and those alloys are mainly used as stationary blades without cavities for cooling.

A new process for the production of ODS nickel base
superalloy turbine blades with cavities for cooling
has been developed in which a couple of blade parts
can be forged precisely and bonded to form a blade.
In forging ODS alloys into blades, the die design
and the thermomechanical process are important.

In this paper, mechanical properties of a newly
developed ODS superalloy TMO-2[4] and a commercial
superalloy MA6000 were investigated before isothermal forging. Post forging recrystallization
behaviors for those alloys were also examined. To
determine the optimum forging conditions, the deformation process during isothermal forging of the
blade parts was simulated by using a rigid-plastic
finite element method (FEM). It is shown that a
plate shaped preform is desirable to obtain uniformity of strain distribution throughout the airfoil. Further, blade parts were forged precisely
and the die shape and the preform shape were improved to avoid cracking.

MECHANICAL BEHAVIORS

In order to estimate the flow stress and the elongation of the developed ODS superalloy TMO-2, tensile tests were carried out in the following conditions: the specimen 4mm $\phi \times$ 15mm, temperature 950 $\sim$1100°C, and strain rate $10^{-4} \sim 10^{-2}$ sec^{-1}. The strain rate was changed in such a way as $1 \times 10^{-4} \rightarrow 3 \times 10^{-4} \rightarrow 1 \times 10^{-3} \rightarrow 3 \times 10^{-3} \rightarrow 1 \times 10^{-2}$ sec^{-1}. Fig.1 shows the relations between flow stress and temperature. The data for MA6000 are also shown in the figure. The flow stress is larger for TMO-2 than for MA6000.

Fig.2 shows the strain rate sensitivity on flow stress. In this figure, the data for P/M-IN100 at 1100°C is also plotted. The m-value for TMO-2 changes from 0.09 to 0.14, which is much smaller than that for P/M-IN100 HIPed billet. The m-value of P/M-IN100 is about 0.5 which means this material shows superplasticity. ODS alloys including TMO-2 as well as MA6000 does not show superplasticity in the test conditions.

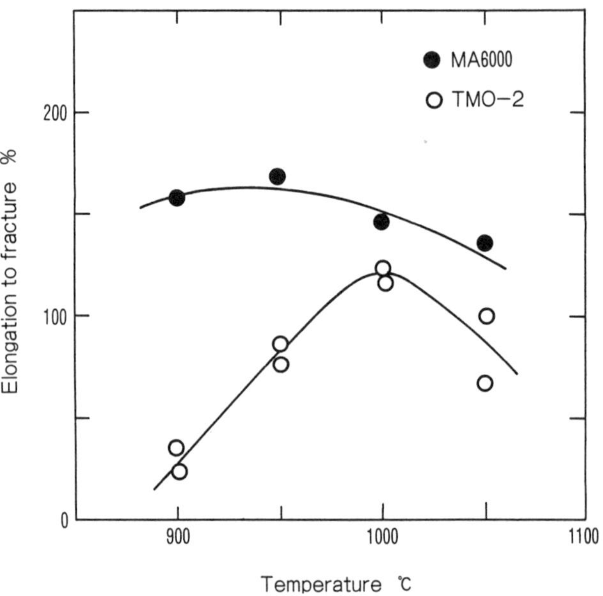

Fig.3 Relations between temperature and elongation to fracture in tensile tests at elevated temperatures for ODS alloy.
(specimen 4mm $\Phi \times$ 15mm, strain rate $1 \times 10^{-4} \rightarrow 3 \times 10^{-3} \rightarrow 1 \times 10^{-3} \rightarrow 3 \times 10^{-3} \rightarrow 1 \times 10^{-2}$ sec^{-1})

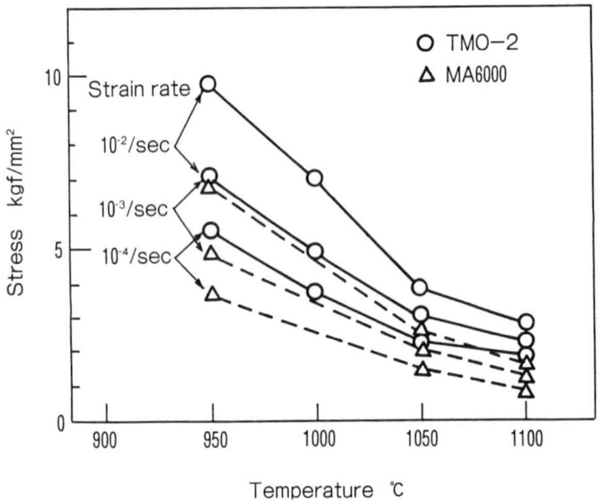

Fig.1 Comparison of flow stresses between TMO-2 and MA6000.

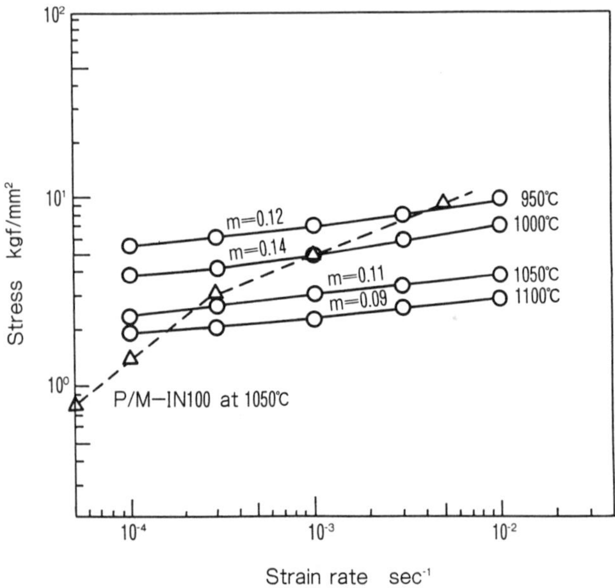

Fig.2 Relation between flow stress and strain rate for TMO-2.

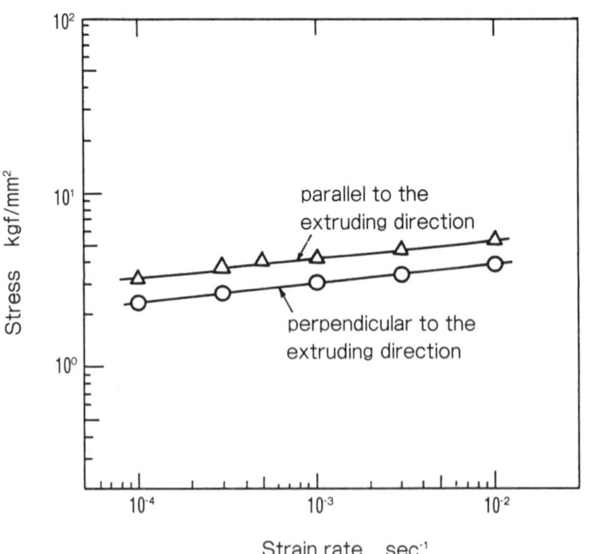

Fig.4 Influence of forging directions on the relation between flow stress and strain rate for TMO-2 at 1050°C

Fig.3 shows the relation between elongation and
temperature. The elongation to fracture for TMO-2
was inferior to that for MA6000. However, it exceeds
more than 100% around 1000℃, and TMO-2 seemed to
have a good forgiability as far as elongation.
To examine anisotropy on flow stress, compression
tests were performed at 1050℃ with strain rate of
$10^{-4} \sim 10^{-2}$ sec^{-1}. The specimen was 8mm $\times$ 8mm $\times$ 10mm.
The results show that flow stress depends on the
forging direction, and that the stress is larger in
the extruding direction than in the transverse direction. In forging of blade parts, the preform may
be cut out in the longitudinal direction. If so,
the forging load should be estimated by the flow
stress in the transverse direction.

RECRYSTALLIZATION BEHAVIOR

To find out the isothermal forging conditions on
grain growth recrystallization of ODS alloys, the
specimen of (thickness)mm $\times$ 10mm $\times$ 50mm were cut from
the MA6000 extruded billet, and forged isothermally
to about 4mm in height.
Fig.5 shows the macrostructures of the specimens
after heat treatment under 1300 ℃ $\times$ 30min. Grain
growth recrystallizations were found when the
natural strain was less than about 0.52 to 0.57,
that is the reduction of thickness was less than
about 40.5% to 43.4%.
Considering of the above results analogically, the
isothermal forging tests were carried out for TMO-2.

The specimen of 9mm $\times$ 7mm $\times$ 20mm were cut from the
extruded billet, and forged isothermally so as to
reduce the thickness within 40% to 55%.
Table.1 shows the test results of the recrystalization characteristics for the TMO-2 billet. Grain
growth recrystalizations were found in all cases in

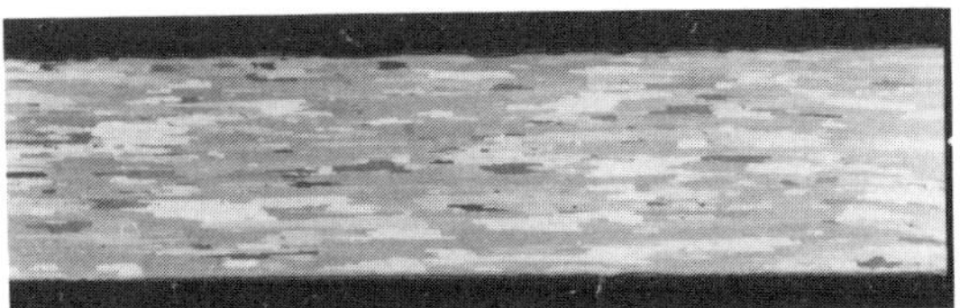

L-direction

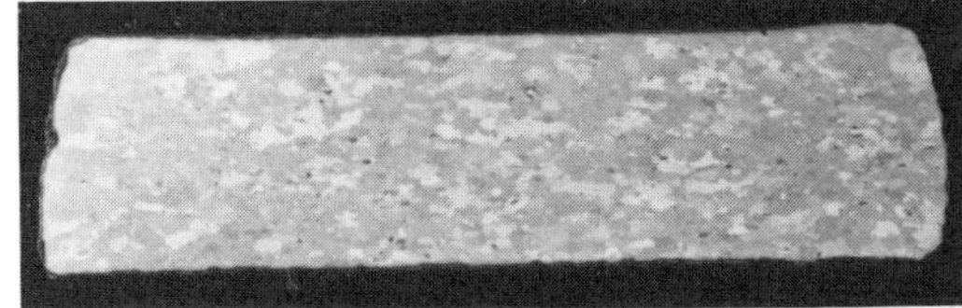

T-direction

(a) Temperature 950 ℃, reduction 40.5%
and strain rate 0.025sec^{-1}.

L-direction

T-direction

(b) Temperature 1000℃, reduction 43.4%
and strain rate 0.026sec^{-1}.

Fig.5 Recrystalization characteristics of tested
MA6000 billet.

Table.1   Recrystalization characteristics of tested TMO-2 billets.

Test Conditions	Forging Temperature ℃	Strain Rate sec^{-1}	Reduction %		
			42.2	48.9	55.6
1	950	10^{-2}	long	long	middle
2	950	5×10^{-2}	long	middle	short
3	950	10^{-1}	middle	long	short
4	1000	5×10^{-2}	long	long	long

'Long' means that the ratio of the elongated grain length to the
specimen size is over 1, 'middle' means between 0.5 and 1, and
'short' means less than 0.5, respectively.

the table. The 1000℃ conditions are more preferable to other cases with regard to the longitudinal
grain sizes.

Fig.6 shows the macrostructures of the specimens
after heat treatment under 1300 ℃×30min as well.
Enlarged grains can be seen even when the reduction
of thickness was about 57%. It seemed to be that
TMO-2 recrystallizes more easily than MA6000.

NUMERICAL ANALYSIS

Two dimentional rigid-plastic FEM was used to

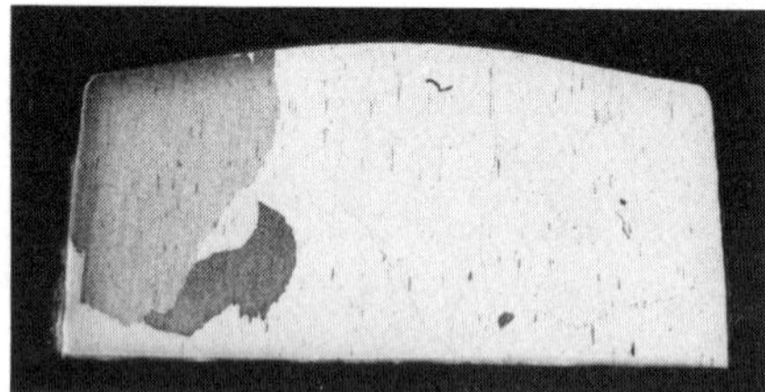

L-direction

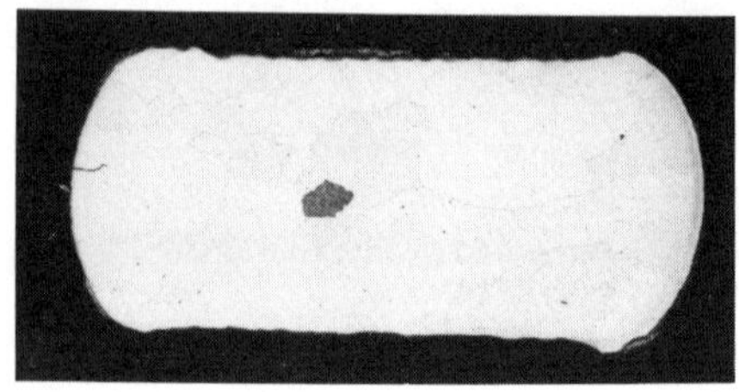

T-direction

(a) Temperature 950 ℃, reduction 52.3%
    and strain rate 0.05 sec⁻¹

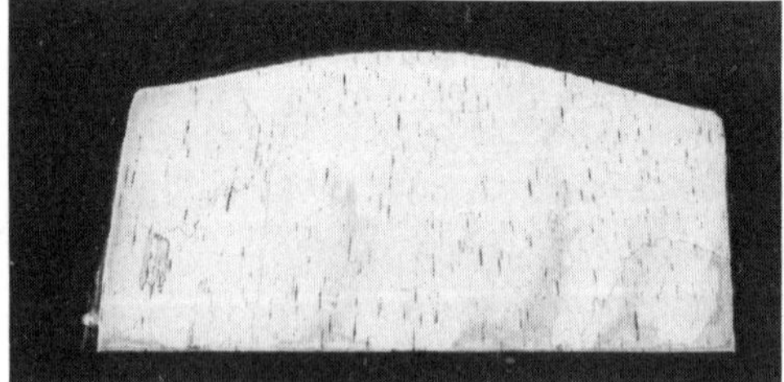

L-direction

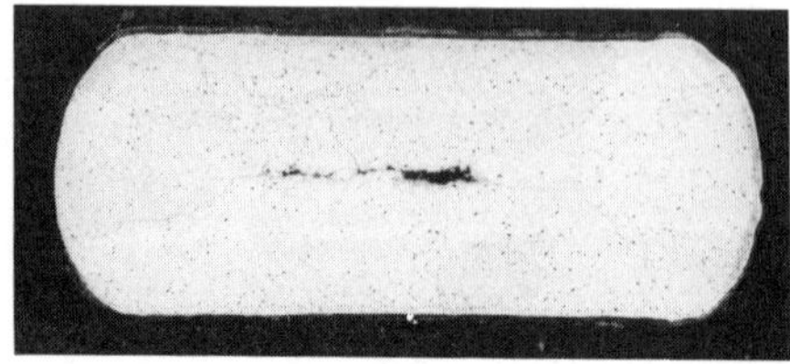

T-direction

(b) Temperature 1000℃, reduction 56.7%s
    and strain rate 0.05 sec⁻¹

Fig.6 Macrostructure after recrystalization for
       tested TMO-2 billets.

calculate the strain distribution and to testify if
the forging conditions satisfy grain growth recrystallization.

The deformation resistance is given by

$$\sigma = C \cdot \dot{\varepsilon}^{\,m} \qquad (1)$$

where

σ : Deformation resistence
$\dot{\varepsilon}$: Strain rate
C : Constant
m : Strain rate sensitivity exponent

The strain rate sensitivity exponent, m-value, for
ODS alloys was about 0.1 as mentioned before, then
plasticine is a good material for simulating the
metal flow. Thus, the simulation tests of forging
blades using plasticine were carried out to compare

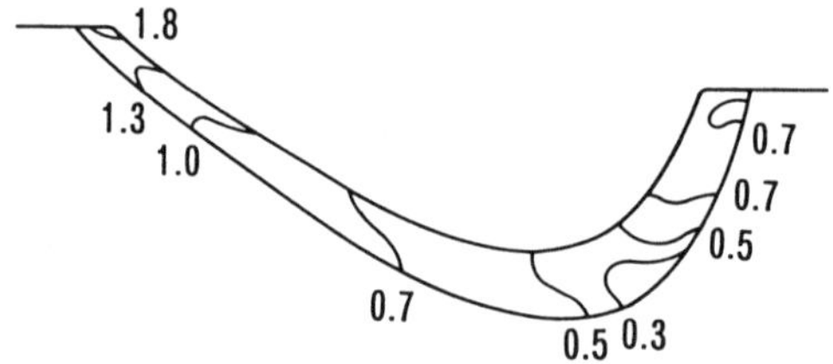

(a) Airfoil cross-section (24mm width)

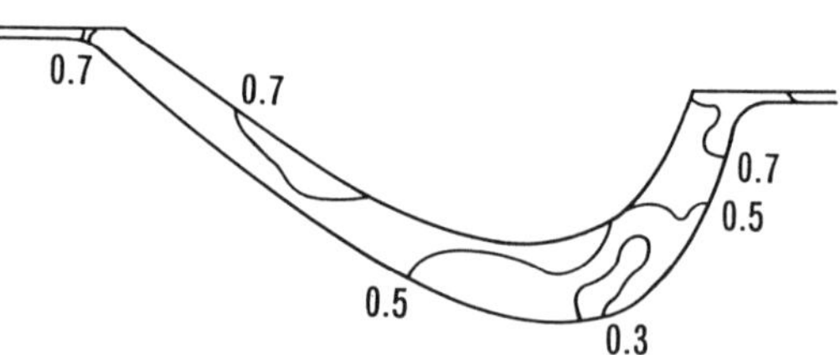

(b) Airfoil cross-section (30mm width)

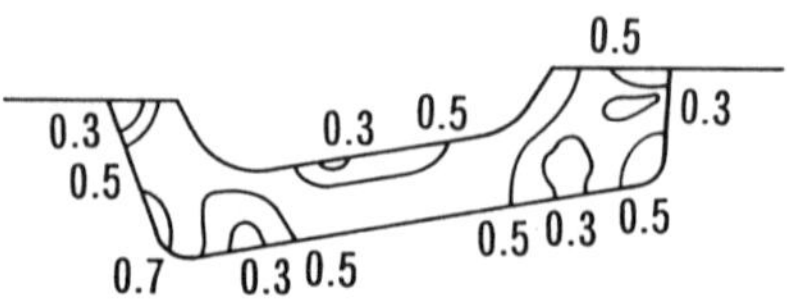

(c) Root cross-section

Fig.7 Equivalent strain distributions in forging of
       blade parts by rigid-plastic FEM.

with the deflection of meshes by numerical analyses.
Both results showed good coincidence.

To determine the preform shape based on the FEM
analysis, the following three cross sections were
considered: the initial cross section 4.1mm ×24mm
for the airfoil section, 4.1mm×30mm for the same
section and 4.6mm×20mm for the root section.
Strain distributions are shown in Fig.7(a)〜(c).
For the section (a) the maximum strain is very high
at the tail end, but for the section (b) the strain
hardly exceeds 0.7 and is almost uniform. The strain
in the root section is also small and uniform.
It was concluded that the optimum preform shape
should be given by the sections (a) and (c).

FORGING OF BLADE PARTS

FORGING EQUIPMENT - A 400tonf hydraustatic press
was used for isothermal forging of ODS alloys. The
ram speed of the press can be controled from 0.1 to
1800 mm/min. Molybdenum alloy, TZM, was used as a
die material, so heating and forging were conducted
in the inert gas atmosphere.

FORGING OF BLADE PARTS WITHOUT PLATFORMS - The
specimen with the initial airfoil section of 4.2mm×
24mm and the initial root section of 4.7mm×20mm was
machined from the TMO-2 extruded billet, and forged
isothermally into a blade part. After forging, the
blade parts were heat-treated at 1300°C×30min.
Fig.8(a) shows a forged blade part, and Fig.8(b)〜
(f) are macrostructures after heat treatment.
Obviously grain growth occurs overall the specimen
even at the tail end of the foil. This might be
caused to the longitudinal metal flow, because the
airfoil width was not perfect.

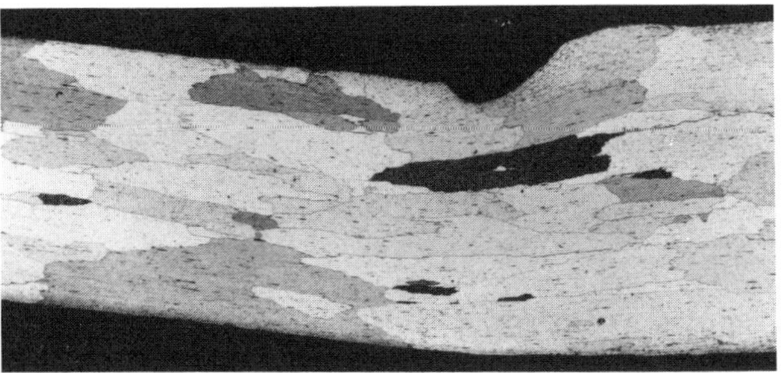

(b) longitudinal cross section at b   ⌊1mm⌋

(c) longitudinal cross section at c   ⌊1mm⌋

(d) longitudinal cross section at d   ⌊1mm⌋

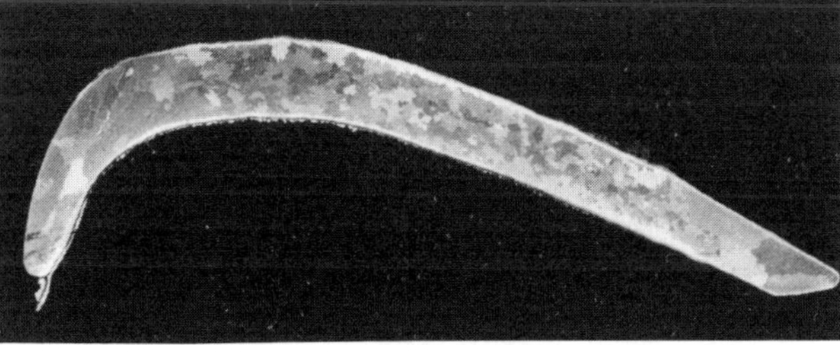

(e) transverse cross section at e   ⌊4mm⌋

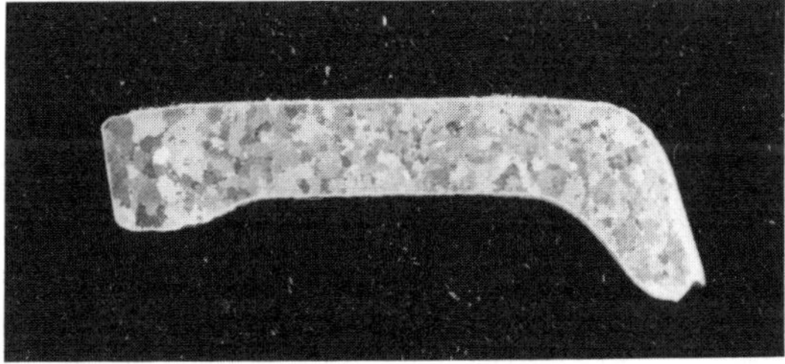

(f) transverse cross section at f   ⌊4mm⌋

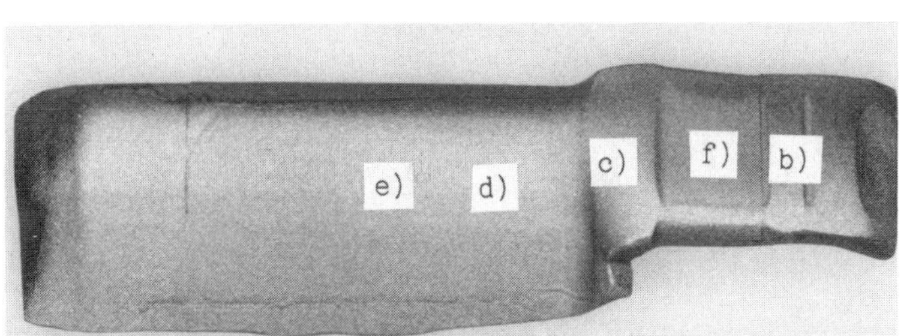

(a) blade part without platform   ⌊10mm⌋

Fig.8 Macrostructures of TMO-2 forged blade parts
after heat treatment.

'EXTRUDING AND FORGING' METHOD - In order to make a near net shape for blade part preforms with platforms, the TMO-2 billet was extruded forward and backward simultaneously. This process was excuted in the isothermal conditions at 1000℃. Dies were TZM to endure the applied high pressure at elevated temperatures.

The inlet of the TZM dies, however, deformed locally when mean pressure became about 30Kgf/mm² at the ram speed 0.01mm/sec. The necessary conditions for recrystallization have not been studied in this case, but the ram speed should be faster than now. Fig.9 is a blade part preform after extrusion, and the sizes of the preforms are shown in table 2. The lengths of airfoils and roots are not constant, because metals were allowed to flow in both forward and backward directions. The platform thicknesses are also different, which might be caused to temperature differences.

FORGING OF BLADE PARTS WITH PLATFORMS - The extruded preforms were forged isothermally into balde parts. Table 3 shows the sizes of blade parts after isothermal forging. The thicknesses of airfoils and roots were mesured at the points on the center line. The flash was thin, and the forging was thought to be almost a near net shape forging process.

Fig.9 Blade part preforms after extrusion.

Table.2 Sizes of extruded blade part preforms.

Specimen No.	Type of blade parts	Airfoil length mm	Root length mm	Platform thickness mm	Force tonf
1-3	Convex type	97	33	6.4	60
1-5	Convex type	86	32	7.0	60
1-6	Concave type	116.5	40.5	5.0	66
1-7	Concave type	95	39.5	5.0	60
1-8	Convex type	100.5	38	4.5	57

Table.3 Sizes of blade parts after isothermal forging.

Specimen No.	Type of blade parts	Thickness mm Airfoil	Root	Flash	Force tonf	Note
2-1	Convex type	3.31	3.40	-	149.5	
2-3	Convex type	3.40	3.46	0.19	154.5	Crack
2-4	Convex type	3.33	3.48	0.23	151	
2-5	Concave type	3.45	3.28	0.30	149	
2-6	Concave type	3.19	3.29	0.45	150.4	

Fig.10 TMO-2 blade parts after isothermal forging.

Fig.10 is a concave type blade part just after isothermal forging, and Fig.11 is a convex type blade
part after cutting and machining, respectively.
No cracks were found on the convex type, while a
hair crack was observed on the concave type. This
is because different bending forces applied to the
concave blade parts in the airfoil section and the
root section.

IMPROVEMENT OF DIE AND PREFORM - Fig.12 shows a
schematic diagram of crack occurences during forging
of blade parts. Cracks happened to occur depending
on the setting method of preforms.

SUMARY AND CONCLUSIONS

In forging of ODS alloy blade parts, the following results are obtained.
(1) The strain rate sensitivity exponent, m-value,
    for ODS alloys is about 0.1 and superplasticity
    can not be expected in the test conditions.
(2) The deformation resistance depends on the
    direction of the extruded billets.
(3) One of the isothermal conditions for grain
    growth recrystallization for TMO-2 is that the
    reduction in height is limitted within 50 ~60%.

(4) A plate shaped preform is desirable to obtain
    uniformity of strain distribution in isothermal
    forging of blade parts.
(5) The elongation for ODS alloys shows anisotropy[5]
    To avoid cracks, it is necessary to improve the
    shape of dies and preforms.
(6) It is possible to forge near net shaped blade
    parts from optimum preforms.

Acknowledgement

This work was performed under the management of the
R & D Association for Future Metal and Composite
Materials as a part of Basic Technology for Future
Industries sponsored by NEDO (New Energy and
Industrial Technology Developement Organization)

References

1) D.C.Wright and D.J.Smith:"Forging of Blades for
   Gas Turbine," Mat. Sci. and Tech.,1986,p.742.
2) E.Grundy:"Structureand Properties of Forged ODS
   Nickel-Base Superalloys,"Int. Conf. on PM Aerospace Materials,1987,p.12.1.
3) B.A.Ewing and S.K.Jain:"Development of Inconel
   Alloy MA6000 Turbine Blades for Advanced Gas
   Turbine Engine Designs," Superalloys 1988,p.131.
4) Y.Kawasaki,K.Kusunoki,S.Nakazawa,M.Yamazaki,
   S.Ochi and K.Mino: Trans. Iron ans Steel Inst.
   Jpn.,26(1986),B-40.
5) G.H.Gessinger:"Powder Metallurgy of Superalloys",
   p.264, p.271

Fig.11   TMO-2 blade parts
         after machining. |10mm|

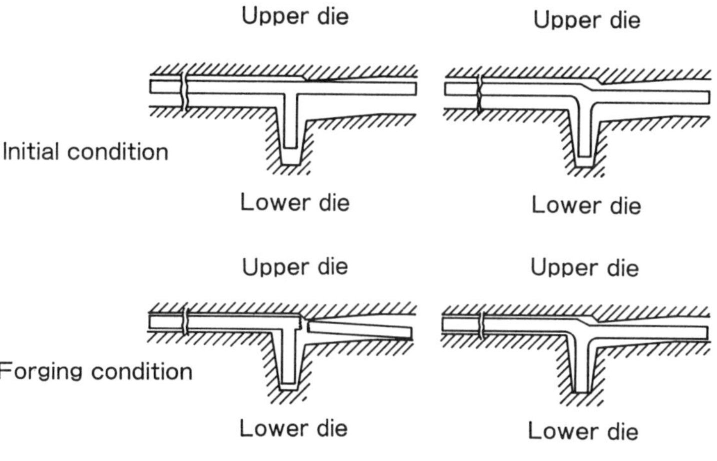

Fig.12   Schematic diagram of crack occurences during
         forging of blade parts.

ISOTHERMAL FORGING AND SUBSEQUENT TREATMENT OF A NEWLY DEVELOPED ODS ALLOY FOR TURBINE BLADE APPLICATIONS

Junji Tsuji, Kazuaki Mino
Ishikawajima-Harima Heavy Industries Co., Ltd.
Aero-Engine & Space Operations
Tokyo, Japan

ABSTRACT

A new oxide-dispersion-strengthened alloy, TMO-2, developed in the MITI project has much higher creep strength than MA6000. Our tasks in the project aimed at developing basic technologies of isothermal forging, directional recrystallization and diffusion bonding of TMO-2 for turbine blade applications. This paper summarizes secondary recrystallization behaviors of TMO-2 and fundamentals of microstructural control during isothermal forging and subsequent zone annealing. Introduction of dislocations by diffusion-brazing was also described, relating to poor joint strength which can be a serious problem for manufacturing an air-cooled blade.

I. INTRODUCTION

THE ADVANCED MATERIALS PROGRAMS sponsored by the Agency of Industrial Science and Technology, MITI have started in 1981. As one of the programs, the cooperative program on advanced ODS alloys for an air-cooled turbine blade was undertaken by the National Research Institute for Metals (NRIM), Sumitomo Electric Industries Ltd. (SEI), Kobe Steel, Ltd. and Ishikawajima-Harima Heavy Industries Co., Ltd. from 1982 to 1988.

NRIM has succeeded in developing a new experimental alloy TMO-2, possessing higher creep rupture strength than MA6000 over a wide temperature range[1]. TMO-2 contains larger amounts of refractory elements such as W and Ta than MA6000 to increase the creep strength. Our task in the program was to investigate directional recrystallization, subsequent to isothermal forging, and diffusion bonding technologies of the newly developed ODS alloy.

The formation of highly elongated grain structure is necessary for ODS superalloys to achieve superior high-temperature strength. Here, the key technology lies in directional recrystallization of isothermally forged blades having a complicated shape to get a uniformly recrystallized structure. Isothermal forging is known to increase a primary grain size by normal grain growth, leading to a loss in ability for secondary recrystallization (SRx)[2]. Details[3] such as its effects on SRx temperature and speed in TMO-2 will be firstly described in the present paper. In addition to these basic knowledges, zone annealing technique of forged parts with a sufficient temperature gradient has to be investigated.

Another problem to be broken through is an improvement of bondability, which is considered to be necessary for manufacturing an air-cooled blade. Diffusion brazing is a convenient bonding method. However, creep rupture strength of butt-joints of ODS alloys is known[4] to be very low as compared to that of oxide-free superalloys. The cause of degeneration of joint strength has not been clarified. In the present paper, a TEM evaluation of diffusion brazing interface was undertaken.

II. MATERIAL

The TMO-2 alloy was developed at NRIM, and prepared by SEI. The alloy was manufactured by the mechanical alloying process described by J.S.Benjamin[5] in which elemental and alloy powders were mixed with yttria powders in a high energy ball mill such as an attritor. Attritted powders were filled in an evacuated steel can and then sealed. Consolidation was performed by hot extrusion at 1020 to 1080°C and a reduction ratio of 15-17:1. The nominal composition of TMO-2 is shown in Table 1. Other alloys used in the present paper are also listed in the table.

Fifteen heats of TMO-2 were supplied for the present study. Their chemical compositions slightly varied from the above-mentioned compositions, but the differences were small. But, it will be noteworthy that an oxygen content was 0.7 to 1.0 wt pct, which was much larger than that in yttria. The extra oxygen

Table 1. Nominal Composition of Base and Insert Alloys (wt pct, Ni:Balance)

Base Alloy	Cr	Co	Mo	W	Ta	Al	Ti	C	B	Zr	Y_2O_3
TMO-2	5.9	9.7	2.0	12.4	4.7	4.2	0.8	0.05	0.01	0.05	1.1
MA6000	15	-	2	4	2	4.5	2.5	0.05	0.01	0.15	1.1
MarM247LC	8	9.5	0.5	9.5	3	5.6	0.8	0.07	0.015	0.015	Hf=1.4

Insert Alloy	Cr	Co	Mo	W	Ta	B	Si	Y	M.P.(℃)
MBF80	(14.68)	-	-	-	-	(4.03)	-	-	1020/1065
N3	6	10	2	12	4	(2.9)	-	0.25	
MBF50	19	-	-	-	-	1.3	8	-	1065/1150

( ): Analized

of 0.5 to 0.8 wt pct will react with 0.6 to 1 wt pct of aluminum or other active elements to form oxides.

Transmission electron micrographs of extruded material are shown in Fig. 1. Slightly elongated grain structures were found in the micrograph of the sample prepared parallel to the extrusion axis. The micrographs revealed a fine grain size of 0.15 to 0.18 μm and oxide dispersoids of 20 to 30 nm in the mean diameter. γ' phase precipitates of about 0.15 μm and about 0.03 μm in size were observed. Extruded round bars of 12 mm in diameter were used in most experiments, but in some cases larger bars of 30 mm in diameter were used.

Fig.1 - Transmission electron micrograph of as-extruded TMO-2

III. SECONDARY RECRYSTALLIZATION BEHAVIOR

AS-EXTRUDED MATERIAL - Small samples of 12 mm in diameter and about 5 mm in height were put into a furnace prefixed at temperatures in the range of 1220 to 1270°C and kept for 20 to 30 min. The macrograin sizes parallel and perpendicular to the extruded axis were measured for each samples heated at various temperatures and shown in Fig. 2. Results for MA6000[2] were also shown in the figure. Largest grains were found in both materials heated at just above the SRx temperatures. Yongenburger and Singer[6] investigated the effects of heating rate on SRx grain sizes of MA6000, and found an interesting fact that larger grains were formed by heating the samples to the SRx temperature at lower rates. When the samples are put into a furnace kept at higher temperatures, they are expected to be heated at higher heating rates to the SRx temperature. Considering that SRx occurs at such a high rate as 0.1 mm/s, they concluded that the grain size dependence of annealing temperatures came from the heating rate.

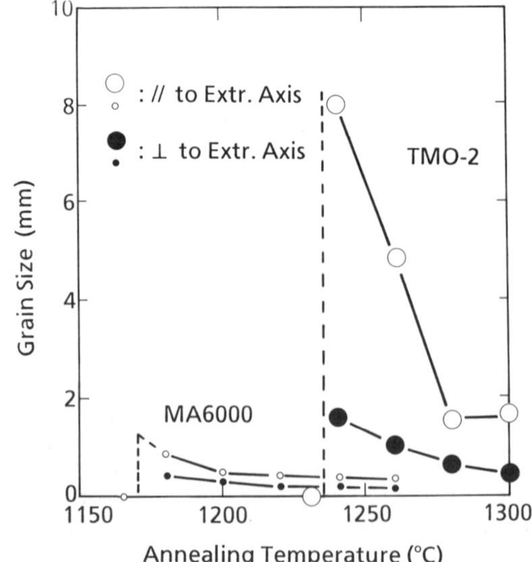

Fig.2 - Relationship between isothermal heating temperature and secondary recrystallized grain size

SRx temperatures of TMO-2 were found to vary from heat to heat in the early stage of the processing works. The hardness of TMO-2 in the extruded condition were plotted in Fig. 3 against the SRx temperature. Effects of an yttria content were also shown. The higher the hardness value, the lower the SRx temperature was observed. The lower yttria version of TMO-2 gave the lower hardness value. The typical value of hardness after SRx was Hv 520 for 1.1 wt pct Y_2O_3 and 490 for 0.6 wt pct Y_2O_3. The hardness dependence of SRx temperatures appears also in preannealing or isothermal forging effects, hence, the details will be explained in the later section.

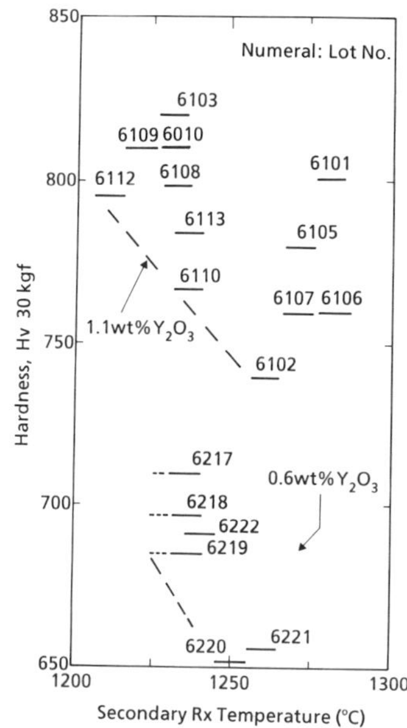

Fig.3 - Relationship between hardness of extruded TMO-2 bars and their recrystallization temperatures

PREANNEALED OR ISOTHERMALLY FORGED MATERIAL - Softening of extruded TMO-2 bars by preannealing at 1050 or 1100°C is shown in Fig. 4. The response to SRx heat treatment at 1280 to 1290°C are indicated by a dark mark (unRx'ed) or a white mark (Rx'ed) for each preannealed materials. The minimum value of hardness required for satisfactory grain growth was about Hv 720.

The relation between the hardness and SRx temperature was examined for one heat. As shown in Fig. 5, the SRx temperature increased with softening, limited by the incipient melting. Although the degradation of SRx'ability by preanneal has been reported[7], our work was the first to show the variations of SRx temperature with preanneal.

Deformation at high temperatures significantly accelerated degeneration of the response to SRx. Cylindrical specimens of 12 mm in diameter and 10 mm in length were deformed to a rectangular shape by compressing in the direction normal to the extrusion axis

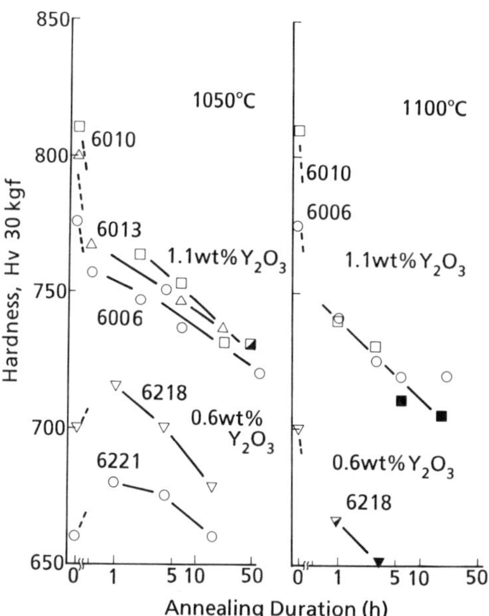

Fig.4 - Softening curve of extruded TMO-2 bars at 1050 and 1100°C

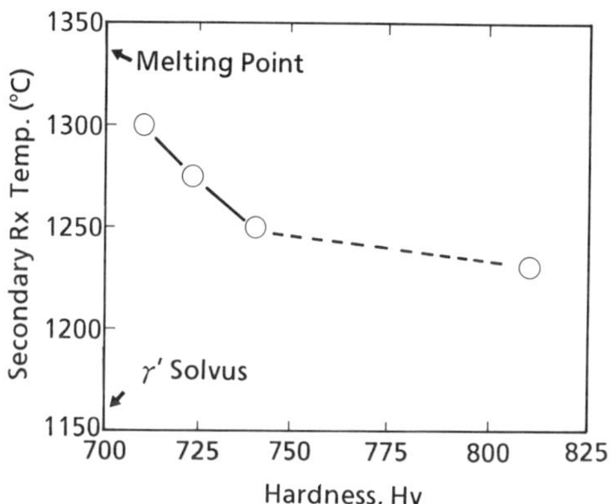

Fig.5 - Relationship between hardness of TMO-2 bars softened at 1100°C for various durations and recrystallization temperatures

at a constant speed. Compressed specimens were subsequently heated at 1280°C and their microstructures were observed to examine the response to SRx. The results are shown in Fig. 6.

Hot deformation softened the specimens as shown in Fig. 7. The hardness profiles were obtained along the direction normal to the press axis. The area indicated by the broken curve has lost the SRx'ability. SRx was not observed when the hardness decreased to Hv 730 or less.

TRIGGGER MECHANISM OF SRx IN TMO-2 - Mechanistic understanding of SRx will be needed to explain the above-mentioned results. SRx is driven by surface energy like a normal grain

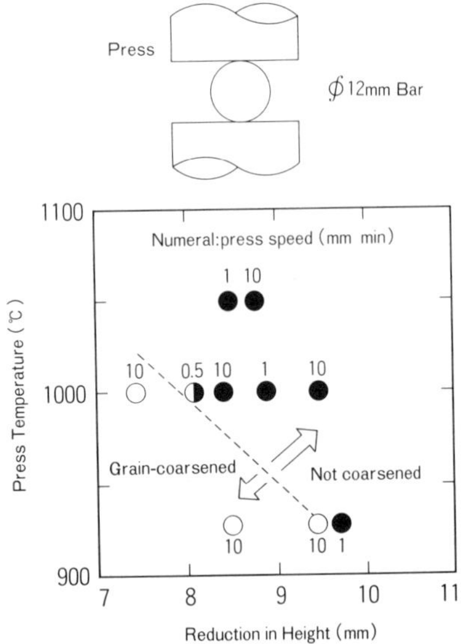

Fig.6 - Relationship between press condition and recrystallization response of TMO-2 at 1280°C

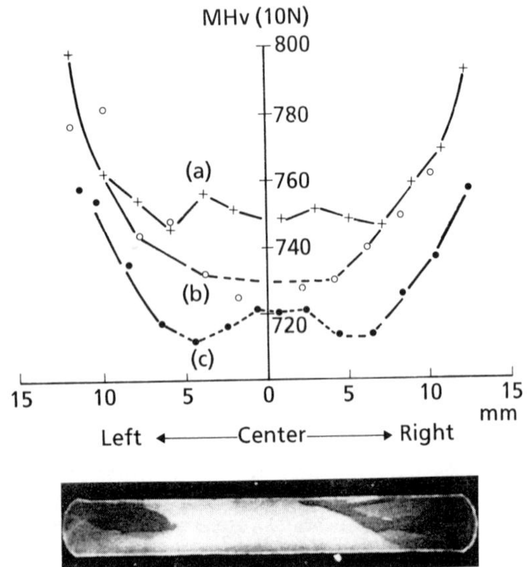

Fig.7 - Hardness distribution of pressed TMO-2 (dotted lines indicate the portion not recrystallized
(a) 930°C, ΔH=9.4mm (b) 930°C, ΔH=9.7mm
(c) 1050°C, ΔH=8.8mm
Photograph shows macrostructure of sample (c) recrystallized.

growth process, but differs from it in the fact that only a small number of grains grow to abnormally large grains. SRx often occurs accompanying dissolution or coarsening was considered to trigger SRx[3]. It was also found that the Hall-Petch type equation as shown in

Fig. 8 existed between the hardness and grain size of extruded materials. Harder materials consist of finer grains, and consequently have larger driving force for grain growth. This leads less oxide coarsening required for abnormal grain growth to occur, i.e. initiation of SRx at lower tempertures.

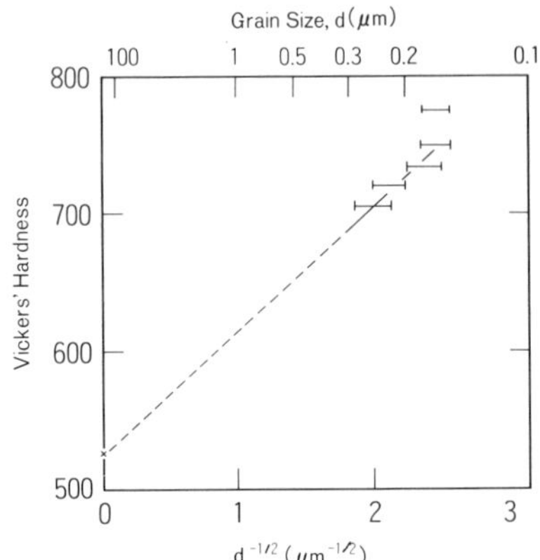

Fig.8 - Relationship between hardness and grain size of TMO-2

Abnormal grain growth was well developed theoretically by Hillert[8]. The average grain size should not exceed an upper limit, which is specific to the material, for abnormal grain growth to occur. This grain size limit varies from the order of 1mm in Fe-3% Si[9] to several μm or less like in Ni-base ODS alloys.

The net driving energy for grain growth per unit volume is given by[9]

$$\Delta F \sim 2\gamma[1/d-(3/8)(f/r)] \qquad (1)$$

where γ: surface energy per unit area

d: average diameter of grains to be consumed

f: volume fraction of dispersoids

r: average diameter of dispersoids.

The minimum value of d to make ΔF positive, d_c, is determined by f and r, and is given by $d_c \sim 8r/3f$. Using the typical values of f=0.04-0.05, r=0.02 μm for TMO-2, d_c was calculated to be about 1.2 μm. The grain size limit, obtained using the preannealed or isothermally forged materials, is about 0.2 μm which corresponds to Hv 710 to 730. The cause of discrepancy has not been clarified.

IV. DIRECTIONAL RECRYSTALLIZATION

ZONE ANNEAL CONDITIONS - No texture has been reported in the primary recrystallization structures of Ni-base alloys with γ' precipitates[10]. Highly elongated grain structures are not likely to be obtained in these alloy systems by isothermal anneal. Zone anneal using RF induction heating is usually applied. The maximum temperature in a zone anneal furnace was set at about 1300°C for isothermally forged TMO-2 parts. This temperature is 30°C below the incipient melting point and 30°C above the SRx temperature.

Higher zone anneal speeds will be desirable for higher production rates. However, the maximum speed is limited by the grain growth rate. Effects of zone anneal speed on the SRx structure and creep rupture time at 1050°C of as-extruded TMO-2 were examined using the round bar specimens of 12 mm in diameter. No significant difference in microstructures similar to those for an isothermally annealed specimen were observed at the central area of the specimens zone-annealed above 300 mm/h (83 μm/s). Shortage of the time to heat up to the center of specimens led the above-mentioned structure. Fig. 9 shows the relationship between the zone anneal speed and the time to rupture at 1050°C, 176PMa. The rupture strength at 1050°C is determined mostly by elongated grain structures.

The response to SRx by zone anneal became worsened in the preannealed or isothermally forged materials even though enlarged grain structures were obtained by isothermal anneal. Fig. 10[3] compares zone annealed macrostructures of preannealed and unannealed specimens of 12 mm in diameter. Maximum temperature in the furnace was about 1290°C and the travel speed was 100 and 50 mm/h. An unrecrystallized area was observed in the center of the preannealed specimen due to the lower heating temperature at the central part than the grain growth temperature. But, elongated grain structures were obtained across the whole section of the unannealed specimen zone-annealed under the same condition and hence the same temperature distribution since the grain growth temperature was lower than that of the annealed specimen. This circumstance was somewhat improved by lowering the travel speed to 50 mm/h; but it partially recrystallized, which significantly differed from isothermally annealed

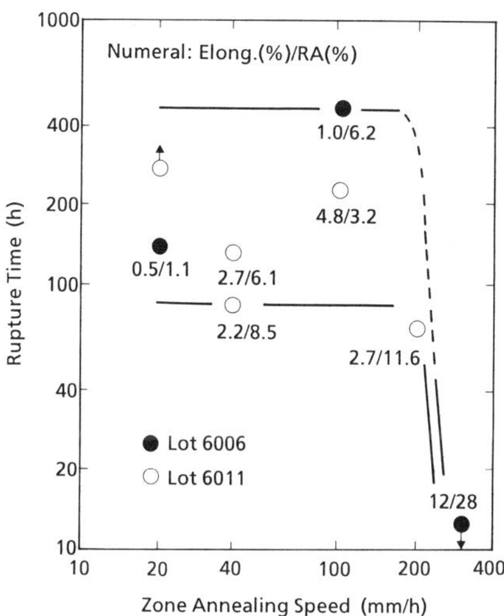

Fig.9 - Relationship between zone annealing speed and creep rupture life tested at 1050°C, 176MPa on TMO-2

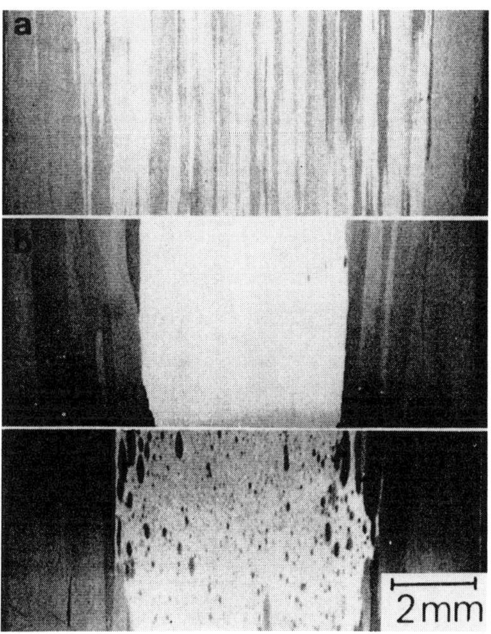

Fig.10 - Longitudinal macrostructures of zone annealed TMO-2 specimens. (a) As-extruded material, zone travel speed 100mm/h. (b) Preannealed at 1100°C for 4h, 100mm/h. (c) same as for (b), 50mm/h. The travel direction is vertical.

macrostructures. This shows that preanneal not only increased the grain growth temperature as shown previously but also decreased the grain growth rate.

ZONE ANNEAL OF ISOTHERMALLY FORGED PARTS - Hot forging tends to induce secondary recrystallized areas with a low grain-aspect ratio or unrecrystallized areas. Too severe working deprives of recrystallizability. On the other hand, keeping high grain-aspect ratio tends to become difficult for forged parts from following two reasons:
(1) Direct induction heating becomes difficult in zone annealing of parts such as a forged turbine blade having a complex configuration. In this case, the temperature gradient across the specimen travel axis has to be sacrificed for overall temperature control.
(2) Secondary recrystallized grains tend to be aligned parallel to the direction of plastic flow rather than that of the imposed temperature gradient.
As a countermeasure to (2), an appropriate design of preforms and/or forged shapes will be needed.

Fig. 11 shows an example of isothermally forged TMO-2 samples. The zone annealed macrostructure is also shown. Transverse grains were found at the bottom and the left-hand side of the top in the root section. Fairly elongated grains were formed in the

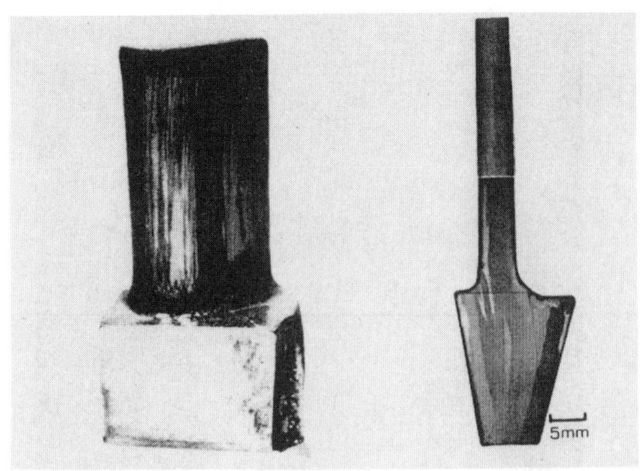

Fig.11 - Isothermally forged TMO-2 and cross-sectional macrostructure of subsequently zone annealed sample.

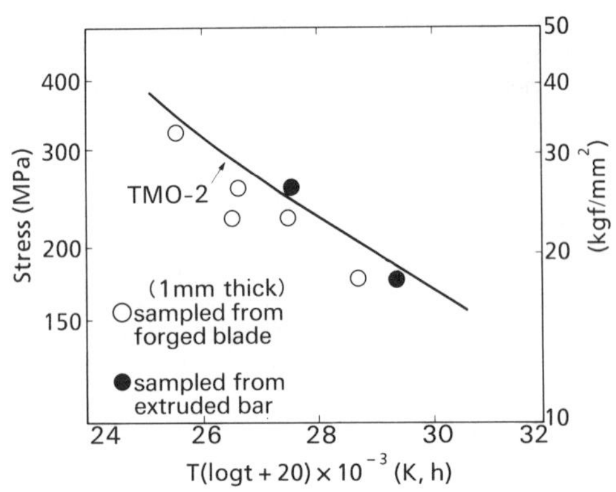

Fig.13 - Degradation of rupture strength of TMO-2 by isothermal forging

airfoil section. Another isothermal forging study, aimed to develop a two-pieces blade, was performed by Kobe Steel, Ltd. using a different preform design to get a forged sample as shown in Fig. 12. An elongated grain structure was observed in the airfoil section.

Creep rupture specimens of a plate form with a gauge section of $1^t \times 4^w \times 10^\ell$(mm) were sampled from the forged and zone anealed specimens as shown in Fig. 12. Fig. 13 shows the result of rupture tests at 900 to 1050°C in air. Specimens having the same size were also machined from the non-forged TMO-2 bar and their test results were shown in the same figure. The forged materials had less strength than the non-forged. The strength reduction was also found not to be caused by the specimen size effect. Defects in directional recrystallization are considered to have led to the reduction in creep rupture strength.

In relation to (1), effects of temperature gradient across the zone travel axis were studied using TMO-2 bars of 12 mm in diameter. The results are shown in Fig. 14. The small temperature gradient of zero to 10°C/cm has led

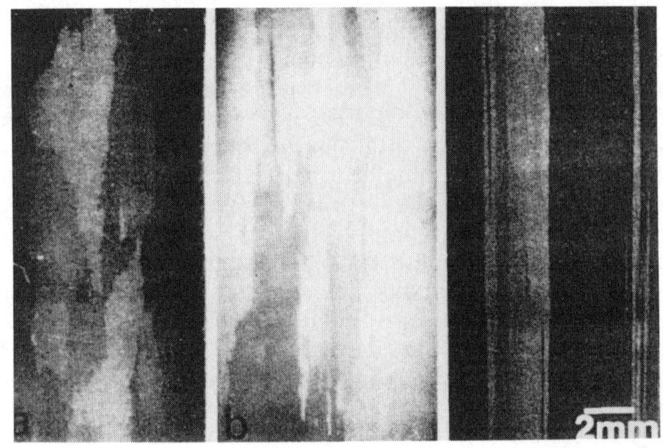

Fig.14 - Effect of temperature gradient on zone annealed macrostructures of TMO-2
a) isothermal, b) ~10°C/cm, c) ~200°C/cm

to incompletely elongated grain structures. It was also found that the temperature gradient required for directional recrystallization was not so large as to be impracticable.

V. DIFFUSION BONDING

Diffusion bonding is necessary for manufacturing air-cooled turbine blades. Bonding of ODS alloys is known to be more difficult than that of conventional superalloys[4].

CREEP RUPTURE STRENGTH OF DIFFUSION BRAZED MA6000 - Butt-joint MA6000 specimens were prepared using filler alloy MBF80 tapes of 40μm in thickness. Other filler alloys MBF50 and N3 tapes were also applied. N3 is our original tape prepared by melt spinning. Bar specimens of 22x25 mm^2 with filler alloys were held together under the stress of 0.2 to 5 MPa and

Fig.12 - Another isothermally forged sample and macro-etched section. A configuration of rupture specimens is shown in the photograph.

heated at 1200°C for 16 to 20h in a vacuum of 1×10^{-4} torr or less.

Creep rupture specimens were sampled from MA6000 bars diffusion-brazed condition. Rupture tests were carried out at 950°C in air using specimens of 3.5 mm in gauge diameter. The specimens were aged at 850°C for 24h in creep furnace prior to loading. The rupture strength of diffusion brazed MA6000 was compared in Fig. 15 with that of the base material in the transverse direction of the elongated structures. The extrapolated stress for 1000 h for causing rupture was about 35MPa; this being about one third the strength of the base material in the transverse direction.

The oxide-free superalloy MarM247LC was diffusion-brazed using MBF80 tapes for purposes of comparison. The stress arising from a difference of thermal expansion coefficients between the test materials and TZM yoke was used as a bonding pressure. The TZM yoke with the specimen was heated to 1200°C and kept for 17h in vacuum. The rupture strength MarM247LC joint was essentially the same as that of the base material. The 1000h-rupture stress at 950°C for the MarM247LC joint was about 120MPa, this being significantly large as compared to the MA6000 joint.

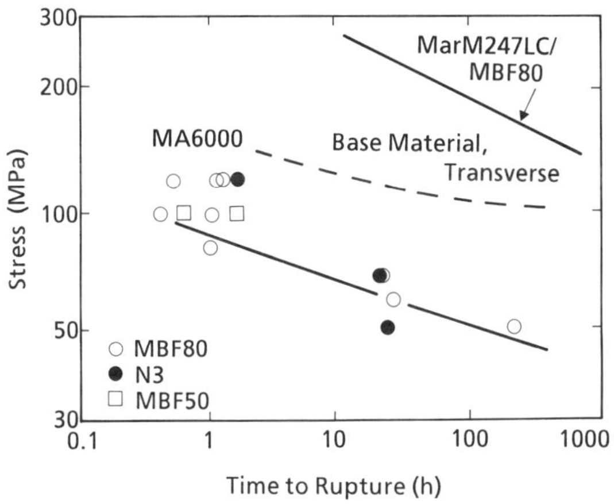

Fig.15 - Stress rupture data at 950°C of diffusion brazed MA6000

TEM INVESTIGATION OF DIFFUSION BRAZED INTERFACE - SEM observations of diffusion brazed MA6000 specimens before and following the stress rupture tests at 950°C for 2h are shown in Figs. 16(a) and (b), respectively. The joints were prepared by the same stressing method as for MarM247LC, but were preheated at 1050°C for 3h and 1150°C for 15h prior to heating at 1150°C or above. The original thickness of the insert alloy was about 40μm, while that of the interlayer decreased by a half following diffusion brazing, as shown in Fig. 16(a). Coarsened oxide layers of Y and Al were found situated about 7μm distance from the center of the interlayer. By comparing Fig.

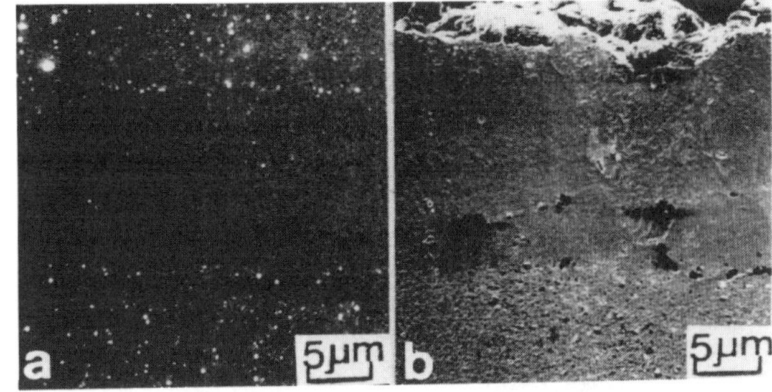

Fig.16 - Scanning electron micrographs of diffusion brazed MA6000 (a) before and (b) after stress rupture tests at 950°C for 2h

16(a) with (b) and assuming the width of the non-dispersed area not to have changed during the stress rupture test, the stress rupture appeared not to have occurred at this oxide layer. Rupture apparently took place at the boundaries between the oxide dispersed and non-dispersed areas situated about 9μm distance apart from the center of the interlayer. Certain regions of the holes in Fig. 16 resulted from spallation of agglomerated oxide particles during metallographic preparation.

Thin foils of brazed specimens were prepared using a twin-jet polisher in a 1:4 mixture of perchloric acid and etanol. Ion milling was carried out following twin-jet polishing of ODS joints since the electrochemical polishing rate was much higher in the ODS area than in the interlayer. A slight difference in foil thickness across the interface could not be avoided.

TEM observations of the diffusion brazed MA6000 interface indicated significant features, as shown in Fig. 17. As mentioned above, the thickness of TEM samples between the interlayer and base material may differ somewhat. Dislocations with extremely high density were observed in the oxide dispersed area and occurred in only a few dislocations in the interlayer. The specimen in Fig. 17(a) was difufsion-brazed at a relatively low temperature of 1150°C. In Fig. 17(b) brazing was conducted at 1200°C and there were observed recrystallized grains of about 1μm in size and rather coarse oxide dispersoids left on the interlayer side. Dispersoids in the recrystallized area were fewer in number than those in the unrecrystallized area. The recrystallized grain size for the specimen diffusion-treated at 1230°C was larger, as shown in Fig. 17(c) and from 1 to 4μm. Large dispersoids with sparsely distributed were evident on the interlayer side and considered to be the coarse oxide layer in Fig. 16(a).

Isothermal solidification of the interlayer starts from a point close to the base material and is completed as the center of the interlayer. Thermal strain induced during

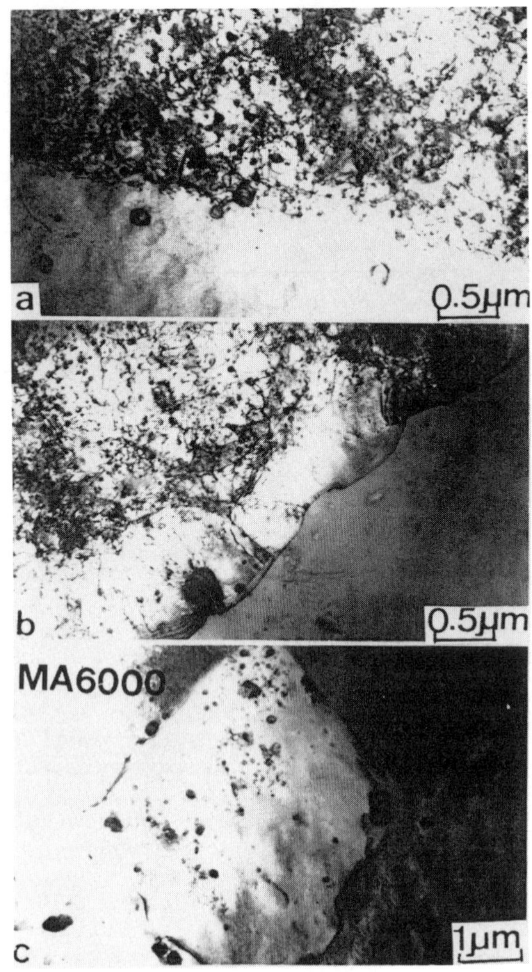

Fig.17 - Transmission electron micrographs of the joint interface of MA6000 diffusion-brazed at (a) 1150°C, (b) 1200°C and (c) 1230°C

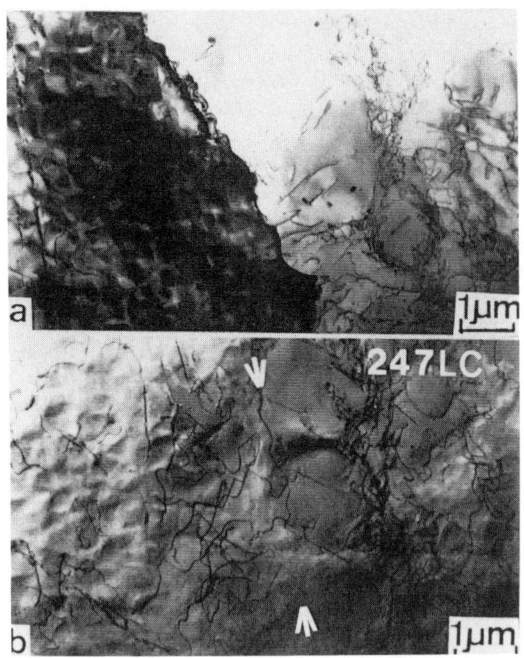

Fig.18 - Transmission electron micrographs of the joint interface of diffusion brazed MarM247LC. Grain boundaries remained in (a), but disappeared in (b).

solidification may possibly give rise to the many dislocations noted in the oxide dispersed area. The features shown in Fig. 17(b) and (c) can be explained on the basis of the data presented by Ashby and Palmer, who reported the following for silica dispersed copper. Recrystallization with high driving force caused no change in the distribution of the dispersoids, while grain growth with low driving force caused dragging of the dispersoids, leaving a dispersoid-free zone and coarse dispersoids. Though strain was introduced at high temperature, the plastic strain energy given to the base material facing the interface was estimated to be very high and primary recrystallization occurred without extensive oxide coarsening. Rather large oxides on the interlayer side should be considered to result from dragging of dispersoids by grain growth while high temperatures are maintained.

Far fewer dislocations were observed near the interface of MarM247LC diffusion-brazed at

1170°C for 17h as shown in Fig. 18. Dislocation density was higher in the γ than γ' phase. Grain boundaries formed at the joining interface remained in most areas (Fig. 18(a)), but were disappeared in others (Fig. 18(b)). Original boundary locations were identified by somewhat coarse γ' phase layers.

As already mentioned, the coarse oxide layer in Fig. 16(a) did not cause degradation of the joint strength of MA6000. This fact requires other causes of the loss in strength by diffusion brazing; the high density of dislocations due to thermal strain during solidification of interlayers, which remain following cooling by oxide pinning, and the recrystallized fine grain structure are considered to lead the poor joint strength.

VI. SUMMARY AND CONCLUSIONS

Understandings of microstructural control of TMO-2 during isothermal forging and subsequent zone anneal have quite been deepened through the MITI project. The present high-strength ODS alloy responded to directional Rx better than MA6000. Its response was well correlated to the hardness, and microstructurally to the grain size. Isothermal forging softened the material, increased the SRx temperature and decreased the migration rate. With proper preform designing, an airfoil part of the isothermally forged blade was found to well develop the directional

Rx structure by subsequent zone anneal. The poor rupture strength of diffusion brazed materials has been a serious concern on ODS materials for many years. Introduction of dislocations and formation of fine grains at the brazing interface, found in the present study, are believed to be developed to establish a countermeasure to the poor joint strength.

ACKNOWLEDGMENT

This work was performed in part under the management of the Research and Development Institute of Metals and Composites for Future Industries as a part of the R&D project of Basic Technology for Future Industries sponsored by the Agency of Industrial Science and Technology, MITI.

REFERENCE

1. Y.Kawasaki, K.Kusunoki, S.Nakazawa, and M.Yamazaki, Tetsu-to-Hagane, 75, 1588-95(1989)
2. R.F.Singer and G.H.Gessinger, Metall. Trans. A, 13A, 1463-70(1982)
3. K.Mino, Y.G.Nakagawa, and A.Ohtomo, Metall. Trans. A, 18A, 777-84(1987)
4. B.Jahnke and G.Dannhauser, "High Temperature Alloys for Gas Turbines and Other Applications", p.175-216, D. Reidel Publishing Co., Dordrecht (1989)
5. J.S.Benjamin, Metall. Trans. 1, 2943-51(1970)
6. C.P.Yongenburger and R.F.Singer, "Advanced Materials and Processing Techniques for Structural Applications", p.339, ASM International, Materials Park, Ohio (1987)
7. R.K.Hotzler and T.K.Glasgow, Metall. Trans. A, 13A, 1665-74(1982)
8. M. Hillert, Acta Metall., 13, 227-38(1965)
9. R.W.Cahn, "Physical Metallurgy", p.1129, North-Holland Publ. Co. (1970)
10. M.Y.Nazmy, R.F.Singer, and E.Török, 7th Intern. Conf. Textures Materials, p.275 (1984)

THE TRANSVERSE CREEP BEHAVIOUR OF
MA 6000 IN TENSION AND COMPRESSION

Russell Timmins, Eduard Arzt
Max-Planck-Institut für Metallforschung
Institut für Werkstoffwissenschaft
Seestrasse 92,7000 Stuttgart 1
FRG

ABSTRACT

Poor transverse strength is a possible limitation for the widespread application of coarse-grained oxide dispersion strengthened (ODS) alloys at high temperatures. This work presents new tensile and compressive creep results, produced with continuous strain measurement, on MA 6000, which reveal the important micromechanisms responsible for the inferior high temperature performance under transverse loading. A model based on an interface reaction controlled diffusional creep mechanism is presented, which is the basis for a better understanding of the transverse creep behaviour of ODS alloys.

1.0   Introduction

Oxide dispersion strengthened (ODS) superalloys are promising materials for critical components operating at elevated temperatures. In particular these alloys, following controlled thermo-mechanical processing, possess excellent creep strength which is primarily due to a fine dispersion of inert oxide particles. The understanding of the creep deformation and fracture behaviour of ODS alloys has advanced considerably in recent years (for a review see Ref.1). A new constitutive equation (1,2) based on the thermally activated detachment of dislocations from inert dispersoids has led to a more consistent interpretation of the creep deformation behaviour in ODS alloys; in particular the high stress exponents and activation energies for creep, which are typical for ODS materials in general, are accounted for in a more natural way.

Efficient strengthening in nickel base ODS superalloys by inert dispersoids is achieved, however, only in combination with an elongated grain structure which suppresses excessive accumulation of creep fracture damage. This role of grain geometry in the creep deformation and fracture behaviour of ODS alloys has also received attention (3-7), but there still remains much debate over the detailed mechanisms involved. Furthermore, the presence of an elongated grain structure, which is a prerequisite for good high temperature creep strength, naturally leads to a far from optimum grain geometry in the transverse direction with a consequent degradation in creep behaviour. Since critical components are rarely subjected to uniaxial loading alone, it is essential to gain a better understanding of the transverse creep properties.

In this contribution emphasis is placed on the transverse creep behaviour of ODS alloys using Inconel alloy MA6000 as a model system. The paper continues in section 2.0 with a short review of the role of grain geometry in the creep behaviour of ODS alloys, and, following the experimental details in section 3.0, the results of new tensile and compressive creep tests on MA 6000 are given in section 4.0. Various models are then presented and discussed in section 5.0 where it is concluded that, for transverse loading, models based on constrained cavity growth are physically unrealistic. It is suggested that a model based on inhibited diffusional creep and cavity growth is more likely to be applicable. The paper closes with a short summary containing the main conclusions in section 6.0.

2.0 The Influence of Grain Geometry on the Creep Behaviour of ODS alloys

Grain boundaries subjected to a tensile stress at elevated temperature accumulate damage in the form of cavities which, after linking up into microcracks, result in material failure. This is the normal failure mode in creep for many polycrystalline alloys in general, but because the matrix strength of dispersion strengthened alloys is so high and the imposed operating conditions of stress and temperature are so severe, the strength differential between matrix and grain boundaries in ODS alloys is particularly marked. It is fortunate therefore that, following controlled thermo-mechanical working, ODS alloys acquire elongated grain structures which minimise the grain boundary area lying normal to a uniaxial tensile stress.

Wilcox and Clauer (3) were the first to demonstrate, in Ni-Cr-ThO$_2$ alloys, the importance of the grain aspect ratio (GAR) rather than the grain size per se for good high temperature strength. The effect was interpreted as being due to the reduction of grain boundary area normal to the maximum principal tensile stress and the suppression of grain boundary sliding. Of course the grain size itself should be large enough to restrict excessive creep deformation by diffusional flow.

Several workers have investigated the influence of GAR on the creep behaviour of ODS alloys. Creep data for MA 6000 reported by Cairns et al. (4) and Arzt and Singer (5) are displayed in Figure 1, which shows how the time to failure increases with GAR. Above a GAR of about 15 fracture occurs transgranularly and the material approaches its "single crystal" limit. Similar observations have been reported for bubble strengthened tungsten (8).

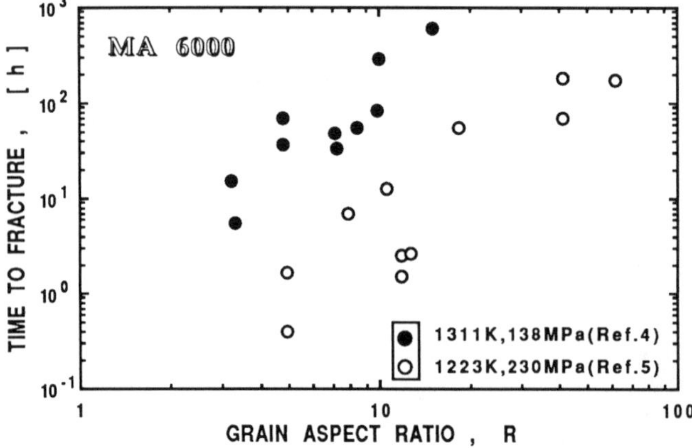

Fig.1: Creep rupture life as a function of GAR in MA 6000.

Various models have been offered as an explanation for these observations, most notably by Arzt and Singer (5), Stephens and Nix (6) and Zeizinger and Arzt (7). All models have a common feature; fracture damage accumulation on grain boundaries normal to the tensile stress is assumed to be constrained by displacements of the surrounding, undamaged, regions. In this way the geometry of the situation results in a relationship between the GAR and the creep deformation rate (or fracture time) . This is understandable since the applied stress is distributed, by an extent depending on the magnitude of the GAR, between the damaged and undamaged regions respectively. Details of the various models differ with respect to the nature of the accommodating creep mechanism. Arzt and Singer (5) attempted an explanation of their results, shown in Figure 1, by assuming cavity growth to be accommodated by grain boundary sliding of the longitudinal grain boundaries, which leads to a natural and direct linear dependence of the time to fracture on the GAR. Since cavity growth and grain boundary sliding are both linearly dependent on stress, a threshold stress for both of the diffusion - controlled processes had to be invoked in order to achieve an adequate correlation with the experimental data. It is now generally accepted, however, that grain boundary sliding alone is not possible in polycrystalline materials without accompanying grain strain accommodation (9,10), a feature that has received general confirmation in ODS alloys (7,11).

The other, more reasonable, explanation considers cavity growth to be constrained by dislocation creep which leads to strong GAR dependence on the rupture time. Models for this have been developed by Stephens and Nix (6) and Zeizinger and Arzt (7), the main aspects of which will be included in section 5.0 on modelling. Stephens and Nix (6) were able to correlate the stress dependence of the strain rate at low stress in MA 754 having intermediate and low GAR structures, where fracture occurred intergranularly, although at the lowest stresses the model consistently underestimated the creep rate. Zeizinger and Arzt (7) developed a similar model, adapted from the constrained cavity growth analysis for equiaxed grain structures given by Cocks and Ashby (12), which showed a good correlation with the results of Arzt and Singer (5).

The models developed so far have not been applied to the important situation arising in transverse loading where the grain geometry naturally results in tensile forces acting across large tracts of grain boundary area. In this case it is questionable whether a grain constraint exists at all.

3.0 Experimental Techniques

3.01 Materials

INCONEL Alloy MA 6000 bar stock with section size 60 mm x 20 mm in the fully heat treated condition (1h 1500 K, 2h 1230 K, 24h 1120 K) was used in the present study. The grain structure, consisting of highly elongated grains, is shown in Figure 2. Longitudinal grain lengths, L_1, reach upto 10 mm; the mean linear intercept length was determined as 3.7 mm. In a transverse section (Figure 3) grain dimensions are rather irregular. The mean linear intercepts of the long transverse (L_2) and short transverse (L_3) grain lengths were determined as 301 µm and 176 µm respectively. This corresponds to a GAR in the transverse direction of 1.7. The GAR in the longitudinal direction, determined from GAR= $L_1/(L_2 \cdot L_3)^{1/2}$, is 16.

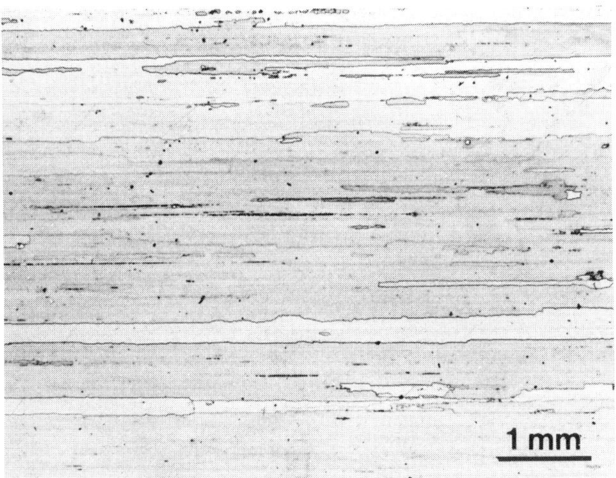

Fig.2: Longitudinal grain structure of MA 6000, GAR=16 (As-received).

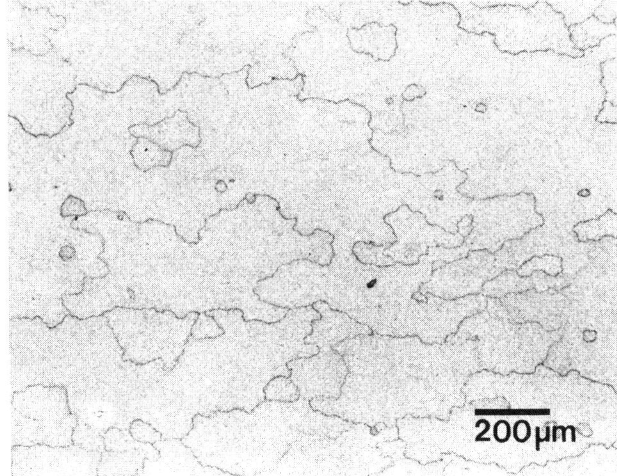

Fig.3: Transverse grain structure of MA 6000, GAR=1.7 (As-received).

In addition a 40 mm diameter bar of fine-grained material was available for producing model microstructures through secondary recrystallisation. Rods of 10 mm diameter spark machined from the bar stock were zone annealed in two stages from opposite ends of the rod under controlled conditions, which allowed grain structures to be produced approaching those of a bicrystal configuration (see section 4.0). Following zone annealing the rods were given the standard heat treatment.

3.02 Creep Testing

Tensile creep specimens with nominal dimensions 30 mm gauge length and 4 mm diameter were machined with the long transverse grain direction lying along the specimen axis. A few specimens were additionally produced from the longitudinal direction. Creep testing under constant load was conducted in an Amsler 5kN creep machine. Thermocouples attached to the specimen enabled the temperature to be controlled to within ± 2 K. Two 5 mm displacement transducers, mounted into a pair of nickel base alloy rod-in-tube extensometers, allowed for continuous displacement-time measurements to be recorded. Tests at the lowest stresses were interrupted before failure; the longest test, at 40 MPa, lasted ~ 1000h.

Compression creep tests, under constant load, were conducted in a Schenck-Trebel 100kN universal testing machine. Tests were performed in the longitudinal, long transverse and short transverse directions using cylindrical specimens 18 mm long and 9 mm diameter. A 1.5 mm displacement transducer connected between two ceramic rods, side mounted onto the specimen surface, allowed for a continuous measurement of displacement. Heating was performed by high frequency induction regulated by three thermocouples which were spot welded with equal spacing along the specimen length. At high stress creep testing was restricted to approximately 2% strain within which the steady state or minimum creep rate could be determined. At low stress a continuously decreasing strain rate was often noted. In these cases final creep rates are reported. Although, in the main, the compression creep tests were of relatively short term, a reasonable estimate of the creep rate could be achieved. All tests, in both tension and compression, were conducted at 1323 K.

4.0 Results

4.01 Tensile Creep Testing

An example creep curve for MA 6000 (65 MPa, 1323 K) tested in the long transverse direction is shown in Figure 4. Following a primary creep range the creep rate reached a steady state and terminated in a tertiary stage prior to failure. All tensile testing in the transverse direction resulted in an intergranular failure, illustrated by the SEM micrograph in Figure 5. The strains to failure were in the range 0.2 - 1.0%.

All tensile creep results are plotted on double logarithmic axes in Figure 6. Although there is considerable scatter, the long transverse data can be reasonably described by a straight line of slope ~7. The longitudinal data agree well with previously published results (full line) taken from reference 13, which can be described using the following constitutive equation:

$$\dot{\varepsilon} = \dot{\varepsilon}_0 \left[\frac{\sigma}{\sigma_0} \right]^n \qquad \text{Eq.[1]}$$

with $\dot{\varepsilon}_0 = 10^{-11}$ s^{-1}, $\sigma_0 = 116$ MPa and n=23.5.

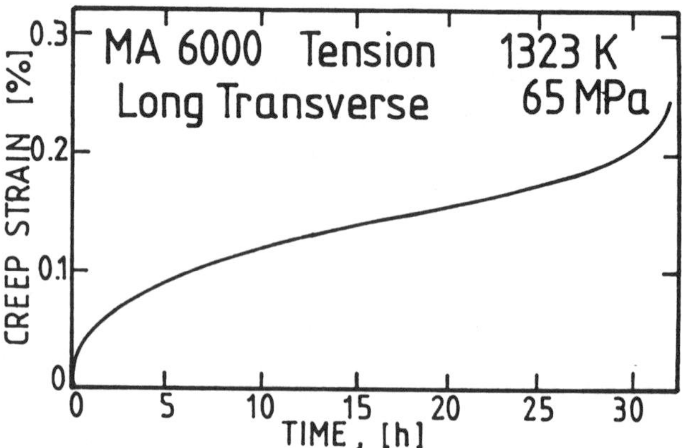

Fig.4: Example creep curve for MA 6000 under transverse loading.

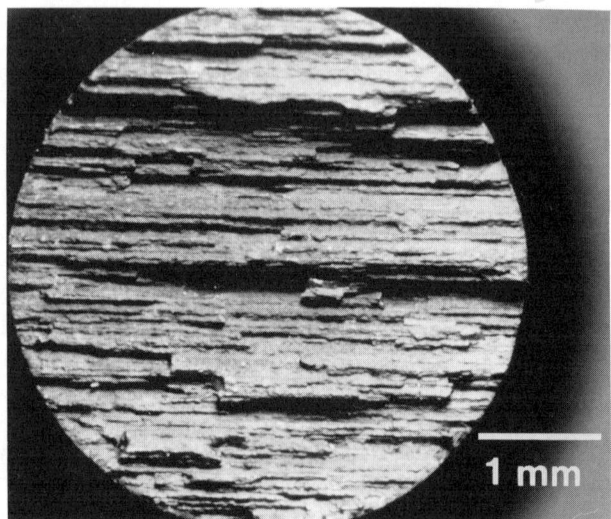

Fig.5: SEM-micrograph showing intergranular fracture in a transverse creep specimen (T=1323 K, σ=65 MPa).

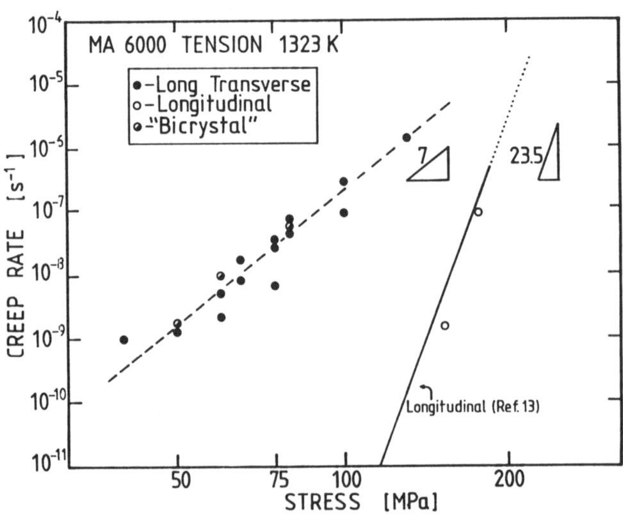

Fig.6: Tensile creep data for MA 6000 in the long transverse and the longitudinal grain direction. The "bicrystal" creep data are also shown.

The tensile creep results confirm the considerable loss in creep strength incurred under transverse loading in comparison to loading in the longitudinal grain direction. Metallographic examination of polished and etched sections taken from fractured specimens revealed profuse grain boundary cavitation lying predominantly on grain boundaries normal to the tensile stress axis (Figure 7). At positions well away from the fracture face individual, spherically shaped, cavities were observed, separated by a white etching zone devoid of γ' precipitates (Figure 8). These observations strongly suggest that diffusional cavity growth mechanisms are involved. The cavity spacing was estimated to lie between 5 and 10μm.

Fig.7: Profuse grain boundary cracking in MA 6000 after creep testing in the long transverse direction (T=1323K,σ=75MPa).

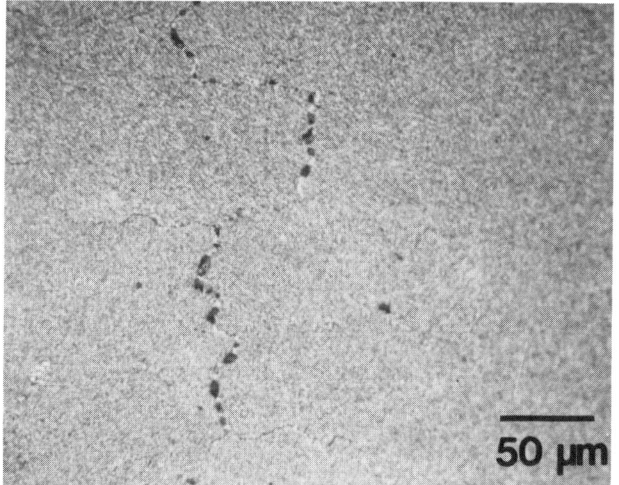

Fig.8: Creep cavities lying on grain boundaries normal to the tensile stress axis (horizontal). Note γ'-denuded zones between cavities.

4.02 Compression Creep Testing

Whereas in the longitudinal grain direction creep is similar in tension and compression (Figure 9), transverse creep in tension is considerably impaired in comparison to creep in compression (Figure 10). Surprisingly, a comparison between the compressive creep behaviour in the longitudinal and transverse directions also revealed significant differences; creep rates are enhanced in the transverse directions (Figure 11). Creep rates in the short transverse direction were found to be, especially at high stress, somewhat faster than in the long transverse direction. The compressive creep data are plotted on double logarithmic axes in Figure 12, which clearly illustrates the enhancement in the transverse creep

rates at low stress in comparison to those in the longitudinal direction.

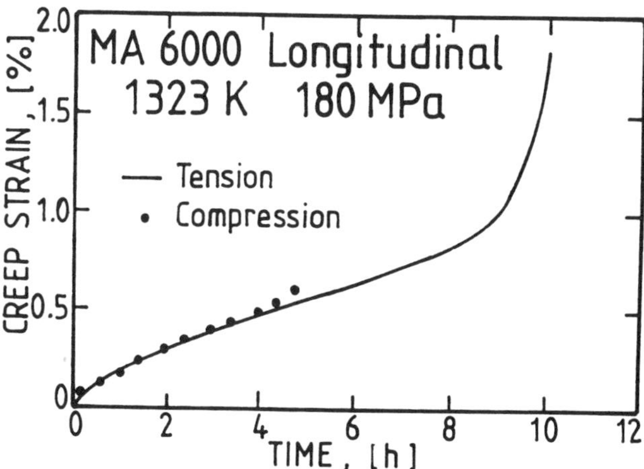

Fig.9: Comparison between tensile and compressive creep in the longitudinal direction.

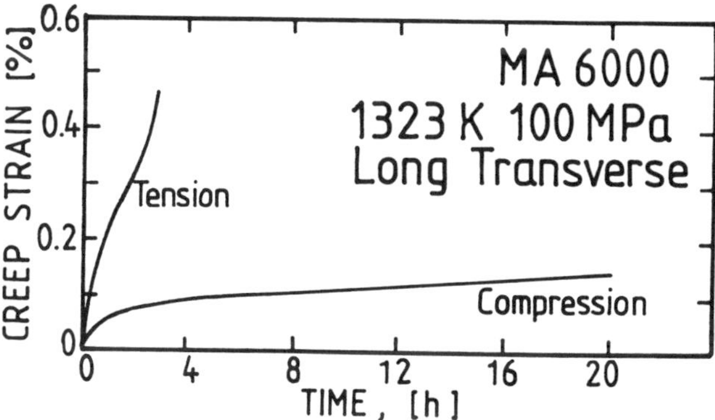

Fig.10: Comparison between tensile and compressive creep in the long transverse direction.

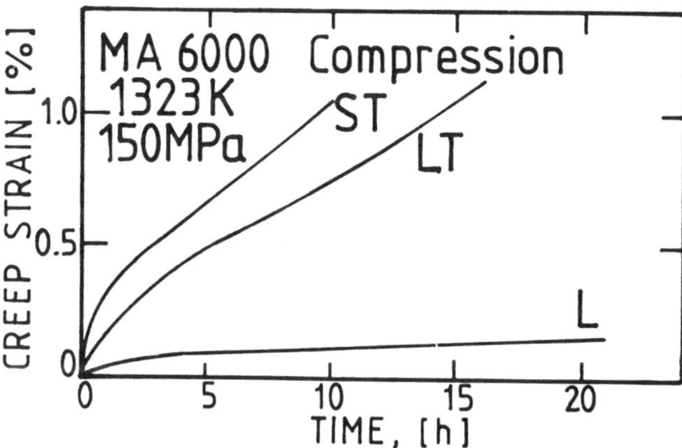

Fig.11: Comparison between compressive creep in the longitudinal, long transverse and short transverse directions.

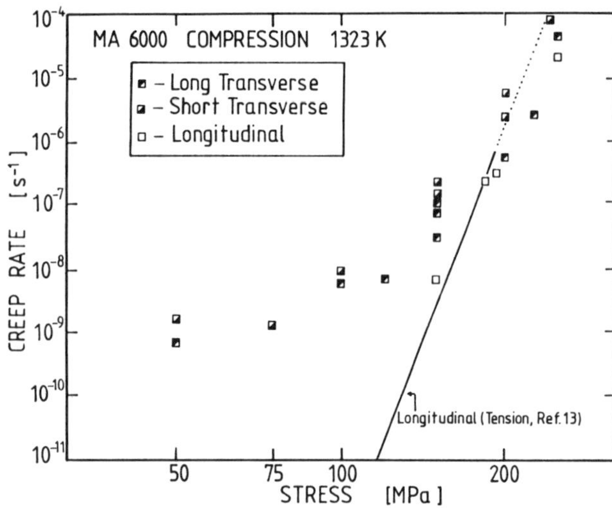

Fig.12: Compressive creep data for MA 6000 in the
transverse directions (longitudinal creep data
are also plotted for comparison).

Metallographic examination of samples tested in
the transverse direction under compressive loading
revealed white etching bands adjacent to grain
boundaries oriented parallel to the compression stress
axis (Figure 13), which have previously been
confirmed as devoid of γ' (14). This feature is strong
evidence for the operation of a stress directed
diffusional creep mechanism. Furthermore, closer
observation revealed local changes in the carbide
particle distribution (Figure 14); within the γ'-depleted
zones a population of ~ 1 µm sized carbide particles,
identified as being W and Mo rich, was observed.
Grain boundaries perpendicular to the compression
stress axis were carbide free. Moreover, these local
changes in microstructure were neither observed in
as-received nor in merely stress-free annealed
material. Attempts have been made, using EDAX in
SEM, to determine the yttrium distribution across the
γ'- denuded zones. A tendency was found for a reduced
Y concentration adjacent to the grain boundary, but
the low resolution of the technique did not allow for a
firm conclusion to be reached. We await further
results from a microprobe analysis.

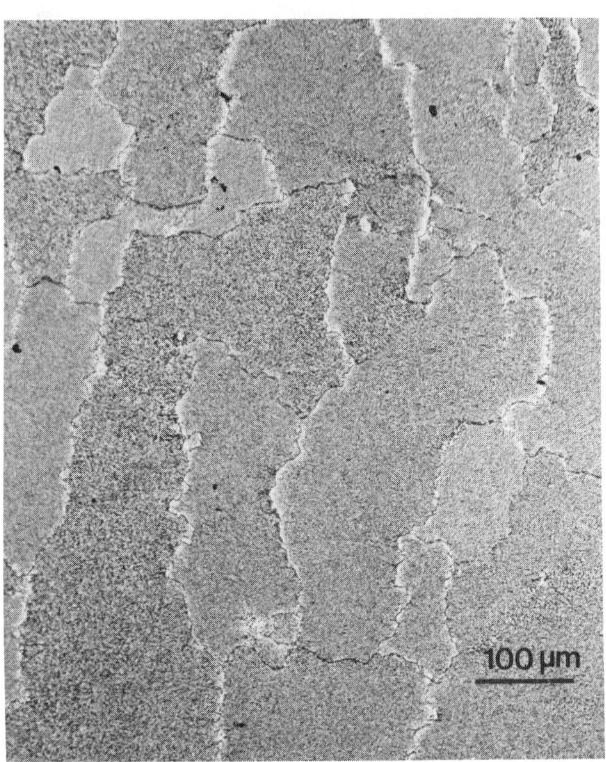

Fig.13: Microstructure following compressive creep in
the long transverse direction. Note γ'-depleted
zones adjacent to grain boundaries oriented
parallel to the compressive stress axis
(vertical). T=1323K, σ=120 MPa.

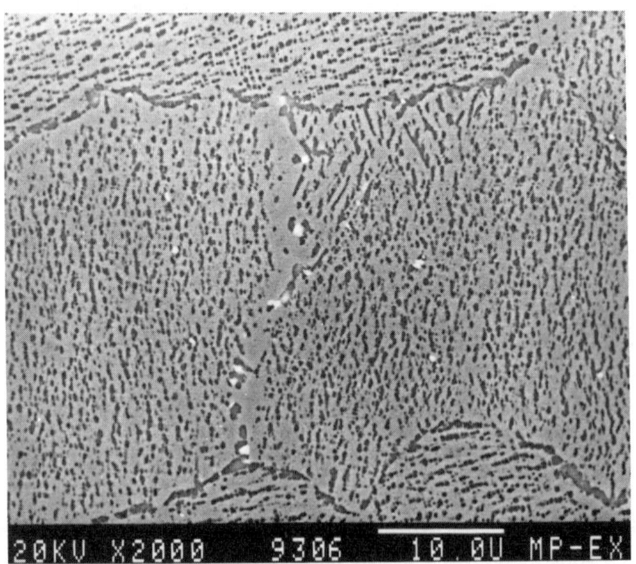

Fig.14: γ'-denuded zones after compressive creep in the
long transverse direction. Note carbide
particles within the zone and absence of
carbides on the other grain boundaries (stress
axis vertical, T=1323K, σ=120 MPa).

4.03 "Bicrystal" Creep Testing

As mentioned in section 3.0 model microstructures having a bicrystal configuration have been produced by sequential zone annealing from opposite ends of specimen rods. The grain structure is depicted in Figure 15 showing how the highly elongated grains meet at a common junction. A few tensile creep tests have been performed on these "bicrystals"; example creep curves, compared with those for the long transverse direction are shown in Figure 16. The times to fracture for the "bicrystal" tests were consistently less than the transverse tests, but the creep rates (also plotted in Figure 6) were, within experimental scatter, surprisingly equivalent. Further reference to these results will be made in the next section.

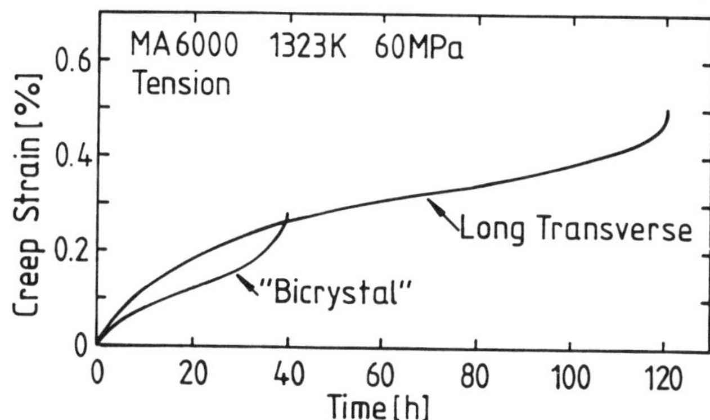

Fig.16: Example creep curves of the "bicrystal" specimens in comparison to the long transverse direction.

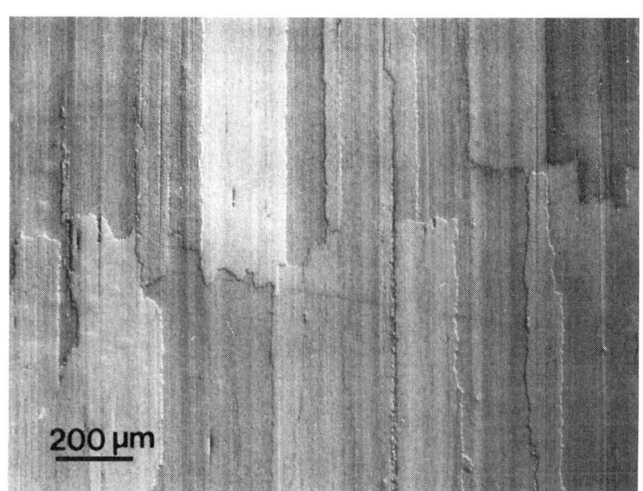

Fig.15: Model "bicrystal" grain structure produced by controlled secondary recrystallisation of fine-grained material.

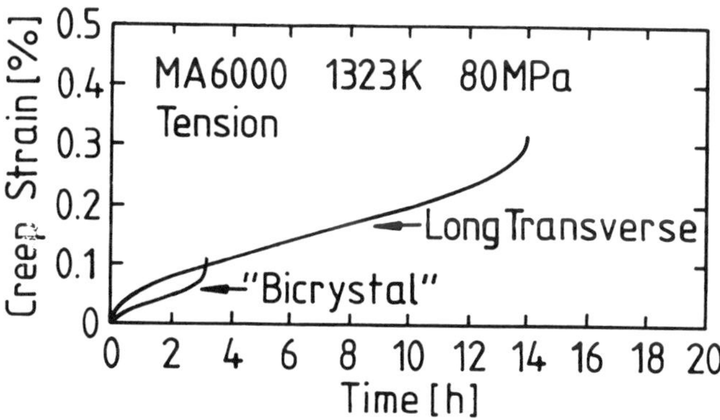

5.0 Discussion and Modelling of the Transverse Creep Behaviour of ODS Alloys

In this section an attempt is made to model the transverse creep behaviour of MA 6000: further details will be published shortly (15).

In view of the success of constrained cavity growth models in describing the longitudinal creep behaviour of ODS alloys as a function of GAR (6,7), it is considered instructive, as a first step, to assess the appropriateness of this modelling approach to the transverse creep behaviour. The main features of these models are contained within the following set of assumptions:

- cavities nucleate on grain boundaries normal to the tensile stress axis and grow by grain boundary diffusion
- void growth is accommodated by dislocation creep in the surrounding, undamaged, grains
- in order to maintain material compatibility the displacement rates in the damaged and undamaged regions are equal.

From these assumptions the following relationship results between the applied stress, σ_∞, and the macroscopic creep rate, $\dot{\varepsilon}_\infty$, including the influence of the grain aspect ratio, R:

$$\frac{\sigma_\infty}{\sigma_0} = \frac{\dot{\varepsilon}_\infty}{A \cdot R} + \frac{R-1}{R}\left[\frac{\dot{\varepsilon}_\infty}{\dot{\varepsilon}_0}\right]^{1/n} \qquad \text{Eq.[2]}$$

where A is a material constant taken from Cocks and Ashby (12),

$$A = \frac{4 D_b \omega \; \Omega \; \sigma_0}{kT \; L \; \lambda^2 \; ln(1/f)} \qquad \text{Eq.[3]}$$

$D_b \omega$ is the grain boundary diffusion coefficient times the grain boundary width, f is the cavitated area fraction, $(r/\lambda)^2$, r is the void radius and 2λ is the cavity spacing. The other symbols have either been given previously or take their usual meanings.

Equation [2] has been evaluated and is plotted, together with the experimental tensile creep results in Figure 17, for both R=1.5, 2λ=5μm and R=2.0, 2λ=10μm, which are reasonable limits for the most sensitive parameters entering into Eq.[2]. L is set equal to the long transverse grain length (L_2) and values for the other parameters are taken from Frost and Ashby (16). The plot shows how the model calculations bound the bulk of the data, but the stress sensitivity of the creep rate is overestimated. The extreme sensitivity of the predicted creep rate to the grain aspect ratio at low GAR is obviously unrealistic. For example at a stress of 50 MPa a small increase in GAR from 1.5 to 2.0 is predicted to raise the strain rate by over four orders of magnitude. The reason lies in the assumption that void growth occurs by classical diffusional mechanisms such that for R=1, which represents a situation where no matrix creep constraint acts (fully cavitated equiaxed polycrystal or bicrystal), a linear stress dependency of the creep rate is predicted. Further model calculations are illustrated in Figure 18 showing how, at low GAR, the stress-creep rate curve changes dramatically on varying GAR by a small, practically insignificant, amount.

The weakness of the constrained cavity growth model lies in the acceptance of the equations for the classical diffusional mechanisms. The microstructural evidence presented in section 4.0 showed clearly that diffusional creep processes do indeed make a significant contribution to the overall creep behaviour in MA 6000. However, it is unlikely that in these dispersion hardened materials diffusional creep rigidly obeys the classical diffusional creep theories. One of the main assumptions of the above model, therefore, cannot be accepted without criticism. Moreover, it is questionable whether the assumption regarding cavity growth constrained by matrix deformation is reasonable for loading in the transverse direction, in which the highly elongated grains extend across virtually the total creep specimen diameter, as already illustrated in Figure 5. Additionally, the model

explicitly predicts a linear stress dependence of the creep rate for testing in the short transverse direction. Creep results, however, have shown that the stress exponent, for the time to fracture at least, remains well in excess of 1 (17). Although some of the problems may be caused by the assumption of an idealised grain geometry used in the modelling, we believe there is little scope for making any significant improvements in the predictions of the constrained cavity growth model when applied to low GAR structures.

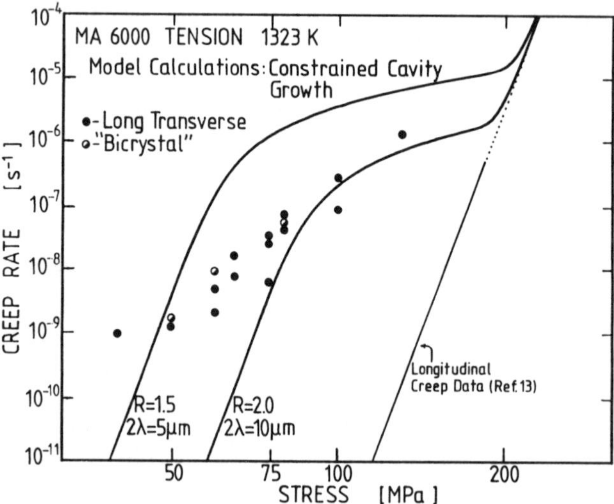

Fig.17: Comparison between model calculations for constrained cavity growth and the transverse tensile creep data.

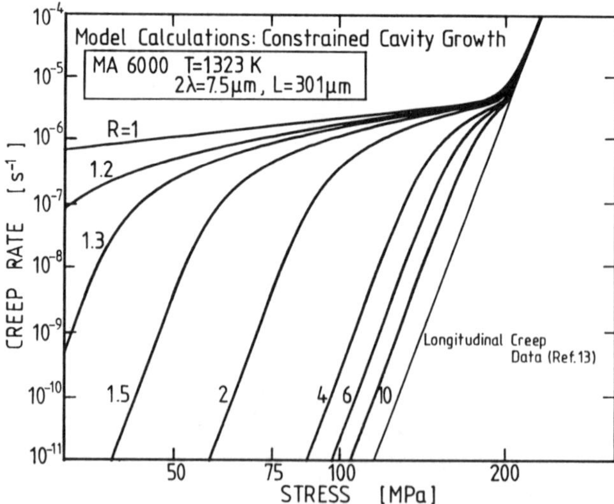

Fig.18: Constrained cavity growth model calculations. Note the extreme sensitivity of the stress-strain rate curve to the GAR at low GAR.

These reasons led to the motivation for conducting the "bicrystal" creep tests, since these would be, in principle, the optimum solution to confirm the applicability of classical diffusional cavity growth theories. The grain structure depicted in Figure 15 reveals that the grains are slightly interlocked, but because of this small degree of waviness we believe such a structure offers negligible matrix constraint to any cavity growth. Hence, the similar strength of the long transverse and the "bicrystal" creep specimens, as described in section 4.0, sheds doubt on the validity of the constrained cavity growth model in the present investigation. They further bring into question the appropriateness of using classical diffusional creep theories alone (see below).

In the following a new model is presented based on interface reaction controlled diffusional creep. In tension, at low stress levels, the creep rates predicted by classical diffusional cavity growth thoery are far greater than the measured creep rates (represented by the R=1 line in Fig.18), and the predicted linear stress sensitivity of the creep rate is not observed. Under compressive loading cavity growth is suppressed, but diffusional creep deformation is not, even though the grains are highly elongated. The reason is (18,19) that, under transverse loading, the vacancy flux adopts a two dimensional form resulting in a creep rate dependent only on the transverse grain dimensions and independent of the long grain length. When lattice diffusion is rate controlling the constitutive equation for transverse Nabarro-Herring creep in an elongated grain structure becomes (18,19):

$$\dot{\varepsilon} = \frac{12\, D_v\, \Omega}{kT} \cdot \frac{\sigma}{L_2^2 + L_3^2} \qquad \text{Eq.[4]}$$

It is well known that in order for diffusional creep processes to continue in an uninhibited manner, the supply of vacancies from the grain boundary must be maintained at the required level as governed by the applied stress. If the creation or annihilation of vacancies is in any way impeded, then the diffusional creep rate will become progressively dependent on the kinetics of the vacancy creation process. It is generally considered that vacancies are created and destroyed via the climb of grain boundary dislocations (20,21), which, therefore, will have to by-pass the dispersoids lying in the grain boundary plane in order to act as efficient sources and sinks for vacancies (22).

We suggest that grain boundary dislocations overcome dispersoids in a similar manner to the way in which matrix dislocations by-pass inert dispersoids in ODS alloys (1,2); dislocations will be strongly pinned at the particle interface as a consequence of an attractive interaction, and thermal activation will be necessary for dislocation detachment. If the stress is too low or the thermal activation is insufficient, then creep due to stress directed diffusion of point defects stops. We realise there is virtually no evidence in the literature for such a mechanism and experimental verification may prove extremely difficult because of the problems involved in imaging grain boundary dislocation - particle interactions in TEM: one report, however, supports this hypothesis (23). In addition, a theoretical analysis has so far not been attempted, but, in analogy with matrix creep behaviour in ODS alloys, we may estimate what sort of features a constitutive equation based on the thermally activated detachment of grain boundary dislocations from inert dispersoids may have. These may include a relatively high stress sensitivity of the dislocation mobility over a large stress range, activation energies for creep far greater than that for the simple diffusional process, and pseudo-threshold stress behaviour.

In general, diffusional cavity growth and diffusional creep superimpose on power law matrix dislocation creep. Therefore three independent strain producing components can be identified:

1) Interface reaction controlled diffusional cavity growth,

$$\dot{\varepsilon}_1 = \dot{\varepsilon}_v \, / \, [1 + (\dot{\varepsilon}_v / \dot{\varepsilon}_m)] \qquad \text{Eq.[5]}$$
where,

$$\dot{\varepsilon}_v = \frac{B\, D_b\, \omega\, \Omega}{kT} \cdot \frac{\sigma}{\lambda^2\, L_2} \qquad \text{Eq.[6]}$$
and $\dot{\varepsilon}_m$ is discussed below.

2) Interface reaction controlled diffusional creep deformation,

$$\dot{\varepsilon}_2 = \dot{\varepsilon}_v \, / \, [1 + (\dot{\varepsilon}_v / \dot{\varepsilon}_m)] \qquad \text{Eq.[7]}$$
where in this case,

$$\dot{\varepsilon}_v = \frac{C\, D_v\, \Omega}{kT} \cdot \frac{\sigma}{L_2^2 + L_3^2} \qquad \text{Eq.[8]}$$

3) Power law creep,

$$\dot{\varepsilon}_3 = \dot{\varepsilon}_0 \left[\frac{\sigma}{\sigma_0} \right]^n \qquad \text{Eq.[9]}$$

Under tensile loading in the transverse direction the creep rate is given by,

$$\dot{\varepsilon}_T = \dot{\varepsilon}_1 + \dot{\varepsilon}_2 + \dot{\varepsilon}_3$$

whereas under compressive loading, where cavity growth does not occur, the creep rate is,

$$\dot{\varepsilon}_C = \dot{\varepsilon}_2 + \dot{\varepsilon}_3$$

In order to proceed further we need a constitutive equation for $\dot{\varepsilon}_m$, which describes the interface reaction. Lacking a detailed thoery for the grain boundary dislocation - dispersoid interaction, we choose to adopt, for the present time at least, an expression for $\dot{\varepsilon}_m$ that simply fits the bulk of the tensile creep data in a reasonable way. We find that a simple power law form for $\dot{\varepsilon}_m$ suffices as follows:

$$\dot{\varepsilon}_m = \dot{\varepsilon}_{m_0} \left[\frac{\sigma}{\sigma_m} \right]^m \qquad \text{Eq.[10]}$$

with $\dot{\varepsilon}_{m_0}=10^{-10}$ s^{-1}, σ_m=40 MPa and m=10.

Using the above relationship, the appropriate combinations of the creep processes 1, 2 and 3 for tensile and compressive loading respectively have been evaluated, and are plotted, together with all the creep data, in Figure 19. The individual uninhibited diffusional creep rates, and the creep rate due to power law creep are also displayed in Figure 19 as dashed lines. Although we have purposely selected a constitutive equation for $\dot{\varepsilon}_m$, we believe the present approach, summarised by Figure 19, is encouraging in two respects:

- firstly, the theoretical creep rates due to both cavity growth and Nabarro-Herring creep predict very well the respective limits of the tensile and compressive creep rate data,
- secondly, through the choice of a single, relatively simple, relationship for $\dot{\varepsilon}_m$, both the tensile and the compressive creep data can be interpreted in a consistent manner.

The assumption that diffusional creep mechanisms in MA 6000 are controlled by an interface reaction is the crucial step in the present modelling approach. In itself this is not unrealistic, but verification of the validity of the constitutive equation for the interface reaction and the details of the mechanism will require further work. An interface reaction process based on a dispersoid - grain boundary dislocation interaction is

certainly plausible. It will be necessary to conduct detailed TEM - investigations in order to ascertain whether such a mechanism indeed occurs in these alloys. In addition a theoretical analysis will be required to support these considerations.

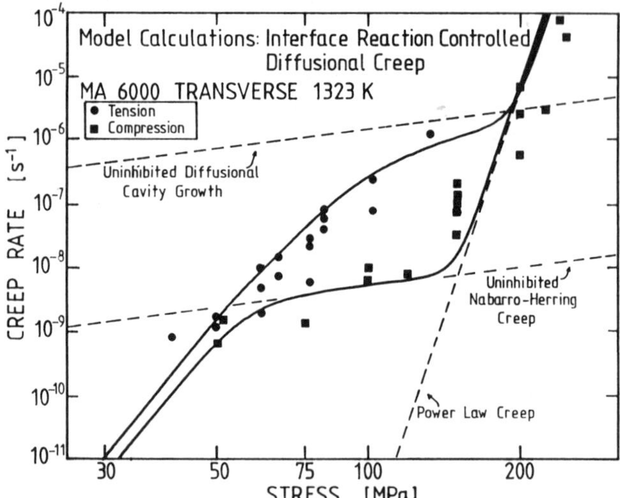

Fig.19: Comparison between model calulations for interface reaction controlled diffusional cavity growth and creep deformation and the transverse tensile and compressive creep data.

It seems clear that the transverse creep behaviour of coarse-grained ODS alloys will, in general, always be inferior to that of the longitudinal direction. Significant improvements in strength would be possible if cavity nucleation could be suppressed. Under the present circumstances this stage of the fracture process occurs relatively easy, and coarse oxide particles are the likely sites for void nucleation. Increased creep strength should be possible through a reduction in the population of the coarse oxide inclusions, which are often present in ODS alloys as stringers located on the longitudinal grain boundaries. Under transverse loading their effect is therefore extremely detrimental.

6.0 Summary and Conlusions

Tensile creep tests on MA 6000 indicate that models based on matrix creep constrained cavity growth are inappropriate for creep in the transverse grain direction. A model based on interface reaction controlled diffusional cavity growth is suggested to be more realistic. Under compression diffusional creep deformation is important which can also be included within the present model framework. Although many details of the micromechanisms involved remain unclear, the present approach can provide a basis for a better understanding of the transverse creep behaviour of ODS alloys. It is suggested that only by reducing the density of the coarse oxide inclusions can significant improvements in the transverse creep strength be acheived.

Acknowledgements
We are especially grateful to Mr K. Lempenauer for producing the "bicrystal" specimens. Particular thanks go to Mr D. Lang, Mrs S. Miller and Mrs C. Elzey for assistance with the creep testing and metallography. We wish to acknowledge the financial support of the Bundesministerium für Forschung und Technologie in the FRG under project number 03 M0013.

REFERENCES

[1]   Arzt, E. (1988) in "New Materials by Mechanical Alloying Techniques", p185 Eds. E. Arzt and L. Schultz, DGM, Oberusel, FRG.

[2]   a) Rösler, J. and Arzt, E. Acta Metall., in press.
b) Rösler, J. and Arzt, E. (1988) Acta Metall., $\underline{36}$, 1043.
c) Arzt, E. and Rösler, R. (1988) Acta Metall., $\underline{36}$, 1053.

[3]   Wilcox, B.A. and Clauer, A.H. (1972) Acta Metall., $\underline{20}$, 743.

[4]   Cairns, R.L., Curwick, L.R. and Benjamin, J.S. (1975) Metall. Trans., $\underline{6A}$, 179.

[5]   Arzt, E. and Singer, R.F. (1984) in "Superalloys 84", p367, Eds. M. Gell et al. TMS-AIME, Warrendale.

[6]   Stephens, J.J. and Nix, W.D. (1986) Metall. Trans. $\underline{17A}$, 281.

[7]   Zeizinger, H. and Arzt, E. (1988) Z. Metall., $\underline{79}$, 774.

[8]   Wright, P.K. (1978) Metall. Trans., $\underline{9A}$, 955.

[9]   Raj, R. and Ghosh, A.K. (1981) Metall. Trans., $\underline{12A}$, 1291.

[10]   Anderson, P.M. and Rice, J.R. (1985) Acta Metall., $\underline{33}$, 409.

[11]   Stephens, J.J. and Nix, W.D. (1985) Metall. Trans., $\underline{16A}$, 1307.

[12]   Cocks, A.C.F. and Ashby, M.F. (1982) Prog. Mat. Sci., $\underline{27}$, 189.

[13]   Guttman, V. (1988) in COST 501 Summary Report by E. Arzt and R. Timmins.

[14]   Timmins, R. and Arzt, E. (1988) Scripta Metall., $\underline{22}$, 1353.

[15]   Timmins, R. and Arzt, E. (1990) to be published.

[16]   Frost, H.J. and Ashby, M.F. (1982) "Deformation-Mechanism Maps", Pergamon Press, Oxford.

[17]   Nazmy, M., Staubli, M. and Ebeling, W. (1987) Mat. Technik, $\underline{15}$, 9.

[18]   Nix, W.D. (1981) Metals Forum, $\underline{4}$, 38.

[19]   Greenwood, G.W. (1985) Phil. Mag., $\underline{51}$, 537.

[21]   Balluffi, R.W. (1980) in "Grain Boundary Sructure and Kinetics", Ed. R.W. Balluffi, ASM, Metals Park, Ohio.

[22]   Arzt, E., Ashby, M.F. and Verrall, R.A. (1983) Acta Metall., $\underline{31}$, 1977.

[23]   Dunlop, G.L., Nilsson, J.-O. and Howell, R.R. (1979) J. Microscopy, $\underline{116}$, 115.

KEYNOTE ADDRESS
MA ALLOYS FOR INDUSTRIAL APPLICATIONS

J. J. Fischer, J. J. deBarbadillo
Inco Alloys International, Inc.
Huntington, West Virginia 25720, USA

M. J. Shaw
Inco Alloys Limited
Hereford, United Kingdom

ABSTRACT

Mechanically alloyed materials have established an outstanding record of performance as the hot section components in military jet engines for more than 12 years. In industrial applications, the acceptance of MA alloys has been slower in developing. This review examines the benefits MA alloys have to offer for demanding high–temperature applications in terms of strength, resistance to environmental attack and other criteria for industrial acceptance. The second part of the paper reviews some specific examples of MA materials being used in industries involved with: thermal processing, glass processing, and energy production.

THE EARLY MECHANICALLY Alloyed (MA) materials were developed primarily with the aerospace market in mind. The first MA alloy to achieve a significant market size was INCONEL® alloy MA 754. This alloy has been used as components in the hot section of military jet engines for more than 12 years.

In contrast to this, the acceptance of MA alloys into more "industrial type" applications has been much slower in developing. There are a number of reasons for this, including:

- reluctance to try new materials
- lack of in–service experience
- availability of the necessary product forms –– sheet, plate, tube, etc.
- higher cost than conventional alloys.

®INCONEL is a registered trademark of the Inco family of companies.

In opposition to these inhibiting factors is the attractiveness of the MA materials performance. The mechanical properties of the MA materials combined with their resistance to high–temperature corrosion/oxidation attack offer a level of performance that cannot be obtained in conventional alloys.

This review is presented in two parts. The first section examines *"What MA Alloys Have to Offer"* in terms of the requirements of industrial applications. The second section reviews various *"Examples of Industrial Applications of MA Materials."*

WHAT MA ALLOYS HAVE TO OFFER

STRENGTH – Mechanically alloyed materials first attracted the attention of aerospace design engineers because of their exceptional strength at high temperatures. The elevated temperature strength of these material is derived from more than one mechanism.

First and of most importance is the role that the oxide dispersoid particles play. These fine particles inhibit dislocation motion in the metal matrix and thereby increase the alloy's resistance to creep deformation.

It is now well known that the mere presence of a refractory oxide is not sufficient to produce the desired results. First of all, the particles must be very fine (5–50 nm). The relatively low temperatures used for MA are ideal for adding and maintaining a controlled particle size. Secondly, the oxide must be uniformly dispersed. Typical particle spacings are of the order of 100 nm. Inadequate milling time, clumping of the oxide particles, powder contamination, improper selection of raw materials, etc., can all lead to less than optimum strength. Finally, the oxide must be stable during service. Early in the the Inco work, yttria was chosen as the most industrially acceptable, stable oxide, although other oxides, such as thoria or lanthana can be used.

Another function of the dispersoid particles is to inhibit the recovery and recrystallization processes. Because of this, many of the mechanically alloyed materials can be made to form a very stable, large grain size. This large grain size is obtained by a secondary recrystallization mechanism that is not fully understood. However, it is known that the large grains are resistant to grain rotation (and grain boundary sliding) during high temperature loading.

It should be noted that for industrial applications where complex fabricated shapes and various mill product forms are required, a relatively random grain orientation and low grain aspect ratio are normally encountered. This is quite different from the situation in textured and/or directionally recrystallized bar used for optimum strength in gas turbine blade and vanes.

Additional benefits to an alloy's strength are obtained from the very homogeneous distribution of alloying elements that is characteristic of mechanically alloyed materials. This uniformity of alloying elements gives both the solid–solution strengthened and precipitation–hardened alloys more stability at high temperatures and overall improvement in properties.

Perhaps a good example of the type of strengthening that may be obtained with M.A. materials is the case of a nickel–base alloy containing 20% Cr. A conventional wrought alloy of this type is NIMONIC® alloy 75. This alloy has a 1000–hour rupture strength of only 2 MPa (0.3 ksi) at 1093°C (2000°F). MA 754, with approximately the same composition, has a 1000–hour rupture strength of 94 MPa (13.6 ksi) at the same temperature. The comparative stress rupture properties of MA 754, NIMONIC alloy 75, INCOLOY® alloy MA 956 and INCONEL alloy 617 (one of the highest strength Ni–base sheet alloys) are shown in Table I. Similar examples of this large increase in high temperature strength can be found in a number of other alloy systems, including copper and aluminum alloys (1,2).

TABLE I

STRESS TO PRODUCE RUPTURE IN 1,000 HOURS AT INDICATED TEMPERATURE

Alloy	871°C(1600°F)		982°C(1800°F)		1093°C(2000°F)		1204°C(2200°F)	
	MPa	Ksi	MPa	Ksi	MPa	Ksi	MPa	Ksi
NIMONIC alloy 75	7	1.0	3	0.5	2	0.3	–	–
INCONEL alloy MA 754	158	22.9	129	18.7	94	13.6	62	9.0
INCOLOY alloy MA 956	83	12.0	67	9.7	51	7.4	31	4.5
INCONEL alloy 617	63	9.2	26	3.8	10	1.5	–	–

RESISTANCE TO ENVIRONMENTAL ATTACK – The mechanically alloyed materials offer more than improved high temperature strength. These alloys are also noted for their excellent oxidation and hot corrosion resistance. The resistance to environmental attack is of prime importance for many applications where long–term exposure is required.

An example of this environmental resistance can be found in the use of MA 754 for gas turbine vanes in General Electric's F110 and F404 engines where the alloy is used in the uncoated condition at operating temperatures above 1000°C (1832°F). Many of the recently developed single crystal alloys are unable to perform in this type of application unless they are plasma–spray coated with MCrAlY alloys to improve the oxidation resistance.

The earliest ODS alloys, such as TD Nickel, were able to provide excellent high temperature strength, but their resistance to environmental attack was quite limited because of the lack of alloying elements required to produce stable adherent oxide scales. the MA process overcame that limitation, and today's alloys of industrial importance contain substantial amounts of chromium and aluminum. The compositions of MA alloys used in

®NIMONIC and INCOLOY are registered trademarks of the Inco family of companies.

industrial applications are given in Table II. In these MA alloys, the increased resistance to oxidation–sulfidation attack is due to the homogeneous distribution of the alloying elements and also the improved scale adherence due to the dispersoid itself. Figure 1 shows the cyclic oxidation resistance of some of the commercial MA materials compared to conventional alloys.

TABLE II

NOMINAL COMPOSITION OF MECHANICALLY ALLOYED MATERIALS

Alloy	Ni	Fe	Cr	Al	Ti	Mo	W	Y_2O_3
INCONEL alloy MA 754	Bal.	–	20	0.3	0.5	–	–	0.6
INCONEL alloy MA 758	Bal.	–	30	0.3	0.5	–	–	0.6
INCONEL alloy MA 760	Bal.	–	20	6.0	–	2.0	3.5	0.95
INCOLOY alloy MA 956	–	Bal.	20	4.5	0.5	–	–	0.5
INCOLOY alloy MA 957	–	Bal.	14	–	1.0	0.3	–	0.25

As shown above, the MA alloys have demonstrated outstanding resistance to oxidation attack. However, real industrial environments are generally much more complex involving other corrosion species, particularly sulfur, halogens, carbon and frequently complex interactions with cyclic or triaxial stresses. Testing in a wide range of environments has shown MA 956 to have the best balance of properties, especially in carburizing or sulfidizing environments. Predicting the performance in actual use is extremely difficult, although in some cases, it can be estimated from coupon exposure tests.

Most environmental degradation at high temperature involves destruction of protective scales or penetration by harmful elements. Exposure to molten glass may present problems caused by the progressive fluxing of the oxide scale. The MA materials have been exposed under a variety of glass conditions with excellent results. An example is shown in Tables III and IV for two types of glass.

TABLE III

LIME GLASS CORROSION OF MA ALLOYS AT 1149°C (2100°F) FOR 240 HOURS

Mass Change for Each Alloy Immersed in Lime Glass*

Alloy	(mg/cm^2)
MA 956	–4
MA 754	–28
MA 758	–42

***Chemical Composition of Lime Glass (wt %)**

SiO_2	Al_2O_3	Na_2O	CaO	MgO	H_2O	Li_2O
73.0	1.7	16.3	4.7	3.1	0.4	0.15

TABLE IV
"C" GLASS CORROSION OF MA ALLOYS AT 1200°C (2192°F)

Alloy	Metal Loss*	
	mm	mils
MA 754	0.04	1.6
MA 758	0.03	1.2

*Based on a five-day immersion test.

COMPOSITION OF "C" GLASS

SiO_2	Al_2O_3	Na_2O	CaO	MgO	B_2O_3
65	4	8.5	14	3	5

AVAILABILITY OF PRODUCT FORMS – In order for a new material to gain acceptance into competitive, industrial markets, it must meet a number of criteria in addition to its mechanical property performance. One of these factors is the availability of various sizes and product forms.

In general, the types of products available must include the conventional forms of: plate, bar (rounds and flats), sheet and small diameter rod. In addition, some applications require tubular or wire products. A list of the product forms available for some of the commercial MA alloys is given in Table V.

TABLE V
PRODUCT FORMS AVAILABLE FOR COMMERCIAL MA ALLOYS

	MA 956	MA 754	MA 758	MA 760	Approximate Size Range
Bar	X	X	X	X	Up to about 250 mm width and thickness
Plate	X	X	X	X	3–25mm thick – widths and lengths over 3m
Sheet	X	X	X		0.1–3mm thick widths up to 3m
Wire	X				0.5mm and larger
Tube	X				5mm OD x 0.5mm wall to 89mm OD x 3mm wall
Forging Stock	X	X	X	X	Similar to bar, except un–recrystallized condition

The Fe–base ODS alloy, MA 956, has excellent fabricability and is amenable to most metal processing techniques. As a result, this alloy is available in the widest range of sizes for all of the common product forms. In particular, sheet material has recently been produced with thicknesses under 0.1mm and tubular products with wall thicknesses below 0.4mm. On the high side, plates over 2.5m wide x 5m long and weighing over 500 kg have also been produced. Through careful processing control, sheet and plate with essentially isotropic properties can now be provided.

Initially, the nickel–base ODS alloys, such as MA 754, were only available as bar with highly directional properties. Subsequently, development led to 6–25 mm plate with isotropic properties.

In recent development, ODS Ni–base alloys have been produced in thin plate and sheet forms and even small diameter wire–rod While the range of forms does not yet match MA 956, this should make ODS nickel–base alloys a possibility for a number of new applications.

The size range in which these product forms can be produced is also of importance to the industrial user. The upper limit of size is usually dependent on the production equipment available, whereas, the lower limit of size (thickness or diameter) may be dependent on the alloy's working characteristics. As with many powder metallurgy

82

products, there is no theoretical upper limit on product weight or section size apart from the equipment limitations. The previously mentioned 500 kg MA 956 plate represents the largest made to date, but larger could be made if needed.

OTHER CRITERIA FOR INDUSTRIAL ACCEPTANCE – Industrial acceptance of new materials also requires that certain other factors be satisfied.

The *performance/cost ratio* must be high. In the case of MA materials which are more costly to produce than conventional cast or wrought alloys, it is particularly important that enhanced performance be obtained. Enhanced performance may include longer component life, more severe operating conditions (higher temperature, more corrosive environment, etc.), or increased productivity (greater load carrying capacity). The MA alloys have demonstrated performance/cost benefits in a number of new industrial applications as described in the next section.

Fabricability of components is an important factor in many applications. Fabricability includes: ease of machining, forming characteristics and joining capabilities. Because the MA alloys are generally much higher strength than conventional alloys, special attention has been paid to developing the methods needed to form these materials. MA 956 is readily formed by most conventional working methods. The alloy also lends itself to cold bending, drawing and stretching, provided it is warmed above the ductile–brittle transition temperature (circa 100°C) (3).

The MA nickel–base alloys are more difficult to form. However, good progress has been made, particularly in the forging area (4,5) and precision forged components are in commercial use. Other forming operations, such as hot shear forming, ring rolling and hot upsetting have been done. For these operations, the relatively low strength and high ductility of the alloy in the fine–grain condition can be exploited to improve fabricability. A grain–coarsening heat treatment is then applied after forming. Properties in the grain–coarsened conditioned are typical of those in recrystallized–mill products.

The machining and joining methods used on MA products have also been examined in some detail and previously reported (6). While machining has not proven to be a problem, the joining of MA materials has been studied to a considerable degree (7). It is clear from these investigations that joining methods utilizing solid state diffusion are preferred, as opposed to fusion welding methods. The reason is due to the fact that the dispersoid particles may be retained in the matrix during solid state joining whereas fusion melting causes the dispersoid particles to be rejected from the molten metal. Nevertheless, sound TIG welds suitable for positioning or non load–bearing joints can be made. Fusion welding processes, which minimize the size of the molten zone, produce welds which are generally stronger. Spot and resistance seam welds with excellent tensile strength can be made. However, it may never be possible to achieve the full stress rupture properties of the base material.

Solid state processes can produce such matching properties. Processes which have been demonstrated include diffusion bonding, explosive bonding and magnetostrictive welding. Brazing is also applied mainly in aerospace industry for attachment of MA components. The joint strength is, of course, limited by the strength of available braze filler materials.

EXAMPLES OF INDUSTRIAL APPLICATIONS OF MA MATERIALS

The number of applications for MA materials in the industrial – non–aerospace – market is increasing rapidly. For many years, certain high–temperature applications have been using the best conventional alloys available and still found their performance lacking. Because conventional alloys often do not provide the necessary life or load carrying capability required for demanding applications, costly down–time and high replacement costs are incurred. The MA materials offer a cost effective way to satisfy some of these stringent operating conditions. A number of examples of MA alloys being used in different industrial environments is given below.

THERMAL PROCESSING INDUSTRY – This industry involves a wide range of applications including furnace fixtures, muffle tubes, furnace transport components (rollers, skid rails, mesh belts, etc.) and even the heat–source materials (electrical resistance windings, burner nozzles, etc.). The thermal environments are extremely varied, including such media as high–temperature air, hydrocarbons, high sulfur compounds, molten glass and various other corrosive gases and particulate matter. In general, a failure in one piece of thermal processing equipment can shut down an entire manufacturing operation. Therefore, material reliability is of prime importance. In addition, scaling can lead to contamination of the article being produced, as in the manufacturing of fine china and electronic chips.

INCOLOY alloy MA 956 is particularly well suited for use in heat processing applications. This iron–base alloy has a melting point of 1482°C (2700°F) which is well above conventional nickel–base alloys. MA 956 has been shown to possess superior oxidation resistance compared to conventional alloys. The carburization and sulfidation resistance are also outstanding and, therefore, allow the alloy to be used in many hostile environments where conventional alloys cannot survive.

The stress–rupture properties of MA 956 are well above those of conventional alloys, such as alloy 617 (Table I). Where dimensional control is important to avoid thermal distortion, MA 956 has a coefficient of expansion that is about 15% lower than nickel–base alloys.

This MA material has already demonstrated its reliability in a number of thermal industry applications. For example, vacuum furnace fixtures made out of MA 956 have shown excellent durability and are able to compete with wrought molybdenum which is also used in these applications (Figure 2). MA 956 offers a number of advantages over molybdenum. The MA alloy is about 30% lower in density than molybdenum which provides both a weight savings and a cost advantage. Also, it has a lower vapor pressure than molybdenum and, therefore, will not coat the inside of the vacuum chamber or the parts being heat treated.

This same alloy, MA 956, has also been used for heat treating baskets (Figure 3) operating in air at temperatures above 1200°C (2200°F). For heat treatments in air, molybdenum cannot be used because of its poor oxidation resistance. Conventional nickel–base alloys cannot be used since they do not have enough strength to support the weight of the parts being heat treated.

GLASS PROCESSING INDUSTRY – A number of years ago, it was recognized that certain MA materials possess exceptional resistance to attack by molten glass (8). The type of molten glass being used will dictate which alloy should be selected. For example, MA 956 is particularly resistant to attack by lime glass as shown in Table III.

Because of this corrosion resistance, MA 956 is being evaluated in a number of molten glass applications.

Figure 4 shows two plunger rods which are used to force molten glass slugs through a nozzle and into molds in order to produce beverage bottles. The glass temperature exceeds 1200°C (2200°F) and the plungers also experience fatigue stresses and thermal shock. Although ceramic rods are currently used in this application, their performance is less than satisfactory. MA 956 has performed well in trials, except for some corrosion attack at the molten glass–air surface line. It is possible that a coating will be needed for this area.

Nickel–base ODS alloys also have shown good performance in certain molten glass applications. Hinze et al (8) have developed a hot–spin forming method to produce components used in the manufacture of fiberglass insulation from mechanically alloyed nickel–base alloys. Their work reports that M.A. materials have excellent resistance against molten glass attack and are capable of producing glass fibers at temperatures of 1315°C (2400°F).

MA 956 has also been found to be particularly resistant to a number of tile and tableware glaze compositions. Because of this corrosion resistance, the alloy is being evaluated for applications such as firing–kiln rollers, muffle tubes and furnace racks.

Other applications for MA materials include molten–glass resistance heaters, thermocouple protection tubes, glass–processing components used in nuclear waste disposal and the bushings used to make single and multi–strand fibers.

ENERGY PRODUCTION – In order to meet the ever increasing demand for power generation, it is often necessary to have additional standby capacity that can be utilized during peak demand periods. In the case of oil fired burners, this means some burners may be held in a non–firing condition even though they are positioned immediately adjacent to operating components. As a result, these components may reach temperatures of 1300°C (2372°F) during these standby periods resulting in a much shortened life cycle of the burner components

One of the earliest industrial applications for M.A. materials was for the flame stabilizers used in these oil–fired burners (Figure 5). D. M. Macdonald (9) has reported on the success of MA 956 in this application. The stabilizer component was examined after 2600 hours of operation at temperatures up to 1230°C (2250°F) in an environment that included sulfur, vanadium sodium, phosphorus and chloride impurities. Although some wastage of material was noted in areas where the impurities were concentrated, the overall condition of the stabilizer was excellent. Additional burner tests were conducted with service exposures of 10,600 hours (14.5 months) and the component was still serviceable.

More recently, M.A. ferritic alloys have been evaluated for use as the fuel cladding in fast–neutron, breeder reactors (10). Conventional austenitic alloys are unsuitable for this application due to the dimensional swelling phenomenon caused by the high neutron fluxes. Alternatively, conventional ferritic steels tend to have lower creep strength in the temperature range desired, 700°C (1300°F).

Because the mechanically alloyed ferritic steels, such as MA 957, have a potentially higher creep resistance, they are currently being evaluated for this application. The fuel–cladding tubes used in the reactors are of relatively small diameter and wall thickness (typically 6 mm dia. x 0.4 mm wall thickness) and require very heavy cold deformation to achieve these dimensions. Thus far, the M.A. ferritic alloys have demonstrated good fabricability in satisfying the cold drawing requirements for these tubes.

M.A. materials are also being evaluated for heat exchanger components in high–temperature gas–cooled reactors (HTGR). In this application, impurities in the helium gas used for cooling can cause severe corrosion problems on the alloys in the system. Results reported by Floreen, et al (11) indicate MA 956 has considerable promise for the heat exchanger tubing material in these reactors.

Because of its high degree of resistance to oxidation and carburization, MA 956 is also being evaluated as a coal–gasification burner nozzle. Although these trials are still in the early stages, MA 956 has shown better performance than conventional alloys in its resistance to attack.

Another application area involving power production is industrial gas turbine components. Recently a new alloy, INCONEL alloy MA 760, has been developed specifically for this market (12). This alloy combines excellent hot corrosion resistance with long–time, high–temperature strength that is beyond the capabilities of conventional superalloys. It is anticipated that this alloy can provide exceptional long–term service for components, such as vanes and blades in turbines used for power generation. The stress rupture properties of MA 760 are given in Table VI.

TABLE VI

STRESS RUPTURE PROPERTIES OF INCONEL alloy MA 760

Stress To Rupture – MPa (ksi) At Indicated Life

Temp°C (°F)	Stress To Rupture – MPa (ksi) at Indicated Life	
	100 h	1000 h
816 (1500)	380 (55)	300 (43)
850 (1562)	320 (46)	240 (35)
982 (1800)	165 (24)	145 (21)
1093 (2000)	115 (17)	105 (15)

FUTURE PROSPECTS – The outlook for growth of MA materials in industrial markets is a good one. Clearly, the availability of alloys that can give both increased load carrying capability and prolonged life at very high temperatures will promote new markets. The fact that MA alloys are now being produced in a variety of shapes and gauges will encourage new users to try these materials. Although much of the past work has been directed at optimizing mechanical properties, it might be expected that in the future, efforts will increase to find lower cost production methods. These efforts will further improve the performance/cost ratio and enable MA materials to compete strongly with conventional alloys.

In the physical metallurgy area, an intensified interest has developed in amorphous materials and nano structures. This research, which involves rapid solidification in addition to mechanical alloying, might be expected to produce a number of new materials with exceptional properties well beyond today's alloys.

REFERENCES

1. I. G. Palmer and G. C. Smith, "Oxide Dispersion Strengthening," p. 253, Gordon and Breach Science Publishers, New York (1968)

2. J. S. Benjamin and M. J. Bomford, Metall. Trans. 8A, 3101 (1977)

3. J. M. Davidson, "Frontiers of High Temperature Materials II," p. 163, Inco Alloys International, Huntington, West Virginia (1983)

4. F. A. Thompson, "Frontiers of High Temperature Materials I," p. 134, Inco Alloys International, Huntington, West Virginia (1981).

5. E. Grundy, Mat. Science and Tech., 3, 782–790 (1987)

6. R. C. Benn, "Frontiers of High Temperature Materials II," p. 37, Inco Alloys International, Huntington, West Virginia (1983)

7. T. J. Kelly, Ibid., p. 129

8. J. W. Hinze, M. L. Robinson and R. D. Lawson, U. S. Patent 4,402,767

9. D. M. Macdonald, "Frontiers of High Temperature Materials II," p. 369, Inco Alloys International, Huntington, West Virginia (1983)

10. S. Nomura, T. Okuda, S. Shikakura, M. Fujiwara and K. Asabe, "Solid State Powder Processing," p. 203, Trans. Met. Soc., Warrendale, Pennsylvania (1989)

11. S. Floreen, R. H. Kane, T. J. Kelly and M. L. Robinson, "Frontiers of High Temperature Materials I," p. 94, Inco Alloys International, Huntington, West Virginia (1981)

12. R. C. Benn and G. M. McColvin, "Superalloys 1988," p. 73, The Metallurgical Society, Inc., Warrendale, Pennsylvania (1988)

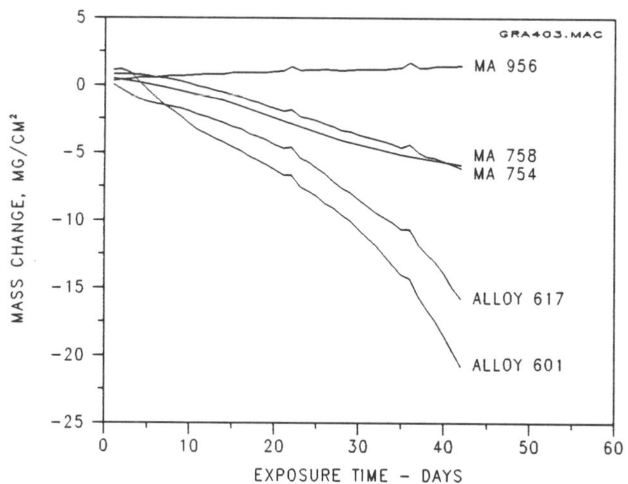

Figure 1. Mass change versus exposure time for various materials exposed at 1100°C in air plus 5% water vapor.

Figure 4. Molten glass plunger rods made of INCOLOY alloy MA 956 operating at 1200°C.

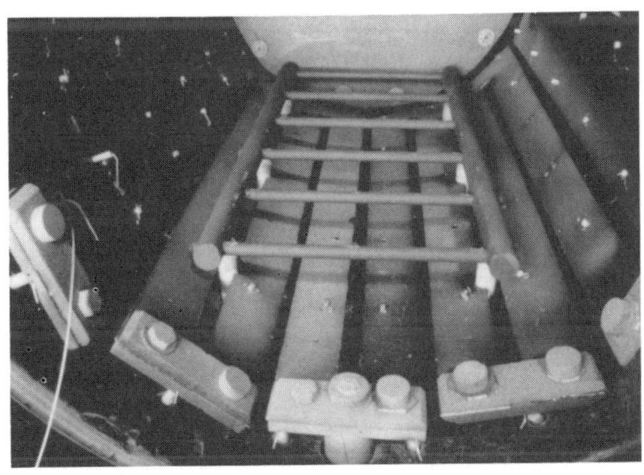

Figure 2. Vacuum furnace hearth rack and other components made of INCOLOY alloy MA 956.

Figure 5. Flame stabilizers from oil–fuel burners made of INCOLOY alloy MA 956.

Figure 3. INCOLOY alloy MA 956 furnace basket.

MICROSTRUCTURE AND TENSILE PROPERTIES OF OXIDE-DISPERSION-STRENGTHENED FERRITIC ALLOYS

A. Alamo, J. Decours, M. Pigoury, C. Foucher
Centre D'Etudes Nucleaires De Saclay
CEREM/DTM/SRMA/SMPX
91191 Gif-sur-Yvette Cédex France

ABSTRACT

Due to their excellent swelling resistance and to their good creep strength, Oxide-Dispersion Strengthened (ODS) ferritic alloys are potential candidate materials for the long-life core of fast breeder reactors. In this paper we investigate the microstructure and the tensile properties of two commercial alloys MA956 and MA957 obtained by mechanical alloying. Nominal compositions of these alloys are respectively Fe-20Cr, 4.5Al, 0.5Ti, 0.5Y$_2$O$_3$ for MA956, and Fe-14Cr 0.3Mo, 1Ti, 0.25Y$_2$O$_3$ for MA957.

With the aim of manufacturing thin tubes for nuclear applications, different experimental fabrication processes at laboratory scale were used. These fabrication routes involved hot-extrusion followed by cold-drawing or HPTR rolling, and eventually intermediate annealing treatments.

10-70% cold-worked tubes have been obtained with different microstructures. These ones consisted of elongated grains of different thickness : fine grains ($\leq 1\mu m$), intermediate size grains (10-50 µm) and coarse grains (> 100 µm). Tensile tests were performed in the range of 20 to 750°C in the longitudinal and transverse directions of the tubes.

Fine grained structures give anisotropic tensile properties : the strength in the longitudinal direction is higher than in the transverse one. Coarse grained structures produce lower values of yield strength compared to the fine grained structures at test temperature below 550°C with a reversed behavior at higher temperature. The best compromise between ductility and strength is obtained for the intermediate grain size structures of the MA957.

FERRITIC AND MARTENSITIC STAINLESS STEELS are now well established as leading materials for fast reactor core components [1]. This choice is based mainly on high void swelling resistance under irradiation but other features as in-reactor creep behavior or sodium compatibility also appear attractive [2, 3].

For fuel pin cladding applications the main limitations of conventional ferritic/martensitic materials compared with the austenitic steels used up to now are their limited tensile and creep strength above 550°C and the occurrence of a ductile-brittle transition temperature which could increase with irradiation [1, 4].

Oxyde dispersion strengthening (ODS) of metals and alloys, made possible by mechanical alloying, is a very effective way to improve high-temperature mechanical properties of ferritic steels to the level of austenitic materials [5]. In this way, two commercial alloys, MA956 and MA957, have been studied.

Nevertheless, high strength materials generally display low ductility which could induce difficulties during cold-working operations as it seems to occur for a similar type of alloys developed in Belgium [6].

The main goal of the present work was to investigate the feasibility of several fabrication processes which could be applied in order to obtain thin tubes.

This paper presents the preliminary results concerning the microstructure and tensile properties resulting from several experimental fabrication routes applied to the MA956 and MA957 alloys. These materials prepared by mechanical

Table 1. Chemical analysis of MA 956 and MA 957 tubes

Elements (wt.%)	Cr	Mo	Ti	Al	Y	O_2	C	N_2	B	P
MA 956	18.7	< 0.05	0.33	4.8	0.35	0.13	0.021	0.035	0.0021	0.007
MA 957	13.7	0.30	0.98	0.03	0.28	0.21	0.030	0.044	0.0009	0.007

alloying have been supplied as rods and hollows in the extruded conditions.

MATERIALS AND FABRICATION PROCEDURES.

Chemical analysis of MA956 and MA057 alloys are reported on table 1. Initial rods (Ø 25mm) and hollows (63/41mm) of both alloys were re-extruded in a press of 6 MN at 950°C with extrusion ratios up to 20. Rods of 8.5mm diameter and hollows of dimensions ranging from 21.3/18.1mm to 7/5.8mm have been obtained.

Several experimental cold forming processes have been applied to extruded hollows : pilgrim rolling, cold-drawing and HPTR rolling. Pilgrim and HPTR rolling differ essentially by a great reduction of the outside diameter in the first case and the thickness in the second one.

All the cold-working procedures listed before can be applied to MA956 ; cold-working ratios (CW) as high as 80% (given by the percentage of area section reduction) can be attained with HPTR and pilgrim rolling.

Contrary to MA956, the MA957 exhibits some difficulties for cold-working. Pilgrim rolling cannot by applied due to the occurrence of instantaneous longitudinal cracking. Maximum CW induced by cold-drawing is about 10% per pass or cumulated passes. Nevertheless, HPTR rolling can be performed by successive passes up to a total area section reduction of 60%.

Bars of 7.5mm diameter have been prepared by cold-drawing with 25% CW without any heat-treatment.

Different fabrication routes have been used to produce tubes of 6.5/5.6mm diameters. The operation sequences applied to the hot extruded hollows are summarized in table 2.

EXPERIMENTAL.

Samples obtained from tubes and bars have been

Table 2. Fabrication routes of MA 956 and MA 957 tubes

Fabrication route	A	B	C	D
	Extrusion at 950°C			
Sequence	Cold drawing 10-15% CW per pass ⇵ annealing 5 min at 1125°C 4 passes	HPTR rolling 10-15% CW per pass 4 passes no annealing	HPTR rolling 10-20% CW per pass 3 - 5 passes Total CW ~ 30-60% ↓ annealing 2h at 1130°C ↓ HPTR rolling 3 - 4 passes	HPTR rolling 10-20% CW per pass 3 - 5 passes Total CW ~ 45-60% ↓ annealing 1h at 1300°C ↓ HPTR rolling 3 - 4 passes
Final state	10-15% CW	70% CW (MA 956) 45% CW (MA 957)	40 - 55% CW	40 - 55% CW
Material	MA 956	MA 956 - MA 957	MA 957	MA 957

examined by optical microscopy on longitudinal and transverse sections. They were polished and chemically etched in Villela reagent.

Vickers hardness measurements have been made under 2kg load ; the reported values are the average of five identations.

Tensile properties of tubes and bars have been measured from 20 to 750°C in the longitudinal and transverse directions with a strain rate of $4.10^{-3}s^{-1}$. The drawings of the tensile samples obtained from tubes are shown in figure 1. Longitudinal and transverse specimens had gauge lengths of 5 and 2mm respectively. In the case of bars cylindrical smooth specimens of 4mm diameter and 20mm gauge length have been used. To characterize the transverse properties of bars, the ring samples shown in figure 1b have been taken.

MICROSTRUCTURE.

MA 956 and MA957 alloys are characterized by an elongated grain structure for all the metallurgical conditions examined here. The grain size depends on the fabrication route performed.

Hot-extruded samples, corresponding to the starting conditions for all the fabrication sequences, present a fine grained structure with a grain size, i. e. a grain thickness of about 1 μm.

Two typical microstructures have been found for the MA956 as shown in figures 2 and 3. Fabrication route A gives rise to the coarse grained structure of the figure 2, characteristic of the occurrence of a recrystallization phenomena. Grain dimensions are about 100-300μm thick and several millimeters long. The heat treatment of 5 minutes at 1125°C after 10% cold-drawing produces the total recrystallization of the hot extruded fine grained structure. Except the first pass, the sequence A is applied on the coarse grained structure.

Fabrication routes of type B involving HPTR rolling with no intermediate heat treatment, produce a very fine grained morphology for MA956 as well as MA957 as shown in figure 3 ; grain sizes seem smaller than the hot-extruded structure.

The fabrication sequences C and D have been only applied to MA 957 alloy, because hollows of MA956 were not available. Sequence C differs from D only on the heat treatment conditions (C : 2h at 1130°C, D : 1h at 1300°C). Both annealings have been performed on tubes having a cold-working ratio ranging from 30 to 60%. The obtained microstructures depend mainly on the CW percentage and less on the heat treatments conditions.

Samples cold-worked to about 60% are totally recrystallized at both annealing temperatures. Structure, as shown in figure 4, is characterized by an intermediate

Table 3. Hardness values and general characteristics of MA 956 and MA 957 tubes

Alloy	Fabrication route	Final cold-working (%)	Grain structure	Grain size (μm)	Hardness	
					LD	TD
MA 956	A	10-15	coarse	100-300	310	300
	B	70	fine	< 1	340	330
MA 957	B	45	fine	< 1	395	385
	C	40	mixed	100-300	340-390	330-380
			(coarse + fine)	< 1		
	C	55	mixed	100-300	350-380	350-380
				< 1		
	C	55	intermediate grain size	10-50	365	350
	D	40	mixed	100-300	220-280	260-320
				< 1		
	D	55	intermediate grain size	10-50	300	290

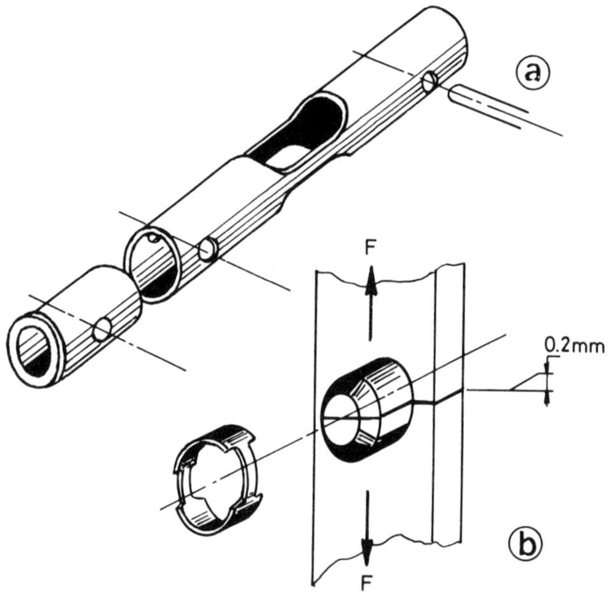

Fig. 1 : Longitudinal (a) and transverse (b) tensile samples obtained from tubes.

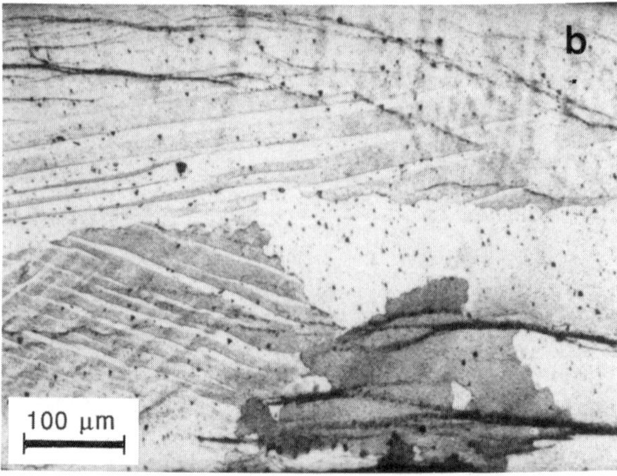

Fig. 2 : MA 956 alloy. Coarse grained structure resulting from the fabrication route A. Longitudinal (a) and transverse (b) sections.

grain size of 10-50μm width and 100-300μm length.

For both sequences tubes cold-worked less than 60% give rise to a mixed structure presented in figure 5, composed of fine grained regions and recrystallized zones near the external surfaces having a coarse grained structure (100-300μm grain width). Very scattered hardness values have been found. The annealing of 1 hour at 1300°C increase the proportion of recrystallized regions compared to 2 hours at 1130°C, but it never leads to a complete recrystallized structure. Neverthless, a systematic decrease of hardness values is obtained for specimens prepared according to the sequence D, compared to samples of sequence C with equivalent cold-working and grain structure.

Table 3 summarizes the hardness values and the structural characteristics of tubes 6.5mm outer diameter 0.45mm thick prepared with the fabrication routes described on table 2. In spite of the great difference on grain structure and cold-working exhibited by different samples, for example MA956 A and B, we can see that the corresponding hardening values are not very distinct and it seems difficult to correlate them with microstructural features. In fact, hardness is not a sensitive parameter to modifications of microstructure as shown in figure 6 where high cold-working percentage induce a slight increase of hardness of fine grained structures.

The recrystallization conditions have been investigated for both materials, but no undoubted conclusions are obtained. The phenomenon is very complex and depends not only on the cold-working ratio but also on the forming process used . Thus, 60% cold-rolled sheet of MA956 and MA957 display recrystallization temperatures of about 1250 and 1400°C respectively (annealing time : 15 minutes). Tubes of MA956 cold-worked at the same ratio recrystallize at 1150°C and at 1100°C after 10-15% cold-drawing. Recrystallization proceeds by an abrupt passage of the fine grained to the coarse grained structure. In the case of MA956, no intermediate grain size morphology have been obtained.

Recrystallization characteristics of MA957 should depend on cold-working percentage. In the particular case of specimens cold-worked by HPTR rolling an intermediate grain size structure seems possible to be obtained at relatively lower temperature (1 hour at 1050°C) if cold-working is about 60%. For lower cold-working percentage complete recrystallization is reached at higher temperature (at least 1 hour at 1350°C). A rapid grain growth takes place resulting in a structure

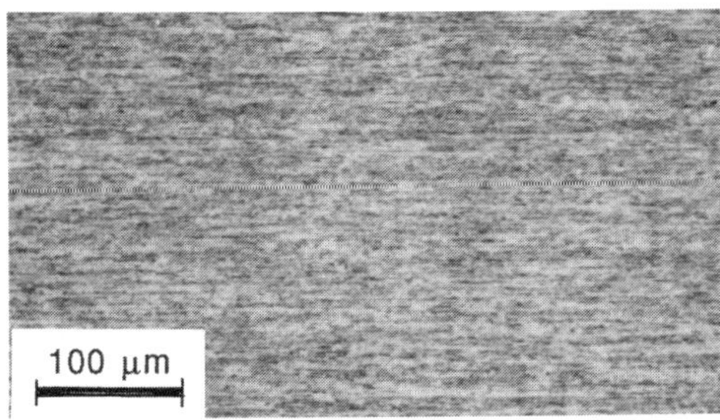

Fig. 3 : MA 956 alloy. Fine grained structure resulting from the fabrication route B. Longitudinal section.

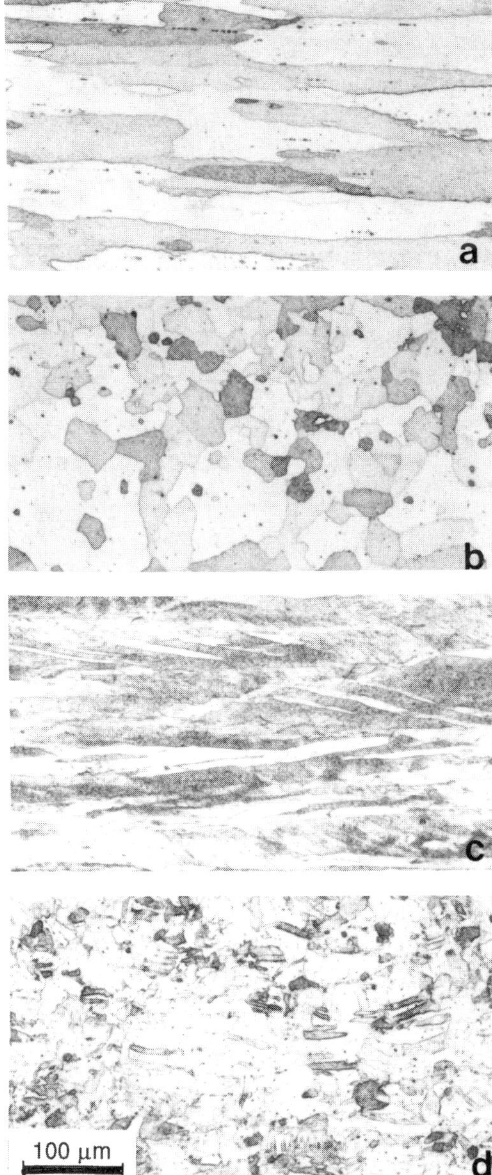

Fig. 4 : MA 957 alloy. Intermediate grain size structure obtained after annealing 2 h. at 1130°C (longitudinal (a) and transverse (b) sections) and 55% CW (longitudinal (c) and transverse (d) sections).

containing coarse elongated grains.

Figure 7 shows the recovery of hardness under isochronal annealings (1 hour) at various temperature for different cold-working ratios. Once again it must be pointed out that hardness is not a good universal measure of condition of microstructure as claimed in ref [7]. The same levels are attained by specimens having different recrystallized region proportions.

Anyway, a systematic study on the effects of hot and cold-forming process and cold-working ratios is necessary in order to have a good description and comprehension of the recrystallization phenomena in this kind of materials.

TENSILE PROPERTIES.

MA956 ALLOY.- Tensile properties have been measured for test temperatures in the range from 20 to 750°C, the last one representing the maximal cladding temperature of fuel pins in normal and incidental conditions.

Typical values of 0.2% yield strength measured at 20 and 750°C are presented on table 4.

Figures 8 and 9 show the plots of 0.2% yield and ultimate tensile strength and total elongation against test temperature for coarse and fine grained structures, obtained with the fabrication routes A and B respectively. Below 550°C, coarse grained structure present lower values of yield strength, compared to the fine grained structure. The behavior is reversed at higher temperature test, especially in the transverse direction as shown in table 4.

Ductility is low (< 10%) for both microstructures below 400°C. Total elongation values attain a maximum at 650°C. Beyond 400°C fine grain size structure displays higher ductility twice higher the coarse grained structure.

Tensile properties of fine grained tubes are not isotropic. Strength and ductility are higher in the longitudinal direction for all the temperature range.

On the other hand, fine grained bars Ø 7.5mm obtained by cold-drawing present also anisotropic characteristics as fine grained tubes. Properties measured in the transverse direction of tubes (~70% CW) and bars (~25% CW) in MA956 are equivalent as shown in table 4 ; in the longitudinal direction the strength of bars is higher showing a low work hardening capability. Therefore, tensile properties seem mainly determined by grain structure. The contributions of cold-

Table 4. MA 956 alloy. Typical values of the 0.2% yield strength [MPa] measured at 20 and 750°C.

Structure	Transversal direction		Longitudinal direction	
	20°C	750°C	20°C	750°C
Fine grained tube 70% CW	925 - 970	55 - 80	1000	165 -180
Fine grained bar 25% CW	850 - 920	50 - 75	1160 - 1170	120 - 130
Coarse grained tube 10-15% CW	790 - 880	135 - 195	845	180 - 190

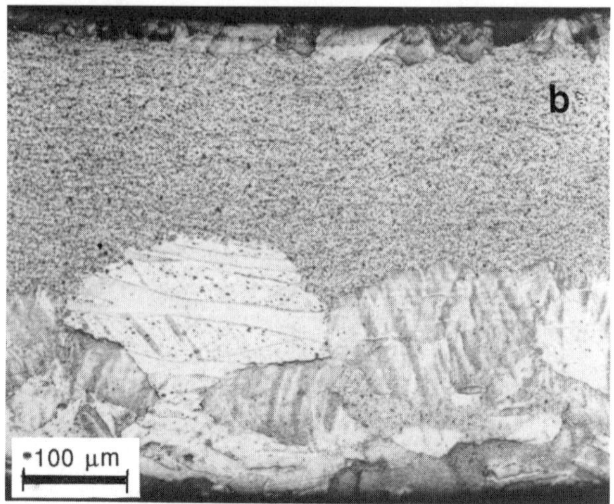

Fig. 5 : MA 957 alloy. Mixed structure : longitudinal (a) and transverse (b) sections.

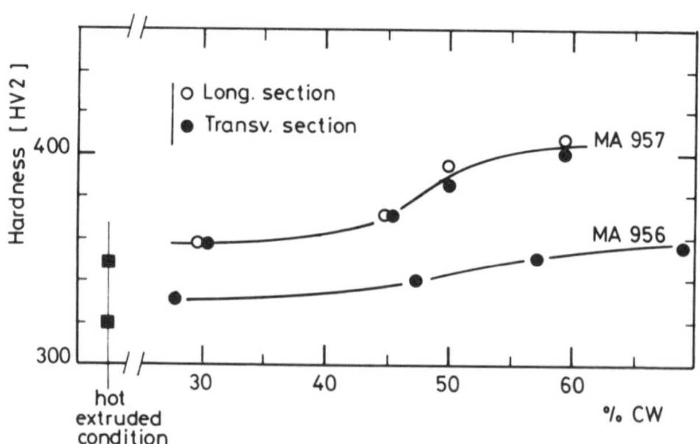

Fig. 6. : Hardness of fine grained structure as function of cold-working percentage.

working percentage and forming process should be less important.

MA957 ALLOY.- As in the former case tensile properties depend strongly on obtained microstructures.

Table 5 summarizes the 0.2% yield strength values of all MA957 samples examined here.

Fine grained bars and tubes present as MA956 anisotropic properties : strength and ductility are higher in the longitudinal direction as shown in figure 10. Bars are slightly more stronger than tubes and display a low work hardening capability in all the temperature test range.

Tubes with mixed structure containing coarse and fine grained regions have similar strength values than fine grained specimens. In contrast, ductility reported on fig. 10b attains the lowest levels over all temperature test range.

The most interesting properties have been found for intermediate grain size structure specimens. They present the higher values of strength especially in the range from 400 to 750°C as can be observed on fig. 11a and table 5.

Fabrication route and structure		B Fine	C Intermediate grain size	C Mixed	D Intermediate grain size	D Mixed	Bars Fine
TD	20	955 -965	970 - 990	830 - 945	840 - 850	570 - 725	1020 - 1050
TD	750	105 - 130	230 - 240	145 - 205	195 - 220	90 - 200	165 - 175
LD	20	1150	975	815 - 1075	730	660 - 920	1180 - 1360
LD	750	255	300	270 - 295	245	205 - 235	280 - 290

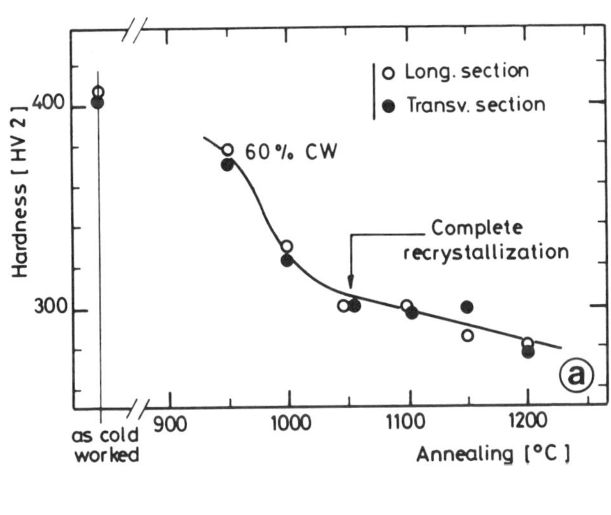

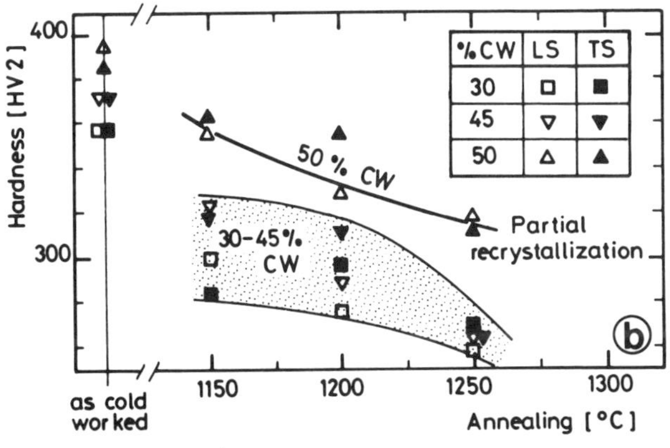

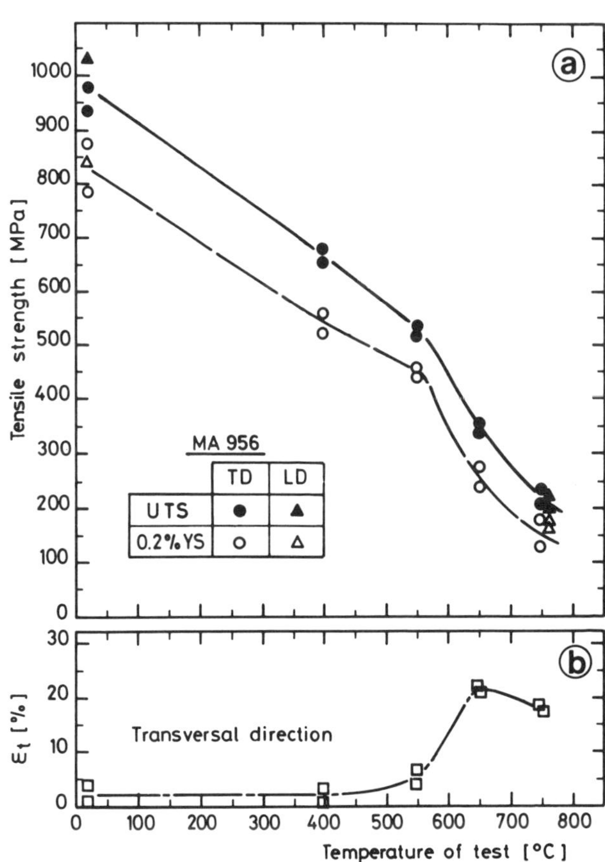

Fig. 7. : Recovery of hardness as function of the isochronal annealing temperature for various initial cold-working : a) 60% CW. b) 30-50% CW.

Fig. 8. : MA 956 alloy. Tensile properties of coarse grained structure tubes.

Ductility seems to be enhanced by annealing 1h at 1300°C performed during the fabrication sequence D. All specimens exhibit an effective increase of the elongation values, compared to the equivalent ones treated 2h at 1130°C.

The most important increase, shown in figure 11b, is produced for the samples having an intermediate grain size structure. The enhancement of ductility is accompanied in all the cases by a slight decrease of strength.

DISCUSSION.

Different fabrication processes have been used in order to obtain tubes. MA956 alloy present nearly the same behavior during hot and cold forming as austenitic materials. In contrast, cold-working of MA957 needs some precautions. Due to the small cold-working which can be made by pass (10-15%), cold-drawing process should involve a great number of passes alternated with annealings, resulting in long and heavy fabrication routes. Neverthless, HPTR rolling is an alternative forming process which leads to relatively high cold-working (up to 60%).

The main problem found during fabrication of tubes and bars was to control the effectiveness of heat treatments in order to obtain the desired microstructure. Heat treatments intended to produce recovering or recrystallization of the structure of cold- rolled sheets, tubes and bars indicate that recrystallization depends strongly on cold-forming process, cold-working percentage, geometry of products and probably also on hot-extrusion parameters. As exemple for equivalent CW amounts, partial recrystallization is observed at relatively low temperatures, 1 hour at 950°C, in HPTR rolled tubes of both alloys, whereas the first indications of recrystallization of sheets appear beyond 1250 and 1400°C for MA956 and MA957 respectively.

However partial recrystallized samples of MA957 need to be treated at high temperature, near 1400°C, to complete recrystallization, except for one particular case of tubes cold-worked to about 60% where total recrystallization occurs at 1050°C.

In both materials, primary recrystallization seems to occur in the range of grain sizes less than 1μm as does in dispersion strengthened nickel-base superalloy IN853 [8]. The recrystallized structures that we detected by optical microscopy corresponds to the stage of secondary recrystallization

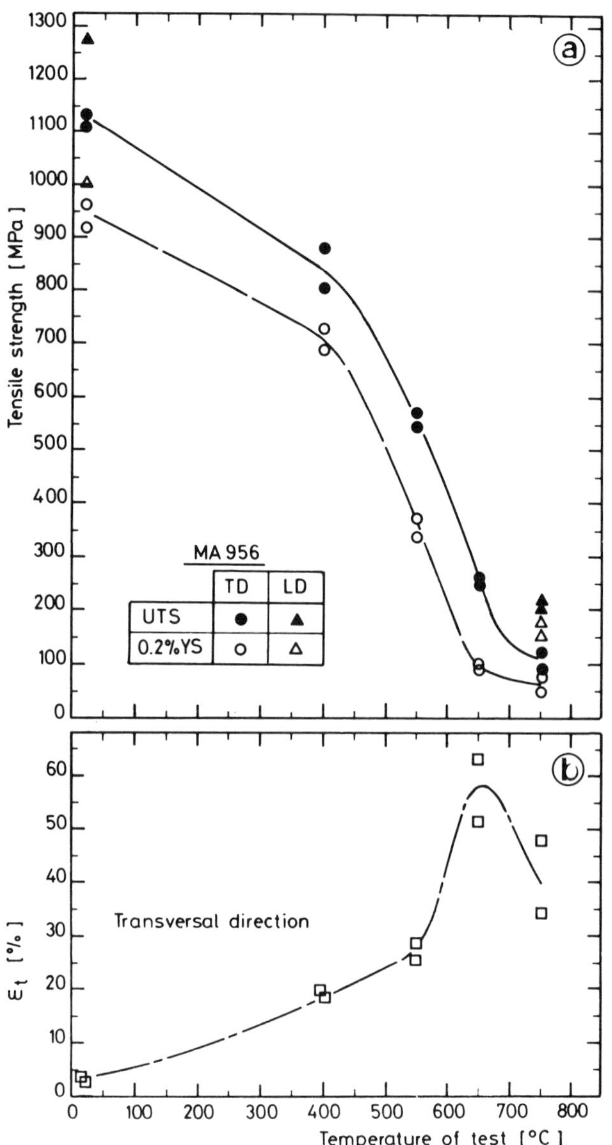

Fig. 9. : MA 956 alloy. Tensile properties of fine grained structure tubes.

characterized by a very coarse grain size, except for the case of intermediate grain size structure.

Consequently, as recrystallization of this kind of materials is a very complex phenomena, a systematic study is now in progress to determine the effects of hot and cold-processing parameters on microstructure and the role of oxide particle distributions during thermo-mechanical treatments.

The different fabrication routes used during this work give rise to a wide variety of microstructures resulting in a wide range of tensile properties.

Cladding materials for FBR fuel pins must present sufficient mechanical properties up to 750°C in order to resist

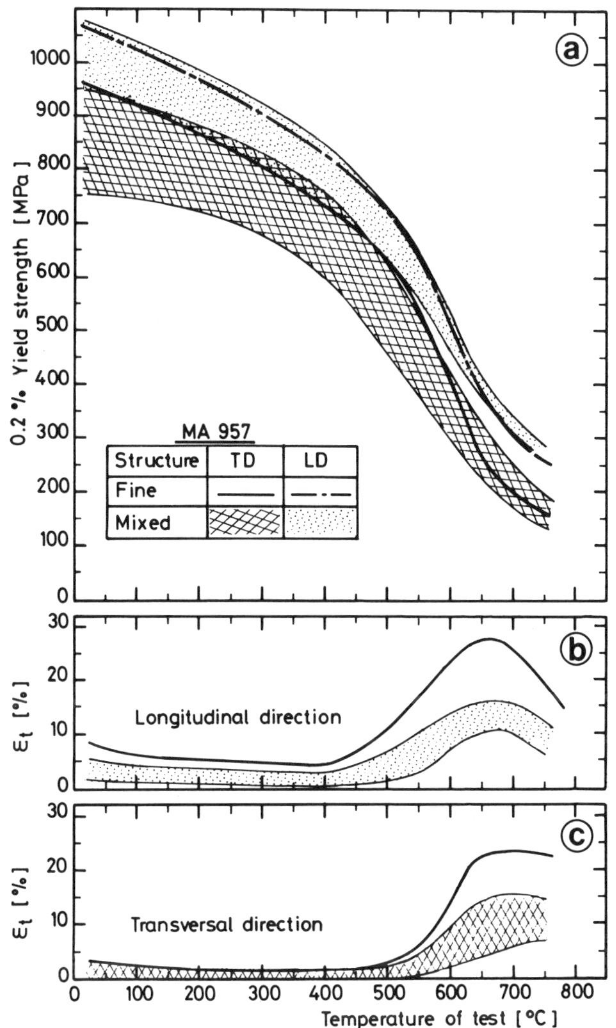

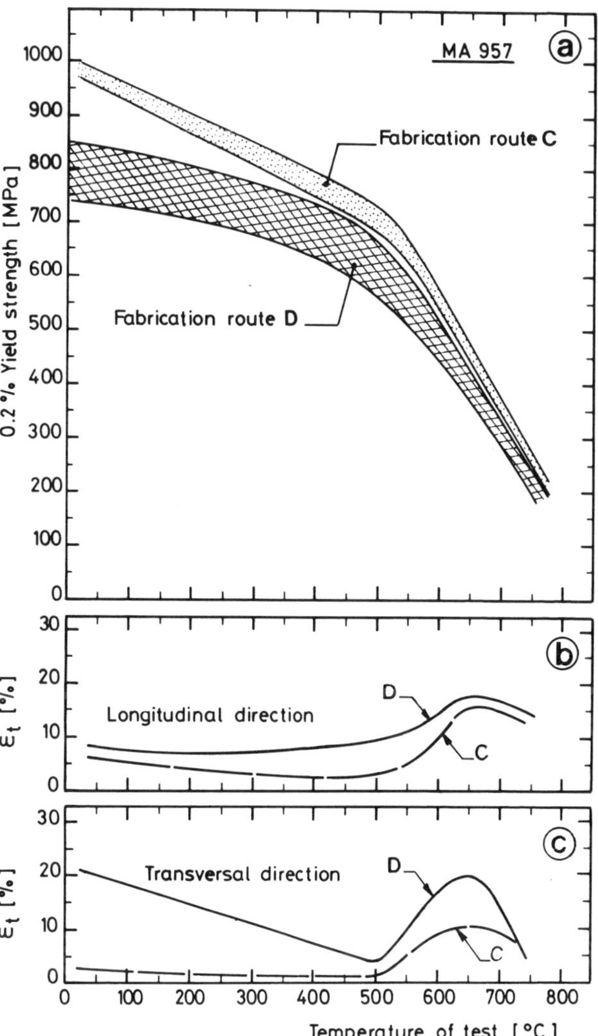

Fig.10. : MA 957 alloy. Typical values of 0.2% yield strength and total elongation of mixed and fine grained structure tubes.

Fig.11. : MA 957 alloy. Typical values of 0.2% yield strength and total elongation of intermediate grain size structure tubes.

to creep deformation induced by internal pressure built up by the fission gases.

The stabilized austenitic stainless steels (AISI 316 Ti CW or 15-15Ti CW) used up to now have adequate mechanical properties at the above mentioned temperatures. Typical values of the 0.2% yield strength are about 675, 560, and 175-300 MPa at 20, 400 and 750°C respectively. So, equivalent levels of strength must be attained by the ODS ferritic materials, especially in the transverse direction, in order to use them for this kind of applications.

The comparison of tensile properties of MA956 and MA957 for equivalent microstructures shows that the last one is a more stronger material, essentially at elevated temperatures.

Tensile properties of these materials are mainly determined by the grain size structure ; the effect of CW percentage should be of the second order.

The higher values of tensile strength are obtained for tubes of MA957 characterized by an intermediate grain size structure (10-50μm width and 100-300μm length). They have been prepared according to the fabrication routes C and D, which differ in the annealing conditions. A heat treatment of 1 hour at 1300°C (D) should probably modify the oxide particle distribution resulting in a slight decrease of strength and an effective increase of ductility. Nevertheless, the strength values of these structures are appreciable higher, especially in the range from 400 to 750°C, than strength displayed by fine-grained and mixed structures. In particular, the last one has the lower strength and ductility in the temperature range interesting for cladding applications.

Therefore, from the only point of view of tensile properties MA957 could be a promising candidate for cladding materials as far as we could control the microstructure and obtain intermediate grain sizes. Moreover fabrication routes must be ajusted to decrease final cold-working and increase ductility.

CONCLUSIONS

Results obtained on MA956 and MA957 allow to draw the following conclusions :

- At laboratory scale, the fabrication of thin tubes in ODS ferritic materials is feasible. Among the cold-forming process, HPTR rolling permit to obtain important CW ratios.
- Grain sizes are very difficult to control. Two extreme structures are found : fine grains (thickness $< 1\mu m$) and coarse grains (100-300μm thickness). Recrystallization of these materials is a very complex phenomenon which need to be systematically investigated.
- Tensile properties depend strongly on grain size structure and less on CW percentage and cold-forming process.
- For equivalent microstructures MA957 is a more stronger material than MA956, essentially at elevated temperatures.
- The higher strength values, in the range interesting for cladding applications (400-750°C), have been found for tubes of MA957 showing an intermediate grain size structure (10-50μm grain thickness). This particular microstructure has been obtained for tubes 60% cold-worked and heat treated at least at 1050°C.

ACKNOWLEDGEMENTS

The auteurs would like to thank Dr. G.A.J. Hack (Wiggin Alloys Limited, Inco MAP, Hereford, U.K.) who supplied the extruded rods and hollows of MA956 and MA957.

REFERENCES.

[1] Bagley, K.Q., E.A. Little, V. Lévy, A. Alamo, K. Erlich, K. Anderko, A. Calzabini. Proc. Int. Conf. on Materials for Nuclear Reactor Core Applications. BNES, London (1987) 37-46.

[2] Johnston W.G., T. Lauritzen, J.H. Rosolowski, A.M. Turkalo, Effects of radiation on materials : Eleventh Conference, ASTM STP 782 (1982), 809.

[3] Dupouy J.M., Y. Carteret, H. Aubert, J.L. Boutard. Proc. Topical Conf. on Ferritic Alloys for use in Nuclear Energy Technologies, Ed. J.W. Davis and D.J. Michel, AIME, Snowbird, Utah (1983) 125.

[4] Gilbon D., J.L. Séran, R. Cauvin, A. Fissolo, A. Alamo, F. Le Naour, V. Lévy, Proc. Int. Symp. on Radiation Effects on Materials, ASTM Andover (1988) to be published.

[5] Huet J.J. and Ch. Lecompte, Proc. Int. Conf. on Ferritic Steels for High Temperature Applications, Ed. A.K. Khare ASM (1983) 210.

[6] De Wilde L., J. Gedopt, S. de Burbure, A. Delbrassine, C. Driesen, B. Kazimierzak. Proc. Int. Conf. on Materials for Nuclear Reactor Core Applications. BNES, London (1987) 271.

[7] Gessinger G.H. and O. Mercier, Powder Metallurgy International $\underline{10}$ (1978) 202.

[8] Cairns R.L. , Met. Trans. $\underline{5}$ (1974) 1677.

[9] Delbrassine A., A. de Bremaecker, L. de Wilde, B. Kazimierzak, C. Driesen, M. Lippens, A. Pay, S. Pilate. Proc. ANS/ENS Int. Conf. Richland U.S., September 1987.

APPLICATION OF MECHANICALLY ALLOYED Ni- AND Fe-BASE ALLOYS TO SKID RAILS FOR WALKING-BEAM-TYPE REHEATING FURNACES

Kenji Tsukuta, Tomohito Iikubo, Susumu Isobe

Daido Steel Co., Ltd.

2-30 Daido-cho, Minami-ku

Nagoya-city, Japan 457

ABSTRACT

The resistance to compressive creep and oxidation of MA754/MA758 and MA956 was proved to surpass a conventional Ni-Co-Cr-W cast alloy at extremely high temperatures up to 1623K. The rails in MA alloys suffered recession and damage far less than castings in the field tests.

INTRODUCTION

Advanced aerospace and petro-chemical industries have required the heat resistant materials with the higher performance and reliability in order to increase output, improve efficiency and curtail maintenance. The operating temperatures of their machines or equipments have been raised steadily year by year. Therefore, even sophisticated γ' - strengthened superalloys won't be able to satisfy their requirements in the near future[1].

The oxide dispersion-strengthened superalloys are candidate materials to replace them in such cases[2]. The mechanical alloying process have already been established by INCO Alloys International, Inc.[3], and some of mechanically alloyed ODS alloys are applied to first-stage turbine vanes of military jet engines and fuel injection nozzles of civil diesel engines[4]. Their MA alloys bearing stable Y_2O_3 are very atractive, because they retain the unrivaled creep and also corrosion resistance at extremely high temperatures.

The steel industries have also searched for new heat resistant materials for skid rails for walking-beam-type furnaces which reheat slabs or billets. Recently they are raising the operating temperatures up to 1623K, and besides are designing the taller rails to eliminate cold spot marks from contact surfaces. Consequently the metal temperatures will be over the capacity of conventional solid-solution hardened cast alloys. On the one hand, skid buttons in advanced ceramics had failed to withstand the erosion due to the intense reaction with built-up scales on tops of them. Therefore, the alternative is to be MA·ODS alloys. The authors intended to prove the superiority of commercially available MA alloys to conventional cast alloys at such extremely high temperatures through accumulating the technical data to design high perfomance skid rails in MA alloys and then the field test results for them at several reheating furnaces.

EXPERIMENTAL PROCEDURE

Table 1 shows the analysis of investigated materials. Alloy MA758 is a high Cr version of MA754 which is derived from Nimonic 75, Ni-20Cr. The ferritic MA956 is originated from Kanthal, Fe-Cr-Al. Alloy A was selected as a reference material which is one of conventional solid-solution strengthened alloys cast into rails. Table 1 also shows the melting points determined by a differential thermal analizer.

All MA alloys were supplied by INCO Alloys International, Inc. where they employ the proprietary mechanical alloying process to prepare them. The extruded bars were fully recrystalized so that they possesed the coarse and elongated macrostructure along with the extrusion direction, which was visible enough with the naked eyes. The temperatures for one hour annealing were reported 1589K for MA758 and MA754 and also 1573K for MA956. Specimens of MA956 were preoxidized at 1373K for 2 hours in air to develop protective α-Al$_2$O$_3$ surface layers before various tests. A reference Alloy A was cast into skid rails.

Figure 1 shows the sampling for creep test specimens from MA alloy bars. The peculiar specimens like bobbins were designed to reduce the area of cross section and then to lessen the load on anvils and a push rod all in ceramics. Specimens of Alloy A were machined out in a

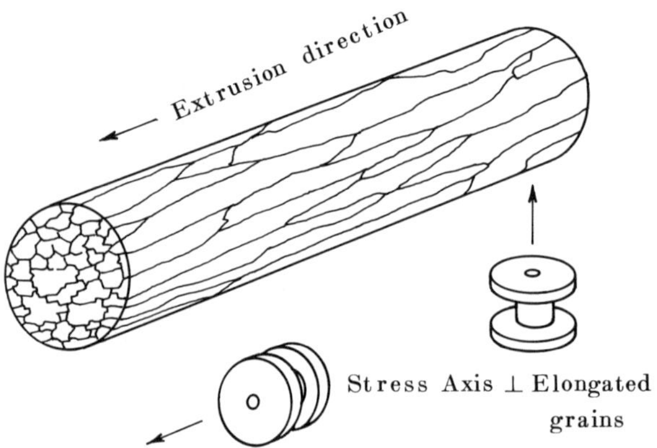

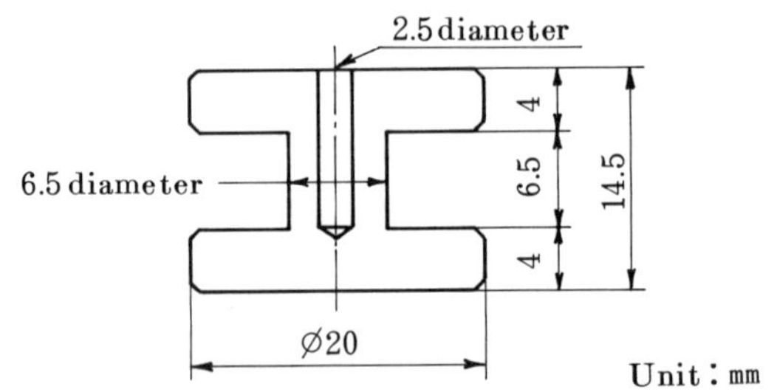

Fig. 1 - Sampling for creep test specimens in MA alloys

Table 1 - Chemical Compositions (wt%) and Melting Points (K) of Specimens

Alloys	C	Ni	Cr	W	Co	Fe	Ti	Al	Y$_2$O$_3$	Solidus	Liquidus
MA758	0.05	Bal.	29.32	–	–	0.67	0.47	0.25	0.60	1646	1672
MA754	0.05	Bal.	19.55	–	–	0.33	0.46	0.33	0.57	1685	1713
MA956	0.01	0.06	19.88	–	–	Bal.	0.38	4.52	0.62	1776	1794
Alloy A	0.13	21.46	31.32	2.07	22.94	(19.9)	–	–	–	1628	1659

parallel direction with columnar grains from cast rails.

The compressive creep test was carried out in the temperature range from 1473 to 1623K up to 50 hours in air with a lever-type constant-loading machine. The recession of gauge was measured with an extensometer attached to the lever under constant or stepwise increased load.

The oxidation behavior was traced by a thermobalance in the range from 1073 to 1588K up to 24 hours. The hot corrosion resistance was evaluated through an immersion and exposure tests. At the immersion test, columnar specimens, 10mm across and 15mm long, were heated at 1573K for 96 hours in Al_2O_3 crucibles containing 45 gr of spalled scales collected at a reheating furnace. The analysis of them was Fe_2O_3-3.8SiO_2-2.1CaO-1.8Al_2O_3-0.9MnO-0.9Cr_2O_3 (in wt%). The unstable Fe_2O_3 transforms at high temperatures to FeO, which would form a compound $FeO \cdot Al_2O_3$ with Al_2O_3 and consequently would deteriorate the oxidation resistance especially of MA956 to be protected by α -Al_2O_3 films. The exposure test was carried out in a reheating furnace fueled the desulfurized oil with low S around 0.16 wt%. The test specimens were sized to columns, 20mm across and 20mm long. The weight loss was measured after removing corrosion products from the surfaces of immersion or exposure test specimens.

RESULTS

COMPRESSIVE CREEP CHARACTERISTICS

Creep Curves for MA758 - Figure 2 shows the creep curves under 49MPa at 1623K, which is 0.98Tm (Tm: solidus). At such an extremely high temperature, the longitudinal specimen exhibited the typical three stages: transient, steady and accelerated creep. On the other hand, the transverse specimen crept rapidly just after loading and collapsed when the strain reached to 0.1, which was analogous to the longitudinal one's. The minimum creep rate of the transverse was 17 times as large as the longitudinal one's.

Figure 3 shows the curves for transverse specimens under a given stress at various temperatures. Even the transverse ones crept in the noticeable three stages at lower temperatures than 1548K(0.94Tm). The creep was accerelated at the strain range from 0.15 to 0.20.

Fracture Mode of MA758 - Photo 1 shows a specimen as machined and the typical ones sheared at the creep test under 48MPa at 1623 or 1573K. The longitudinal specimen collapsed with the significant slipping off between upper and lower

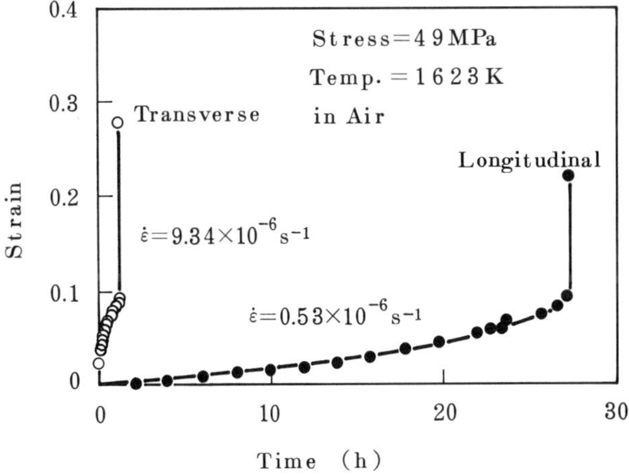

Fig. 2 - Compressive creep curves for MA758

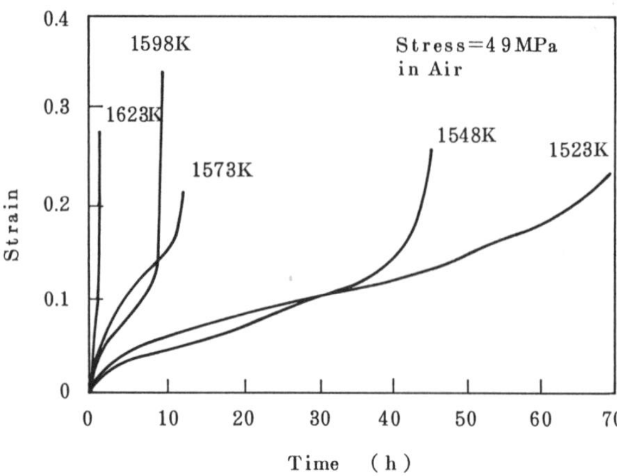

Fig. 3 - Compressive creep curves for transverse specimens of MA758

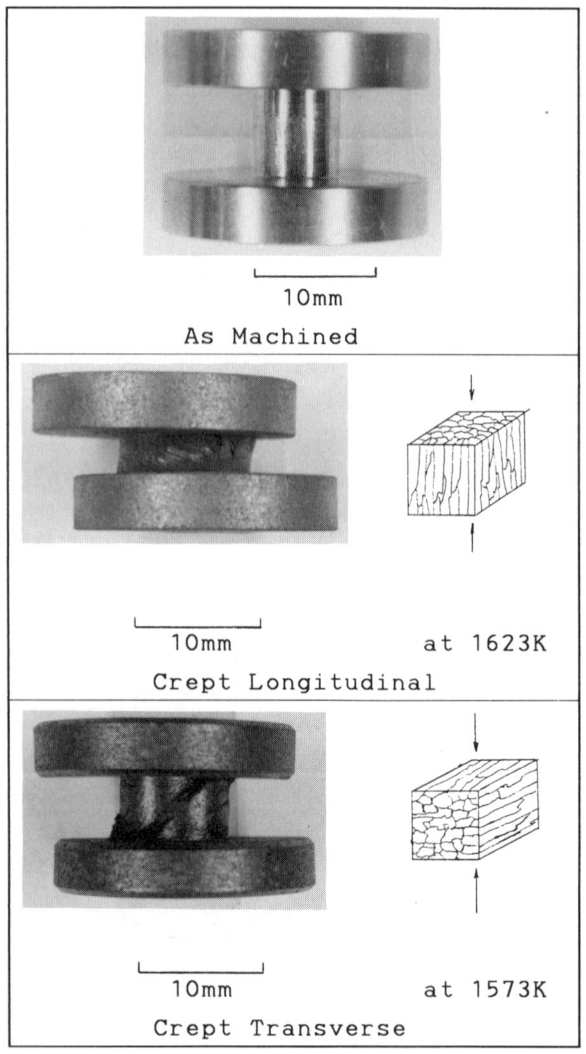

Photo 1 - Test specimens of MA758 fractured under 48MPa

flanges, while the transverse one sustained the slight gliding. Such a shear mode of both specimens was a contrast to that of conventional Alloy A castings which broke down after the concentric swelling into barrels.

Photo 2 shows the optical and the scanning electron micrographs of the above specimens. In the longitudinal one, the bulging of elongated grains in the bamboo structure caused the separation of grain boundaries, where voids grew and clustered into cracks. On the contrary, the transverse one bore the sliding at grain boundaries at an angle to the applied stress, which generated voids and then cracks.

The Steady-State Creep Rates for MA758 - Figure 4 shows the stress dependence of steady-state creep rates. The stress exponent, n, was derived from linear fitting after the equation, $\dot{\varepsilon} = A\sigma^n$. As for the longitudinal specimens, the exponent was low and almost constant about 2 at different temperatures. On the other hand, the transverse ones were more sensitive to applied stress and so the exponent was high up to 6. At 1573K the strain rates for both directions became very close under the lower stress around 30MPa.

Figure 5 shows the Arrhenius plot of the steady-state creep rates at various tempertures. The apparent activation energy, Q, was calculated from the gradient of fitting lines with the equation, $\dot{\varepsilon} = B\sigma^n \exp(-Q/RT)$. The values for both directions were approximately the same.

Comparison of Creep Strength between MA Alloys and A Conventional Alloy - The compressive creep

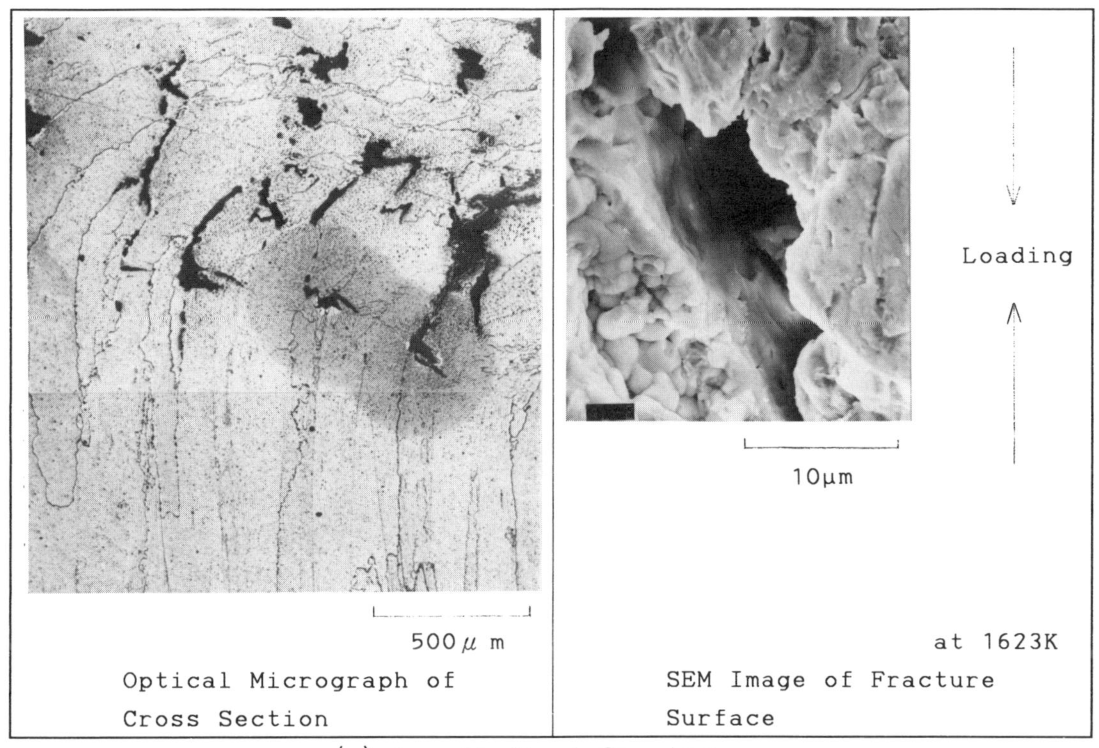

(a) Longitudinal Specimen

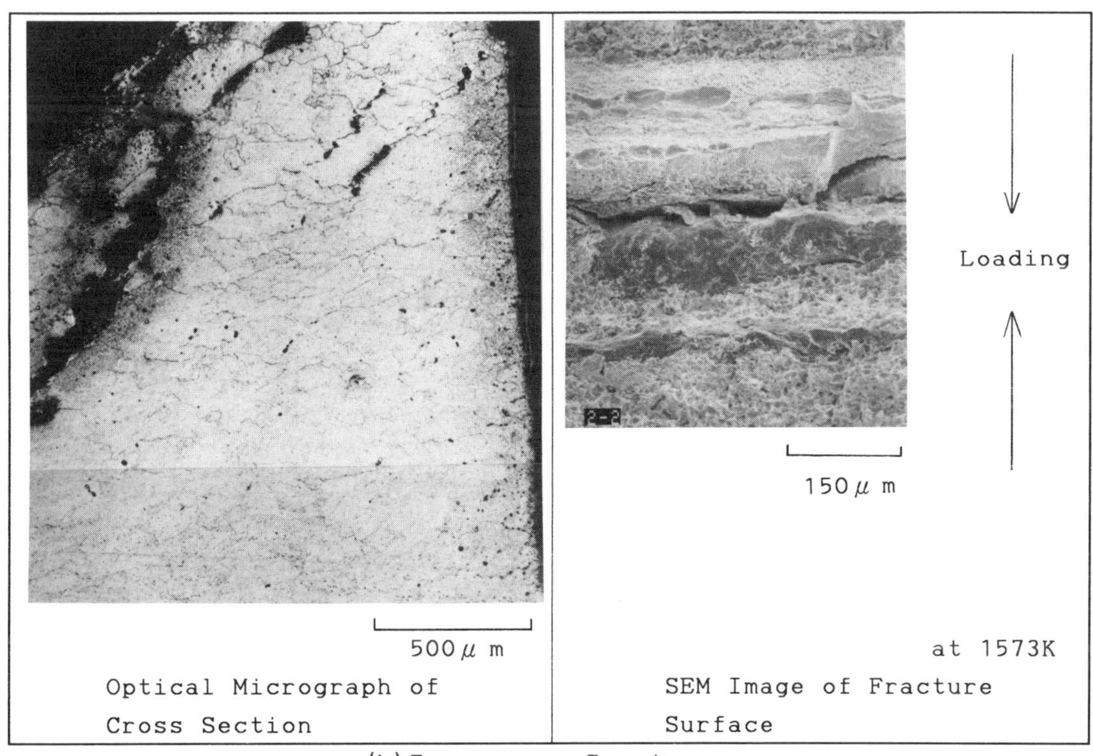

(b) Transverse Specimen

Photo 2 – Microstructures and surface cracks of MA758 fractured
under 48MPa

characteristics of other two MA alloys and a cast alloy were also obtained in the same manner as MA758. Table 2 lists the constitutive equations for steady-state creep rates of longitudinally loaded specimens. The equations give the strain rates in terms of applied stress and service temperature.

The fabricators generally adopt the stress for 0.025%/hour creep as allowable one to design skid rails. Figure 6 shows the allowable stresses calculated according to the above equations for MA alloys and a cast one. It reveals that Ni-base MA alloys maintain the creep strength more than 12 times as large as a conventional alloy over the wide range up to 1623K. And also it is noteworthy that a ferritic MA956 with the higher melting point surpasses Ni-base alloys at the high temperature beyond 1600K.

The above comparison was based on the longitudinal creep strength of MA alloys. As mentioned before,

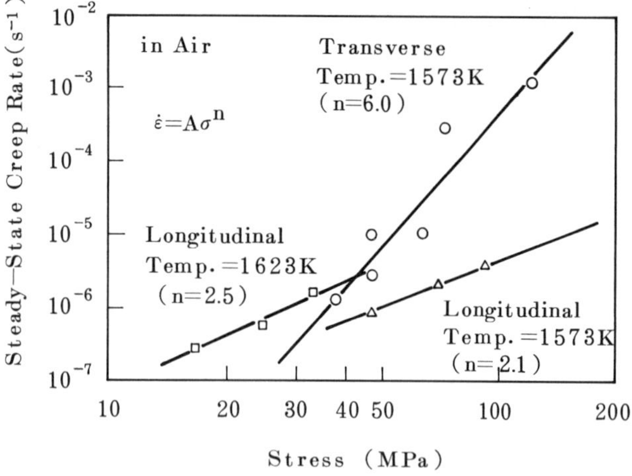

Fig. 4 - Steady-state creep rates vs. applied stress for MA758

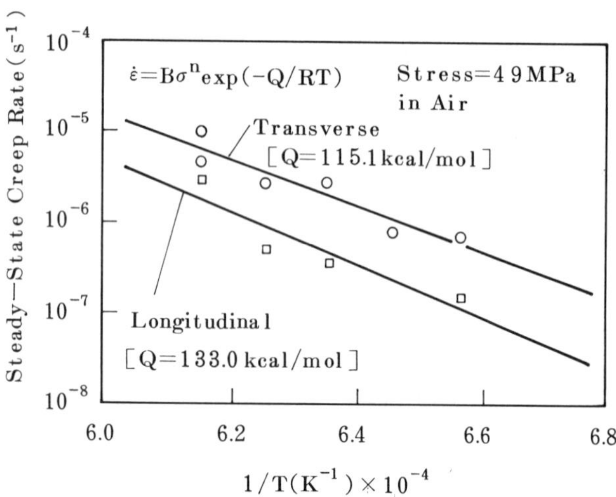

Fig. 5 - Steady-state creep rates vs. temperatures for MA758

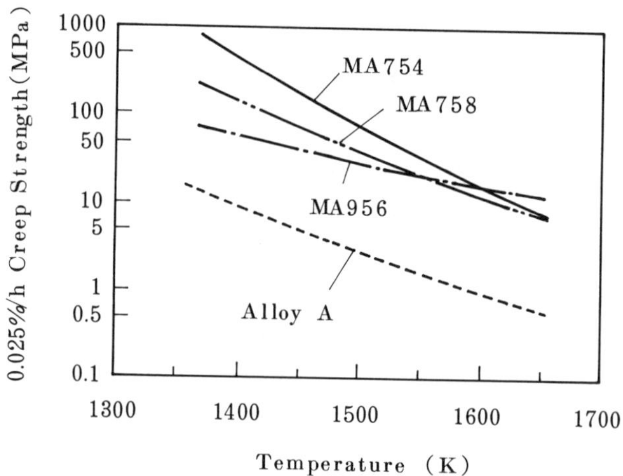

Fig. 6 - Creep strength at elevated temperatures

Table 2 - Constitutive Equations for Steady-State Creep Rates

Alloys	Equations
MA758	$\dot{\varepsilon} = 2.053 \times 10^8 \sigma^{2.46} \exp(-133.0/RT)$
MA754	$\dot{\varepsilon} = 1.276 \times 10^{10} \sigma^{1.99} \exp(-144.1/RT)$
MA956	$\dot{\varepsilon} = 1.264 \times 10^2 \sigma^{3.59} \exp(-99.8/RT)$
Alloy A	$\dot{\varepsilon} = 2.207 \times 10^7 \sigma^{2.19} \exp(-106.0/RT)$

the transverse strength is inferior to the longitudinal. Figure 7 shows the creep curve of transverse MA758 compared to Alloy A sampled in a parallel direction with columnar grains. It affirms that even under the transverse loading the MA alloy deforms less and survives longer than a conventional cast alloy.

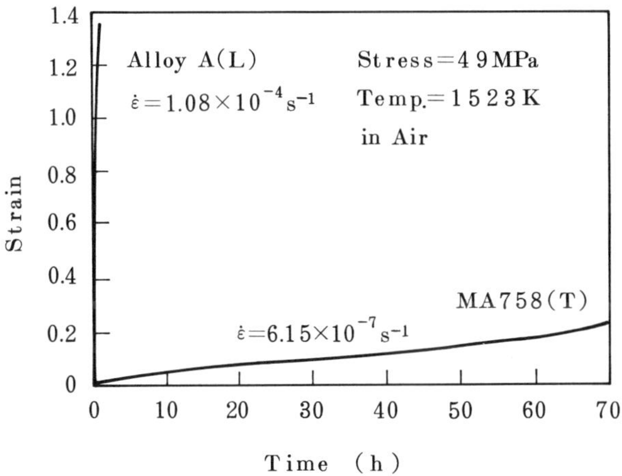

Fig. 7 - Compressive creep curves for transverse MA758 and longitudinal Alloy A

CORROSION RESISTANCE - Figure 8 shows the weight gain measured by a thermobalance for MA758 and a conventional Alloy A in still air at 1588K. Comparing with alloy A, MA758 possessed the smaller parabolic rate constant and also the longer incubation period to breakaway. Figure 9 compares the rate constants for various alloys. It reveals that a ferritic MA956 with high Al is the most oxidation resistant among them, and Ni-base MA758 bearing higher Cr is the next. In the early stage of oxidation under parabolic law, the oxidation resistance of MA754 is inferior to Alloy A at lower temperatures but comparable with it at higher temperatures over 1600K.

The activation energies derived from the slopes were categorized into two groups: large values for higher Cr (Cr_2O_3) or Al (Al_2O_3) and small values for lower Cr ($NiCr_2O_4$) [oxides in parentheses were dominant surface products].

Figure 10 shows the corrosion loss in collected scales. The corrosion attack on Alloy A was the severest and that on MA956 was the slightest. The higher Cr MA758

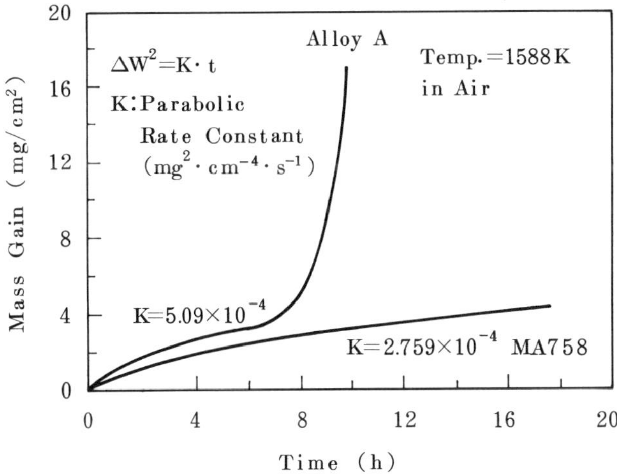

Fig. 8 - Mass change of MA758 and Alloy A in still air at 1588K

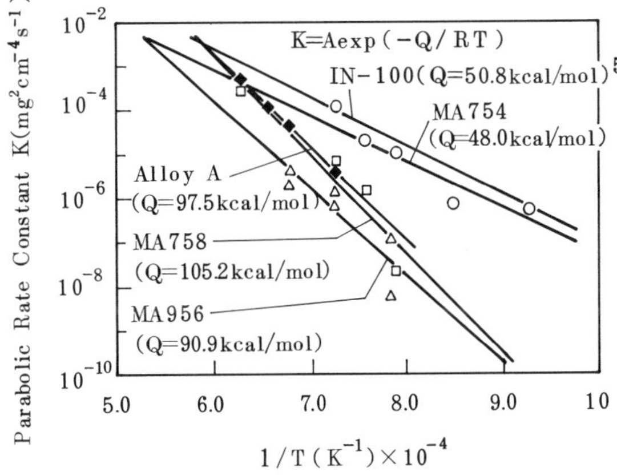

Fig. 9 - Parabolic rate constants and activation energies for oxidation

endured better than MA754. Figure 11 shows the weight loss of specimens placed at the bottom of crucibles in a typical reheating furnace where the low S oil was burned. The environment temperature in the furnace was controlled from 1273 to 1573K through the exposure duration extended to 3 months. The weight loss of MA956 was negligibly small compared with a conventional Alloy A. In the service condition, the corrosion resistance of lower Cr MA754 was equivalent to higher Cr MA758. Both alloys suffered less than a half of Alloy A's loss.

Field Tests of MA Alloy Rails in Reheating Furnaces - The skid rails in MA758, MA754 and MA956 have been under evaluation tests in

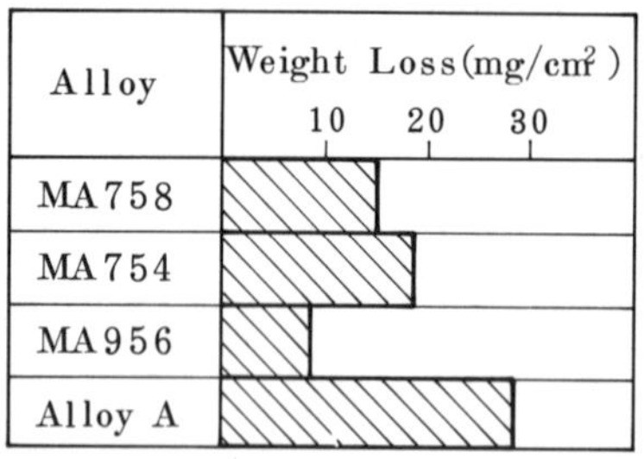

Alloy	Weight Loss (mg/cm^2) 10 20 30
MA758	
MA754	
MA956	
Alloy A	

45g Scale/6.3 cm^2 T. P./1573K/96h/in Air

Fig. 10 - Weight changes in collected scales

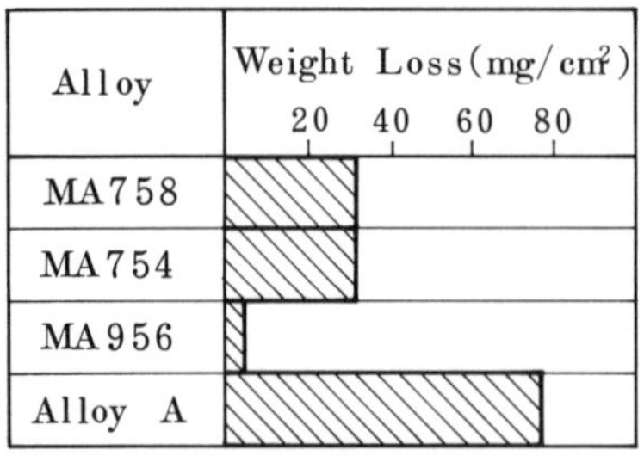

Alloy	Weight Loss (mg/cm^2) 20 40 60 80
MA758	
MA754	
MA956	
Alloy A	

1273−1573K/2057. 1h

Fig. 11 - Weight changes of specimens in a reheating furnace

several kinds of reheating furnaces. There has been no trouble except for the extensive corrosion damage experienced on rails in Ni-base MA754 and MA758. In this exceptional case, they burn the fuel oil with very high S around 1%, which attacked rails heavily. Most of furnaces are fueled coke oven gas or desulfurized oil with low S so that the corrosion damage is not fatal to MA alloys.

The stationary skid rails in MA758 are operating now successfully at a soaking zone in a reheating furnace since July, 1988. They burn the coke oven gas with low S from 200 to 300ppm and maintain the environmental temperature from 1553 to 1613K. The interim checkout after one year from the installation proved that the corrosion damage was negligibly small and the recession was only 1mm, which was 1/9 of conventional rails in a cast alloy similar to Alloy A.

The competition program at some furnace revealed that MA alloy rails are tougher than even a type of ceramic/metal composites.

DISCUSSIONS

Steady-State Creep of MA Alloys - Table 3 shows the tensile creep data reported by Nix et al. on longitudinal MA754 specimens[6]. The divide stress for stress exponents was attributed to its yield stress at 1473K. In this work the applied stress was always set below the yield stress, so the exponents for the stress under 100MPa are to be refered here. Although the data were obtained through tensile tests, they are comparable enough to ones applied to the equations for

Table 3 - Longitudinal Tensile Creep Characteristics for MA754[6]

Test Condition	Stress Exponent		Activation Energy (kcal/mol)
	36 to 46	2.5 to 5	95.7 to 167.4
Stress (MPa)	>100	< 100	62 to 204
Temp. (K)	1473		1300 to 1473

compressive creep rates in Table 2.

The stress exponents for all alloys in Table 2 were greater than unit. It means that they did not deform in the diffusional creep even at extremely high temperatures. The activation energy for all alloys was larger than self-diffusion energy, 62.9kcal/mol for Ni and 57.2 to 67.2 kcal/mol for Fe. It is noteworthy that ODS Ni-base alloys possessed the greater energy than solid solution Alloy A. Such a phenomenon was observed in ODS Al alloys[7].

Stability of Oxide Dispersoids - Photo 3 shows the TEM micrographs of MA758 exposed to 1623K for 200hr without loading. Even at extremely high temperature, 0.98Tm (Tm: solidus), the size and the number of Y_2O_3 combined with Al_2O_3 stayed almost intact as received. Table 4 also shows the change in hardness, the drop of which was minimal 6 or 7%. The above experiment proved that the oxide dispersoids in MA alloys are very stable up to their melting points.

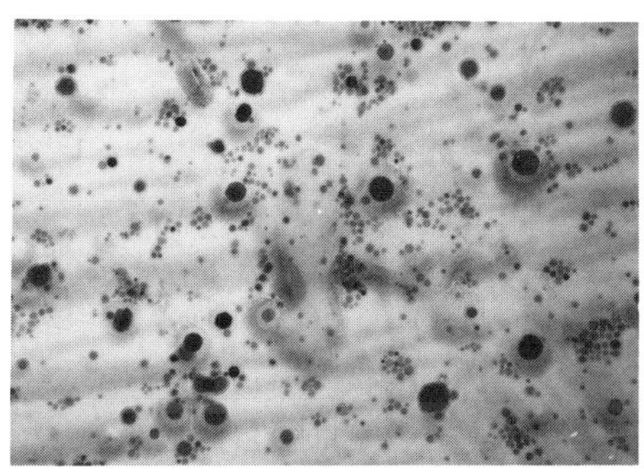

100nm

(a) As Received

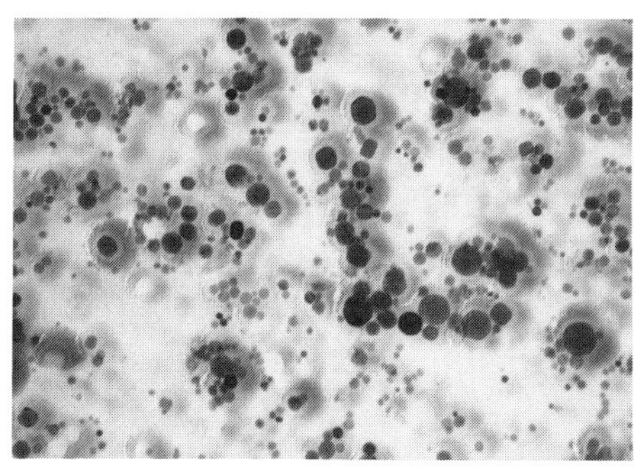

100nm

(b) Exposured to 1623K for 200hr

Photo 3 - TEM micrographs of MA758

Table 4 - Hardness of MA758 before and after Exposure to 1623K for 200hr

Test Specimen	HV (Load: 1kg) (%Drop)	
	As Received	Exposured
Longitudinal	303	284 (6)
Transverse	306	284 (7)

CONCLUSIONS

In order to apply ODS Ni-base MA758 and MA754, and Fe-base MA956 to skid rails for reheating furnaces, their compressive creep and hot corrosion resistance was evaluated at extremely high temperatures comparing with a conventional cast alloy, Fe-Ni-Co-Cr-W. The results are as follows.

1. Up to 1623K, MA alloys maintain the 0.025%/hr creep strength more than 12 times as high as a conventional alloy.

2. The elongated and coarse grain structure of MA alloys originates their anisotropic strength, but the creep rate in the weaker transverse direction still is much smaller than a conventional alloy.

3. The hot corrosion resistance of MA alloys is far better than a conventional alloy.

The perfomance of skid rails in some MA alloys is satisfactory and exceeds that of rails in conventional cast alloys at several field tests.

REFERENCES

1. Donachie, M. J, Jr., "Super-alloys Source Book", p. 3-19, American Society for Metals, Metals Park, Ohio (1984)

2. Curwick, L. R., "The Mechanical Alloying Process: Powder to Mill Product", Published in Proceedings from Frontiers of High Temperature Materials Conference, New York City, 3-10 (1981)

3. Hack, G. A. J., "Fundamentals of Mechanical Alloying", Published in Proceedings from Frontiers of High Temperature Materials II Conference, London, 3-18 (1983)

4. Wilson, R. K. and F. L. Perry, Ind. Heat. 51 [5], 27-30 (1984)

5. Wasielewski, G. E., "Nickel-Base Superalloy Oxidation", AFML-TR-67-30 (1967)

6. Stephen, J. J. and W. D. Nix, Met. Trans. 16A, 1307-1324 (1985)

7. Ansell, G. S. and J. Weertman, Trans. AIME. 215, 838-843 (1959)

HIGH TEMPERATURE CORROSION RESISTANCE OF HEAT RESISTANT MECHANICALLY ALLOYED PRODUCTS

Gaylord D. Smith, Pasha Ganesan
Inco Alloys International, Inc.
Huntington, West Virginia 25720, USA

ABSTRACT

The high temperature corrosion performance of the heat resistant, mechanically alloyed, oxide dispersion strengthened alloys, INCOLOY® alloy MA 956, INCONEL® alloy MA 754, INCONEL alloy MA 758, INCONEL alloy MA 760 and INCONEL alloy MA 6000 is described in this paper. Oxidation, carburization, molten glass, and oxidation-sulfidation data for a range of temperatures and environmental conditions are presented, along with comparative data on six conventional wrought alloys. Scale types are related to performance.

THROUGHOUT THE HISTORY of alloy development, there has been a need for ever increasing strength while at the same time providing for adequate oxidation and high temperature corrosion resistance (1). Oxide dispersion strengthened (ODS) alloys, made possible by the invention of mechanical alloying, is one of the latest chapters in this advancing technology (2). These alloys combine unique strength characteristics with high temperature corrosion resistance superior to that found in their cast or conventional wrought counterparts. This paper describes the oxidation and burner rig corrosion resistance of five commercial mechanically alloyed (MA) products: INCOLOY alloy MA 956, INCONEL alloy MA 754, INCONEL alloy 758, INCONEL alloy MA 760 and INCONEL alloy 6000. Similarly, INCOLOY alloy MA 956, an advanced combustor material, is compared with five conventional wrought alloys.

It has been known for nearly two decades that finely dispersed rare earth oxides including yttria improve the high temperature corrosion resistance of heat resistant alloys. The mechanisms by which the rare earth oxides enhance protective oxides on high temperature alloys have been recently reviewed by Moon (3) and others (4–6). The key to enhancing high temperature corrosion behavior is the diffusion of the rare earth elements into the scale and its subsequent segregation to high diffusivity pathways and interfaces.

Several investigators have examined the oxidation and sulfidation behavior of INCOLOY alloy MA 956 (7–9). These workers conclude that the alumina scale that forms on INCOLOY alloy MA 956 grows by both inward diffusion of oxygen and the outward diffusion of metal cations. The addition of the yttrium oxide into the scale suppresses the outward scale growth by the mechanisms described by Moon. Thus the incorporation of yttrium oxide into the alumina scale leads to improved scale adherence, decreased scale growth and enhanced oxidation resistance for the alloy. Similar scale behavior has been reported for nickel base alloys that form protective chromia oxides (10–13). This paper will characterize the oxidation, carburization, molten glass, sulfidation and burner rig (oxidation/sulfidation) corrosion data obtained for these yttria–containing mechanically alloyed products.

EXPERIMENTAL PROCEDURE

MATERIALS – Eleven high temperature alloys were selected for this study: five ODS alloys, INCOLOY alloy MA 956, INCONEL alloy MA 754, INCONEL alloy MA 758, INCONEL alloy MA 760 and INCONEL alloy MA 6000 and six conventional wrought alloys, INCONEL alloy 617, INCO® alloy HX, NIMONIC® alloy 86, HAYNES® alloys 188 and 214 and INCOLOY alloy 800HT.

All of the alloys were produced by Inco Alloys International except the last two alloys which were purchased from Haynes Alloys International. The alloys are nickel–based with the exception of iron–based INCOLOY alloy MA 956 and cobalt–based HAYNES alloy 188. The nominal composition of the materials evaluated in this study is given in Table 1.

Table 1. Nominal Compositions of the Alloys Studied (wt. %)

Alloy	Ni	Co	Fe	Cr	Mo	W	Al	C	Y_2O_3	Others
MA 754	Bal	–	–	20.0	–	–	0.3	0.05	0.6	0.5 Ti
MA 758	Bal	–	–	30.0	–	–	0.3	0.05	0.6	0.5 Ti
MA 760	Bal	–	–	20.0	2.0	3.5	6.0	0.05	0.95	0.15 Zr
MA 956	–	–	Bal	20.0	–	–	4.5	0.05	0.5	0.5 Ti
MA 6000	Bal	–	–	15.0	2.0	4.0	4.5	0.05	1.1	2.5 Ti
617	Bal	12.5	1.5	22.0	9.0	–	1.2	0.07	–	0.3 Ti
86	Bal	2.0	5.0	25.0	10.0	–	–	0.05	–	0.03 Ce
HX	Bal	1.5	19.0	22.0	9.0	0.5	–	0.1	–	–
IN–738	Bal	8.5	–	16.0	1.75	2.6	3.4	0.13	–	3.4 Ti, 1.75 Ta 0.85 Cb 0.12 Zr
IN–100	Bal	15.0	–	10.0	3.0	–	5.5	0.18	–	4.7 Ti, 1.0 V
800HT	Bal	–	40.0	21.0	–	–	0.4	0.08	–	Al + Ti .85 to 1.2
188	22.0	Bal	3.0	22.0	–	14.0	–	0.01	–	0.04 Ce
214	Bal	–	2.5	16.0	–	–	4.35	0.02	–	0.01 Y

Table 2. Metal Loss and Depth of Maximum Attack Data, 1008 hr in Air Plus Water Vapor

Alloy	900°C		1000°C		1100°C	
	Metal Loss (μm)	Max. Attack (μm)	Metal Loss (μm)	Max. Attack (μm)	Metal Loss (μm)	Max. Attack (μm)
617	10	38	20	64	28	107
86	8	8	13	13	18	18
HX	5	20	13	43	33	97
MA 956	0	0	0	0	5	5
188	5	20	10	33	20	107
214	5	13	0	0	5	5

Table 3. X–ray Diffraction Analyses of Corrosion Scales, Air plus 5% Water Vapor, 1100°C/1008 hr

Alloy	Phases Present	
	Major	Minor
617	Cr_2O_3	M_3O_4
86	Cr_2O_3	Fe_3O_4
HX	Cr_2O_3; $NiCrO_3$	M_3O_4
MA 956	Al_2O_3	$Y_3Al_5O_{12}$; Al_2O_3
188	Cr_2O_3	--
214	Al_2O_3; $FeAl_2O_4$	--

Five of the conventional wrought alloys and INCOLOY alloy MA 956 are candidate combustor alloys. INCONEL alloy MA 754 is a commercially–used first stage vane alloy and INCONEL alloys MA 760 and MA 6000 are candidate blade alloys. INCONEL alloy MA 758 is being commercially evaluated for use as skid rails in ferrous ingot soaking furnaces and as bushings and spinners for manufacturing glass wool. A description of the mechanical alloying process is presented in a number of sources (14–16). Most of the alloys of this study are chromia scale formers, while INCOLOY alloy MA 956, INCONEL alloys MA 760 and MA 6000 and HAYNES alloy 214 will form an alpha alumina scale under most oxidizing conditions.

The performance of these alloys was evaluated in five environments: oxidation, carburization, molten glass, oxidation–sulfidation and typical burner rig conditions. The test procedures for each environment are described below.

OXIDATION TEST PROCEDURE – The oxidation resistance of the MA alloys was measured at 1000°C, 1100°C and 1200°C for 1008 hours (42 days) using an air plus 5% water vapor mixture. The testing was conducted in electrically heated 100 mm diameter mullite tube furnaces. Furnace temperatures were controlled to ± 5°C. The water vapor was maintained at 5% by passing saturated air through a condenser maintained at constant temperature (33.1°C), which condenses out any excess moisture. A dewpoint hygrometer was used to ensure that the desired water vapor content was achieved and maintained. Specimens were weighed after each cycle (24 hours or 100 hours) in order to obtain mass change data as a function of time. Prior to testing, the test pins (7.6 mm diameter x 19.1 mm long) were degreased in Acetone, measured and weighed. Upon completion of each test, the alloys were sectioned and metallographically examined.

OXIDATION–SULFIDATION TEST PROCEDURE – INCOLOY alloy MA 956 and a number of typical combustor alloys were screened in an atmosphere of oxygen and 4% sulfur dioxide flowing through an electrically heated 100 mm diameter mullite tube. Duplicate pins (7.6 mm diameter x 19.1 mm long) of each alloy were exposed for 1008 hours (42 days) at 704°C. The pins were cycled to room temperature at approximately 100 hour intervals. To minimize air oxidation during thermal cycling, the specimens were moved from the hot zone of the furnace to the cold end under argon by means of a pusher rod without breaking the atmosphere seal of the tube. The specimens were removed from the furnace after cooling for a minimum of 1 hour.

Selected alloys of this study were evaluated in an aggressive hydrogen, 45% carbon dioxide, 1% hydrogen sulfide atmosphere at 816°C and at 982°C for various times to 3720 hours (155 days). Exposures were made in an electrically heated 100 mm diameter mullite tube furnace. Directly upstream of the test specimens was a 2.5 mm thick disc of cordierite catalyst carrier coated with platinum and gasketed with silica wool. Behind this was placed 50 mm of platinum–coated catalyst pellets (0.3% platinum on alumina). Next was a 75 mm bed of alumina bubble insulation to serve as a preheater. The bed was held in place by a second cordierite disc. This preheated catalyst bed reliably catalyzed the inlet gas mixture to achieve equilibrium at the temperatures and flow rates used. The dry inlet mixture of hydrogen, carbon dioxide and hydrogen sulfide was prepared by metering the individual gases with electronic flow controllers at a total flow rate of 8.33 mL/s. The compositions of the inlet mixture and the test atmosphere were monitored by gas chromatography and dewpoint hygrometry. The measured concentrations of

hydrogen, carbon dioxide, water and hydrogen sulfide agreed well with those anticipated from equilibrium calculations. The test pins were cycled to room temperature for periodic weighing using the same procedure as with the oxygen–4% sulfur dioxide specimens. At the conclusion of each oxidation–sulfidation test, every alloy was sectioned and metallographically examined.

BURNER RIG TEST PROCEDURE – The hot corrosion testing was performed using a G.E.–Lynn designed low velocity burner rig. A schematic diagram of the burner rig is presented in Figure 1. The MA alloy evaluations were conducted using pins 3.2 mm in diameter and 38.1 mm in length. Prior to

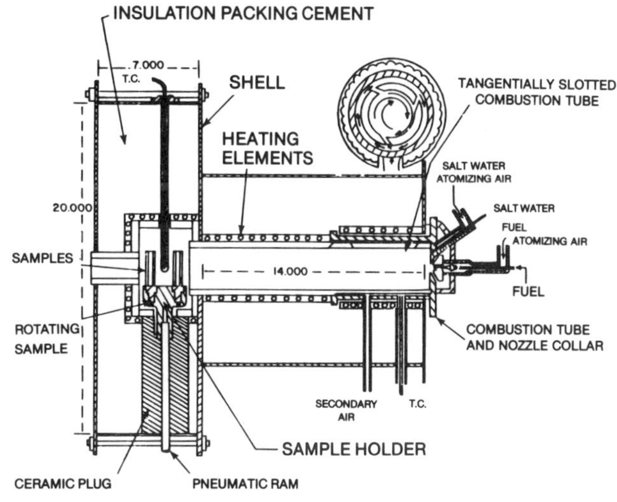

Figure 1. A schematic of the G.E.–Lynn burner rig.

exposure, the specimens were coded, measured, degreased with Acetone and cemented into a metallic cup using alumina cement. Duplicate specimens of each alloy were exposed to the following conditions:

Temperature	927°C
Air Fuel Ratio	30 to 1
Velocity	approximately 55 meters/minute
Sea Salt	5 ppm
Fuel	JP–5 as–received (0.08% sulfur)

The burner rig tests were run for 500 hours (21 days). The specimens were exposed in the hot zone for 58 minutes and then for 2 minutes in a blast of room temperature air. Various test parameters such as temperature, fuel flow, cooling water flow, air flow, air atomization pressure, up/down cycle, and flame condition were checked and recorded three times during each 8–hour period. The specimen chamber temperature was recorded continuously on a strip chart recorder. Upon completion of each test, pins were evaluated as described below.

SPECIMEN EVALUATION PROCEDURE – The principal measure of corrosion resistance was the mass change per unit area as a function of time. As mentioned earlier, in the oxidation and oxygen–sulfidation tests, the samples were periodically cycled to room temperature and weighed to obtain mass change data as a function of exposure time. To determine metal loss and the depth of attack after the completion of the tests, transverse sections cut from the specimens were mounted and polished using standard metallographic procedures. Samples were then evaluated in the unetched condition.

A schematic cross section of a specimen is shown in Figure 2. D is the original diameter of the specimen measured with a

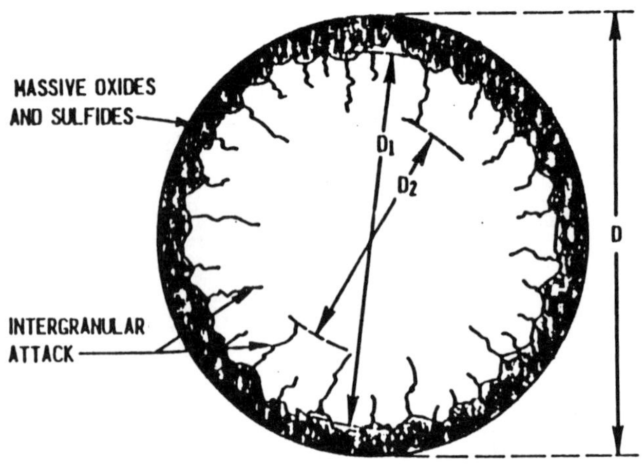

Figure 2. Method of measuring hot corrosion attack.

micrometer before testing. D_1 is the diameter of structurally useful metal (measured at 200X, and D_2 is the diameter of metal unaffected by oxides and/or sulfides (also measured at 200X). Metal loss was defined as $D - D_1$, the loss in diameter due to oxide and sulfide scale formation; and depth of attack was defined as $D - D_2$, the loss in diameter due to internal attack. At least three diameter traverses of D_1 and D_2 were taken. The three readings were averaged to give one value each for metal loss and depth of attack.

In addition, phase analyses of the corrosion scales were performed using x-ray diffraction. SEM/EDX analyses were conducted on most of the specimens to correlate with the XRD data and to characterize the corrosion morphologies.

CARBURIZATION TEST PROCEDURE – The carburization resistance of the five MA alloys of this study were compared to INCOLOY alloy 800HT in two carbonaceous atmospheres at 1000°C. The initial inlet gas mixture was hydrogen–1% methane (carbon activity of 1) and the second inlet gas mixture was hydrogen–5.5% methane–4.5% carbon dioxide [–log P_0 (atm.) was 20.57 and a carbon activity of 1]. The gas mixtures were fed into a mixing chamber using electronic

mass flow controllers at a total flow rate of 8.33 mL/s, and then into an electrically heated furnace with a mullite tube containing a reformer catalyst to equilibrate the gases. The gases then flowed to a separate furnace of similar design, passed over a carbon bed, and then over the test specimens. The test specimens (7.6 mm diameter x 19.1 mm long) were cleaned in alcohol and Acetone, placed in a cordierite boat and placed in the cold zone of the furnace. The furnace was sealed and purged with argon for 30 minutes, then the boat was pushed into the hot section with a feed–through pusher rod. The gas mixtures were admitted when the furnace regained temperature. The specimens were pulled into the cold zone under argon and allowed to cool to room temperature prior to removal from the furnace and weighing. The specimens were initially exposed to 24–hour cycles for 10 days, then weekly cycles thereafter.

MOLTEN GLASS TEST PROCEDURE – INCOLOY alloy MA 956 and INCONEL alloy MA 758 were exposed to molten glass in static crucible tests. An INCOLOY alloy MA 956 specimen (7.6 mm diameter x 19.1 mm long) was placed upright in an alumina crucible and 75% submerged in lime glass. The composition of lime–glass is given in Table 9. The crucible was placed in an electrically heated furnace at 1150°C exposed to air for 120 hours. The specimen was then cooled to room temperature, the glass removed by gentle grit blasting using –400 mesh alumina powder, and weighed for mass change. The INCONEL alloy MA 758 specimen was tested similarly in C–glass (see Table 10 for composition) at 1200°C.

RESULTS

The oxidation resistance in air plus 5% water vapor vs. time to 1008 hours (42 days) at 1000°C, 1100°C and 1200°C for INCOLOY alloy MA 956, INCONEL alloy MA 754, INCONEL alloy MA 760 and INCONEL alloy MA 6000 are depicted in Figures 3 through 5, respectively. Sulfidation resistance in hydrogen–45% carbon dioxide–1% hydrogen sulfide was determined for times to 3720 hours (155 days) at 816°C and 982°C for selected alloys. The mass change data vs. time are shown in Figures 6 and 7. Burner rig results for the MA alloys of this study are presented in Figure 8 for 500 hours (21 days) at 927°C.

INCOLOY alloy MA 956 and five candidate combustor alloys were evaluated in air plus 5% water vapor at 900°C, 1000°C and 1100°C for 1008 hours (42 days) and the mass change data vs. time are shown in Figures 9 through 11. Table 2 presents the metal loss and depth of attack data for this test series. Table 3 defines the phases present in the scales of these alloys after exposure at 1100°C for 1008 hours (21 days) and in Table 4 for four ODS MA alloys at 1100°C and 1200°C for varying times. Metal loss and depth of attack data are presented in Table 5 and x-ray diffraction analyses of the corrosion scales in Table 6 for specimens exposed to oxygen plus 4% sulfur dioxide at 704°C for 1008 days (42 days). Table 7

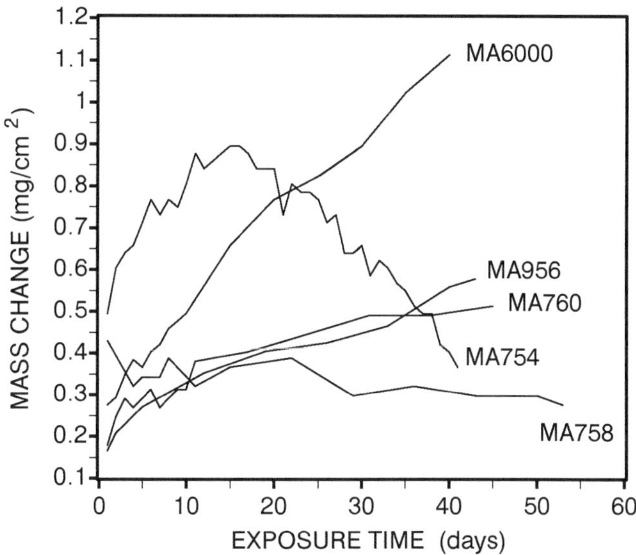

Figure 3. Mass change versus exposure time results for five mechanically alloyed materials exposed at 1000°C in air plus 5% water vapor.

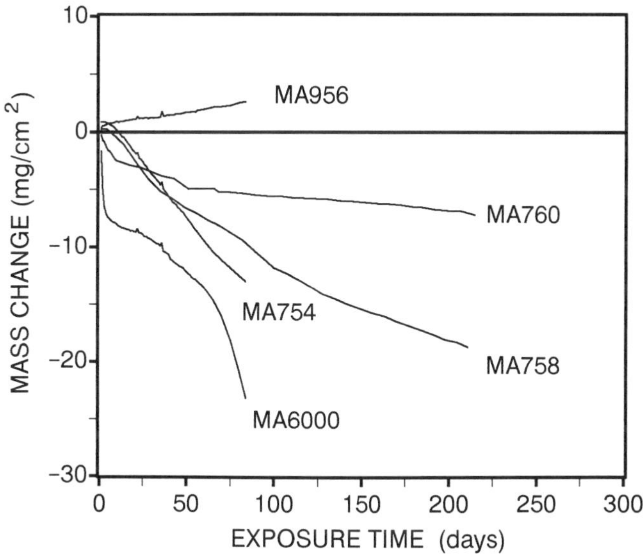

Figure 4. Mass change versus exposure time results for five mechanically alloyed materials exposed at 1100°C in air plus 5% water vapor.

records the metal loss and depth of attack for these alloys following 500 hours (21 days) of burner rig testing at 927°C. The x-ray diffraction analyses of the scales present on these alloys at the conclusion of burner rig testing are presented in Table 8.

The mass change data for the five ODS MA alloys of this study for two carburizing atmospheres are presented in Figures

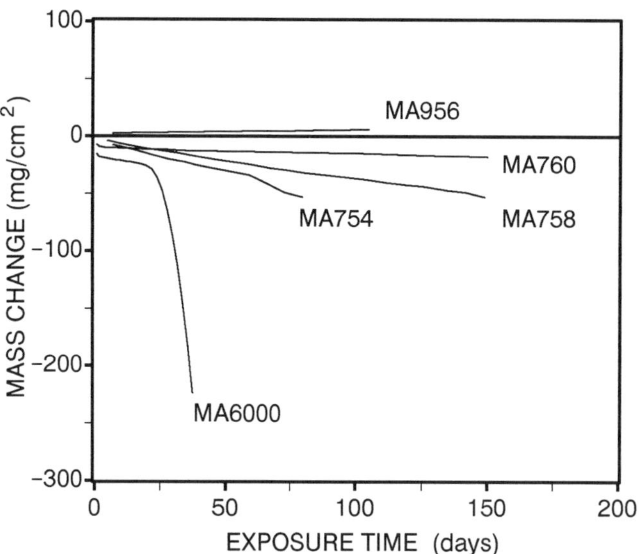

Figure 5. Mass change versus exposure time results for five mechanically alloyed materials exposed at 1200°C in air plus 5% water vapor.

12 and 13. Molten glass corrosion data for INCOLOY alloy MA 956 in lime glass and INCONEL alloy MA 758 in C–glass are given in Tables 9 and 10.

DISCUSSION

OXIDATION ENVIRONMENTS – At 900°C and 1000°C in air plus 5% water vapor, all the ODS alloys exhibited positive mass changes suggesting adherent protective scale formation (Figures 3, 9 and 10). However, at 1000°C, the oxidation rate for INCONEL alloy MA 754 and INCO alloy HX did become negative during the test period probably due to volatilization of chromium. At 1100°C, INCONEL alloy MA 6000, NIMONIC alloy 86 and INCONEL alloy 617 showed similar but acceptable mass loss behavior (Figures 4 and 11). INCONEL alloy MA 6000 displayed extensive mass loss behavior at 1200°C (Figure 5). This rapid mass loss can be attributed to the relatively low chromium level and high refractory element content of the alloy. INCOLOY alloy MA 956, INCONEL alloys MA 754 and 760 and the five conventional wrought alloys yielded satisfactory performance at all test temperatures to which they were exposed. The metal loss and depth of attack data at 900°C, 1000°C and 1100°C after 1008 hours (42 days) in air plus 5% water vapor for INCOLOY alloy MA 956 and the five conventional wrought alloys are given in Table 2. For these same alloys, the x-ray diffraction analyses of the corrosion scales are presented after exposure to air plus 5% water vapor at 1100°C for 1008 hours (42 days) in Table 3 and in Table 4 for four ODS MA alloys at 1100°C and 1200°C for varying times. As expected, INCONEL alloy MA 758, INCONEL alloy 617, NIMONIC alloy 86, INCO alloy HX and HAYNES alloy 188 formed essentially pure chromia scales, while INCOLOY alloy

MA 956, INCONEL alloy MA 760 and HAYNES alloy 214 formed basically an alpha alumina scale.

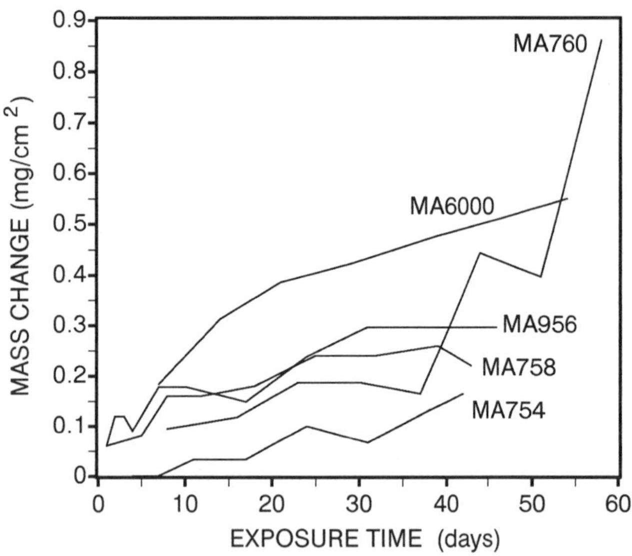

Figure 6. Mass change versus exposure time results for five mechanically alloyed materials exposed at 816°C in hydrogen– 45% carbon dioxide–1% hydrogen sulfide.

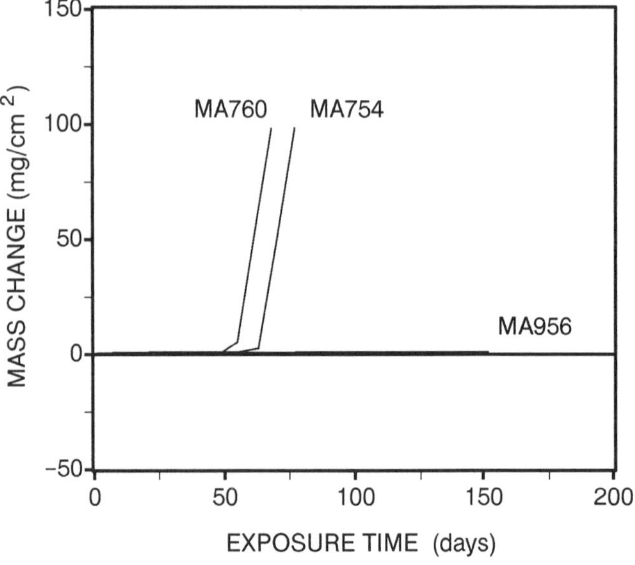

Figure 7. Mass change versus exposure time results for three mechanically alloyed materials exposed at 982°C in hydrogen– 45% carbon dioxide–1% hydrogen sulfide.

OXIDATION-SULFIDATION ENVIRONMENTS – In the hydrogen–45% carbon dioxide–1% hydrogen sulfide test at 816°C after 1008 hours (42 days), INCOLOY alloy MA 956 and INCONEL alloys MA 754 and MA 6000 showed only minimal mass gains, whereas INCONEL alloy MA 760 displayed breakaway oxidation–sulfidation after about 864 hours (35 days) (Figure 6). In the same test environment at 982°C, both INCONEL alloys MA 754 and 760 exhibited breakaway oxidation–sulfidation within 1440 hours (60 days), whereas INCOLOY alloy MA 956 showed no mass gain even after 3720 hours (155 days) (Figure 7). In oxygen plus 4% sulfur dioxide at 704°C for 1008 hours (42 days), INCOLOY alloy MA 956 and the five conventional wrought alloys performed well with the exception of HAYNES alloy 214 which experienced extensive mass loss. See Table 5. This behavior is explained by the tendency of this alloy to form a non–protective nickel oxide corrosion scale in this environment as shown by the x–ray diffraction analyses data presented in Table 6. Similarly, INCOLOY alloy MA 956 failed to develop its alpha alumina scale but was protected by a chromia scale as were the other alloys that were evaluated in this test environment.

Table 4. SEM analysis of oxidation scales in air plus 5% water vapor of MA alloys exposed at 1100°C and 1200°C for varying lengths of time.

| | Phases Present | | | |
| | 1100°C | | 1200°C | |
Alloy	Duration Hours	Major	Duration Hours	Major
MA 956	2,016	Al_2O_3	1,008	Al_2O_3
MA758	5,256	Cr_2O_3	3,576	Cr_2O_3
MA 760	10,000	Al_2O_3	10,344	Al_2O_3
MA 6000	2,016	Al_2O_3	1,008	NiO Cr_2O_3 Outer Al_2O_3 Inner

Table 5. Metal Loss and Depth of Maximum Attack Data, 1008 hr in Oxygen Plus 4% Sulfur Dioxide at 704°C

Alloy	Metal Loss (μm)	Depth of Maximum Attack (μm)
617	3	3
86	3	3
HX	3	10
MA 956	0	0
188	5	5
214	206	206

Table 6. X–ray Diffraction Analyses of Corrosion Scales, Oxygen plus 4% Sulfur Dioxide, 704°C/1008 hr

Alloy	Phases Present	
	Major	Minor
617	Cr_2O_3	--
86	Cr_2O_3	$NiMnCrO_4$
HX	Cr_2O_3; $NiCrO_3$	--
MA 956	Cr_2O_3	$Y_3Al_5O_{12}$; Al_2O_3
188	Cr_2O_3	--
214	NiO; Cr_2O_3	--

alloys of this study are presented for burner rig exposure at 927°C for 500 hours (21 days) in Figure 8. All four ODS alloys performed well with INCONEL alloy MA 6000 experiencing the largest depth of attack. Table 7 presents the metal loss and depth of attack data for INCOLOY alloy MA 956 and the five conventional wrought alloys under the same burner rig conditions. INCO alloy HX and HAYNES alloys 188 and 214 showed depths of attack similar to that of INCONEL alloy MA 6000. The x–ray diffraction data presented in Table 8 show that these alloys failed to develop their customary protective oxidation scales in the burner rig environment. INCO alloy HX and HAYNES alloy 188 formed a major scale of spinel and HAYNES alloy 214 developed a predominant mixture of nickel oxide and chromia.

Table 7. Metal Loss and Depth of Maximum Attack Data, Burner Rig Test 927°C/500 hr

Alloy	Metal Loss (μm)	Depth of Maximum Attack (μm)
617	25	53
86	43	43
HX	25	84
MA 956	0	0
188	23	99
214	23	99

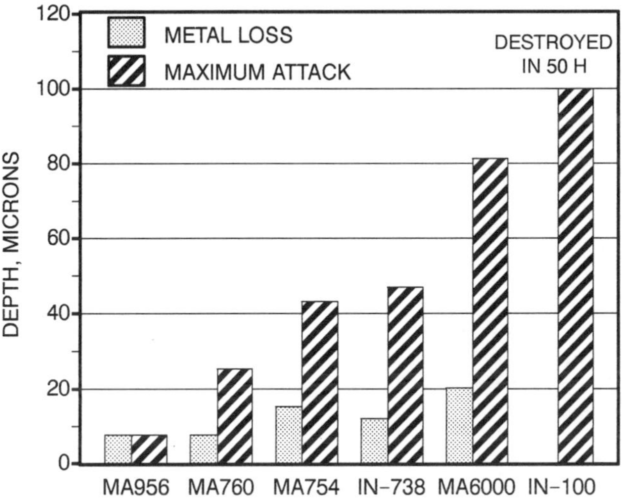

Figure 8. Metal loss and depth of attack data for four mechanically alloyed materials, IN–738 and IN–100 following burner rig exposure at 927°C for 500 hours.

BURNER RIG ENVIRONMENT – Burner rig testing is a typical method of evaluating the oxidation sulfidation resistance of gas turbine alloys under cyclic thermal conditions. Because a portion of each sample is cemented into the specimen holder, it is not possible to obtain meaningful mass loss data. Consequently, metal loss and depth of attack are metallographically obtained. These data for the four ODS

Table 8. X–ray Diffraction Analyses of Corrosion Scales, Burner Rig Test 927°C/500 hr

Alloy	Phases Present	
	Major	Minor
617	Cr_2O_3; NiO	--
86	NiO	Cr_2O_3
HX	M_3O_4	Cr_2O_3; NiO
MA 956	Al_2O_3	$Y_3Al_5O_{12}$
188	M_3O_4	Cr_2O_3
214	NiO; Cr_2O_3	Al_2O_3; $NiCrO_3$

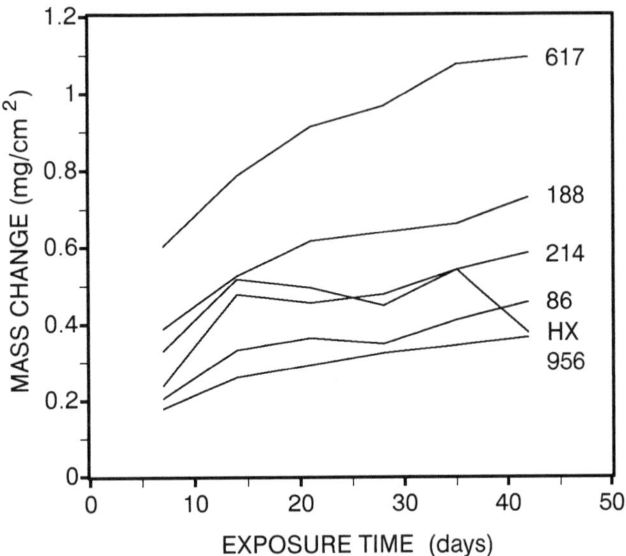

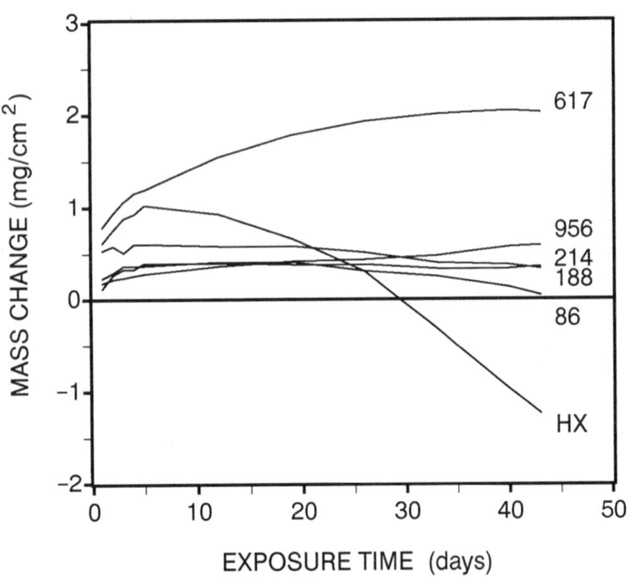

Figure 9. Mass change versus exposure time results for INCOLOY alloy MA956 and five conventional wrought alloys exposed at 900°C in air plus 5% water vapor.

Figure 10. Mass change versus exposure time results for INCOLOY alloy MA956 and five conventional wrought alloys exposed at 1000°C in air plus 5% water vapor.

CARBURIZATION RESISTANCE – The five ODS MA alloys of this study displayed excellent resistance to carburization in both carboneous test atmospheres. In Figures 12 and 13, data are presented on carburization resistance for times up to 450 days at 1000°C. Results are similar for each

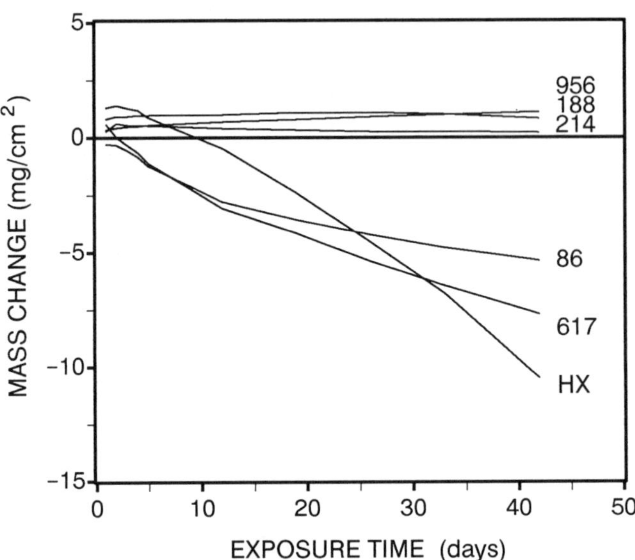

Figure 11. Mass change versus exposure time results for INCOLOY alloy MA956 and five conventional wrought alloys exposed at 1100°C in air plus 5% water vapor.

MA alloy in both atmospheres, attesting broad range of oxygen partial pressures. INCONEL alloy MA 760 was somewhat more resistant to carbon pick–up in the lower oxygen partial pressure atmosphere. The alumina scale formers, INCOLOY alloy MA 956, INCONEL alloys MA 760 and MA 6000 did significantly better than the chromia scale formers, INCONEL alloys MA 754 and MA 758 and INCOLOY alloy 800HT. MA alloys should perform exceptionally well in commercial heat treating service pyrolysis and steam methane reforming atmospheres.

MOLTEN GLASS RESISTANCE – INCOLOY alloy MA 956 and INCONEL alloy MA 758, as demonstrated in Tables 9 and 10, possess excellent resistance to lime glass and C–glass respectively. Fiberglass bushings and spinners, made from INCONEL alloy 758, are currently undergoing commercial evaluation. This alloy offers the strength of ODS alloys as well as enhanced corrosion resistance over presently used cast alloys for rotating glass–manufacturing hardware.

Table 9. Lime glass* corrosion resistance of INCOLOY alloy MA 956

Alloy	Test Temperature (°C)	Test Duration (Hours)	Mass Change (mg/cm2)
MA 956	1150	120	–4

*Chemical composition of lime glass (wt %) – 73.0% SiO_2, 1.7% Al_2O_3, 16.3% Na_2O, 4.7% CaO, 3.1% MgO, 0.4% H_2O, 0.15% Li_2O.

116

Table 10. C–Glass* corrosion resistance of INCOLOY alloy MA 758

Alloy	Test Temperature (°C)	Test Duration (Hours)	Metal Loss (mm)
MA 758	1200	120	0.03

*Chemical composition of C-glass (wt %) – 65.0% SiO_2, 4.0% Al_2O_3, 8.5% Na_2O, 14.0% CaO, 3.0% MgO, 5% B_2O_3

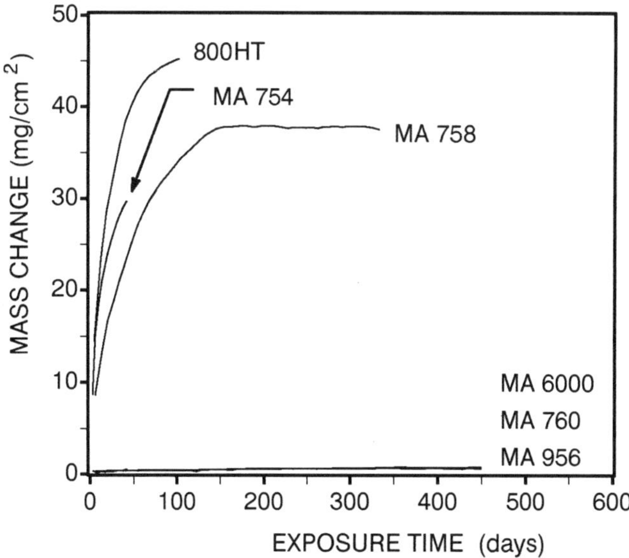

Figure 13. Mass change data for five MA alloys and INCOLOY alloy 800HT exposed at 1000°C in H_2 – 5.5% CH_4 – 4.5% CO_2.

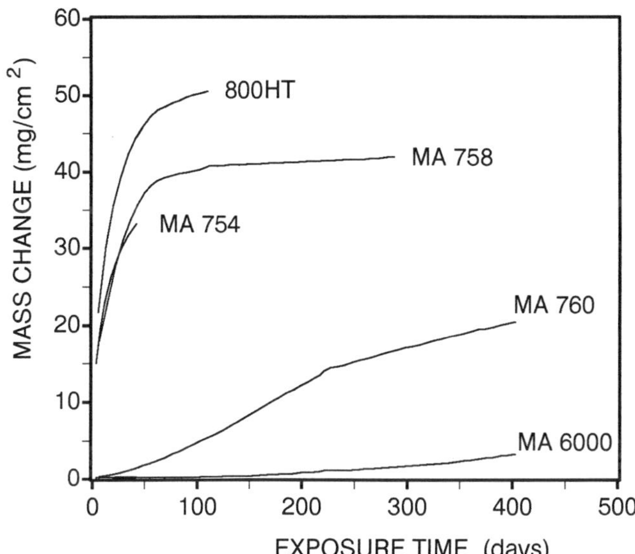

Figure 12. Mass change data for four MA alloys and INCOLOY alloy 800HT exposed at 1000°C in H_2 – 1% CH_4.

SUMMARY

The performance of the four ODS alloys, INCOLOY alloy MA 956, INCONEL alloy MA 754, INCONEL alloy MA 760 and INCONEL alloy MA 6000 has been evaluated in this study for exposure in air environments to 1200°C and in oxidation–sulfidation environments to 982°C. The performance of INCOLOY alloy MA 956 was exceptional in all test environments. INCONEL alloy MA 754 displayed satisfactory oxidation resistance to 1200°C, performed well in the burner rig environment, but did exhibit breakaway oxidation–sulfidation corrosion under severe laboratory test conditions. INCONEL alloy MA 760 exhibited oxidation and burner rig test results similar to those of INCOLOY alloy MA 956 but displayed breakaway corrosion behavior in the more severe laboratory oxidation–sulfidation tests. INCONEL alloy MA 6000 has adequate oxidation resistance to 1100°C but would be unsuited for service above that temperature without a coating. The alloy performed reasonably well in the burner rig environment and in laboratory oxidation–sulfidation exposure at 816°C. INCOLOY alloy MA 956 displays excellent resistance to molten lime glass as does INCONEL alloy MA 758 to C-glass . All five ODS MA alloys of this study exhibited exceptional carburization resistance.

REFERENCES

1. Fawley, R. W., The Superalloys, Sims, C. T. and Hagel, W. C., eds., John Wiley and Sons, New York, 1972, pp. 1–29.

2. Benjamin, J. S., Met. Trans. 1, 1970, pp. 2943–51.

3. Moon, D. P., Mater. Sci. Technol. 5, 1989, pp. 754–64.

4. Whittle, D. P. and Stringer, J., Philos. Trans. R. Soc., A295, 1980, pp. 309–29.

5. Hindam, H. and Whittle, D. P., Oxid. Met., 18, 1982, pp. 245–84.

6. Bennett, M. J., in Proc. Conf. on High Temperature Corrosion, Rapp, R. A., ed., NACE-6, Houston, TX, 1983, pp.145–54.

7. Kane, R. H., in Proc. Conf. on Frontiers of High Temperature Materials II Incomap, London, 1983.

8. Weber, J. K. R. and Hocking, M. G., in Proc. Conf. on High Temperature Alloys–Their Exploitable Potential, Joint Research Centre, Petten, The Netherlands, 1985.

9. Ramanarayanan, T. A., Raghavan, M. and Petkovik–Luton, R., J. Electrochem. Soc., 131, 1984, pp. 923–31.

10. Braski, D. N., Goodell, P. D., Cathcart, J. V. and Kane, R. H, Oxid. Met. 25, 1986, pp. 29–50.

11. Yurek, G. J., Przybylski, K. and Garrett–Reed, A. J., J. Electrochem. Soc., 134, 1987, pp. 2643–44.

12. Michels, H. T., Met. Trans. A., 7A, 1976, pp. 378–88.

13. Michels, H. T., Met. Trans. A., 8A, 1977, pp. 273–78.

14. Gilman, P. S. and Benjamin, J. S., Ann. Rev. Mater. Sci., 13, 1983, pp. 279– 300.

15. Fleetwood, M. J., J. Inst. Met., 1986, pp. 1176–82.

16. Weber, J. H., in Proc. Conf. of The 1980's––Payoff Decade for Advanced Materials, Vol. 25, Science of Advanced Materials and Process Engineering Series, S.A.M.P.E., P. O. Box 613, Azusa, CA, 1980, pp. 752–64.

MECHANICAL PROPERTIES OF NEW MECHANICALLY ALLOYED Fe-BASE ODS ALLOYS

M.A. Daeubler
Motoren-und Turbinen-Union München GmbH
D-8000 München 50, FRG

D. Froschhammer
Polytechnical University München
D-8000 München 2, FRG

Abstract

Microstructural and mechanical properties of two newly developed ferritic oxide dispersion strengthened (ODS) alloys for high temperature applications, containing 0.5 and 1.0% Y_2O_3, respectively, have been examined. The influence of chemical composition, production parameters and recrystallization conditions have been included. The recrystallized microstructure is characterized by elongated grains with an aspect ratio greater than 10:1 and a relatively small content of porosity. Results from tensile and stress rupture tests were superior to those given by the commercial ODS alloy MA 956. Due to the promising results, commercialization of both alloys is expected in the near future.

Introduction

Oxide dispersion strengthened (ODS) superalloys exhibit excellent creep properties. Ni- and Fe-base ODS alloys are already finding use in high temperature applications. Apart from their good stress rupture properties it is generally the remarkable mechanical strength of Ni ODS alloys, and the high oxidation resistance of ferritic ODS alloys, that make them especially attractive.

The high temperature strength is achieved by fine, homogeneously distributed oxides in the matrix and a recrystallized elongated grain structure, where the properties of the ODS alloys are importantly influenced by the dispersoid content and the manufacturing process, especially mechanical alloying, the deformation of the material and the recrystallization treatment.

Ferritic ODS alloys have been gaining increasing acceptance since they became available in semifinished forms (sheet and bar materials). Ferritic ODS alloys make especially suitable materials for aero engine combustion chambers, gas turbines (1, 2) and parts in the hot section of Diesel engines (3). Ferritic ODS alloys take their corrosion resistance at elevated temperatures from sufficient Cr and Al contents (approximately 20 wt.% and 5 wt.%, respectively). Their mechanical strength essentially derives from their Y_2O_3 content and defined grain structure (aspect ratio).

The newly developed ferritic ODS alloys PM2000 and PM2010 were investigated with respect to microstructure and mechanical properties. These new ODS superalloys are presently undergoing testing at MTU Munich for their value as combustion chamber shingles for aero engine gas turbines.

The ferritic ODS alloys PM2000 and PM2010 under investigation were jointly developed under a collaboration program between universities and industrial partners. The material under study was manufactured by Metallwerk Plansee GmbH. The chemical composition of the alloy is shown in Table 1.

Alloy	Fe	Al	Cr	Ti	Y_2O_3
PM2000	balance	5.5	20	0.5	0.5
PM2010	balance	5.5	20	0.5	1.0
MA956	balance	4.5	20	0.5	0.5

Table 1: Chemical compositions of the alloys PM2000 and PM2010 in weight %

PM2000 and PM2010 differ from the commercial ODS alloy MA956 by essentially the higher Al content in both alloys and the higher Y_2O_3 content in PM2010. The alloys were manufactured from powders consisting of single elements mixed with binary prealloyed materials (identifier E) and from blended master alloys (identifier B), respectively. These alloys were manufactured using conventional PM techniques (mechanical alloying, encapsulation and thermomechanical treatment, decapsulation) followed by forming and recrystallization annealing. For further details, see Table 2.

During the development phase the rolling temperatures were varied for an estimate of their impact on structure and mechanical strength properties. So far, bars (circular and square) and also 3 mm, 2.5 mm and 1 mm thick sheets have been rolled from these novel alloys.

Microstructural Analysis

For microstructure analysis and crack path characterization, light microscopy was used.

Mechanical Testing in Air

To assess the mechanical properties, tensile tests at 980 ^{o}C and creep rupture tests at 900 ^{o}C were conducted on round at flat test specimens in rolling direction. Investigations at higher temperatures were made at partners' facilities. To test the bar material, cylindrical specimens 6 mm in gage diameter and 30 mm in gage length (A5 specimen) were taken, while for the sheet material, flat specimens 7 x 3 mm^2, 7 x 2.5 mm^2 and 13 x 1 mm^2 in cross-sectional area were used.

Thermal Fatigue Tests

These new ferritic ODS alloys were also tested for their thermal fatigue properties, considering that these alloys will be exposed to temperature variations depending on operating conditions. The thermal fatigue behavior has so far been tested only on the ferrritic ODS sheet material. For these investigations, use was made of either thermal fatigue specimens (Fig. 3) wedge-shaped with a tip radius of 0.5 mm, or of 1 mm sheet strips 65 mm long and 22 mm wide, where the long edges were also given an 0.5 mm radius. In the thermal fatigue test the central section of the longitudinal specimen edge was cyclically heated to 1050 ^{o}C in a hot gas stream and cooled down to room temperature in an air stream. At the gage edge a 10 mm hot spot was produced causing a sharp temperature gradient. The duration of a heating/cooling cycle was 60 seconds. As a measure of thermal fatigue strength the number of thermal cycles to the first discernible crack (crack length 0.2 mm approx.) was used. Since at these test conditions the material is stressed with rather high temperature gradients, the experiments permit the simulation of thermal fatigue behavior in extreme service conditions, e.g. for aircraft turbine engines. Straight-forward comparison of results from thermal fatigue tests are permitted only if test conditions, especially temperature gradient and specimen geometry, are comparable.

Microstructure Analysis

For the unidirectionally rolled ferritic ODS alloys PM2000 and PM2010 the typical structure is characterized by elongated grains of appropriately high grain aspect ratio. Fig. 1 illustrates the ferritic ODS structure of PM2000 bar material (12×12 mm^2) in longitudinal and transverse directions. Owing to the rolling operation the grains forming in the surface zone of bar material are typically smaller than in the core, although no appreciable variation in grain aspect ratio is noted. The typical structure of cross-rolled ODS sheets features relatively small flat equiaxed grains. The 1 mm sheets from the PM2000 ODS alloy exhibit a uniform grain structure over the entire thickness (Fig. 2). Porosity is modest in the ferritic ODS alloys PM2000 and PM2010; apart from small isolated pores (under 0.1 mm dia.) no other porosity was noted in the structure.

Tensile Properties

The averaged results (of at least two tests) of the tensile test at 980 ^{o}C in air are given in Figs. 4 and 5, where the results for MA956 are added for comparison. The tensile strength of PM2000E (elemental) and PM2000B (blended) bar material (Fig. 4) is the same, at about 110 MPa, whereas the elongation A5 of the PM2000B alloy, at about 11%, runs some 2% higher than the PM2000E figures. The 3 mm PM2000E sheet material gave appreciably superior high temperature strength. In the rolling direction this sheet reached a strength of 160 MPa at an elongation of nearly 5%. At about 140 MPa the bar material in the ferritic PM2010 ODS alloy (Fig. 5) clearly surpassed PM2000 in strength, but this plus was penalized by a drop in elongation to values between 7% and 8%. No differences were noted between PM2010E and PM2010B in mechanical properties.

Creep Rupture Tests

The results of the creep rupture tests at 900 ^{o}C in air are given in Figs. 6 and 7. At a test load of 100 MPa the bar material in the ferritic ODS alloy PM2000B (Fig. 6) attained a life of at least 40 hours. PM2000E even reached a minimum of 80 hours at the same 100 MPa. The time to rupture figures were clearly superior to MA956 for the ferritic ODS sheet material PM2000E, where at 100 MPa the time to rupture exceeded 600 hours. The time to rupture of the PM2010E (Fig. 7) bar material compares with that of the bar material in PM2000E (Fig. 6). The blended bar material PM2010B exhibits appreciably superior creep rupture strength over the elemental bar material PM2010E and also over the PM2000B and PM2000E bar materials (compare Figs. 6 and 7). This is attributed to structural features discussed in more detail below. Typical structures of broken creep rupture specimens are represented in Fig. 8.

The relationship between time to rupture t_B and grain aspect ratio l/d is shown in Fig. 9 for PM2000 and PM2010. As it will readily become apparent the time to rupture increases when the grain aspect ratio does. Shown schematically in Fig. 10 is the time to rupture versus rolling temperature (dashed line), with supplemental readings given. The times to rupture are a' maximum at rolling temperatures between 700 ^{o}C and 750 ^{o}C.

Thermal Fatigue Tests

In the thermal fatigue tests first discernible cracks (about 0.2 mm) were noted in PM2000 and PM2010 specimens after 12 to 15 thermal cycles. For a closer look into the pattern and propagation behavior of the thermal cracks, further thermal cycles (about 5 to 8) were run after the first cracks had been noted. Shown in Fig. 11 is the structure with the typical thermal fatigue cracks originating at the gage edge of the specimen after totally 20 thermal cycles. From the first discernible crack size (0.2 mm approx.) upon 12 thermal cycles the thermal cracks grew in length to a maximum 2 mm, and simultaneously widened, in the course of another 8 thermal cycles. The main crack obviously propagates away from the gage edge, preferably following the grain boundaries, and also branches out in again preferred orientation along the grain boundaries. In the cracked area the thermally fatigued structure is interspersed with a plurality of small pores.

Discussion

Compared with the commercial ODS alloy MA956 the new ferritic ODS alloys PM2000 and PM2010 show improved high temperature tensile strength (Figs. 4 and 5) and stress rupture properties (Figs. 6 and 7). For the PM2000 bar material (Fig. 4), only minor improvement in tensile strength at 980 $^{\circ}$C is noted, whereas its elongation, at 8% to 11%, is markedly higher than that of MA956. The ductility increase is attributed to the 1% higher Al content, which improves the oxidation resistance and so keeps surface embrittlement down, which in turn benefits elongation. When the grain aspect ratio is very high on account of extensive deformation, as in the case of PM2000E sheet, the tensile strength grows to 160 MPa with elongation dropping to about 5%. The strength increase is attributed to the more pronounced directionality of the grains (rolling texture), and the drop in elongation might be explained by a reduction of the active slip systems. Compared with the PM2000 bar alloy (Fig. 4) the ODS bar alloy PM2010 (Fig. 5) gives appreciably improved high temperature strength at elongations still ranging above 5%. The improved mechanical performance of the PM2010 bar material may conceivably derive from the higher Y_2O_3 content.

The higher Y_2O_3 content does not significantly reflect on the stress rupture live at 900 $^{\circ}$C. This becomes readily apparent from a comparison of the times to rupture of sheet materials PM2000E (Fig. 6) versus PM2010B (Fig. 7), since both alloys exhibit the same grain aspect ratio of about 30:1 (Fig. 9). A primary observation made was that time to rupture was linked with grain aspect ratio (Fig. 9). Grain aspect ratios being the same, the sheet achieves longer times to rupture than the bar material. It is assumed that the greater amount of deformation undergone by the sheet material causes a more pronounced intensity of texture, which then leads to longer times to rupture. The structures in the fracture area of stress rupture specimens here shown in Fig. 8 are typical of ODS alloys. In the fracture area of the stress rupture specimens the structure exhibits conspicuously asymmetric reduction patterns. Supplemental SEM examinations reveal small dimples on the crack initiation fracture surface, and large dimples in the rupture area. Since the mechanical properties of the ODS alloys, as witnessed by the present study, are largely determined by the microstructure, thermomechanical treatment plays a correspondingly important part. A vital consideration next to the judicious matching of recrystallization temperature and time, is the exact definition of the severity and temperature of deformation in the manufacture of the semifinish. The time to rupture, therefore, can be correlated as a function of rolling temperature, a process parameter which at approximately constant deformation determines the dislocation density and, together with further given recrystallization conditions, establishes the grain aspect ratio which in turn determines the time to rupture (Figs. 9 and 10). The times to rupture for PM2000 (Fig. 10) are a maximum at a rolling temperature in the 700 $^{\circ}$C to 750 $^{\circ}$C range, and the structure of the stress rupture specimens here exhibits a high grain aspect ratio. In PM2000B a rolling temperature of 750 $^{\circ}$C produces an approximate grain aspect ratio of 20:1, which at 600 $^{\circ}$C drops to a mere 15:1 approximately. The times to rupture of the ODS alloys PM2000 and PM2010 increase with growing grain aspect ratio (Fig. 9), but the time to rupture does not extend rapidly until the grain aspect ratios have risen to a very high figure in excess of 20:1. It is noted that raising the grain aspect ratio from 20:1 to 30:1 will approximate quadruple the time to rupture, while at a constant grain aspect ratio of, e.g., 20:1 a rise in Y_2O_3 content from 0.5% to 1% does not show a significant effect on creep rupture performance. The noted rapid increase in time to rupture at growing grain aspect ratios - the averaged pattern is indicated by the broken line (Fig. 9) - cannot conclusively be charged to grain structure. We suspect that only part of the increase in time to rupture can effectively be attributed to the grain structure, while the remaining portion is brought about by thermomechanical process effects, such as greater deformation in the semifinish during the rolling operation. This is a likely proposition inasmuch as the PM2000E sheet material with its great amount of deformation relative to bar material, achieves a still longer time to rupture at the same grain aspect ratio. A plausible explanation of the steep rise in time to rupture can presently not be offered; it is prevented by the persisting lack of insight into the interaction of the various thermomechanical process paramters involved in the manufacture of the ODS alloys.

In the thermal fatigue test defined for aero engine gas turbines, with its extremely high temperature gradients, the targeted minimum 100 thermal cycles without discernible crack initiation have, as yet, not been attained. In tests featuring lower temperature gradients - such as they may be typical of a host of other applications - the new ODS alloys gave entire satisfaction, and at these conditions the thermal fatigue behavior was graded acceptable (4). A new batch of PM2000 material with recrystallization conditions further optimized is scoring a remarkable improvement with up to 70 crackfree thermal cycles also under the present rigid test routine.

Conclusions

- High temperature tensile and stress rupture properties of PM2000 and PM2010 are superior to those of commercial ODS alloy MA956.

- Stress rupture life increases with rising grain aspect ratio.

- Increasing the Y_2O_3 content primarily boosts the high temperature tensile strength, while the stress rupture performance does not improve significantly.

- Thermal fatigue properties are still below expectation when tested at high temperature gradients. Efforts to improve by further optimizing recrystallization conditions are under way.

- Due to the promising results, commercialization of both alloys PM2000 and PM2010 is expected in the near future.

Acknowledgement

This project was sponsored by the German Federal Minister for Research and Technology (BMFT). Partners in the joint project are:

Metallwerk Plansee GmbH, Lechbruck
Metallgesellschaft AG, Frankfurt
Daimler-Benz, Stuttgart
MTU, München
ABB, Mannheim
MPI, Stuttgart
KFA, Jülich

References

1 Benn, R.C.
"Frontiers of High Temperature Materials II",
J.S. Benjamin and R.C. Benn, eds., INCO Alloy International, 1983, pp. 37-71.

2 Tank, E.
"Frontiers of High Temperature Materials II",
J.S. Benjamin and R.C. Benn, eds., INCO Alloy International, 1983, pp. 251-271.

3 Hedrich, H.D.
"New Materials by Mechanical Alloying Techniques",
E. Arzt and L. Schultz, eds., DGM Informationsgesellschaft-Verlag, 1988, pp. 217-230.

4 Hedrich, H.D.
Neue ODS-Superlegierungen für heißgasbeanspruchte Komponenten,
BMFT-Matfo-Verbundprogramm, 7. Report (1989), Daimler-Benz, Stuttgart, Germany.

Alloy	Alloy Production	Consolidation Techniques	Rolling Temeprature	Product Dimensions	Recrystallization Conditions
PM2000E	elemental powder + binary blended powder + Y_2O_3	forging press	800 ^{o}C to 600 ^{o}C	sheet 3 mm thick	1300 ^{o}C/1h/air
PM2000E	elemental powder + binary blended powder + Y_2O_3	hot extrusion	700 ^{o}C, 750 ^{o}C 950 ^{o}C, 1000 ^{o}C	12 mm square bar	1350 ^{o}C/2h/air
PM2000B	fully blended powder + Y_2O_3	hot extrusion	600 ^{o}C 700 ^{o}C	12 mm square bar	1350 ^{o}C/1h/air
PM2010E	elemental powder + binary blended powder + Y_2O_3	hot extrusion	700 ^{o}C	12 mm square bar	1350 ^{o}C/1h/air
PM2010B	fully blended powder + Y_2O_3	hot extrusion	700 ^{o}C	12 mm square bar	1350 ^{o}C/1h/air

Table 2: Tested charges of ferritic ODS olloys PM2000 and PM2010

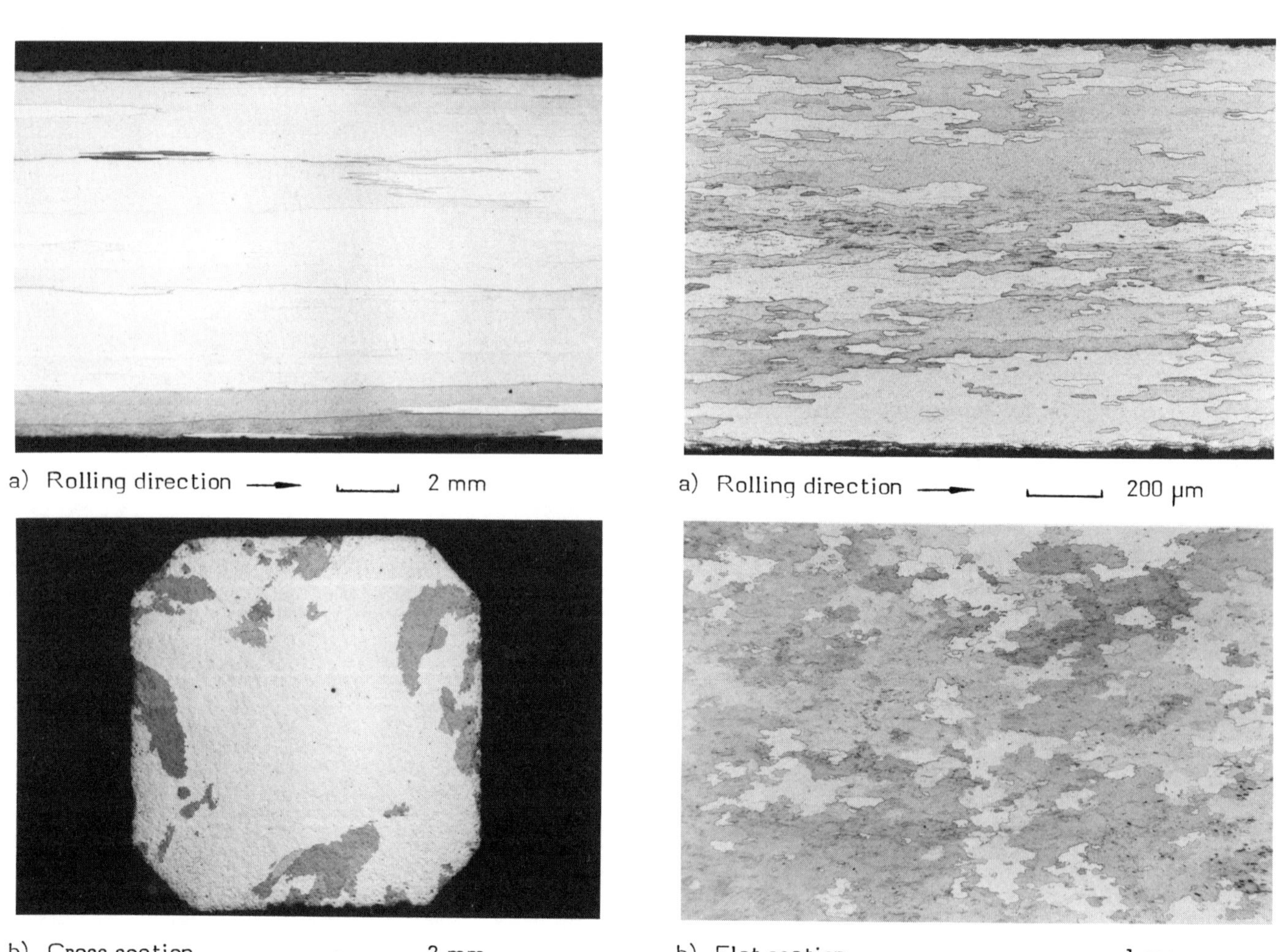

a) Rolling direction ⟶   2 mm

b) Cross section   2 mm

Fig. 1: Microstructures of unidirectionally rolled ferritic ODS supperalloy PM2000

a) Rolling direction ⟶   200 μm

b) Flat section   1 mm

Fig. 2: Microstructures of cross rolled ferritic ODS supperalloy PM2000

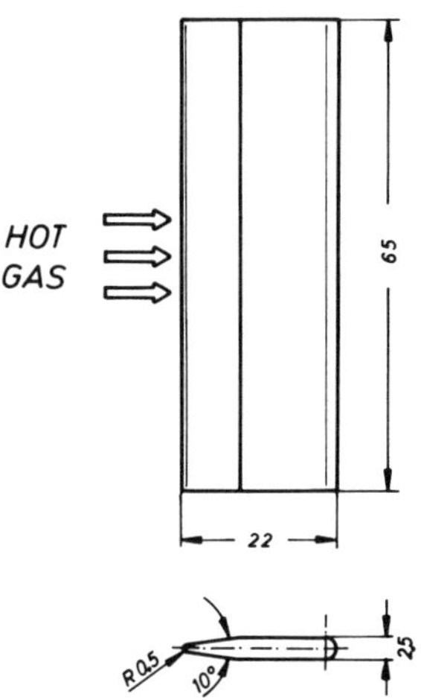

Fig. 3: Geometry of thermal fatigue test specimen

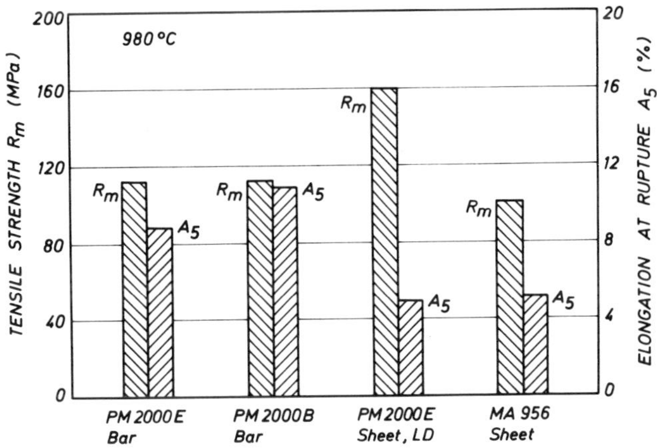

Fig. 4: Tensile properties of PM2000 in comparison to MA956 tested at 980 °C

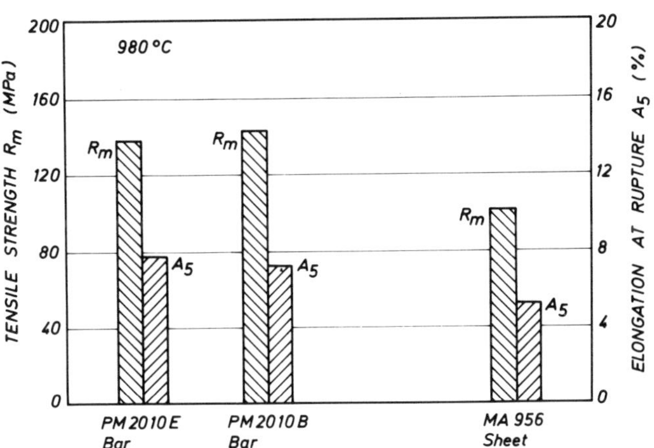

Fig. 5: Tensile properties of PM2010 in comparison to MA956 tested at 980 °C

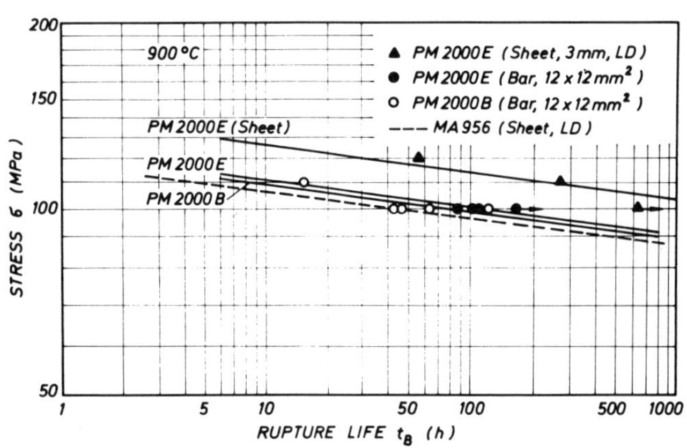

Fig. 6: Creep rupture life for PM2000 tested in longitudinal direction at 900 °C in air

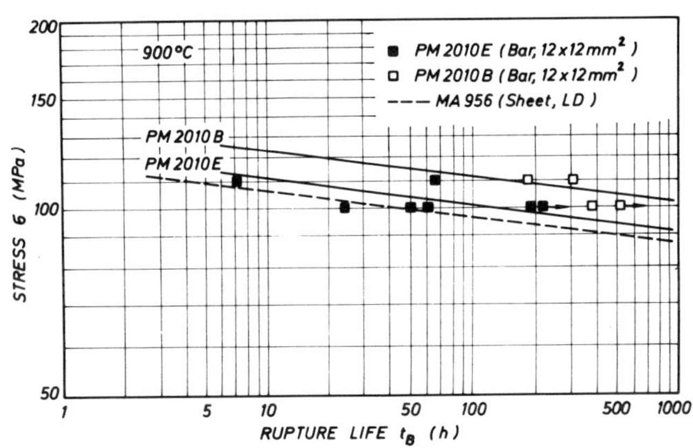

Fig. 7: Creep rupture life for PM2010 tested in longitudinal direction at 900 °C in air

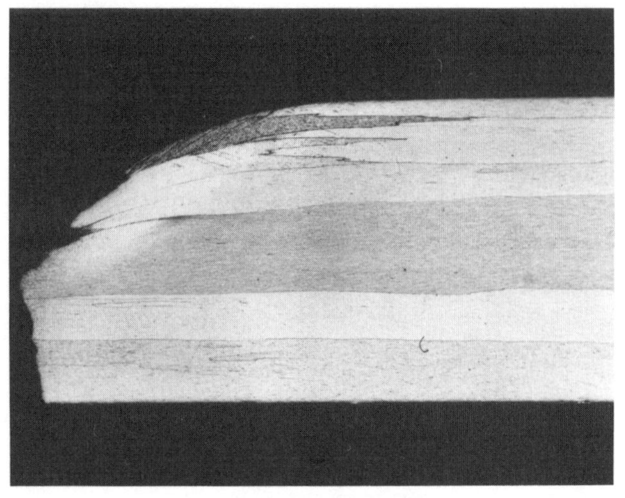

a) l/d = 20:1, rupture life 65 hrs   1 mm

b) l/d = 30:1, rupture life 307 hrs   1 mm

Fig. 8:   Fracture patterns of PM2010 after creep rupture tested at 900 °C and 110 MPa

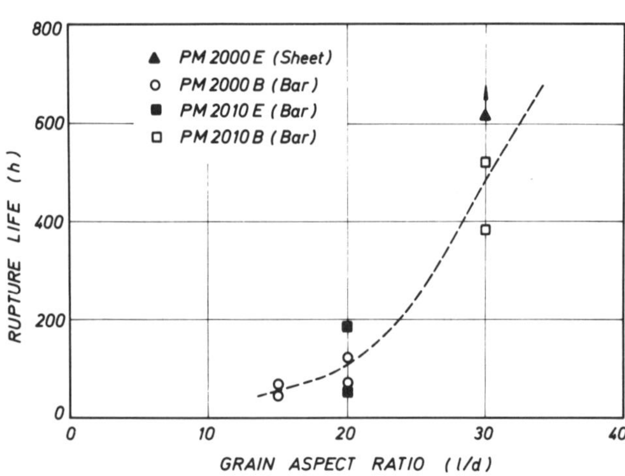

Fig. 9:   Creep rupture life time (900 °C, 100 MPa) as function of grain aspect ratio.

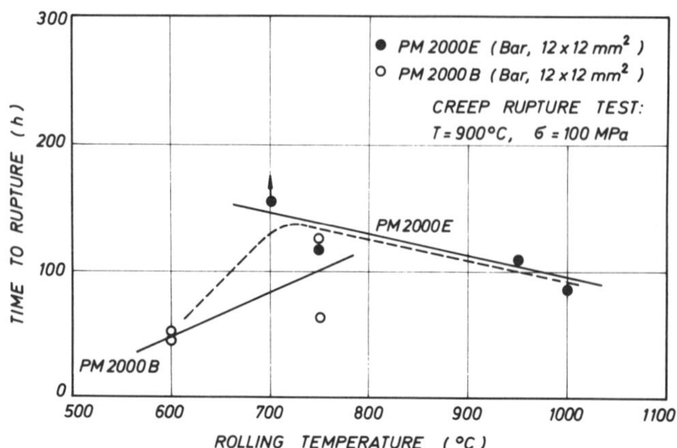

Fig. 10:   Creep rupture life as function of rolling temperature for PM2000 (deformation rate nearly constant)

a) Overview   2 mm

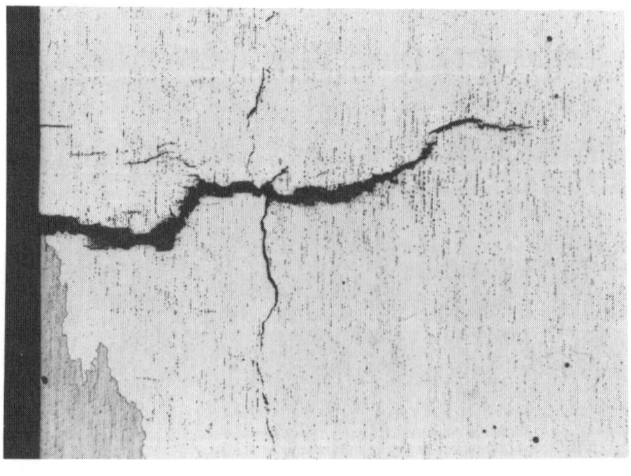

b) Detail   200 µm

Fig. 11:   Microstructures of PM2010 showing thermal fatigue cracks after 20 cycles

EVOLUTION OF STRUCTURE AND PHASE COMPOSITION DURING MECHANICAL ALLOYING OF NICKEL WITH ALUMINIUM

P. A. Vityaz, A. A. Kolesnikov, G. P. Pimenova, A. A. Stefanovich
Byelorussian Powder Metallurgy Association
Minsk, 220600, USSR

ABSTRACT

The paper represents the results of research works on formation of structure and phase composition of nickel alloyed with aluminium (70%Ni-30%Al). The process was carried out in an attritor. The investigation technique developed combines different methods of physical-chemical investigation. During the initial stage of treatment, up to 8 hours, the growth of powder pellets stops, their structure and content stabilize. The microstructure represents alternating layers of nickel and aluminium. Substructural level reveals chaotically oriented formations, including parallel needles of long size, the latter being nickel aluminides of various stoichiometry in microcrystalline state. Further treatment results in formation of dark grey phase on the pellets' surfaces or in the areas with fine layered structure. X-ray phase analysis reveals formation of intermetallide NiAl, volume share of which increases with the time of treatment. A large number of small particles appear, being chippings of the dark grey phase from the pellets' and milling balls' surfaces. Generally, electron diffraction patterns display diffusion halation, which is the sign of amorphization of the intermetallic compound NiAl. Phenomenological description of the relationships during formation of the material structure is being presented.

During the past years an increased interest to aluminides of transition metals as the base for advanced refractory alloys is being expressed. Nickel aluminides are the most advanced and promising for operation under corrosive, oxidizing media, high temperatures and high mechanical stresses /1/. The technique of mechanical alloying can be successfully used for producing powders of intermetallides /2,3/.

The mixture of nickel powder with 30% by weight of aluminium powder was taken as a tested sample. The process was conducted in the attritor in argon atmosphere with preliminary evacuation. The process duration was changed within 0.5 to 64 hours.

The investigation was carried out according to the developed technique, when products of mechanical alloying were studied by a microhardness tester, a light, scanning and transmission electron microscopes, X-ray diffractometer and a microanalyser. In the beginning of the process in the attritor, formation and growth of pellets take place, and in 8 hours it stops. With the treatment time increasing the pellets break into separate particles, the process proceeding most intensively within the time period from 16 to 24 hours. Further mechanical alloying does not result in noticeable reduction in size of the particles, and in 64-hour treatment the average particle size reaches 4 μm.

In the initial stage of the treatment the pellets are not uniform by structure. There are particles of grain, coarse and fine laminated structures. On the interlayer boundaries in some areas one can observe pores. With treating time increase the pores in the pellets disappear, the quantity of laminated constituent increasing. Two phases different by microstructure are observed: Ni-base grey one and Al-base white phase with a few patches of the grey (Fig.1a). It is confirmed by X-ray phase analysis, which identified only the lines of Ni and Al.

In the initial state Ni has microhardness 0.75 GPa, aluminium particles – 0.32 GPa. With further mechanical alloying microhardness increases, and in 8 hours it constitutes 0.9 GPa (white phase) and 2.2 GPa (grey phase). Treatment during 24 hours leads to growth of microhardness of the white phase up to 1.2 GPa, and microhardness of the grey phase increases slightly.

Study of the fine structure has shown that in the initial stage, within 16 hours of treatment, intensive plastic deformation results in sharp reduction of coherent scattering area. Further treatment leads to lower rate of reduction of blocks' sizes and after 64 hours it constitutes 12.4 nm for Ni and 16.0 nm for Al. Lattice parameters for Ni and Al during treatment remain unchanged and correspond to pure metals parameters.

X-ray point microanalysis of the pellets has shown that concentration of Ni in aluminium ingradient is much higher than of Al in nickel part. The difference is most noticeable in areas with coarse layered structure. It is worth mentioning that the thinner nickel and aluminium layers the higher concentrations of Ni and Al in them. The same behaviour of concentration change of the elements is observed with treating time increase. In the result of mass transfer one can observe formation of the third dark grey phase with microhardness 6.0-6.5 GPa in the pellets' microstructure after 16-hour treatment and on the tumbling balls' build-ups after 8-hour treatment.

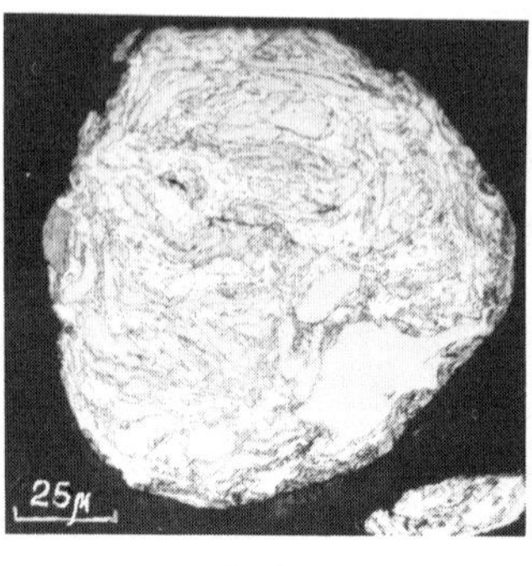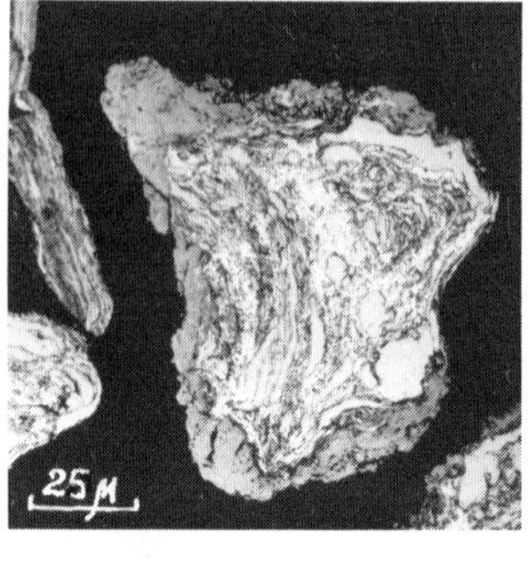

(a)   (b)

Fig.1. Microstructure of powder after treatment in an attritor:(a) 8 hours; (b) 24 hours.

The third phase is usually formed in the pellets' near-surface areas or in fine laminated structures. X-ray photographs display, besides nickel and aluminium lines, the lines corresponding to intermetallide NiAl. Quantitative X-ray microanalysis has determined the elements content change in the dark grey phase within 48-52 at.%, corresponding to β-phase of NiAl with a wide zone of homogeneity.

Further mechanical alloying leads to the growth of the dark grey phase, situated on the pellets' surfaces and in thin layers build-ups, the thickness of which becomes larger (Fig.1b). The dark grey phase as if grows through the aluminium layers inside the pellets or build-ups. A large number of chippings of the grey phase and pellet fragments appear from the pellets' surfaces and build-ups. Stereological analysis has shown that after 64-hour treatment the powder comprises up to 70% of the dark grey phase.

X-ray analysis has stated that the width of NiAl intermetallide line (100) in the dark grey phase does not change noticeably with duration of the treatment, and has approximately the same values on the chippings of the tumbling balls (Table I). It means that blocks of coherent scattering area do not get smaller, and their size remains the same.

Table I. The width of NiAl intermetallide line (100) in powder vs. mechanical alloying time.

Time of mech.alloying, hours	B, rad.
24	0.062
32	0.044
48	0.043
64	0.044
64 (chippings)	0.050

Analysis with a transmission electron microscope has revealed that generally during 1-hour mechanical alloying a fine structure is formed in nickel and aluminium. The structure consists of fragments of 50-70 nm in size, which in their turn consist of finer blocks of 10-30 nm (Fig.2a). Diffuse halation on the electron diffraction patterns may evidence about the possible presence of the amorphous constituent of the material just on the initial stage of the treatment (Fig.2b). Formation of new phases during 1-hour mechanical alloying is confirmed by electron diffraction analysis: monocrystals and crystallites of $NiAl_3$ and Ni_3Al phases are detected.

With the treatment time increasing up to 16 hours, substructure of the pellets undergoes noticeable changes, while going to random-oriented formations from parallel long needles. Fig.2c illustrates a pellet consisting of layers and needles (Fig.2c). The needle's diameter is up to 0.1 μ and the length from 1 to 5 μ. The needles in their turn consist of finer subgrains-blocks of 10-20 nm in size (Fig.2d). Electron diffraction pattern of such areas displays haloes with fine X-ray interferences, corres-

ponding to microcrystalline state of
the material. Electron diffraction patterns have revealed NiAl, Ni$_2$Al$_3$, NiAl$_3$,
Ni$_3$Al phases in the needles.

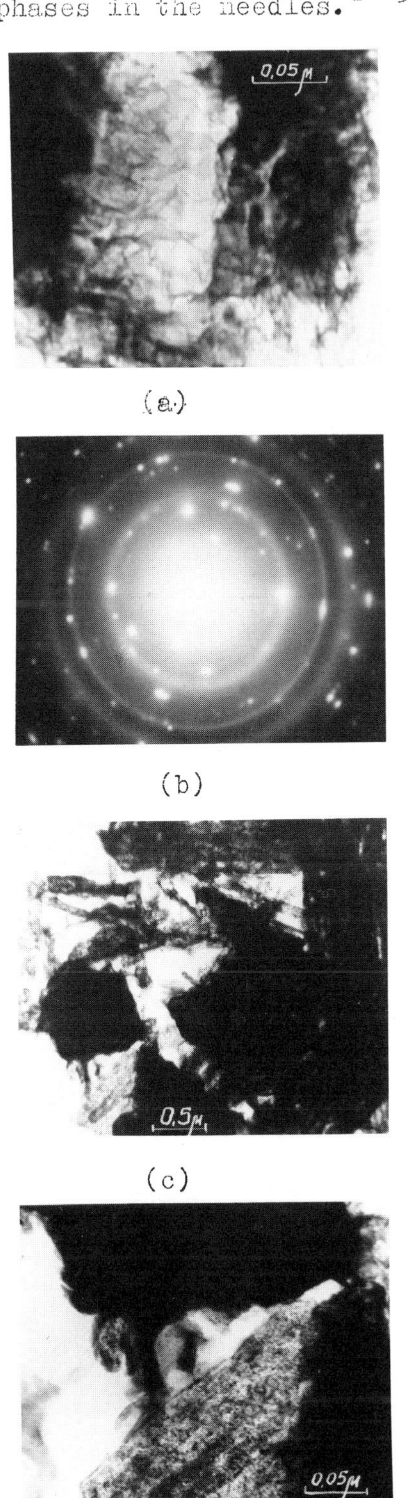

(a)

(b)

(c)

(d)

Fig.2. Electron diffraction patterns of
powder after treating in attritor (a)-
1 hour (substructure);(b)-1 hour; (c)-
16 hours (substructure); (d)-16 hours
(substructure).

After 64-hour mechanical alloying
electron diffraction patterns of the particles generally reveal only diffuse halation (Fig.3a,b), indicating amorphization of NiAl intermetallide.

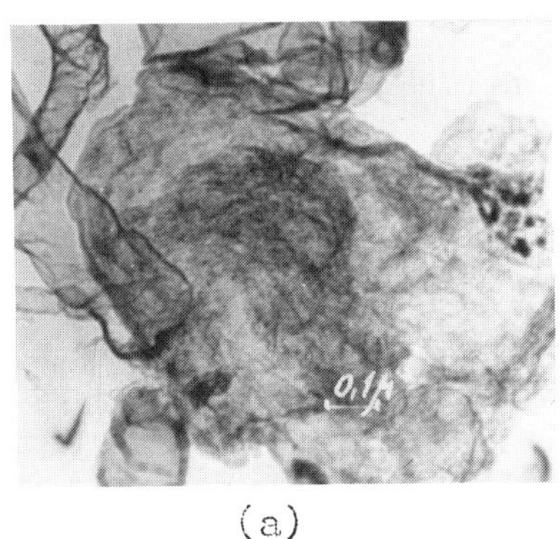

(a)

(b)

Fig.3. Electron microscopic patterns of
powder after treatment in attritor during 64 hours: (a)- substructure, (b)-
electron diffraction pattern.

The analysis of the investigation
results allows one to explain the transformations taking part during treatment
of nickel and aluminium powders'mixture
in an attritor.

Strong action of the tumbling balls
results in deep plastic deformation of
the powder particles. Under such conditions in the attritor working chamber
one can observe the whole spectrum of
stressed-deformed states on particle-
to-particle and particle-to-ball boundaries, shear effect being predominant.
/3/. The scheme of stressed-deformed
state is realized, pressure-shear-type,
which enables hardening of particles
with each other and with the tumbling
ball. Pellets are formed, consisting of
deformed viscous initial particles. The
process of repeated welding and milling
of the particles and pellets is going
on /2/.

With the treatment time increasing
more and more particles are involved in
plastic deformation. Initial nickel particles with polycrystalline structure,
and those with close contact to other
particles, under hindered plastic de-

formation, are more intensively affected by deformation strengthening as compared to small individual particles, because most dislocations in them remain
after interactions related to plastic
deformation. In the result, in large nickel particles dissipative structures
are formed. At the repeated actions on
such particles deformation becomes less
and is realized, as a rule, due to the
spare surface, for example, on particle
boundaries in the pellets or on the particle surfaces, thus leading to their
so-called "cold sintered state". Thus,
plastic deformation of two particles
with dissipative structures leads to
little plastic deformations on the boundary between them. This is not enough
for obtaining rather large area of hardening, and moreover, after unloading accumulated elastic energy will cause separation of the particles. Obviously,
pellets' growth under such conditions
occurs due to introduction of new portions of nickel particles and aluminium
particles with large margin of plasticity. Hence, characteristic structure
of mechanical alloying with alternating
layers of hard (nickel) and plastic (aluminium) materials is obtained. With the
quantity of plastic particles decreasing the growth of pellets becomes lower and the process of mechanical alloying is stabilized by structure, content
and microhardness.

Practically all the involved materials are treated by multiple plastic
deformation. Assuming interaction time
between two tumbling balls 1 ms, the period between the interaction about 10
ms (it comes from the mean relative
velocities of the balls and their diameters), and a particle is affected one
time per 10 interactions, then the average period between such interactions
constitutes about 100 ms. Hence, it follows that during all the period of mechanical alloying the particle is affected more than 10^6 times. Under such conditions the material is in excited
state. Relaxation of the excited state
is realized by the release of defects,
i.e. voids, dislocations, etc. These
processes go most intensively in plastic aliminium, where mass transfer velocities are rather high. Interaction
on nickel-aluminium boundary takes place, being more noticeable in the pellets with laminated structure. In the
result of mass transfer, especially in
aluminium constituent, supersaturated
solid solution is formed in the local
areas. Continuous and stable changing
of the material structure leads to its
mechanical activation, increases its reactivity and enables formation of che

mical compounds of different composition
in certain crystallographical planes at
low temperatures of micro- and crystalline states. With mechanical alloying
time increasing volume fraction of aluminides in the tested alloy increases and
the thickness of aluminium layers becomes less, thus shortening the ways of
nickel mass transfer in aluminium. At
definite critical thickness of aluminium layers concentration of nickel reaches the values enough for formation of
continuous NiAl intermetallide interlayer
with high hardness. Aluminium layers become more hardened with released aluminides and reach microhardness level of
the nickel constituent. Joint plastic
deformation begins. As far as the velocity of deformation strengthening of the
pellets' surfaces is much higher than
that of the inside one, a layer with
denser (more) defects is formed on the
pellets' surfaces. In the result of
more intensive mass transfer NiAl intermetallide is formed on the pellets'
surfaces, its thickness getting more
with further treatment in the attritor;
then it begins to spall. As intermetallide hardness is higher than that of the
pellets, further stages of plastic deformation go deeper in the pellet, near
aluminide-matrix boundary. The growth of
intermetallide constituent takes place.
NiAl phase spallations undergo plastic
deformation, amorphous particles of
about 4 μm size are formed. Amorphization process is completed, nickel and
aluminium constituents absent in the
powder mixture treated.

CONCLUSIONS

Kinetics of structural and phase
transformations in Ni-30%Al system has
been studied during treatment in the attritor. Formation of nickel aluminides
of different stoichiometric composition
versus mechanical alloying duration has
been stated. The evolution of transformation of a metal powder into microcrystalline and amorphous states has been
described. Phenomenological description
of structure formation relationships
has been presented.

REFERENCES

1. K. Portnoy, V. Babushkin, B. Zakharov,
   A. Sharykov, J. Poroshkovaya metallurgiya, 2 , 33, (1980)
2. Benjamin J., Volin T., J. Met. Trans, v.5,
   No.8, 1929-1934 (1974)
3. Steinberg A., Kolesnikov A., Properties
   and application of dispersion powders
   (in Russian), p.78, J. Navukova Dumka,
   Kiev (1986)

DISPERSOIDS IN MECHANICALLY ALLOYED ALLOY

I. S. Polkin, E. V. Ivanova, B. P. Matyhin
All Union Institute of Light Alloys
Moscow, USSR

High level of properties in the dispersion - strengthened alloys are managed to realize when the size of oxide phase particles does not exceed several tens nm and the distance between the particles - several hundreds nm. It is difficult to determine the optimum parameters of the dispersoid provided they are not connected with the whole thermomechanical previous history and end purpose of semi-finished material so long as the dispersoids have appreciable influence on the material behaviour during recristalization, formation of texture and dislocation substructure and so on. But for the known oxide dispersion - strengthened (ODS) nickel alloys designed for work at $1100°C$ the size of oxide phase 15-30 nm is indicated as preferable one [1,3-6] .

In the time of the mechanical alloying the powder of yttrium oxide is introduced in attritor together with the powders of metals and master alloys. It is precisely at this stage of the technology the parameters of strengthening phase are laid down in the alloy.

If sufficient quantity of the papers has been devoted to research of the dispersoid in the consolidated metal [1, 2] then the information about the influence of the initial powders size on the parameters of dispersoid in the alloy is absent.

In this work the attempt has been undertaken to connect the initial powder size of yttrium oxide introducing in ODS nickel alloy on the base of the $Ni-Cr-Al-W-Y_2O_3$ system with the parameters of oxide strengthening phase in the alloy and to observe how the type of oxide phase changed at the various stages of the technology.

The yttrium oxide powder of the different sizes was obtained by calcining of carbonate on oxalate of yttrium by change the conditions of the compound precipitation and the temperature decomposition.

Examination of the powder particles was carried out by transmission electron microscopy, and an object was prepared in the following way: alcohol suspension of the yttrium oxide powder was put on the carbon replica - substrate. The size of particles was determined by the dark - field images.

Features of the oxide phase in the consolidated metal were examined by analitical transmission electron microscopy and X-ray diffraction analysis.

The mean sizes of the powder particles of yttrium oxide introduced in the alloy under the mechanical alloying and the sizes of the oxide phase in the rod after recrystallization annealing at $1300°C$, 5 hours are given in Table 1.

The size distribution and the mean sizes of the oxide strengthening phase in the bar are similar to the distribution and the mean sizes of the initial particles.

It is obvious that the more dispersed initial oxide powder the less the size of the strengthening phase in the alloy. It is known that oxide phase in the compact metal does not exist as Y_2O_3 but forms with Al_2O_3 mixed oxide compounds [2,5] . The formation of the complex oxide phases has to lead most probably to enlargement of the particles especially after heat treatment. It can be supposed that at the stage of mechanical alloying some decrease of the size Y_2O_3 takes place owing to the fracture of the coarse particles and agglomerates and then in the bar - enlargement owing to the formation of the mixed oxide compounds.

Formation of the mixed oxide compounds takes place at the stage of alloy consolidating. The energy dispersive analysis of oxide particle and matrix obtained under examination of the foils after hot extrusion are given at fig.1. The particle contains Al and Y. During mechanical alloying 1.1% (mass)Y_2O_3 are introduced in the alloy, with that the oxygen content in the compact metall is 0,4% (mass) and Y_2O_3 introduces 0.23% (mass) from that. "Free" oxygen($\sim$0.17%) reacts with Al forming just at the stage of the mechanical alloying approximately 0.36% mass Al_2O_3. It is probably that under the thermal influence dissociation alumina (Al_2O_3), diffusion and interaction with Y_2O_3 in the nickel matrix take place. The Y_2O_3 - particles are more thermodynamically stable and so the nucleation places of the complex oxides occur. The 0.17% oxygen content which is not connected with Y_2O_3 in the alloy corresponds to the region of coexistence of 2 phases: $2Y_2O_3 \cdot Al_2O_3$ and $3Y_2O_3 \cdot 5Al_2O_3$ at the phase diagram of $Al_2O_3 - Y_2O_3$ system (fig.2).

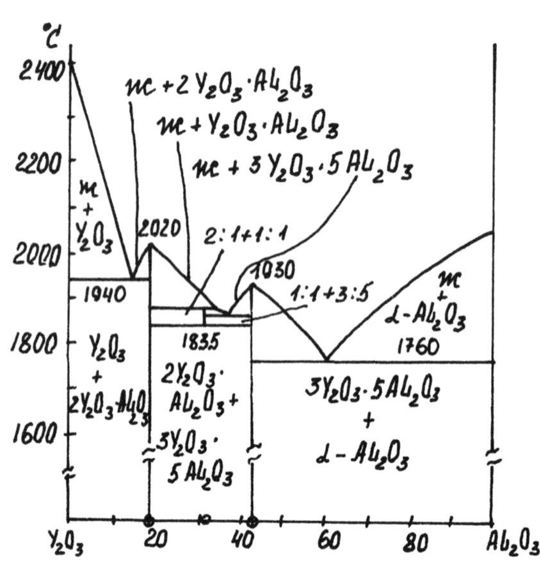

Fig.2- Y_2O_3-Al_2O_3 phase diagram.

Phase analysis of the extracts precipitated from the alloy allow to suppose that after extrusion the mixture of phase takes place in the alloy and the most probable phases are the following: $Y_4Al_2O_9$, $2Y_3Al_2(AlO_4)_3$ and $M_{23}C_6$ carbide (table 2). After the thermal treatment (1300°C, 5h) according to the electron difraction patterns the oxide phase in the alloy

was identified as $Y_4Al_2O_3$ (table 3,fig.3)

Results of X-ray diffraction analysis of an extract confirmed that after the heat treatment the phase $Y_4Al_2O_9$ and carbide coexist in the alloy (table 4).

It must be noted when mixed oxide compounds $Y_4Al_2O_9 (2Y_2O_3 \cdot Al_2O_3)$ are formed the volume part of the strengthening phase increases from 1.88% vol. for Y_2O_3 to 2.45% vol. for $Y_4Al_2O_9$ and the size of the oxide phase in the alloy must increase insignificantly as compared with the initial size of the particles.

CONCLUSIONS

1. It **has** been established that the correlation between the size of the yttrium oxide powder introduced in the alloy by mechanical alloying and the size of the mixed oxide compounds existed in the consolidated metal structure, the less initial powder Y_2O_3, the less strengthed oxide phase was by size.

2. It has been shown that yttrium oxide introduced in the alloy does not exist as Y_2O_3 but forms with Al and oxygen mixed oxides compounds. The formation of the complex oxides takes place at the stage of bar extrusion; in this condition the phases $Y_4Al_2O_9$ and $2Y_3Al_2(AlO_4)_3$ are appeared in the alloy. After heat treatment the phase $Y_4Al_2O_9$ exists in the alloy.

   Most probably the type of complex oxide and its stability in the alloy depend first of all on oxygen content.

Table 1 – Influence of the Initial Powder Size
of Yttrium Oxide on the Size of Phase in the Alloy

Sample No.	Average size of the as received powder Y_2O_3 , nm. (size range observed)		Average size of the strengthening oxide phase in the heat treated bar, nm. (size range observed)	
1	27	(20-40)	23	(19-57)
2	41	(28-87)	53	(17-90)
3	300	(100-2200)	132	(30-177)

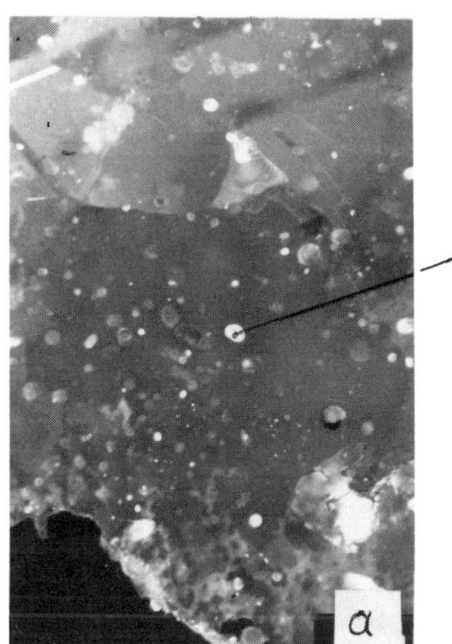

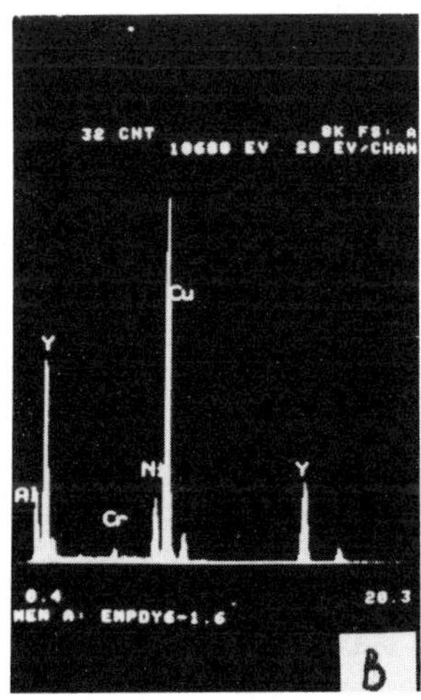

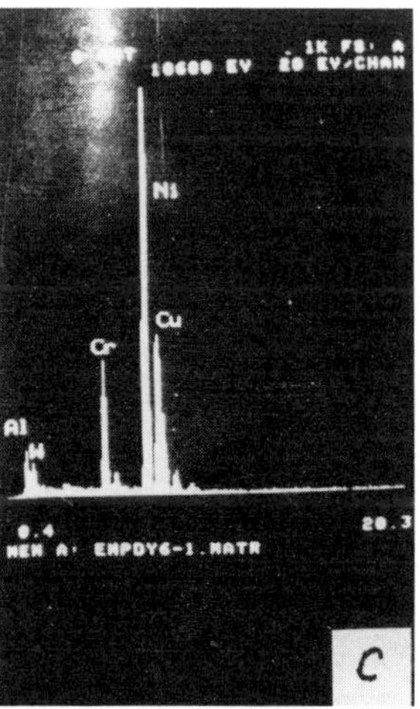

Fig.1 – Dark field electron micrograph showing fine oxides (a)
x30000, energy dispersive analysis of oxide dispersoid (b)
and matrix (c) in as-extruded bar.

Table 4 - X-ray Diffraction Analysis of an Extract
from the Alloy after the Thermal Treatment.

No.	$\frac{d}{n}$	$\frac{d}{n}$ $Y_4Al_2O_9$	HKL	$\frac{d}{n}$ $M_{23}C_6$	HKL
1.	4.67	4.69	012,021	–	–
2.	3.70	3.71	022	–	–
3.	3.27	–	–	–	–
4.	3.03	3.01	$22\bar{1}$	–	–
5.	2.97	2.908	023,032	–	–
6.	2.86	2.884	$2\bar{1}3,220$	–	–
7.	2.65	2.615	040	–	–
8.	2.57	2.559	202	–	–
9.	2.37	–	–	2.37	024
10.	2.17	–	–	2.168	224
11.	2.09	2.09	$240,24\bar{2}$	–	–
12.	2.04	–	–	2.04	333,115
13.	1.97	1.981	$241,12\bar{5}$	–	–
14.	1.88	–	–	1.878	044
15.	1.83	–	–	1.795	135
16.	1.77	–	–	1.770	006,244
17.	1.64	–	–	1.62	335

REFERENCES

1. J.S.Benjamin,M.S.Bomford
   Metal.Trans.,1974,v.5,N 3,p.615-
   -621.
2. J.S.Benjamin,T.E.Volin
   High Temperature - High Pressures,
   1974,v.6,p.443-446.
3. W.L.Kimmerle,V.C.Nardone,J.K.Tien
   Met.Trans.,1987,v.18A,N 6,p.1029-
   -1033.
4. W.Hoffelner,R.F.Singer
   Met.Trans.,1985,v.16A,N 3,p.393-
   -399.
5. J.D.Whittenberger
   Met.Trans.,1984,v.15A,N 9,p.1753-
   -1762.
6. T.E.Howson,D.A.Mervyn,J.K.Tien
   Met.Trans.,1980,v.11A,N 9,
   p.1609-1615.
7. N.A.Toropov,I.A.Bondar,F.J.Galahov,
   H.C.Nikogosjan,N.V.Vinogradova
   Izvestija AN SSSR,Serija Chimiches-
   kaja,1964,N 7,p.1158-1164.

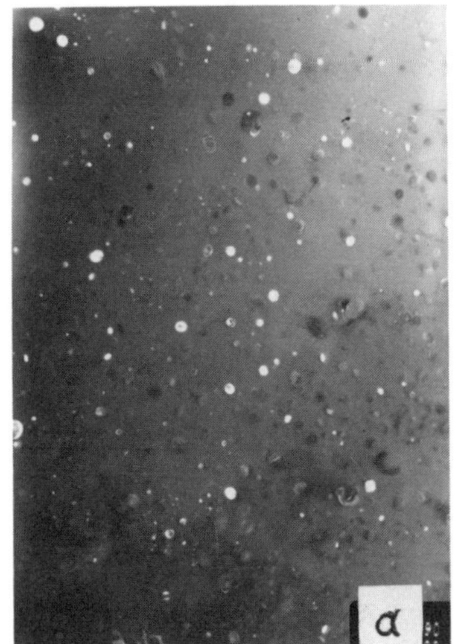

Fig.3 — Dark field electron micrographs oxide dispersoids in heat treated bar (a), diffraction pattern from oxide dispersoids shown in (a), (b).

Table 2 — X-ray Diffraction Analysis of Extracts.

No.	$\frac{d}{n}$ exp.	$\frac{d}{n}$ $Y_4Al_2O_9$	HKL	$\frac{d}{n}$ $2Y_3Al_2(AlO_4)_3$	HKL	$\frac{d}{n}$ $M_{23}C_6$	HKL
1.	3.22	−	−	3.21	321	−	−
2.	2.93	3.01	$22\bar{1}$	−	−	−	−
3.	2.86	2.908	023,032	−	−	−	−
4.	2.71	−	−	2.69	420	−	−
5.	2.49	−	−	2.45	422	−	−
6.	2.34	−	−	−	−	2.37	024
7.	2.23	2.291	231	−	−	−	−
8.	2.14	−	−	−	−	2.168	224
9.	2.10	2.09	$240,24\bar{2}$	−	−	−	333
10.	2.02	−	−	−	−	2.04	115

Table 3 — Calculation of the Circled Diffraction Pattern at Fig.3

No.	$\frac{d}{n}$ exp.	$\frac{d}{n}$ $Y_4Al_2O_9$	HKL
1.	7.48	7.41	011
2.	5.23	5.26	002
3.	4.62	4.69	012,021
4.	3.49	3.71	022
5.	3.20	3.01	$22\bar{1}$
6.	2.86	2.908	023,032
7.	2.71	2.62	040

DEVELOPMENT OF FERRITIC ODS TUBES
FOR HEAT EXCHANGERS OPERATING
ABOVE 1100°C

B. Kazimierzak, J. M. Prignon
Dour Metal
Belgium

F. Starr
British Gas
United Kingdom

L. Coheur
Cen/SCK
Belgium

C. Lecomte, D. Coutsouradis, M. Lamberigts
CRM
Belgium

ABSTRACT

Recent developments in ferritic ODS materials for applications above 900 °C in very stringent environments are presented.

Results available today feature outstanding thermal creep and corrosion resistance behaviour. Although further research have been undertaken, present composition and process route allow to launch commercial material ODM" 751.

1. INTRODUCTION

The MANUFACTURE OF FERRITIC ODS materials is one of the most promising applications of mechanical alloying. These alloys combine the following advantages :

- Oxidation resistance due to their Fe Cr Al matrix

- Excellent high temperature mechanical resistance due to the dispersion of oxide particles.

Their ability to develop coarse grain structures allows them to feature an outstanding creep behaviour at high temperatures.

The purpose of this paper is to present the encouraging properties obtained on ODS ferritic alloys developed at Dour Metal in cooperation with CEN/SCK and CRM.

---

* ODM - DOUR METAL's Trade Mark
  [Oxide Dispersion Microforged materials]

A considerable input has been given by British Gas, London Research Station where Dour Metal's materials amongst other alloys have been evaluated for an advanced heat exchanger system.

2. GENERAL ASPECTS

It has been shown that to reach a high level of properties above 1050°C (1920°F) ODS ferritic materials should feature :

- a regular and homogeneous dispersion of fine oxide particles;

- a coarse grain structure;

- a favourable grain texture (with high shape factor in the load orientation);

- a stable and adherent surface layer of protective oxides.

It has also been shown that defects causing premature failure are linked to the occurrence of :

- fine recrystallization areas within coarse grains, even when these areas are very small [1];

- films with disturbed recrystallization patterns at grain boundaries [2];

- Thermal induced porosity [3,4];

- Defects in surface protective layer.

3. RESEARCH BACKGROUND

The Belgian Nuclear Research Center, CEN/SCK, started in the late 60's on the development of ODS ferritic material for the use as canning tube in fast breeder reactors.

Since the early 80's DOUR METAL's team has been involved in this development and since 1985, DOUR METAL supplied ODS tubes for the french Phenix reactor, the UK's PFR and some samples to the US DOE. These were based in a Fe-13Cr material.

At the same time, work was carried out mainly in the framework of the European cooperative action COST 501, Round 1 to develop an ODS ferritic material for non nuclear high temperature uses.

To give the oxidation resistance required for this new range of applications, a composition with 3 % Al was developed.

Dour Metal's team conducted for more than 10 years an extensive program to master hot deformation processes which are now applied to the manufacture of ODS materials.

In parallel to this, Dour Metal have studied powder microforging phenomena and put together a skilled team as well as dedicated equipment to produce iron based ODS materials as tubes, bars and plates.

Since 1988, Dour Metal has been taking part to the Round 2 of the Cost 501 action and has been working together with British Gas to develop Iron based ODS tubing for an advanced heat exchanger operating at temperatures in excess of 1100 °C (2012 °F).

4. DEVELOPMENT OF ODM 751 TUBING FOR HEAT EXCHANGER APPLICATIONS

As the pressurized heat exchanger application required multiaxial high temperature strength, this new avenue of research aimed principally at improving our mastery of the metal structure.

Both grain orientation and recrystallization phenomena need to be addressed.

It is well known that the occurrence and the development of dynamic secondary recovery of b.b.c. metallic materials depend on several parameters including those related to the extrusion, during which the dynamic recovery is depending on strain-rate compensated temperature written :

$$Z = \dot{\epsilon} \exp \Delta H/GT \quad [5]$$

Where : $\dot{\epsilon}$ = strain rate
ΔH = activation energy
G = universal gas constant
T = temperature in Kelvin

This expression also applies to ODS ferritic materials, though their "recrystallization" is linked to other phenomena.

In particular, dispersoids tend to block grain growth and thus higher activation energy levels are required.

Extensive research was performed at Dour Metal to determine the influence of process parameters on "recrystallization".

The understanding of these phenomena had to lead to the optimization of grain texture for pressurized tubing.

Fig. 1 - some of the grain morphologies obtained by varying process parameters.

Another important criterium which influences the macro recrystallization is the orientation texture of micrograins produced at the extrusion.

After extrusion grain size is less than one micron.

X-ray examinations show that micrograins $\langle 1, 1, 0 \rangle$ axis is parallel to the extruded product axis. In the case of tubes, only slight variations can be observed on this overall orientation.

It should be stressed that after "recrystallization" treatment, resulting macrocrystals feature a different orientation, their $\langle 1,1,1 \rangle$ axis being parallel to the tube axis.

Macro-recrystallization is a phenomenon which strongly depends on the temperature level.

Macrograin formation can be compared to a diffusion phenomenon as it appears on the partially recrystallized specimen that are observed under polarized light on tube sections (figure 2).

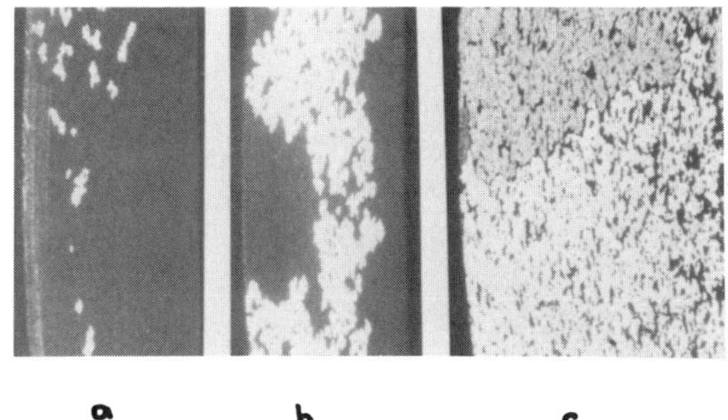

Fig. 2 - Etched sections. Polarised light. Early stages of "macrorecrystallisation" (a) (b) coherently oriented incipient grains (c) misorientation between adjacent emerging macrograins.

Once the secondary recovery is mastered, it is quite easy to obtain a coarse grain structure in one direction, that most often corresponds for bars and sheets to the load orientation during product operation.

For pressurized tubes, the maximum stress is observed in the hoop direction of the tube and hence transverse grains growth is important to be obtained.

Figure 3 shows how we have managed to produce macrograins elongated in the transverse direction around the circumference of the tube.

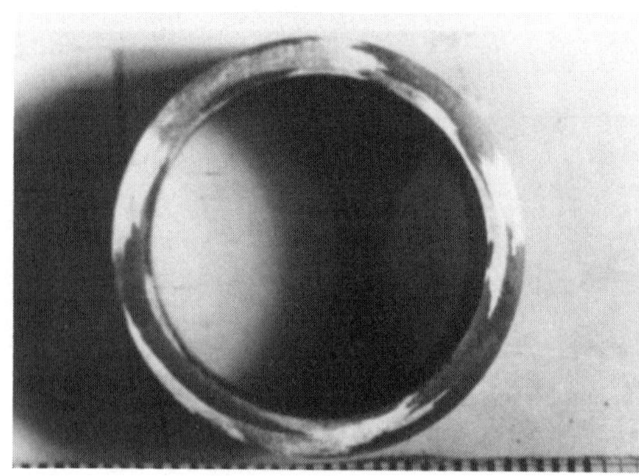

Fig. 3 - Transverse tube section with peripheral elongated macrograins.

Fig. 5 - Cyclic long term oxidation tests on tubes preoxidised for 1 hour at 1150 °C.

In parallel to the development of appropriate grains structure for pressurized tubes, another issue was the resistance of the material in oxidation at temperatures in the 1050 °C to 1250 °C range.

Therefore the five compositions shown in figure 4 with varying Al and Cr contents were produced and tested for oxidation. (Fig. 5)

NAME	Cr	Al	Mo	Ti	Y_2O_3	Fe
ODM 331	13	3	1.5	0.6	0.5	bal
ODM 031	20	3	1.5	0.6	0.5	bal
ODM 061	20	6	1.5	0.6	0.5	bal
ODM 361	13	6	1.5	0.6	0.5	bal
ODM 751	16.5	4.5	1.5	0.6	0.5	bal

Fig. 4 - Nominal compositions produced at Dour Metal

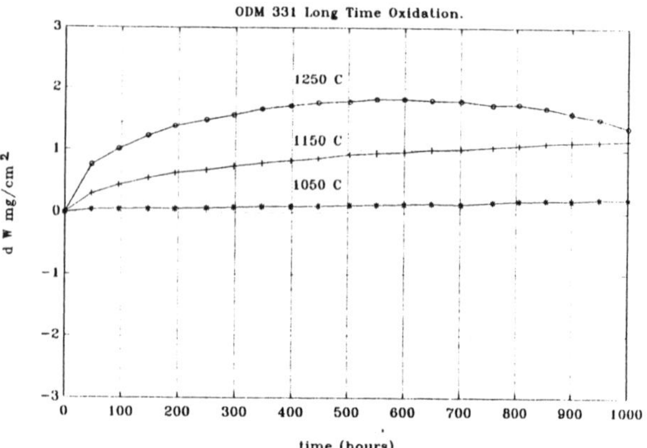

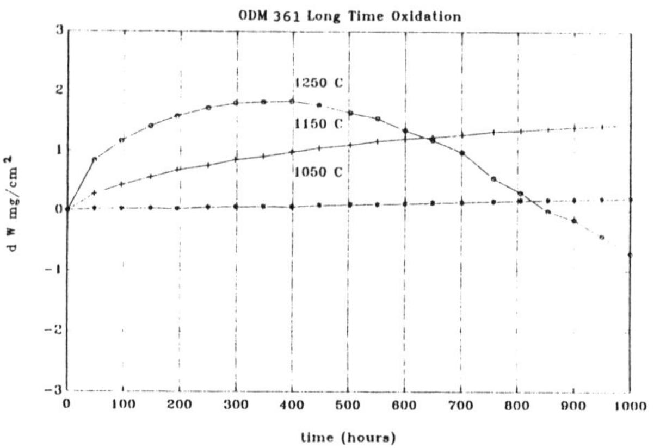

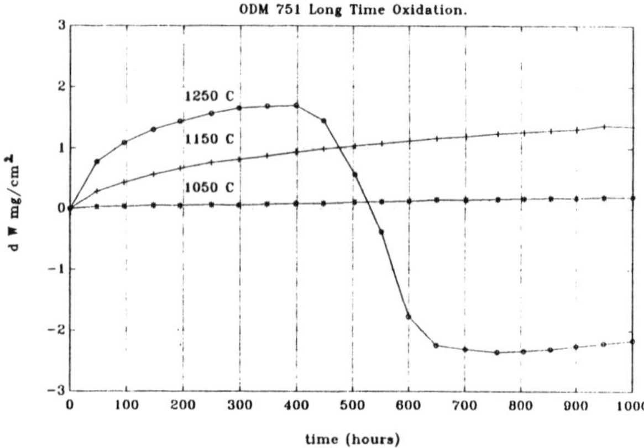

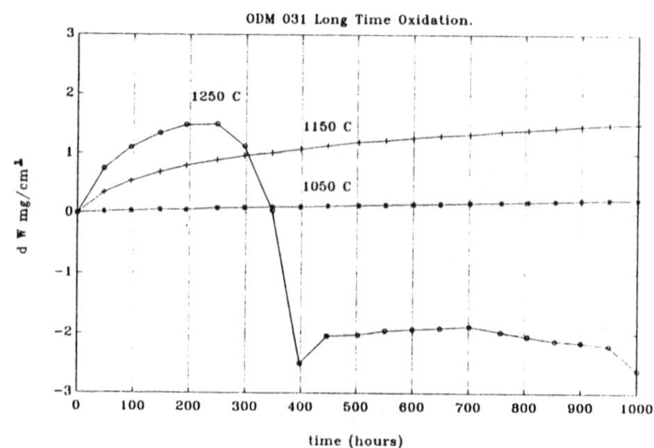

In addition to corrosion resistance, consideration had to be given to the workability of the recrystallized material. In particular sufficient ductility was required to allow ends swaging as well as the explosive welding onto a nickel based refractory material.

A composition was selected as offering a satisfactory compromise : ODM 751.

Finally to get the best out of the structures obtained, considerable attention had to be paid to eliminate the various defect sources above mentioned (see point 2).

It has indeed been observed that these defects tend to produce extremely random results both for the creep resistance and the corrosion behaviour.

Aiming at eliminating these defects gave us an extremely tedious job as it required special care and control for the various parameters for each step of the fabrication process.

Similarly extreme attention had to be paid when conducting the various tests which allowed us to confirm that ruptures originated from typical defects sources such as porosities, grain boundaries, unhomogeneities, surface defects, etc.

In a few words, thorough examination of the various fabrication steps allowed us to :

- eliminate one of the identified sources of thermal induced porosity;

- reduce the frequency of unrecrystallized areas within the coarse grains.

which causes defective deckle-edge ruptures (Fig. 6).

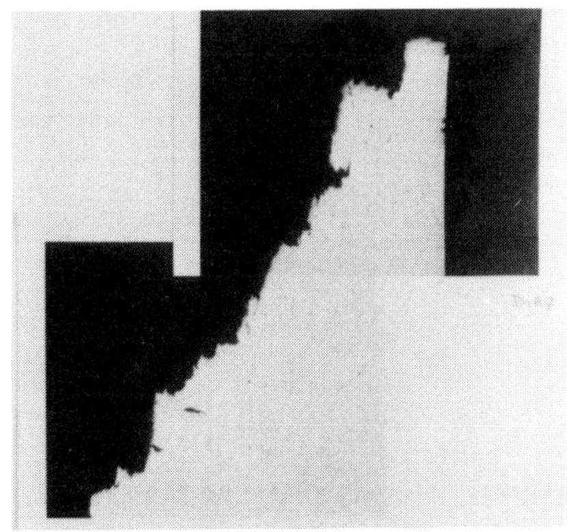

Fig. 6 - Deckle edge transgranular rupture.

It is well known that single crystal materials feature outstanding creep properties and thus it can be assumed that major weakness causes usually appear at the grain boundaries.

It was observed that intergranular ruptures (Fig. 7) appeared whenever the material could not sustain higher stress level. Hence, we thus aimed at identifying the conditions to obtain transgranular rupture patterns instead of intergranular patterns.

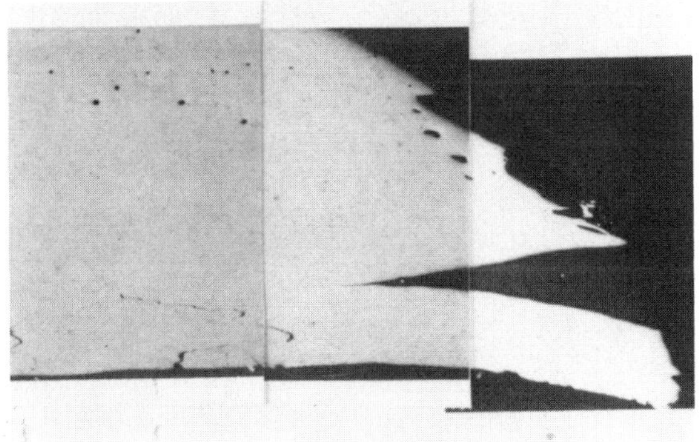

Fig. 7 - Intergranular rupture by macrocrystal decohesion.

Creep experiments carried out lately indicated the impact of these factors as the achieved strength level was outstanding. (Fig. 8)

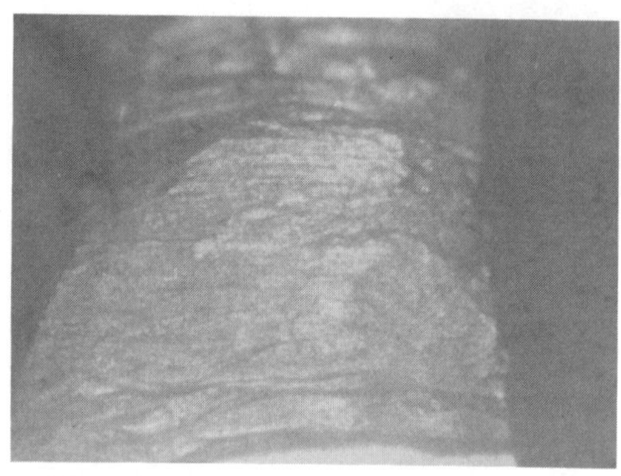

(a) smooth surface of rupture

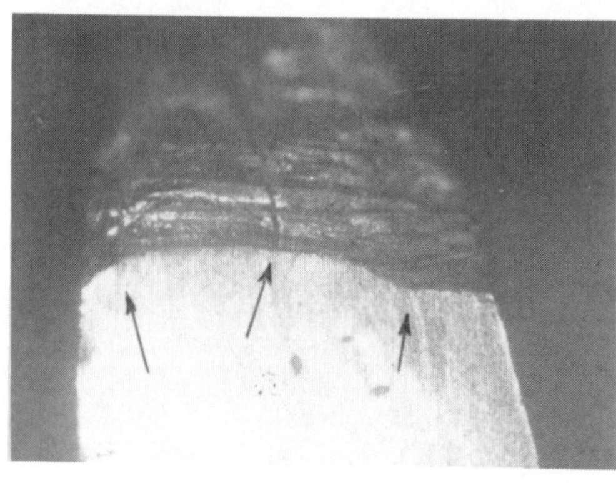

(b) unaffected grain boundaries.

Fig. 8 - Transverse section and surface of well processed tube ruptured after creep test at 1100 °C/65 MPa.

Indeed, typical lifetime in creep experiments at 1100 °C (2012 °F) and stress level of 65 MPa (9400 psi) exceeds 1000 H and for a stress level of 70 MPa (10100 psi) it exceeds 300 H.

These results, obtained repeatedly, are well above any data earlier released on Fe ferritic ODS materials.

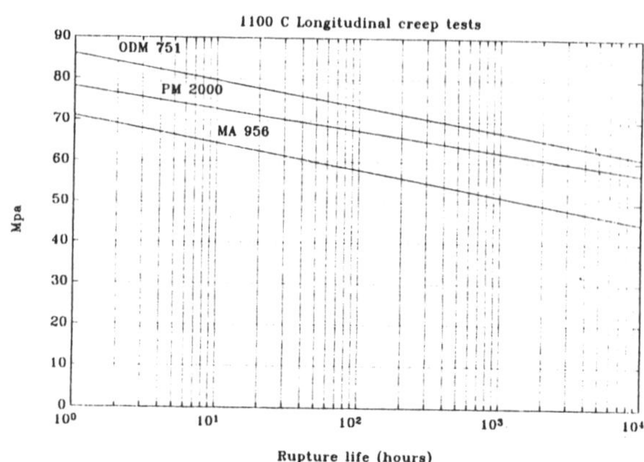

Fig. 9 - Creep properties of ODM 751

The strength level of ODM 751 approaches some of Ni ODS materials at 1100 °C and could probably surpass them at higher temperatures.

Extensive work is now probably being carried out within the European Cost 501 action to characterize further ODM 751.

Although these current results are satisfactory, it is considered that even higher creep values could be reached.

Our future work will therefore consist in improving further materials properties by identifying and eliminating other defects sources.

5. INDEPENDENT TESTS ON TUBE MATERIALS AND COMPARISON WITH CONVENTIONAL ALLOYS (F. Starr)

Such is the rate of development of these alloys, so far only an earlier alloy has been subject to tube burst and long term oxidation tests. This earlier alloy, ODM 331, differs from ODM 751 in having a lower aluminium and chromium contents. These are 3.0 and 13 % respectively. Furthermore, the process route was not then fully optimised. Nevertheless, the long term tests are already showing the superiority of this variant of ODS tubing to the best conventional wrought alloys.

As was mentioned earlier, ODS tubing differs from conventional tubing in that the transverse stress rupture strength, that is in the critical hoop direction, is usually inferior to the longitudinal strength. Hence, it is necessary to subject the tubes to long term pressure or burst test at temperature to obtain data value to the heat exchanger designer.

Another way in which ODM materials are different is that the alloy exhibits high stress sensitivity. Hence, the short term resistance to stress and pressure can appear quite disappointing. However, in the longer term, at times of real interest to the commercial operator, there is only a small fall off in strength.

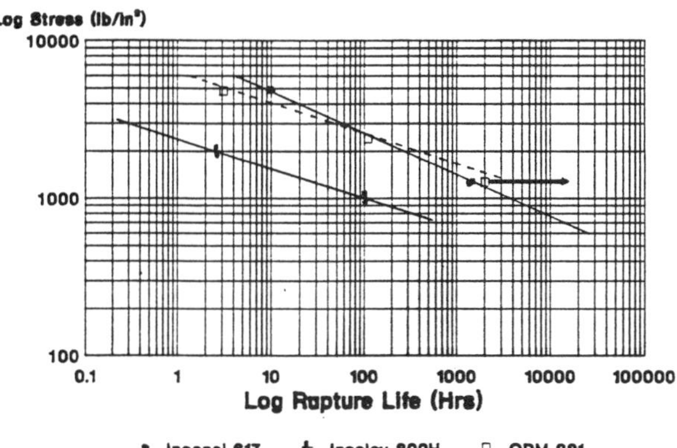

Fig. 10 - "Stress Rupture 1100 °C".

These features are well displayed in a Fig. 10 which compares the hoop strength of two typical high temperature wrought alloys, Inco 800H and Inco 617 with ODM 331. At the relatively high stress of 30.3 MPa, the ODM tube failed after 3 hours at 1100 ° C. Manufacturers data suggests that at the same stress Inco 617 would have lasted about 10 hours. However, at a reduced stress level of 8.7 MPa, the Inco 617 would have failed after some 1500 hr. However, in tests the ODM tubing survived this time and after 2000 hours shows no indication that it is about to burst. It is planned to continue this test for a further 3000 hours.

It is recognised that expensive tests of this type will be needed on each manufacturers ODS products even though the chemical compositions of different alloys may be very similar. This is because the processing route dominates the stress rupture properties.

Turning now to oxidation resistance, even ODM 331 with its comparatively low alloy content again is showing its superiority over conventional alloys, providing the test duration is sufficiently long and the conditions realistic.

Much commercial heat recovery equipment operates at steady working temperatures for long periods of time. Hence the need is for long term oxidation resistance data. Information from short term cycling tests, where the aim is to assess spalling resistance, is almost useless for design purposes.

The European Community tests are planned to be of long duration. Some have today reached 1500 hours at 1100 and 1150 ° C. During the early stages of these test, samples were removed every 300 hour for weighing and inspection. However, slow cooling and heating were employed in keeping with typical operational practice.

Fig. 11 shows the comparison between two ODS alloys, ODM 331 and Inco MA956 and four conventional alloys Inco 617, Haynes 230 and Haynes HR 160 (see table for compositions).

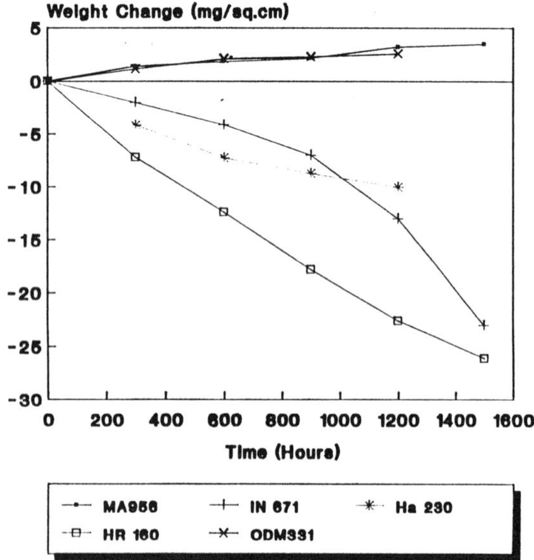

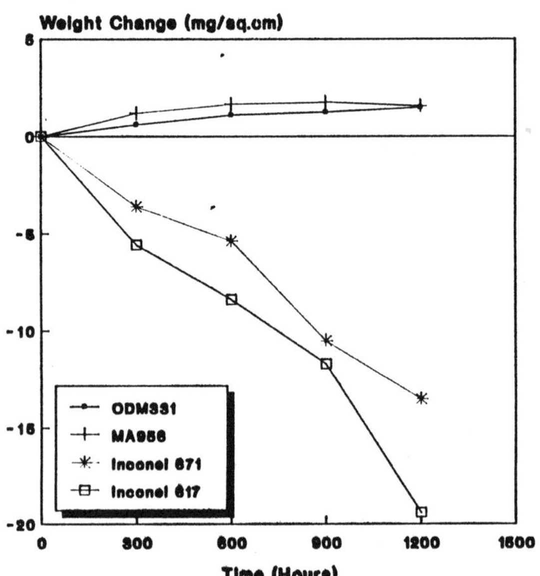

There is little different between the two ODS alloys despite differences in composition. Both suffered very slight spalling, but the weight changes and visual appearance are suggestive of good long term performance.

However, as the tests progressed all the conventional alloys suffered marked weight changes which were clearly associated with chromia volatilisation. Furthermore, despite the lack of real cycling, the scales of these were tending to spall quite badly.

Materials tested

Material	Composition
ODM 331	Fe - 13 Cr - 3 Al - 1.5 Mo- 0.6 Ti - 0.5 Y_2O_3
Incoloy MA956	Fe - 19 Cr - 4.5 Al - 0.6 Ti - 0.5 Y_2O_3
Inco 671	Ni - 48 Cr - 0.4 Ti
Inconel 617	Ni - 22 Cr - 12 Co - 9 Mo - 1.2 Al - 0.4 Ti - 0.2 Si - 0.07 C
Haynes 230	Ni - 22 Cr - 14 W - 5 Co - 3 Fe - 2 Mo - 0.4 Si - 0.3 Al - 0.1 C - 0.02 La
Haynes HR160	Ni - 28 Cr - 27 Co - 4 Fe - 2.75 Si - 0.05 C

6. CONCLUSIONS

This paper showed that the ODM 751 as now meets many of the requirements for high temperature materials to be applied above 1000 °C.

Dour Metal therefore decided to freeze the current process as it is and to introduce ODM 751 as a commercial material available in tubes, rods and plates.

It should be stressed that Dour Metal performs in house the whole manufacturing process, starting from elementary powders up to the final product.

Each step is thoroughly controlled in order to supply a consistent product and to give it the adequate properties.
The dimensions currently available are :

<u>Tubes</u> : outer dia 25 to 40 mm, wall thickness 2.5 to 5 mm, length up to 4 m.

<u>Rods</u>   : dia 10 to 40 mm, length up to 6 m.

<u>Sheets</u> : thickness from 1 up to 20 mm, width 250 mm, length up to 10 m.

REFERENCES

1. DM Elzey and E. Arzt. "ODS superalloys : the role of grain structure and dispersion during high Temperature low cycle fatigue"
Superalloys 1988 The metallurgical Society pp 595-604

2. H. Zeininger and E. Arzt "The role of grain boundaries in high temperature creep fracture of an ODS superalloy"
Z. Metallk. Bd 79 (1988) H12 pp 774-781.

3. G. Korb "Problems related to processing of ferritic ODS Superalloy" "New materials by mechanical alloying technique" DGM 1989 pp 175-182.

4. H.D. Hedrich "Properties and application of Iron Base ODS Alloys"
"New materials by mechanical Alloying Technique" DGM 1989 pp 217-230.

5. T. Sheppard "Metallurgical principles and the control of properties during the extrusion process"
Extrusion, Scientific and Technical Developments. DGM 1981

6. Incoloy alloy MA 956 INCOMAP mechanically Alloyed Products MP1 MP2.

KEYNOTE ADDRESS
MECHANICALLY ALLOYED ALUMINUM ALLOYS
FOR AIRCRAFT APPLICATIONS

J. H. Weber
IncoMAP Light Alloys
Inco Alloys International, Inc.
Huntington, West Virginia, USA

D. J. Chellman
Lockheed Aeronautical Systems Company
Burbank, California, USA

ABSTRACT

Mechanically alloyed (MA) aluminum alloys have been provided to the aircraft market for several years, and additional alloys are under development. Prior to widespread acceptance of these products, knowledge of durability and damage tolerance properties is mandatory. The present work addresses these issues by examining tensile, fracture toughness, fatigue, crack growth, and corrosion resistance of IncoMAP® alloy AL–905XL forgings. Improvements in structural efficiency compared to competitive or alternative materials are estimated by using a methodology pioneered by Lockheed. Progress on the development of high temperature mechanically alloyed Al–Ti alloys is also addressed using the same methodology.

THE PAST DECADE HAS SEEN THE EMERGENCE OF MECHANICAL ALLOYING (MA) as a technology for the production of materials with consistent properties attractive for high performance applications in aircraft. While the earliest materials were nickel- and iron-base alloys directed toward the hottest parts of the propulsion systems, the current IncoMAP product line includes aluminum–base alloys which are directed more toward structural applications (1). These aluminum–base materials are in various stages of product life. IncoMAP alloy AL–905XL, an Al–Mg–Li alloy, developed for use in structural forgings is in commercial production. By contrast, MA Al–Ti alloys are in the development and production scale–up stage.

Prior to the widespread acceptance of the MA aluminum materials, a knowledge of the durability and damage tolerance properties is mandatory. In addition to expanding the data base, techniques such as the ABVAL–G (2) and the Lockheed developed   methodology for evaluation of weight savings (3,4)

®IncoMAP is a registered trademark of the Inco family of companies.

can be used to determine the usefulness of new materials in various airframe components.

The purpose of the present work is three–fold: (a) to provide the results of property evaluations of forged IncoMAP alloy AL–905XL and of the developmental MA Al–Ti alloys, (b) to compare these properties with those of current and competitive materials, and (c) to project potential benefits for weight savings when using these MA products in airframe structural components.

EVALUATION OF IncoMAP ALLOY AL–905XL

BACKGROUND. Aerospace industry needs for aluminum alloys which can reduce weight in structures have resulted in considerable emphasis on alloys containing lithium. Several ingot metallurgy alloys have been developed and are available as plate, sheet, and/or extrusions. However, their usefulness as forgings has been limited due to their need for cold working after solution treatment but prior to aging for the development of optimum properties.

The mechanical alloying process has also been used to develop a lightweight Al–Li forging alloy. The design goal for IncoMAP alloy AL–905XL was to develop an alloy with mechanical properties equivalent to AA7075–T73 in forgings (5,6). Composition selection was focussed by: using only alloying elements which reduce density, limiting addition levels to minimize precipitation hardening, and strengthening via dispersoids resultant from the MA process. The nominal composition of the alloy developed is: Al–4.0 Mg–1.3 Li–1.1 C–0.4 O. This alloy represents an ideal forged product due to its excellent combination of strength, fracture toughness, and corrosion resistance compared to AA7XXX alloys. Due to the design of alloy AL–905XL, conventional solution heat treatment and aging are not required after the final forging operation.

MATERIALS. This study evaluated IncoMAP alloy AL–905XL in the forged condition. These 38 mm thick products were forged at approximately 345°C from 89 mm diameter rod. All forged material was used in the as–forged or –F temper for the property characterizations. The measured mechanical properties were compared to those of competing AA7075 forgings.

DENSITY AND MODULUS PROPERTIES. Alloy AL–905XL provides a 7 to 8 percent density reduction, due to the magnesium and lithium additions, compared to conventional AA7075 alloy, as shown in Figure 1. Elastic

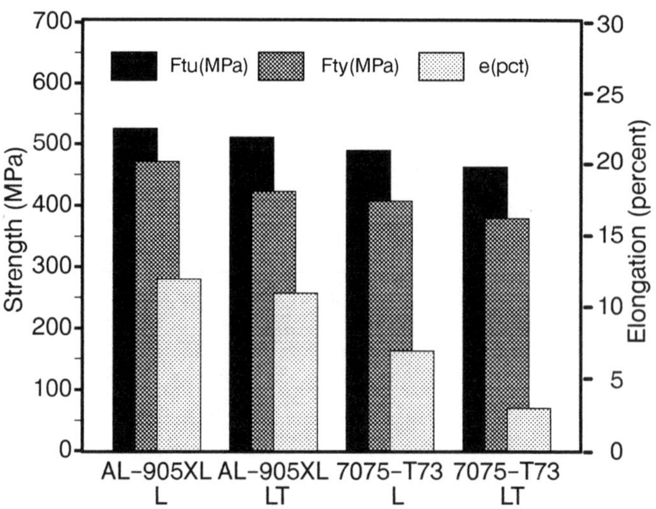

Figure 2. Tension properties as a function of orientation for alloy AL–905XL and AA7075–T73 hand forgings.

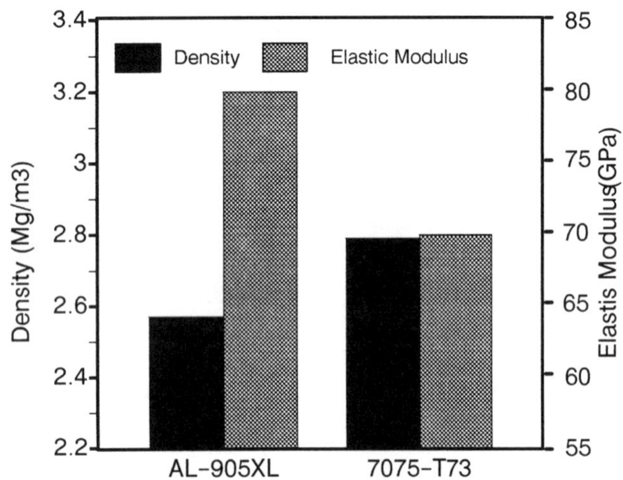

Figure 1 – Density/Modulus comparison of alloy AL–905XL with AA7075–T73.

modulus values obtained from tension stress– strain curves are also given in the figure. The 12 to 14 percent modulus improvement is associated with the lithium alloying, and contributes to a significant specific modulus and strength advantage for alloy AL–905XL.

TENSILE PROPERTIES. The results of room temperature tension tests conducted on the alloy AL–905XL hand forgings are given in Figure 2. Property comparisons for the most appropriate baseline aluminum forged product, AA7075–T73, are also shown in the figure. Testing was performed in the three orthogonal directions of the 38 mm thick hand forged blanks. Significantly higher yield and tensile strengths were obtained for alloy AL–905XL, with approximately 14% and 10% improvements, respectively, compared to AA7075–T73. Nearly isotropic properties in the L (longitudinal) and LT (long transverse) orientations were shown for the MA forgings, with only modest strength decreases of 4 to 7 percent in the ST (short transverse) direction. Excellent strength and ductility property combinations were exhibited for the alloy AL–905XL hand forgings, generally exceeding current AA7XXX forged parts.

FRACTURE TOUGHNESS. Plane strain fracture toughness measurements were conducted on the MA forgings.

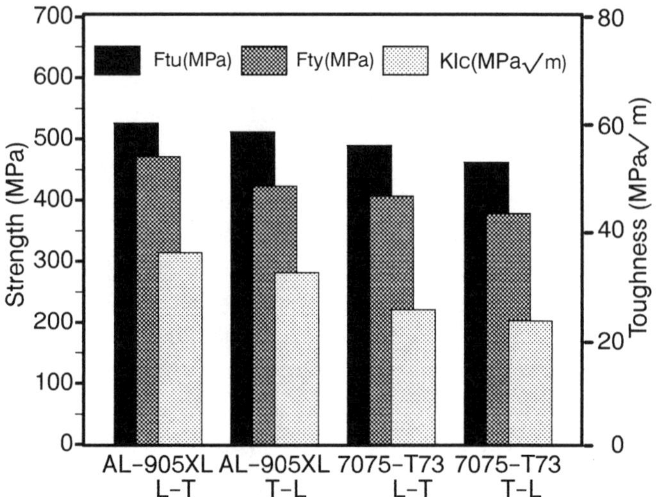

Figure 3. Strength–Toughness relationship for alloy AL–905XL and AA7075–T73 hand forgings in the L–T and T–L orientations.

Compact tension (CT) type specimens of 19 mm thickness were machined and tested according to established ASTM E399 procedures. Provisional K_q values for the L–T and T–L orientations are summarized in Figure 3. The fracture toughness values did not conform with all validity criteria, although the findings are judged to be representative of the overall toughness behavior. Data reported earlier on similarly sized hand forgings (7) using nominally 25.4 mm. thick CT test samples showed valid fracture toughness, K_{Ic}, values of 34.0 to 38.9 MPa.m$^{1/2}$ in the T–L orientation. These data corroborate the results obtained in the current study. In combination with the advantages in tensile strength properties, the alloy AL–905XL forgings were observed to be superior to the static properties for the baseline AA7075–T73 products. Additional testing is under way to establish the influence of thermal exposures on the toughness properties of the alloy AL–905XL forgings.

AXIAL S–N NOTCHED FATIGUE. Fatigue initiation behavior is an important consideration in structural applications designed to primary durability and damage tolerance (DADT) criteria. Preliminary results on the fatigue initiation resistance of alloy AL–905XL hand forgings are presented in terms of S–N curves in Figure 4 for the L

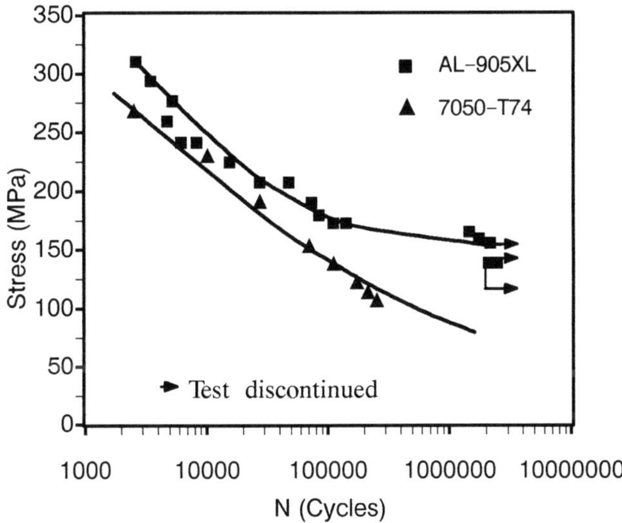

Figure 4. S–N notched fatigue behavior of alloy AL–905XL hand forgings for $K_t = 3.0$ at $R = +0.1$.

orientation. Testing conditions were $K_t = 3.0$, $R = +0.1$, and $f = 20$ Hz, in line with the extensive alloy database available on aluminum alloy products. Trend lines for both the alloy AL–905XL and the comparative AA7050–T74 (8) forged products are also given on the S–N plot. The notched fatigue behavior of alloy AL–905XL is clearly superior to AA7050–T74 throughout the high–cycle and low–cycle regions, and exhibits relatively low scatter. The fatigue preliminary design properties were estimated from these findings at a cyclic lifetime of 1.0 x 10⁵ cycles.

FATIGUE CRACK GROWTH PROPERTIES. Preliminary results on the fatigue crack growth resistance of alloy AL–905XL hand forgings were obtained using CT specimens, tested under constant amplitude loading and relative humidity in excess of 90%. The specimens were evaluated in both the L–T and T–L orientations at $R = +0.1$ conditions. The test results are plotted in the customary manner of da/dN versus ΔK in Figure 5, and compared to AA7050–T74 plate (8). The alloy AL–905XL data reflect slightly higher fatigue crack propagation rates, particularly in the lower range of ΔK values. Previous research on powder metallurgy aluminum alloys indicated that the observed behavior is primarily due to the fine sub–grain and grain sizes inherent in this processing route. Microstructural examinations are under way on the CT specimens to assess the influence of grain size and morphology

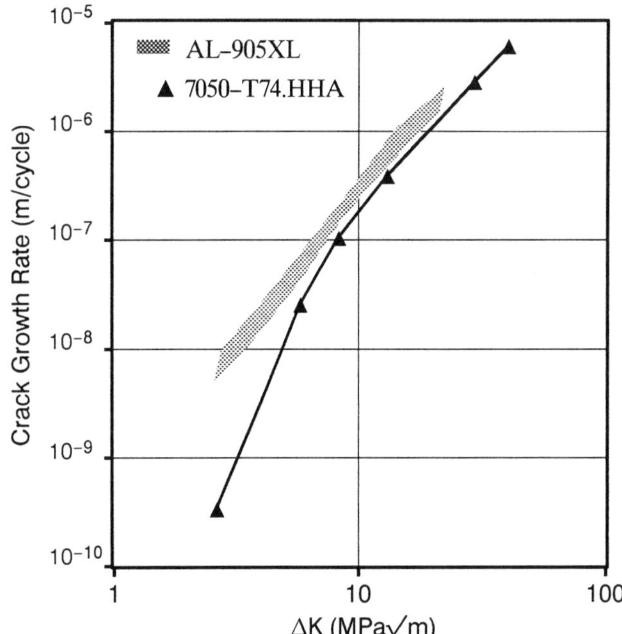

Figure 5. Fatigue crack growth behavior for alloy AL–905XL at constant amplitude, $R = +0.1$, and $f = 20$ Hz.

on crack growth rates. In addition, further tests will be conducted to determine typical fatigue crack growth trends as a function of frequency, environment, and loading conditions. Both constant amplitude and spectrum loading studies will be performed on die forged alloy AL–905XL products during the 1990 time frame.

STRESS CORROSION CRACKING BEHAVIOR. The stress corrosion cracking susceptibility of the alloy AL–905XL hand forgings was evaluated according to ASTM G44 procedures in both 3.5 percent NaCl and seacoast environments. Threshold values were estimated by orienting the specimens in the most susceptible S–T direction. Triplicate time–to–failure tensile specimens were machined from the forgings and tested at stress levels between 172 and 310 MPa. Results of the SCC tests after 20–day alternate immersion exposures are shown in Figure 6. No failures were experienced in either the 3.5 percent NaCl or the seacoast exposures. Based on these findings, the threshold values are estimated to exceed 310 MPa. The SCC behavior of alloy AL–905XL is judged to be at least equivalent to that of the incumbent AA7075 and AA7050 forged parts.

EXFOLIATION CORROSION BEHAVIOR. Resistance to surface corrosion attack typified by exfoliation corrosion is also an important material characteristic for airframe applications. Several exfoliation corrosion environments were examined for the IncoMAP alloy AL–905XL, including EXCO, MASTMAASIS, and seacoast. Although alloy AL–905XL exhibited excellent resistance to the accelerated EXCO medium, exfoliation behavior for other Li containing alloys has been the subject of some controversy (9). A comparison of the

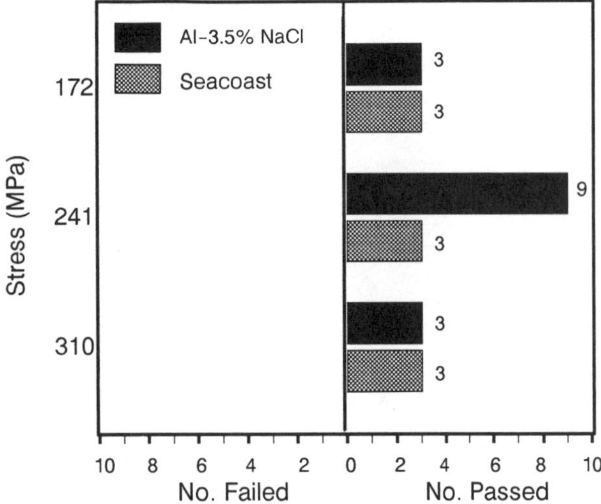

Figure 6. Stress corrosion cracking results for alloy AL–905XL –F temper for laboratory and seacoast environments.

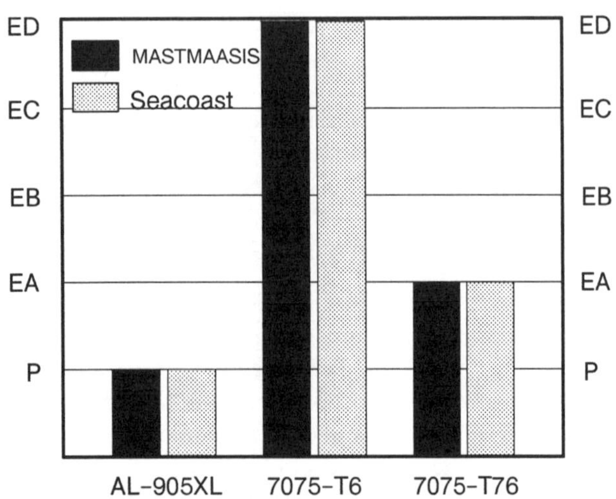

Figure 7. Exfoliation corrosion results for alloy AL–905XL hand forgings.

alloy AL–905XL results with those of two tempers of AA7075 are shown in Figure 7 for MASTMAASIS and seacoast exposures. Only shallow pitting attack was observed for alloy AL–905XL, with the assignment of a semi–qualitative P rating compared to the AA7075 products. The laboratory tests appear to provide an adequate measure of the long–term exposures experienced at the Pt. Loma, CA seacoast site.

ABVAL–G STATISTICAL ANALYSIS. The results of the mechanical property tests for tension and compression behavior were compiled in terms of specimen replicates for each grain direction. An input file representing the individual data points was created for statistical analysis using the ABVAL–G subroutine. The ABVAL–G calculations provide

an inexpensive and statistically significant means of estimating preliminary design properties from typical property data. The –G revision of the computer subroutine originally developed by Boeing (2) allows for simpler presentation and graphical handling of the output results. Preliminary B–basis design properties were calculated for the alloy AL–905XL hand forgings to facilitate an accurate assessment of the weight savings potential based on replacement of the incumbent AA7050–T74 alloy. The 3p–Wiebull statistical properties appear to be the most appropriate for these comparisons, with the individual values shown in Table I.

Table I. ABVAL–G Generated Preliminary Design Properties, B–basis, for IncoMAP alloy AL–905XL Hand Forgings.

ORIENTATION	L	LT
F_{tu} (MPa)	475	460
F_{cy} (MPa)	425	385
E_t (GPa)	80	80
DADT (MPa)	159	--
Density (Mg/m3)	2.58	

WEIGHT SAVINGS ESTIMATES. The preliminary design properties established in the previous section were used to estimate the potential weight savings for structural applications of alloy AL–905XL forgings. The weight savings methodology developed by Lockheed for comparing various advanced aluminum alloys was employed for specific failure modes (3,4) in aircraft structures. Weight saving advantages for ten primary failure modes were calculated from the basic material properties, see Table II. The weight savings analysis includes consideration of the margins of safety (MS) for various failure modes that can influence the size of the aircraft structure. Results of the weight savings benefits associated with alloy AL–905XL forgings are compared in Figure 8 for a baseline of AA7050–T74. Positive weight advantages are observed for all failure modes, with increases in excess of 10 percent for criteria incorporating the elastic modulus benefits of alloy AL–905XL. The results of the analysis are applicable to an advanced fighter and carrier–based patrol aircraft, where the total weight savings range up to approximately 10 to 15 percent.

EVALUATION OF MA AL–TI MATERIALS

BACKGROUND. The development of high temperature aluminum alloys is driven by the material needs for advanced aerospace vehicles. The high temperature strength and stability requirements for these materials greatly exceed the

Table II. Critical Failure Modes from LASC Developed Metholology to Determine Structural Efficiency of Advanced Materials

Mode	Design Failure Criteria
1	Tensile Strength
2	Compressive Strength
3	Crippling
4A	Compression Surface, Column/Crippling, N/L < 300
4B	Compression Surface, Column/Crippling, N/L 300–800
4C	Compression Surface, Column/Crippling, N/L > 800
5	Buckling, Compression/Shear
6	Aeroelastic Stiffness
7	Durability and Damage Tolerance Assessment Cut-Off (DADT)
8	General Instability, Compression/Shear
9	Minimum Gauge
10	Thermal Stress

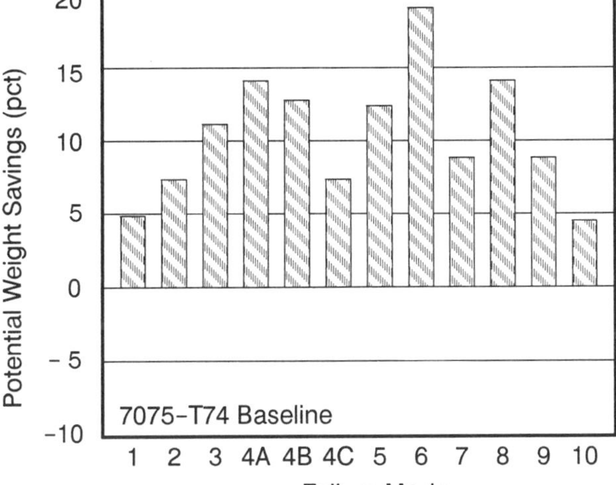

Figure 8. Calculated weight savings for alloy AL–905XL compared to AA7050–T74 baseline using Lockheed methodology.

capabilities of conventional AA2XXX and AA7XXX alloys. Aluminum alloy development has thus focussed on compositions strengthened by fine dispersions of incoherent intermetallic phases, carbides, and oxides, produced by powder metallurgical practices. Such alloys normally involve transition element additions such as iron or titanium. Rapid solidification techniques have been used to produce Al–Fe–Ce (9–13) and Al–Fe–V–Si (14,15) alloys with enhanced elevated temperature strength and stability. By contrast, Al–Ti alloys have been produced primarily by mechanical alloying (16,17).

Additions of titanium to aluminum greatly enhance high temperature capability because titanium: (a) forms a stable, high melting point intermetallic, Al_3Ti, (b) causes only minimal increase in density, and (c) diffuses very slowly in aluminum. The incorporation of titanium into aluminum alloys has not found much use in conventional aluminum alloys, other than as a grain refiner, because it greatly increases the melting temperature and expands the temperature range of the liquid + solid phase regime. This leads to the formation of large, blocky Al_3Ti particles which can be deleterious to alloy properties. Use of mechanical alloying allows large additions of titanium to be made while still controlling the Al_3Ti particle size.

Aluminum–titanium alloys containing up to about 12 percent titanium have been produced by MA and found to exhibit attractive combinations of room and elevated temperature strength, elastic modulus, and ductility (16). These materials have also been found very stable after extended exposures at elevated temperatures. While these MA Al–Ti materials are still under development, preliminary property evaluations and analysis using the Lockheed weight savings methodology have been performed to assess their potential.

MATERIALS. Evaluations of the developmental MA Al–Ti compositions, containing about 6 to 12 percent titanium, were performed on laboratory produced extruded product (16,17). These materials were made using varying extrusion conditions. The extrusion temperature was typically between 400°C and 450°C. The extrudates were rods 12.7 to 22 mm diameter or 12.7 x 51 mm flats. All material was evaluated in the as–extruded or –F temper for the property characterizations. The measured mechanical properties were compared to those of competing high temperature monolithic aluminum alloys, including data previously reported in the technical literature, e.g. Ref. 11,12,13, and 18.

SPECIFIC STIFFNESS. The measured densities of the MA Al–Ti alloys ranged from 2.80 to 2.89 Mg/m3. Modulus measurements yielded values of about 86 to 103 GPa. Thus, the MA Al–Ti alloys show a significant advantage in specific stiffness at ambient temperature compared to either the competing Al–Fe–X type aluminum materials or Ti–6Al–2Sn–4Zr–2Mo (Ti–6242). Figure 9 shows that this advantage exists to temperatures as high as 371°C to 427°C. This behavior results from two factors: the presence of significant amounts of Al₃Ti which increases stiffness, and the minimal increase in density resultant from the Ti additions. Interestingly, the specific stiffness of these Al–Ti materials compares favorably with Al–Li alloys known to have excellent stiffness combined with low density. At the higher Ti levels, the measured modulus values approach those of some Al–base metal matrix composites.

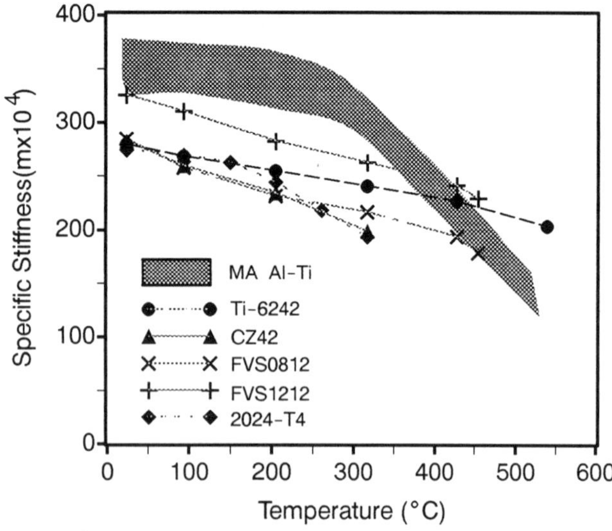

Figure 9. Specific stiffness for MA Al–Ti alloys compared to current and competitive elevated temperature alloys.

SPECIFIC STRENGTH. Room and elevated temperature tension tests were conducted on the MA Al–Ti materials and the results compared with the competitive materials on a density corrected basis. Figure 10 shows the specific strength, i.e., tensile strength–to–density ratio, for the MA Al–Ti alloys as well as the comparison alloys from ambient temperature to

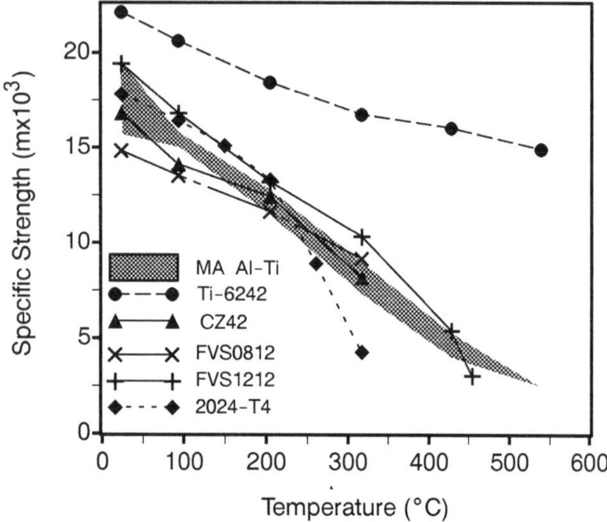

Figure 10. Specific strength for MA Al–Ti alloys compared to current and competitive elevated temperature alloys.

538°C. Although the MA Al–Ti alloys compare well with the other dispersion strengthened aluminum alloys, it is apparent that the specific strength values are significantly lower than current Ti–base alloy products. The conventional aluminum alloy, AA2024–T4, separates from the band about 250°C. Similarly, the Al–Fe base materials show indications of degradation at about 400°C. By contrast, the MA Al–Ti alloys show a linear behavior to at least 538°C, 0.85 T_{MP}. However, the specific strength differences observed at room temperature increase at higher temperatures and represent an even greater penalty for all of the Al–base alloys compared to Ti–6242 as service temperature increases.

CREEP/RUPTURE PROPERTIES. Comparison of the specific 0.2% creep properties of the MA Al–Ti alloys with current and competitive elevated temperature alloys using the Larson–Miller parameter is given in Figure 11. The data for the

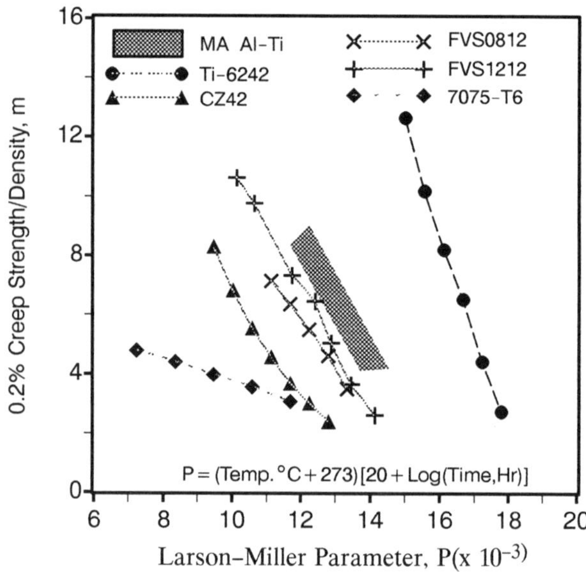

Figure 11. Specific 0.2% creep strength vs. Larson–Miller parameter for MA Al–Ti alloys compared to current and competitive elevated temperature alloys.

MA Al–Ti alloys has been combined with that of other work reported earlier (18,19,20). Clearly, these monolithic aluminum base alloys are not competitive with Ti–6242 in the temperature range evaluated. The MA Al–Ti binary alloys do compare favorably with the current and competitive aluminum materials, both conventional and PM Al–Fe base materials. Evaluations to obtain more data are under way to allow a more quantitative definition of the apparent benefit shown here for the MA Al–Ti alloys.

WEIGHT SAVINGS ESTIMATES. An initial weight savings survey, as described in the previous section on alloy AL–905XL, was conducted on the MA Al–Ti alloys. Due to the limited data on these emerging high temperature aluminum alloys, typical property values instead of preliminary design properties were employed for the weight savings calculations. The potential weight savings by failure mode is shown in Figure 12 for two cases involving ambient and elevated temperature

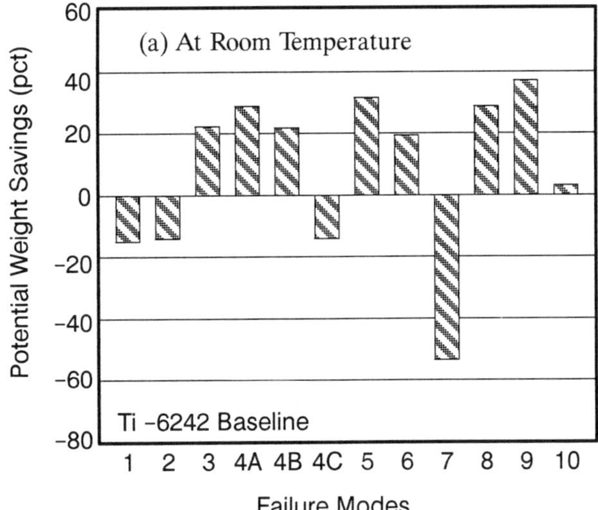

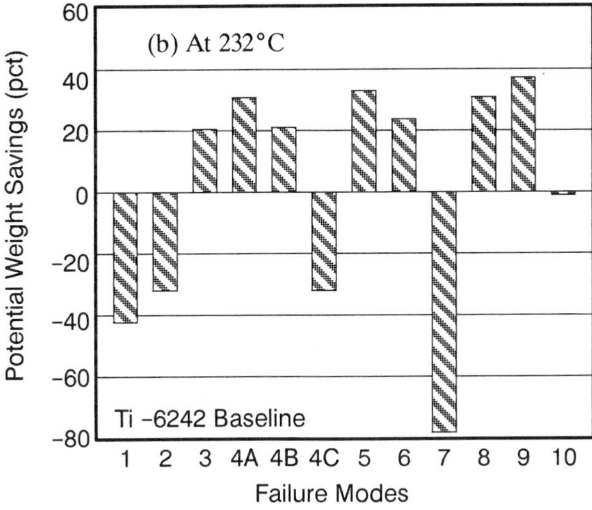

Figure 12. Calculated weight savings for MA Al-9Ti compared to Ti-6242 baseline using the LASC methodology.

design criteria. For the specific strength and specific modulus failure modes, the relative weight benefits are similar to the behavior described for alloy AL-905XL. In combined loading situations involving design criteria based on strength and stiffness (13), the MA Al-Ti alloys appear to offer weight savings advantages of 20 to 30 percent. These structural weight reductions are realized at ambient and 204°C temperatures, although deterioration might be anticipated at higher temperatures for all Al-base alloys.

SUMMARY/CONCLUSIONS

The static and cyclic properties of IncoMAP alloy AL-905XL hand forgings were obtained and compared to those of aluminum alloys currently used, namely, AA7075-T73 and AA7050-T74. Alloy AL-905XL provides improvements in density and modulus of 8% and 13%, respectively. Strength properties compare favorably with AA7075-73 and appear to be nearly isotropic. While crack initiation resistance is good, crack propagation was found slightly faster than in the current alloys. Corrosion properties were also found to be quite good. Based on these properties, positive weight savings were found for all failure modes. Advantages exceeding 10% were found when the criteria involved the contribution of elastic modulus to structural applications.

Data from limited evaluations of developmental MA Al-Ti alloy extrusions were obtained for comparison with current and competitive materials. The mechanical behavior with respect to tension properties appears to be attractive for operating temperatures exceeding 300°C. Significant weight savings potential can be realized at both ambient and elevated temperatures in design situations dominated by the specific stiffness benefits of the MA Al-Ti alloys. Additional work is required to establish the durability and damage tolerance performance of this emerging class of high temperature Al alloys.

ACKNOWLEDGEMENTS

The valuable technical contributions of the following personnel are gratefully acknowledged: A. S. Watwe and D. J. Cometto at Inco Alloys, and Glenn T. Mandigo for specimen testing and analysis, and Susan L. Lim for corrosion behavior characterization and interpretation at Lockheed.

REFERENCES

1. J.H. Weber and R.D. Schelleng, "Mechanical Alloying of Dispersion Strengthened Aluminum: A Retrospective Review", Dispersion Strengthened Aluminum Alloys, ed. Y-W. Kim and W.M. Griffith, TMS, Warrendale, PA, 1988, p. 467–482.

2. R.A. Jones and F.W. Scholz, "A- and B-Allowables for the Three Parameter Weibull Distribution", Boeing Computer Services Company, Technical Report No. 10, October 1983.

3. J.C. Ekvall, J.E. Rhodes, and G.G. Wald, "Methodology for Evaluating Weight Savings from Basic Material Properties", ASTM STP 761, Philadelphia, PA, 1982, p. 328–341

4. M.D. Goodyear, S.L. Langenbeck, and D.J. Chellman, "Aluminum–Lithium Alloy Applications

on Military Aircraft", presented at the 1988 Aluminum–Lithium Symposium (WESTEC/ASM International), March 1988.

5. S.J. Donachie and P.S. Gilman, "Microstructure and Properties of Al–Mg–Li Alloys Prepared by Mechanical Alloying", Aluminum–Lithium Alloys II, ed. T.H. Sanders, Jr. and E.A. Starke, Jr., TMS/AIME, Warrendale, PA, 1984, p. 507.

6. R.D. Schelleng, A.I. Kemppinen, and J.H. Weber, "Low Density Al–Mg–Li Alloy for Aerospace Forgings" Space Age Metals Technology, vol. 2, ed. F.H. Froes and R.A. Cull, SAMPE, Covina, CA, 1988, p. 177.

7. R.D. Schelleng, A.I. Kemppinen, and J.H. Weber, "IncoMAP alloy AL–905XL for Aerospace Forgings – Production Status", Aluminum–Lithium Alloys V, vol. III, ed. T.H. Sanders and E.A. Starke, Materials and Component Engineering Publications Ltd., Birmingham, UK, 1989, p. 1577.

8. L.M. Mueller, Alcoa Aluminum Alloy 7050, Green Letter No. 220, Aluminum Company of America (ALCOA), Third Revised Edition, October 1985.

9. E.L. Colvin, "Exfoliation and Stress Corrosion Cracking Performance of Al–Li Alloys", Aluminum Lithium Alloys – Design, Development, and Applications Update, ed. R.J. Kar, S.P. Agrawal, and W.E. Quist, ASM International, Metals Park, OH, 1988, p. 273–290.

10. Y.–W. Kim, "Advanced Aluminum Alloys for High Temperature Structural Applications", Industrial Heating, May 1988, p. 31. Kim, 1988

11. J.C. Ekvall, R.A. Rainen, D.J. Chellman, R.R. Flores, and M.J. Gersbach, " Elevated Temperature Aluminum Alloys for Advanced Fighter Airframe Structures", AIAA/ASME/ ASCE/AHS/ASC 30th Structures, Structural Dynamics, and Materials Conference, AIAA Paper No. 89–1407, Mobile Alabama, April 1989.

12. D.J. Chellman, J.C. Ekvall, and R.R. Flores, "Elevated Temperature Al Alloys for Advanced Aircraft Applications", P/M in Aerospace and Defense Technologies", comp. F. H. Froes, MPIF, Princeton, NJ, 1990.

13. D.J. Chellman, J.C. Ekvall, and R.A. Rainen, "Elevated Temperature PM Aluminum Alloys for Aircraft Structure", Metal Powder Report, vol. 43, No. 10, October 1988.

14. P.S. Gilman, S.K. Das, D. Raybould, M.S. Zedelis, and J.M. Peltier, "Applications of High Temperature/High Stiffness Aluminum Alloys", Space Age Metals Technology, vol. 2, ed. F.H. Froes and R.A. Cull, SAMPE, Azusa, CA, 1988, p. 189.

15. P.S. Gilman and S.K. Das, "Rapidly Solidified Aluminum Alloys for High Temperature/High Stiffness Applications", Metal Powder Report, v. 44, September 1989, p. 616.

16. P.K. Mirchandani, R.C. Benn, and K.A. Heck, "Structure–Property Relationships in Elevated Temperature High Stiffness Mechanically Alloyed Al–Ti Based Alloys", Light–Weight Alloys for Aerospace Applications, ed. E.W. Lee, E.H. Chia, and N.J. Kim, TMS, Warrendale, PA, 1989, p.

17. P.K. Mirchandani, D.O. Gothard, and A.I. Kemppinen, "Elevated Temperature Al–Ti Alloy by Mechanical Alloying", 1989 Advances in Powder Metallurgy, vol. 3, comp. T.G. Gasbarre and W.F. Jandeska, MPIF, Princeton, NJ, 1989, p. 161.

18. J.A. Hawk, J.K. Briggs, and H.G.F. Wilsdorf, "Creep in Mechanically Alloyed Dispersion Strengthened Alloys", ibid., p. 285.

19. S.L. Langenback, W. M. Griffith, G.J. Hildeman, and J.W. Simon, "Development of Dispersion Strengthened Aluminum Alloys", Rapidly Solidified P/M Aluminum Alloys, ASTM STP 890, ed. M. E. Fine and E. A. Starke, Jr., 1986, p. 410–422.

20. D. J. Chellman, J. C. Ekvall, and R. A. Rainen, "Elevated Temperature Aluminum Alloy Development", NADC Symposium on Advanced Aluminum Alloys for Naval Aircraft, Valley Forge, PA, October 1988.

MICROSTRUCTURE AND PROPERTIES OF MECHANICALLY ALLOYED Al-Mn

Pierre Le Brun, Ludo Froyen, Luc Delaey
K.U. Leuven
Department MTM
Heverlee, Belgium

ABSTRACT

An Al-8wt%Mn alloy has been mechanically alloyed in a planetary ball mill starting from elemental powders. The mechanically alloyed powders consist of manganese particles finely dispersed in the aluminium matrix. During consolidation by hot extrusion, formation of aluminides occurs, the equilibrium phase being Al$_6$Mn. The extruded material exhibits a high strength up to 300°C at least, with an acceptable level of ductility. The E-modulus of the alloy is increased in comparison with Al, but remains below the predictions. This is explained by porosity formation in the intermetallic particles.

MANGANESE IN ALUMINIUM ALLOYS – Manganese is a common alloying element for wrought aluminium alloys: the serie 3xxx benefits of the increase in strength, hardness and modulus caused by the manganese atoms in solution [1]. The drawbacks are an increase of the density and a reduction of the ductility. The latter is mainly affected by the intermetallic phase formed (Al$_6$Mn). Manganese is also an efficient strengthening element at elevated temperatures. Finally, the good corrosion resistance is due to the almost equal electrochemical potential of Al and Al$_6$Mn. Nevertheless, manganese contents higher than 2% (all weight percentage unless mentioned) in castings are deleterious as large plate-like primary intermetallics will form upon cooling (cf phase diagram : figure 1)[2].

A renewed interest for Al-Mn alloys arises from the scientific community since the discovery of icosahedral phases in those alloys [3].

POWDER METALLURGICAL AL-MN ALLOYS – The properties of Al-Mn alloys are very sensitive to the nature of the manganese atoms present either in solid solution or as intermetallics. Mn atoms can easily be retained in supersaturation and fast cooling techniques (mainly PM techniques) have thus been applied to Al-Mn alloys in order to increase their properties [4,11].

The tensile properties of several ternary Al-Mn-X alloys are presented in table 1 and 2. Ternary alloying elements like Mg,Fe,Si have been tested. A common drawback for the PM Al-Mn alloys is their low ductility. Magnesium in solid solution is the most powerfull strengthening element at room temperature (RT), but loses its effectiveness at temperatures higher than 150°C due to precipitation of intermetallic phases. The highest elevated temperature properties are obtained with a quaternary alloy containing titanium as fine dispersoids forming element (Al$_3$Ti).

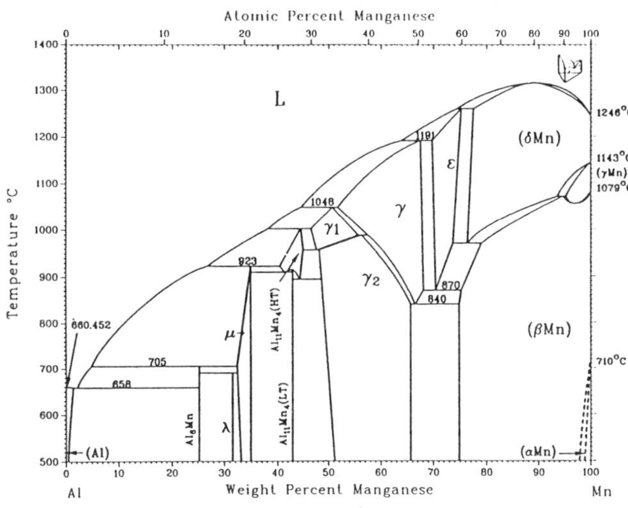

Fig.1 – Phase diagram of Al-Mn [2]

Table 1 : Tensile properties of Al–Mn–X alloys at room temperature

Alloy	Preparation Technique	$\sigma_{0.2}$ (MPa)	σ_{max} (MPa)	A (%)	E (GPa)	Ref
Al–10Mn	Gas Atomisation + Splat Cooling	308		5	100	[4]
Al–9.25Mn	Inert Gas Atomisation	300	370			[5]
Al–7.5Mn–5.3Fe	Air Atomisation	405	460	2.5		[6]
Al–9.25Mn–1.4Mg	Inert Gas Atomisation	300	420			[5]
Al–7.00Mn–6.13Mg	Air Atomisation	550	550	0.2	83.5	[7]
Al–2.00Mn–4.5Mg	High Pressure Casting	160	310	13		[8]
Al–4Mn–4Mg–2.5Zr	Inert Gas Atomisation	470	550	8		[9]
Al–9.8Mn–2.4Si	Air Atomisation	240	320	3.5	90	[10]
Al–10Mn–1.6Si–2.0Ti	Inert Gas Atomisation (Ar)	420		10		[11]
Al–10Mn–1.6Si–2.0Ti	Inert Gas Atomisation (He)	460		10		[11]

Table 2: Elevated temperature properties of Al–Mn–X alloys ($\sigma_{0.2}$ in MPa)

Alloy	RT	150°C	200°C	250°C	300°C	Ref
Al–10Mn	300		220		140	[12]
Al–7.5Mn–5.3Fe	405	325		195	150	[6]
Al–7Mn–6.13Mg	550	360		110		[7]
Al–9.8Mn–2.4Si	204		175		100	[10]
Al–10Mn–1.6Si–2.0Ti	420			232		[11]

MECHANICAL ALLOYING – Mechanical Alloying (MA) is a technique allowing to disperse initial powder constituents on a fine scale by repeated welding and fracturing of powders [13]. The MA process results in materials with submicronic grain sizes and an homogeneous and fine dispersion of the constituents. In the case of aluminium alloys, inert dispersoids are formed by the reaction of aluminium with the oxygen from the atmosphere and the carbon and oxygen present in the organic additives. Their size ranges from 10 to 100 nm.

The aim of the present work is to produce MA Al–Mn powders with a fine dispersion of elemental manganese within the aluminium powders. During further processing, these Al–Mn composite powders are heated which can lead to a reaction between dispersed manganese particles and the aluminium matrix. It is expected that a uniform distribution of fine aluminides will be formed, due to the short diffusion distances and a highly defected structure.

EXPERIMENTAL PROCEDURE

Elemental powders of aluminium and manganese have been mechanically alloyed in a planetary ball mill (Pulverisette 5 from Fritsch). The weight composition is Al–8Mn. Organic additives needed in order to limit the welding of the powders have been added. The size of the powders has been determined by use of a laser diffractometer (Coulter LS–100), using an He–Ne source. The kinetics of the reaction (formation of aluminide) inside the MA powders has been analysed by differential thermal analysis (DTA). The peak temperature (T_m) has been determined in function of the heating rate (Φ), for rates up to 50 K/min. The consolidation of the powders is done following a traditional PM route, involving cold compaction, degassing, preheating and extrusion. The mechanical properties have been determined at RT, 200 and 300°C on an Instron tensile testing machine. The long term stability of the extruded material is assessed by means of microhardness Vickers indentation under a 0.1 N force. At the different stages of the process, the microstructure is evaluated by electron microscopy (scanning (SEM) and transmission (TEM) on selected samples), and the phases have been analysed by X-Ray diffraction (XRD).

EXPERIMENTAL RESULTS

The morphology of the powders is strongly influenced by the MA process. The elemental aluminium powders are elongated, the manganese powders are angular and the MA powders are composed of several layers welded together (figure 2 a,b,c respectively). The mean size of the Al, Mn and MA powders is 125, 13 and 70 μm respectively.

The microstructure of the powders consists of fine Mn particles (<1 μm) embedded in the Al matrix (figure 3). The kinetics of the reaction between Al and Mn is determined by DTA. In table 3, the values of the peak temperatures (T_m) are listed in function of the heating rate (Φ).

Table 3 : DTA measurements on Mechanically Alloyed Al-8Mn

Heating Rate Φ (K/min)	Peak temperature T_m (K)
5	725
10	751
25	762
50	785

The microstructure of the extruded material consists of manganese containing intermetallics and very small dispersoids probably Al_2O_3 and Al_4C_3 uniformally distributed in the Al matrix. Although most of them are submicronic (figure 4 a), a few are up to 5 μm in diameter (figure 4 b). The aluminium grain size ranges between 0.2 and 0.5 μm with fine Al_2O_3 and Al_4C_3 dispersoids (figure 5 a). In some areas with less inert dispersions, larger grains (1 μm) are found (figure 5 b).

The evolution of the phases has been analysed by XRD (figure 6). After MA, pure aluminium and manganese are detected (bottom curve) but no formation of intermetallics can be detected. After extrusion, the formation of Al_6Mn is evident (top curve). The peaks of Al_6Mn are indicated by dotted lines, the first peak of Mn is indicated by a continuous line. The two most intense peaks are the (111) and (200) peaks of Al.

The yield strength of the MA(Al-8Mn) extruded at 400°C reaches 408 MPa at RT; the ultimate tensile strength (UTS) is 465 MPa. The ductility is about 6% and the Young's module is 74 GPa. The evolution of the mechanical properties at elevated temperature are given in Figure 7. At 200°C, the UTS is still 320 MPa and 242 MPa at 300°C.

In order to assess the long term stability, the extruded bar has been annealed for 100 h at temperatures up to 500°C (figure 8). The microhardness remains stable up to 400°C at a value 1.30 GPa. At 500°C, the microhardness falls down to 1.13 GPa.

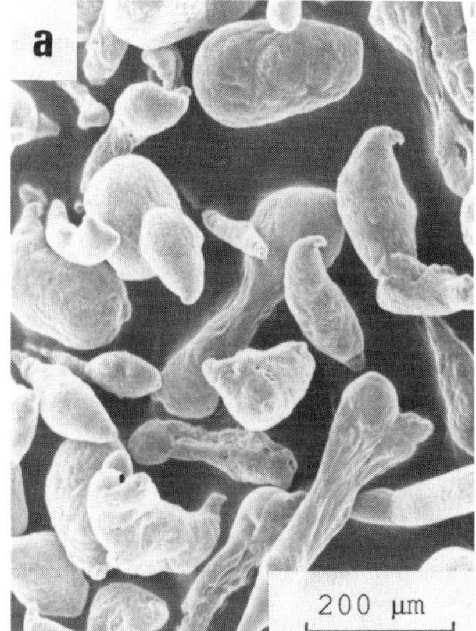

Fig.2 – Morphology of the elemental powders (a : Al; b : Mn, c : MA)

Fig.3 – Microstructure of MA powders

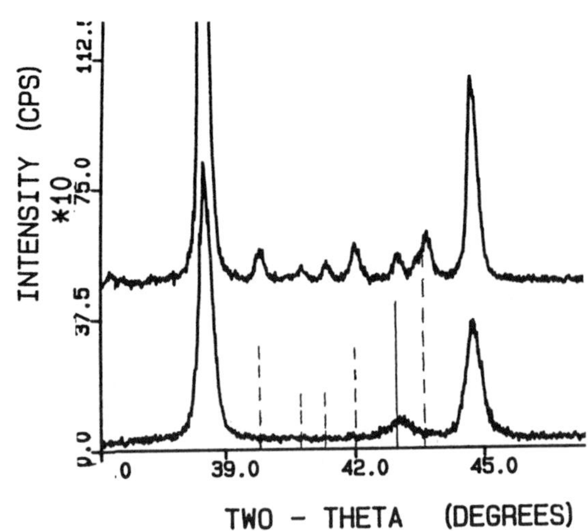

Fig.6 – XRD of the MA powders (bottom) and of the extruded (top)

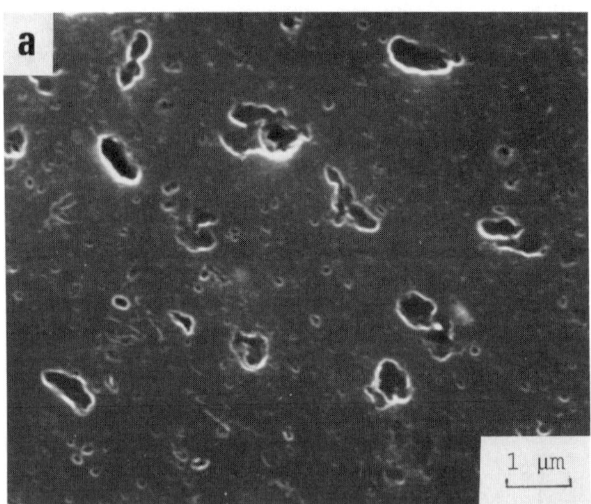

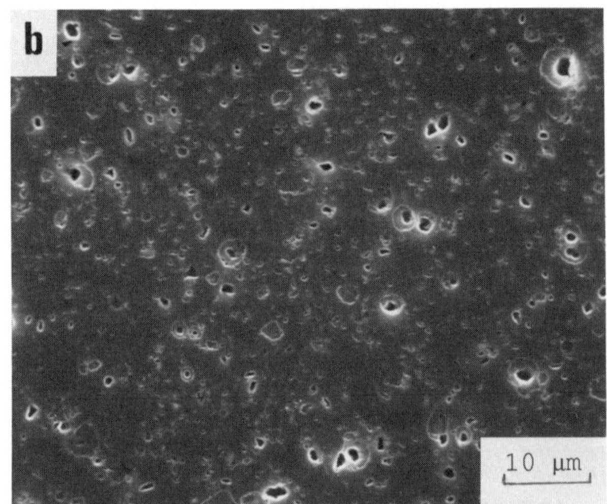

Fig.4 – SEM of the extruded alloy

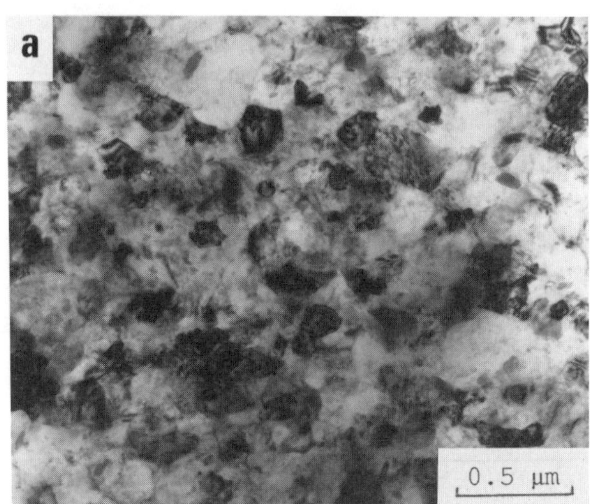

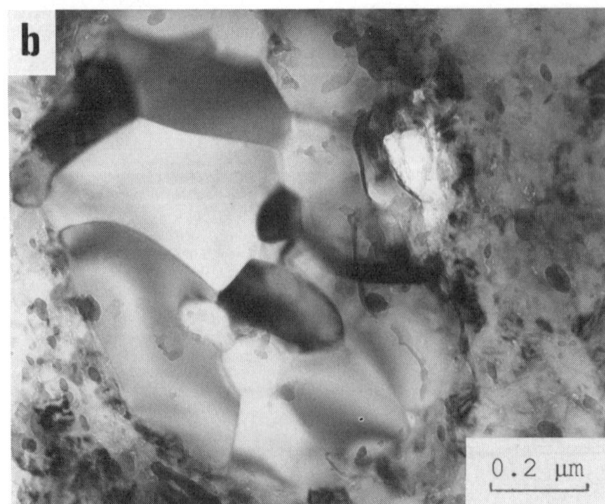

Fig.5 – TEM of the extruded alloy

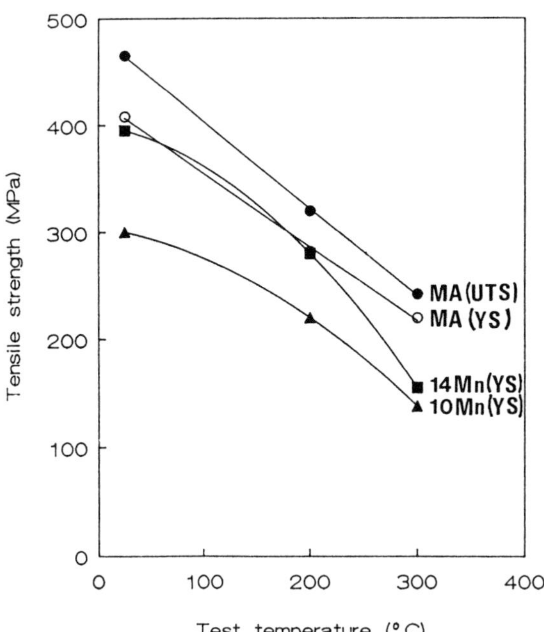

Fig.7 –   Tensile strength of MA and rapidly
           solidified Al–Mn alloys at elevated
           temperature

DISCUSSION

MICROSTRUCTURE AND PHASE ASSESSMENT –
After mechanical alloying, the second phase is
uniformly distributed inside the Al matrix and
XRD analysis does not provide the evidence of
the formation of intermetallics at this stage
of the process. However, manganese particles
are very finely distributed throughout the Al
matrix. Indeed, manganese is detected in the
smallest volume that can be analysed by
microprobe (about 1 μm^3). The weight content is
lower than expected on base of the composition
of the alloy. This is due to some large
particles of pure manganese subsisting in the
alloy.

The total content of the oxides and
carbides is about 5 wt %, based on a full
reaction of C with Al. Some larger
recrystallised grains are found, probably due
to a local low content of dispersions or their
insufficient homogeneous distribution. During
degassing and extrusion, Al and Mn are in close
contact and interdiffusion occurs. A reaction
layer is formed between the pure metals. In
order to ensure full reaction, both Al and Mn
have to diffuse through the layer. Several
intermetallics based on Al and Mn can be
formed. The formation of Al_6Mn is unambiguously
detected. With a longer annealing time, the
amount of Al_6Mn increases. The study of the
formation of intermediate intermetallics is in
progress, possibly Al_4Mn could be formed before
Al_6Mn.

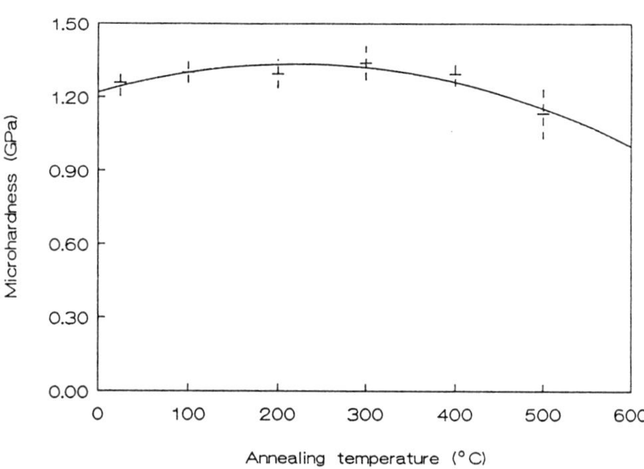

Fig.8 –   Microhardness of the MA alloy after
           annealing for 100 h at temperature

DTA ANALYSIS – Amongst the transition metals,
Mn is known to be a fast diffuser in Al [14].
The atomic volume of Al is larger than that of
Mn (0.0166 nm^3 for Al against 0.0128 nm^3 for
Mn). The diffusion constants of Al in Mn are
not available but it is anticipated that Mn
could diffuse faster in Al than Al in Mn due to
the size effect. This would account for the
porosity formation observed inside the
intermetallic particles (fig. 9).

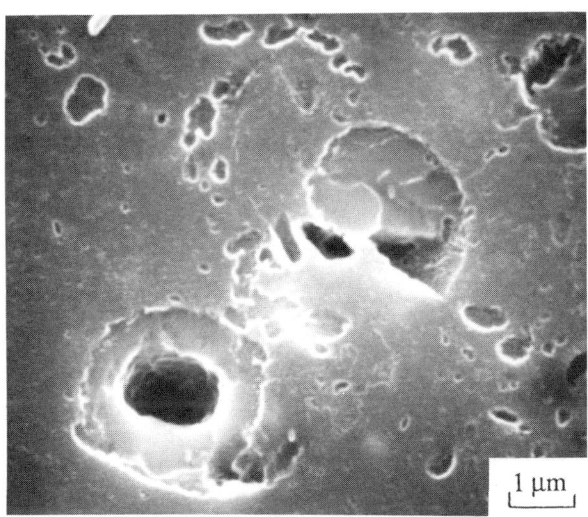

Fig.9 –   Porosity formation inside the
           intermetallic particles

The rate of the reactions can be monitored by DTA. The rate of a first order reaction can be expressed by [15]:

$$dx/dt = A(1-x)e^{-E/RT} \qquad (1)$$

The temperature T_m at which the rate is maximum can be derived from eq. (1) [15]. The variation of T_m as a function of the heating rate is given by :

$$d(\ln(\Phi/T_m^2))/d(1/T_m) = -E/R \qquad (2)$$

By plotting the values of $\ln(\Phi/T_m^2)$ as a function of $1/T_m$, a first order reaction will be represented by a straight line, from which the activation energy can be calculated. Figure 10 gives such a plot for the MA alloy. The correlation coefficient for a straight line is 0.98, the activation energy is 176 kJ/mole. This value needs to be considered carefully. During the heating, interdiffusion occurs and a reaction layer is formed between Al and Mn. This could result in a variation of the order of the reaction.

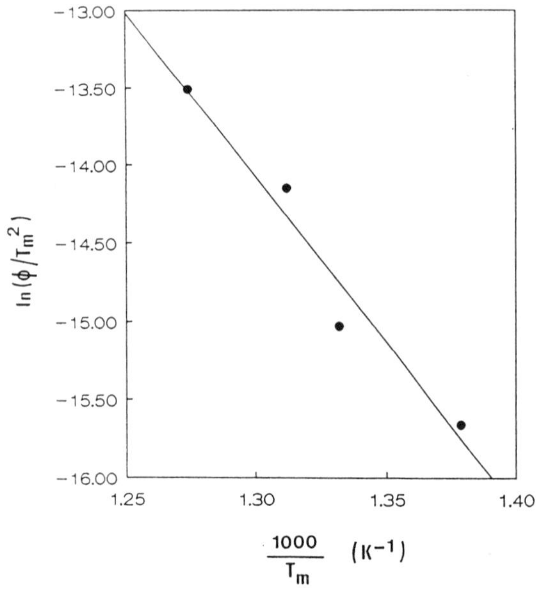

Fig.10 – Relation between peak temperature (T_m) and heating rate (Φ)

MECHANICAL PROPERTIES – As mentioned in the introduction, Mn influences strongly the mechanical properties of Al alloys. The MA alloy with 8%Mn reaches the same level of strength at RT than a splat cooled alloy containing 14%Mn [12]. This is believed to be mainly due to the fine grain size of Al and the homogeneous distribution of the intermetallics inside the matrix. Al_6Mn can be considered as totally undeformable and will therefore strengthen considerably. In the splat cooled

alloy mentioned in figure 7, Mn retained in supersaturation is responsible for an increased strength. As the temperature is increased, precipitation occurs and the strength is decreased. In contrast, the MA alloy still has a high level of strength at elevated temperatures.

The influence of the intermetallic phase is expected to be much more pronounced on the E-modulus. A computation of the theoretical modulus has been done, based on the theory of Hashin and Shtrikman for composite materials. The values used are summarised in table 4 (σ stands for the Poisson's coefficient). The values for the Al_6Mn are taken from [16]. In figure 11, the dotted lines indicate the theoretical lower and upper bounds (LB and UB). The continuous line represents the expected module based on a linear increase of the module by 2.34 GPa for each wt%Mn [17] and corresponds fairly well to the upper bound of the composite. The measured modulus of the MA alloy is lower than the lower bound of the theory which can be attributed to the porosity formed in the alloy during the reaction between Al and Mn.

Table 4 : Prediction of E-modulus

Composite	E_{Al}	:	62 GPa	LB: 78.5 GPa
	σ_{Al}	:	0.33	HB: 81.2 GPa
	E_{Al_6Mn}	:	150 GPa	
	σ_{Al_6Mn}	:	0.25	
Linear	dE/dMn	:	2.34 GPa	E: 80.7 GPa

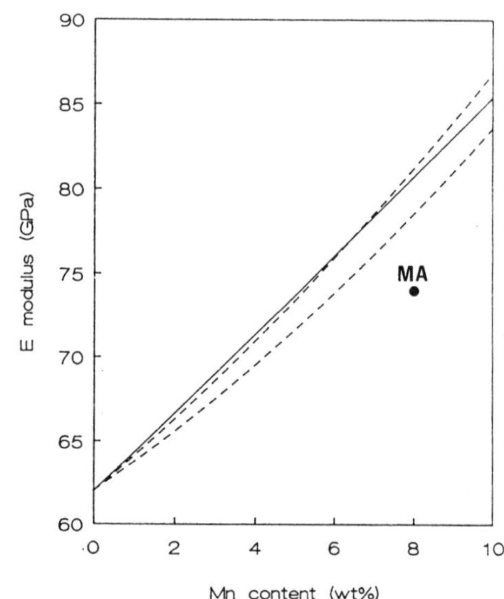

Fig.11 – Calculated and experimental E-modulus of Al–Mn alloys

The properties after a long term annealing are evaluated based on microhardness measurement. The value is slightly increasing up to 300°C, probably due to the transformation of intermetallics. At 400°C, the hardness starts to decrease due to coarsening of the structure.

In a previous paper [18], the preparation of Al–5Fe–4Mn by MA is reported. The tensile strength of MA(Al–8Mn) and MA(Al–5Fe–4Mn) are about the same in the temperature range considered, but MA(Al–8Mn) has a higher ductility (6% against 3%). Al–8Mn has been degassed, the Fe containing alloy not. As the two alloys have comparable alloying contents, it is anticipated that the degassing treatment can explain the difference in ductility.

CONCLUSIONS

1. An Al–8Mn alloy has been successfully produced by MA in a planetary ball mill.
2. The formation of the intermetallics has been investigated by differential thermal analysis and X-Ray diffraction. The calculated activation energy fot the formation of the equilibrium phase Al_6Mn is 176 kJ/mole.
3. The extruded material has a submicronic grain size with a complex distribution of intermetallics. Some of them are up to 5 μm in size.
4. The mechanical properties are improved by comparison to alloys with almost the same Mn content, especially at elevated temperature (up to 300°C at least). The ductility is acceptable (6%) and the E–modulus is increased to 74 GPa. This remains below the predictions for composite materials, probably due to porosity formation.
5. Further work is needed for a better understanding of the evolution of the intermetallics during thermal treatment of the MA powders.

ACKNOWLEDGEMENTS

The authors are gratefull to J. Eyssen and Niu Xiaoping for their help in the practical work. This work has been partially done under the EURAM contract MAIE–0013–C and under the IUAP–4 programme.

REFERENCES

[1] Mondolfo L.F., "Manganese in Aluminium Alloys", The Manganese Centre.

[2] Massalski T.B.,"Binary Phase diagrams", ed. Murray J.L., Bennett L.H., Baker H., ASM , vol.1, 131–133 (1986).

[3] Shehtman D., Bleh I., Gratias D., Cahn J.W., Phys. Rev. Lett., 53, 1951 (1984).

[4] Terlinde I., Peters M., Lütjering G., Williams J.C., Z. Metallkde, 78, H8, 607–612 (1987).

[5] Ahrens T., Gysler A., Lütjering G., "Horiz. Powder Metall.", vol 2, ed. W.A. Kaisser, W.J. Huppman, W.J. Schmid, 973–976 (1986).

[6] Zaidi M.A., Robinson J.S., Sheppard T., Mat. Sci. Tehn., 1, 737–742 (1985).

[7] Marshall G.J., Ioannidis E.K., Sheppard T., Powder Met., 29, 57–64 (1986).

[8] Nishi N., Kami S., Takahashi Y., Komoto H., Conley J.G., "Dispersion Strengthened Aluminium Alloys", edt. Y.–W. Kim and W.M. Griffith, 451–464, TMS (1988).

[9] Helm D., Gysler A., Lütjering G., Becker J., Fisher G., "Light–Weight Alloys for Aerospace Applications", edt. E.W. Lee, E.H. Chia, N.J. Kim, 99–109, TMS (1989).

[10] Ahrens T., Gysler A., Lütjering G., Z. Metallkde, 76, H.6, 391–396 (1985).

[11] Saal B., Albreht J., Lütjering G., Beker J., Fisher G., "Light–Weight Alloys for Aerospace Applications", edt. E.W. Lee, E.H. Chia, N.J. Kim, 3–13, TMS (1989).

[12] In–Gyu Park, Gysler A., Lütjering G., Z. Metallkde 80, H.5, 345–351 (1989).

[13] Gilman P.S., Benjamin J.S., The Annual Reviews, 279–300 (1983).

[14] Ping Liu, "Development of Microstructure in Cast and Rapidly Solidified Aluminium Alloys", Ph.D. Thesis, Chalmers University of Technology, Goteborg, 31 (1987).

[15] Kissinger H.E., J. Nat. Bureau of Standards, 57, 217–221 (1956).

[16] Van Aken D.C., Miller D.J., Kurath P., Fraser H.L., "High Strength Powder Metallurgy Aluminum Alloys 2", TMS, 387–402 (1986).

[17] Dudzinski N., J. Inst. Metals, 81, 1414, 49 (1952–1953).

[18] Le Brun P., Niu Xiaoping, Froyen L., Munar B., Delaey L., "Solid State Powder Processing", edt. Clauer A.H. and deBarbadillo J.J, TMS, 273–283 (1990).

MECHANICAL ALLOYING OF ELEVATED
ALUMINUM ALLOYS

Jörg Kumpfert, Günter Staniek
DLR, German Aerospace Research
Köln, West Germany

Wolfgang Kleinekathöfer
Daimler-Benz AG
Stuttgart, West Germany

Manfred Thumann
ABB Corporate Research
Baden, Switzerland

INTRODUCTION

This contribution reports results from a cooperative effort of German industrial and government research laboratories and the universities of Hamburg-Harburg and Vienna with the objective of developing elevated temperature (ET) aluminum alloys by mechanical alloying (MA), including reaction milling (RM). Alloy selection was guided by specifications obtained from service conditions of the parts proposed for application. Table 1 shows material property requirements for jet engine compressors, automotive gas turbines and combustion engine components manufactured from ET-Al alloys. Figure 1 gives an overview of temperature ranges and alloys under investigation compared with the conventional alloy 2618 and the aims of alloy development. Three groups of alloy bases, represented by three different engine components, were investigated with respect to fabrication by MA:

AlFeCe base for compressor wheels and other components in jet engines

AlSi base for compressor wheels in automotive gas turbines

AlSi base for connecting rods in combustion engines

---

*   DLR, German Aerospace Research, Köln, West-Germany

**   now with ABB Corporate Research, Baden, Switzerland

*** Daimler-Benz AG, Stuttgart, West-Germany

EXPERIMENTAL PROCEDURE

Gas atomized alloy powder was produced by Eckart-Werke (Fürth, West-Germany) and Metalloys Limited (Birmingham, Great Britain). Milling was performed in 10 kg batches at Eckart-Werke. Powder processing, consolidation and property evaluation was carried out at DLR. (Köln, West-Germany, including cold isostatic pressing (CIP), canning, degassing and extrusion in a 250 ton direct extrusion press. Alloys were extruded from 55 mm diameter billets to 12 mm diameter rods, corresponding to a reduction in area by a factor of 21. Tensile testing was performed on round bar specimens of 63 mm length.

MILLING OF RAPIDLY SOLIDIED ALLOY POWDER

Elevated temperature properties of aluminum alloys have been considerably increased by means of rapid solidification. Maximum service temperatures of conventional ET-Al alloys such as 2618, 2219 or 2024 are in the range of 180 to 200°C. However AlFe-based alloys are being produced by melt spinning or gas atomization that are thermally stable up to approximately 350°C. Most AlFe-based alloys contain 8 to 12wt.-% Fe that is responsible for a high content of intermetallic phases in fine dispersion, if solidified rapidly. Additions of alloying elements such as V, Mo, or Ce appear to increase the elevated temperature strength by stabilizing the AlFe-phases [1].

Further increases in service temperature up to 450-500°C was achieved by mechanical alloying or reaction milling of aluminum alloys. Milling of aluminum powder with graphite and a subsequent heat treatment leads to a fine distribution of Al_3C_4 and Al_2O_3 dispersoids which are stable up to the melting temperature of aluminum [2]. Unfortunately, these alloys have low room temperature strength compared to conventional ingot or PM alloys as shown for Dispal^{TM} in figure 2.

The additive combination of rapidly solidified Al8Fe4Ce powder with reaction milling was an attempt to improve room temperature properties and to simultaneously extend the temperature range for thermal stability. Hardening by rapidly solidified intermetallic phases was intended to be superimposed with dispersion hardening through carbides and oxides produced by mechanical alloying.

For this purpose, Al8Fe4Ce powder, atomized in a mixture of helium and argon by Metalloys Ltd. and sieved less than 45μm was milled in an attritor with 0.5 wt.-% stearic acid for 90 minutes in air. As atomized and milled powder was processed according to the specifications given before and tensile properties were measured from room temperature up to 450°C. In figure 2, the MA Al8Fe4Ce is compared with the as atomized powder extrusions, RM Al2C1O (Dispal^{TM}) and Al2Cu1.5MgNiFe (IM 2618). Extrusion temperature of Al8Fe4Ce alloys was 400°C.

The Al8Fe4Ce yield strength is superior to Dispal and IM 2618 in the whole temperature rang. Milling further increases elevated temperature strength compared to the as atomized alloy and even extends the temperature range of application. For a more detailed view of coarsening, hardness was measured at 450°C as a function of time (fig.3). Extrusions produced from as-atomized powder show a decrease in hardness of about 50 Vickers hardness after 250 hours, whereas with milled powder, the

hardness is generally higher and the total decrease is only 35 Vickers hardness.

The increase in thermal stability and strength at elevated temperatures and the reported higher creep resistance [3] are attributed to two effects taking place during mechanical alloying or milling. Milling introduces a high number of defects into the alloy in addition to oxides and carbides generated during the process. These dispersoids and defects are expected to harden the matrix and to stabilize the structures. Transmission electron microscopy (TEM) showed that no considerable annealing or coarsening occurs up to elevated temperatures (Fig. 4). A detailed description of the oxide and carbide particles on the creep behavior is given elsewhere [4, 5].

Oxides and carbides can act as obstacles as described above, but they may also affect the process of milling by their presence as a hard phase. The process of particle deformation may be affected and the defect structure introduced by mechanical milling should also depend on the presence of dispersoids.

As a measure of the volume percent of oxides and carbides in the alloy, the oxygen and carbon content of extrusions were analyzed and are shown in figure 5 for varying milling conditions. Powders were milled in air and argon with different contents of stearic acid. As expecte milling in air produces the highest oxygen content. Under argon, oxygen content is generally lower, but oxygen from the stearic acid also contributes to the total content and increases with the amount of stearic acid. Carbon is lowest with milling under argon and small additions of stearic acid. With increasing amounts of stearic acid and graphite as milling agents, the carbon content of the alloy likewise increases.

The ultimate tensile strength of extrusions as a function of oxygen and carbon content is shown in figure 6. It was assumed that all oxygen and carbon detected during the measurement originated from oxides and carbides, respectively. The total volume content of oxides and carbides was calculated from their densities and the analyzed percentage of oxygen and carbon. No assumption was made on dispersoid sizes and distributions. At all test temperatures, the ultimate tensile strength (UTS) increases with dispersoid content up to approximately 7 vol.-%. At room temperature, a maximum in UTS occurs just below 8 vol.-%. At higher vol.-% the UTS decreases.

A maximum may also occur at higher temperatures and at a higher dispersoid content. The fracture elongation of the MA (milled) RS alloy AL8Fe4Ce is low at a high volume content of dispersoids, where the strength is high. This behavior is nearly independent of test temperature. At lower dispersoid contents, the elongation is reasonable but the increase in strength by MA is still small. This is demonstrated in figure 7 where the fracture elongation of the corresponding UTS data from figure 6 are plotted versus the dispersoid content.

Low ductility is a general problem with this family of dispersion strengthened alloys, as already known from the SAP-powder. Although ductility becomes a problem with higher dispersoid contents, there are means to improve ductility by minimizing other causes of low ductility. Especially in mechanical alloying, impurities from mill materials must be avoided.

One of the most important effects on ductility in mechanical alloying is presented by the role of hydrogen during milling. As-atomized and mechanically alloyed Al8Fe4Ce powder was degassed and the evolving gases were analyzed in a quadrupole mass spectrometer. The hydrogen signals are plotted versus temperature in figure 8 at two different scales in order to show the orders of magnitude of gas volumes and details in the dependence on temperature. Mechanical alloying with stearic acid as a milling agent increases the hydrogen content of the powder by two orders of magnitude compared to the as atomized powder. Graphite leads to hydrogen content one order of magnitude higher. Mass spectrometry gives values of gas composition with good reproducibility.

For determination of gas content in the powder alloy and in the extrusions, melt extraction analyses were performed (fig. 9). Qualitatively, melt extraction analysis confirms the results obtained by mass spectroscopy. Mechanical alloying does increase the hydrogen content drastically. Degassing after mechanical alloying reduces the hydrogen content to some extent. However, most surfaces containing hydrogen are incorporated into the interior of the mechanically alloyed particles and cannot be reached by a degassing treatment.

The main sources for hydrogen introduced by milling are particle surfaces and milling atmosphere. Depending on the atmosphere (humidity) during atomization and storage of powder, surfaces may contain a high volume content of hydrogen in the outer layers of the surface in the form of hydroxides. From these hydroxides, hydrogen is released during heating probably because of continuing oxidation of the aluminum alloy surfaces that uses up the oxygen from the hydroxide skin. This process occurs repeatedly during mechanical alloying whenever new aluminum surfaces are formed that will immediately oxidize. Hydrogen may also be released from milling agents, specifically from stearic acid and is then incorporated into the particles during mechanical alloying.

Before mechanical alloying, particle surfaces can be treated by degassing in vacuum in order to reduce the hydroxide skin and, thereby, minimize a potential source for hydrogen. Hydrogen introduced during mechanical alloying can only be removed by degassing from surfaces that can be reached by a degassing treatment. Therefore the better way of reducing the hydrogen content is the selection of the proper milling agent.

From our results, graphite leads to a lower hydrogen content than stearic acid. However, all benefits from the milling agents in respect to the properties of the final product have to be considered for the choice of the milling agent.

MECHANICAL ALLOYING OF ALLOY POWDER WITH Si

A second class of elevated temperature alloys based on the AlSi system find potential applications in the automotive industry. There are specifications for pistons and connecting rods in combustion engines or compressor wheels in an automotive gas turbine that cannot be fulfilled by conventional AlSi-based alloys. A combination of sufficient elevated temperature strength up to 300°C and thermal expansion below $16 \cdot 10^{-6} K^{-1}$ are difficult to achieve. For connecting rods, fatigue strength of $\sigma_a = 150$ MPa at 150°C is postulated in addition to elevated temperature strength up to 150°C. (See again Table 1).

A high Si-content of 20 wt.-% is necessary for a low coefficient of thermal expansion; however, rapid solidification has to be applied in order to obtain a fine distribution of second phases in the alloy and good mechanical properties at elevated temperatures. Silicon additions lower the coefficient of thermal expansion of the alloy linearly up to approximately 25 wt.-% [6]. Solidification rates achieved by gas atomization alloys allow to be produced with a maximum of about 17 wt.-% Si and a sufficiently fine microstructure. Above this concentration of Si, primary crystals appear in coarse distribution and deteriorate the mechanical properties of the consolidated powders [7].

Mechanical alloying was applied to produce AlSi-based alloys with high Si-content and a fine distribution of second phases. The powder alloy Al20Si5Mn2Ti was chosen [7] and produced in three different ways. Al20Si5Mn2Ti and Al12Si5Mn2Ti were gas-atomized in an argon and oxygen mixture and screened to powders with a particle size < 71 µm. The alloy containing 20 wt.-% Si exhibited primary Si-particles as shown in figure 10. For a low coefficient of thermal expansion, 8 wt.-% Si was added to the rapidly-solidified Al12Si5Mn2Ti alloy by mechanical alloying with elemental Si-powder of 5 µm mean particle size. Figure 10 shows the powder microstructure after mechanically alloying 8 wt.-% Si. The Al20Si5Mn2Ti alloy powder was also milled in an attritor with stearic acid as a processing agent. All three Al20Si5Mn2Ti alloy powders were compacted and extruded at 360°C and the microstructure of powders and extrusions and tensile properties were determined at room temperature and elevated temperatures.

Figure 12 shows the yield strength of the alloy produced in the three different ways. Compacts of as atomized Al20Si5Mn2Ti exhibit the lowest strength values at all test temperatures. Milled powder compacts have considerably increased strength as already observed before with the AlFeCe alloy. Nearly the same result is reached by mechanically alloying the Al12Si5Mn2Ti alloy with 8 wt.-% Si.

This method allows the combination of rapid solidification up to the concentration limits of an alloying element where a fine distribution of phases can be obtained, with mechanical alloying to further increase the concentration range of the alloying element. By optimization of processing, it should be possible to not only to reach the same results as with milling of a relatively coarse rapidly solidified microstructure, shown in figure 10, but even to exceed the properties.

All three alloys show low ductility with an elongation of the order of magnitude of 1 %. With these low values, scatter is high and no effect of processing method was detected.

Improvement of the mechanical alloying with Si is expected if Si powder with small particle size is used. It was observed by electron microscopy that the Si particles are not broken during the process of mechanical alloying. Milling should be performed with the smallest particle size possible. However, agglomeration has to be avoided by the use of suitable activators as they are known from electrolytical codeposition of metals and oxide particles.

SUMMARY

1. Mechanical alloying by milling of rapidly solidified powders with processing control agents generally increases elevated temperature strength and reduces ductility. For a possible application of this process, means to increase ductility have to be evaluated. Experiments are being carried out where milling is combined with a degassing treatment and chemical treatment during the milling process.

2. Mechanical alloying of rapidly solidified powders with Si opens the possibility to extend the composition range of Si above what can be achieved by gas atomization. With the alloys investigated, ductility was generally low and was not significantly affected by mechanical alloying.

ACKNOWLEDGEMENT

This work was carried out at the German Aerospace Research Establishment, Köln and Daimler Benz AG, Stuttgart and is a part of a cooperative project. Other partners were: MPI für Metallforschung, TU Hamburg-Harburg, Eckart-Werke, Erbslöh-GmbH, Sintermetallwerk Krebsöge, Otto Fuchs Metallwerke and Porsche AG. Funding was granted from the German Ministry of Research and Technology (BMFT) and is gratefully acknowledged.

REFERENCES

[1] I.J. Polmear, M.J. Couper, M.J. Bannister, "Characterization of Precipitation Reactions in Rapidly Solidified Powders Based on the Al-Fe and Al-Cr-Systems", Metal Forum, 12, 1988, pp. 54-61.

[2] V. Arnhold, K. Hummert, "Properties and applications of dispersion strengthened aluminum alloys" in "New materials by mechanical alloying techniques" eds. E. Arzt, L. Schulz, DGM Informationsgesellschaft, Oberursel, 1989, pp. 263-278.

[3] M.L. Övecoglu, W.D. Nix, "Elevated temperature characterization and deformation behavior of mechanically alloyed rapidly solidified Al8Fe3.5Ce alloy" in "New materials by mechanical alloying techniques" eds. E. Arzt, L. Schulz, DGM Informationsgesellschaft, Oberursel, 1989, pp. 263-278.

[4] E. Arzt, J. Rösler, Acta Met. 32 1988, pp. 1053-1060.

[5] J. Rösler, E. Arzt, Acta Met. 32, 1988, pp. 1043-1051.

[6] Changynan Gau, "Reibung, Verschleiß und thermische Ausdehnung von Al-Si-Legierungen", Fortschr.-Ber. VDI Reihe 5, Nr. 150, VDI-Verlag, Düsseldorf, West-Germany, 1988.

[7] B. Saal, J. Albrecht, G. Lütjering in Proc. "Light-Weight Alloys for Aerospase Applications", ed. E.W. Lee, E.M. Chia, N.G. Kim, TMS, 1989, pp. 3-13.

components	temperature range [$^\circ$C]	yield strength at $^\circ$C	[MPa]	fatigue strength at $^\circ$C	[MPa]	thermal expansion coefficient [1/K]	
jet-engine compressor	RT to 400	RT	460	**120**	**130**		elongation > 5%
compressor-wheel in automotive gas turbine	RT to 300	RT **300**	340 **330**	RT 300	150 130	< 18*10^{-6}	disc diameter 160 mm
connecting rod	- 50 to 150	150	270	**150**	**150**	< 16*10^{-6} (- 50 to 200°C)	

RT: room temperature

Tabl. 1:   Required mechanical properties for specific components

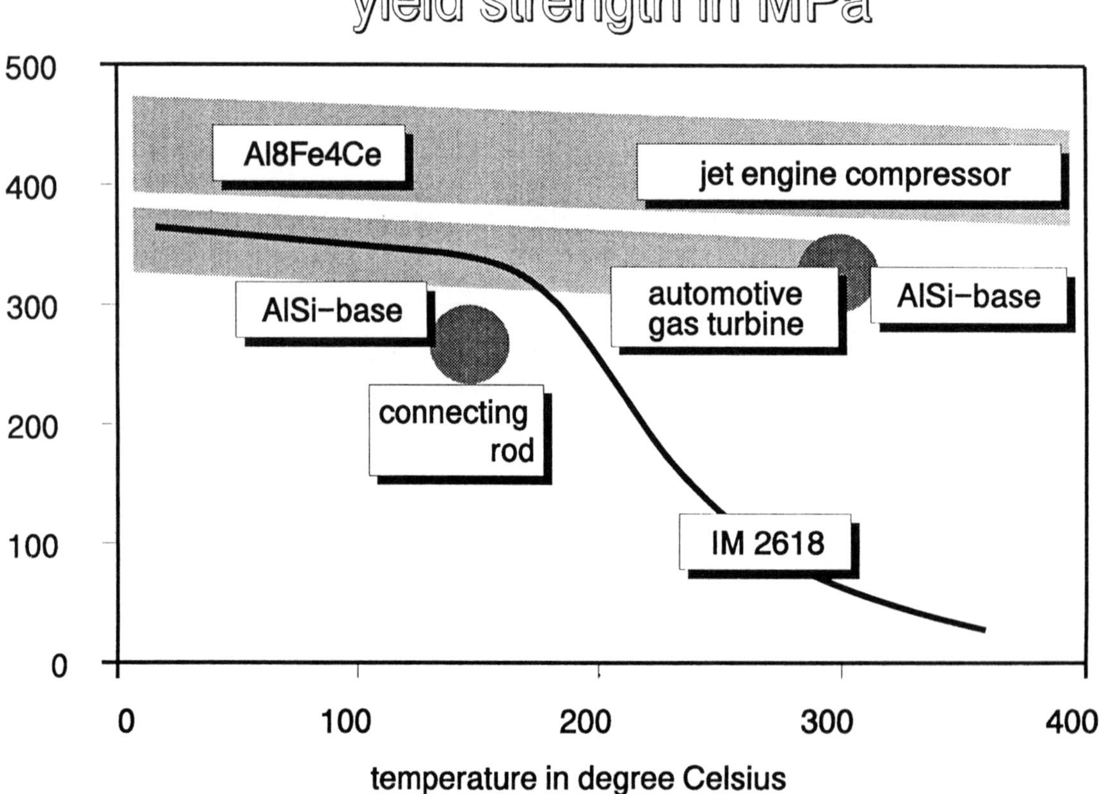

Fig. 1: Yield strength requirements of different applications compared with 2618.

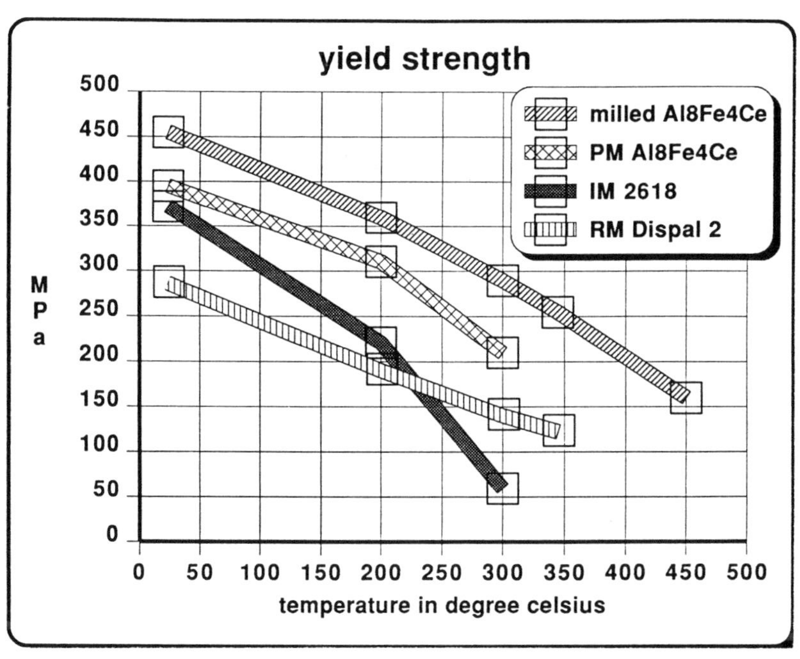

Fig. 2: Comparison of Al8Fe4Ce and Dispal with 2618 yield strength.

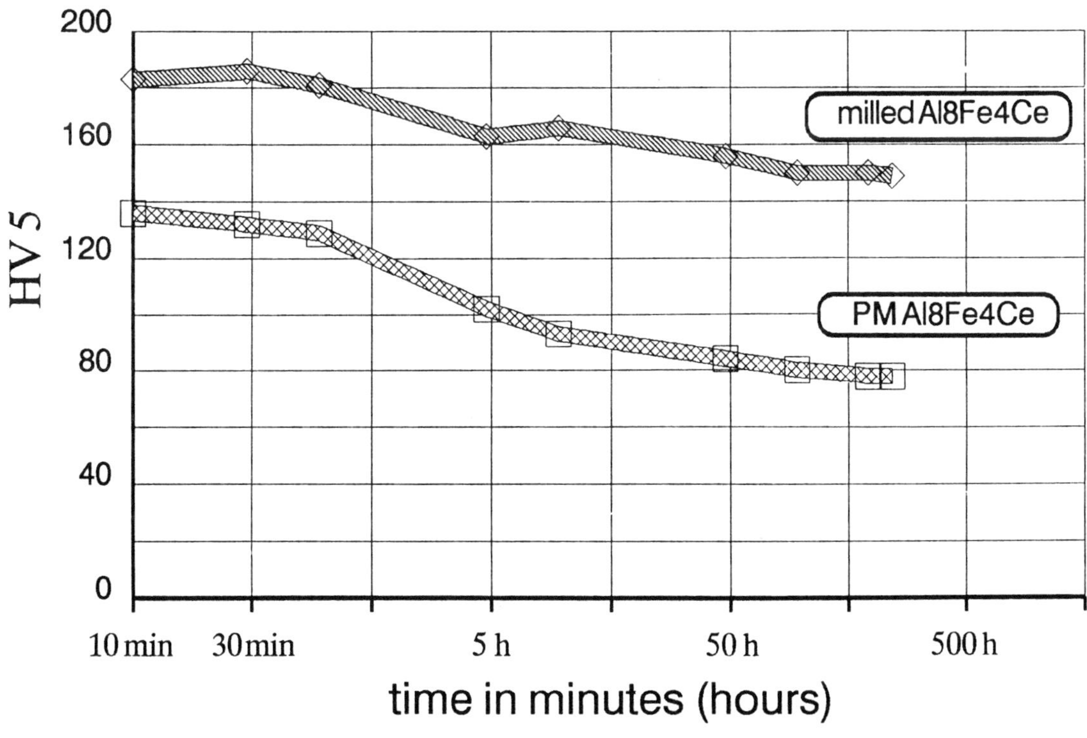

Fig. 3: Hardness of as-atomized and milled Al8Fe4Ce powder extrusions
after ageing at 450°C.

Fig. 4: TEM micrograph of consolidated Al8Fe4Ce powder annealed at 450°C, 570h.

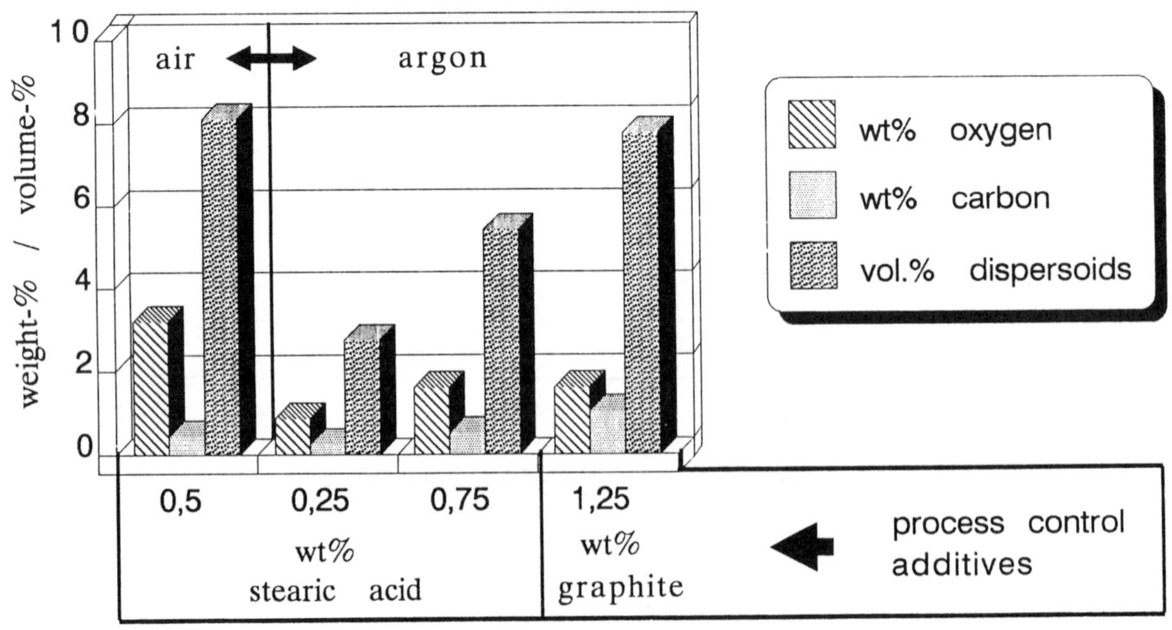

Fig. 5: Oxygen and carbon content of milled Al8Fe4Ce powder.

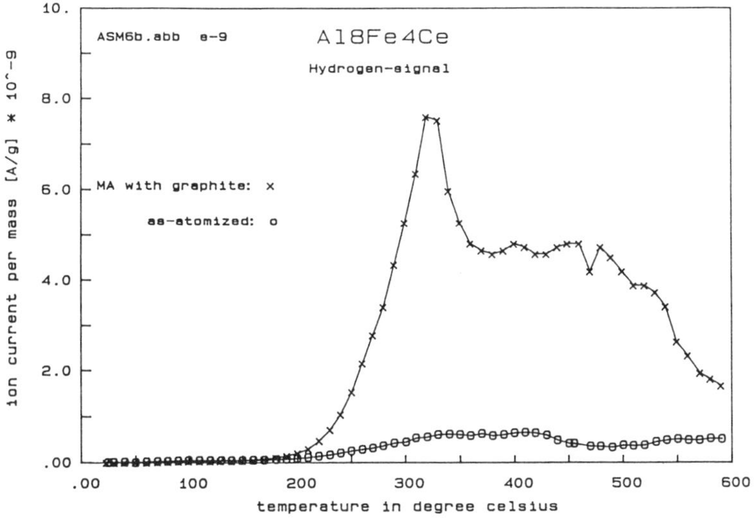
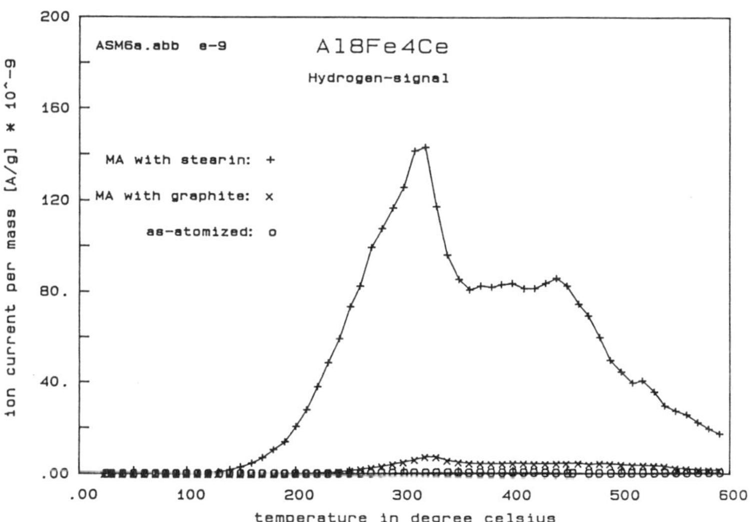

Fig. 8:
Mass spectroscopy of rapidly solidified and mechanically alloyed aluminum alloy powder (note the different scale in b).

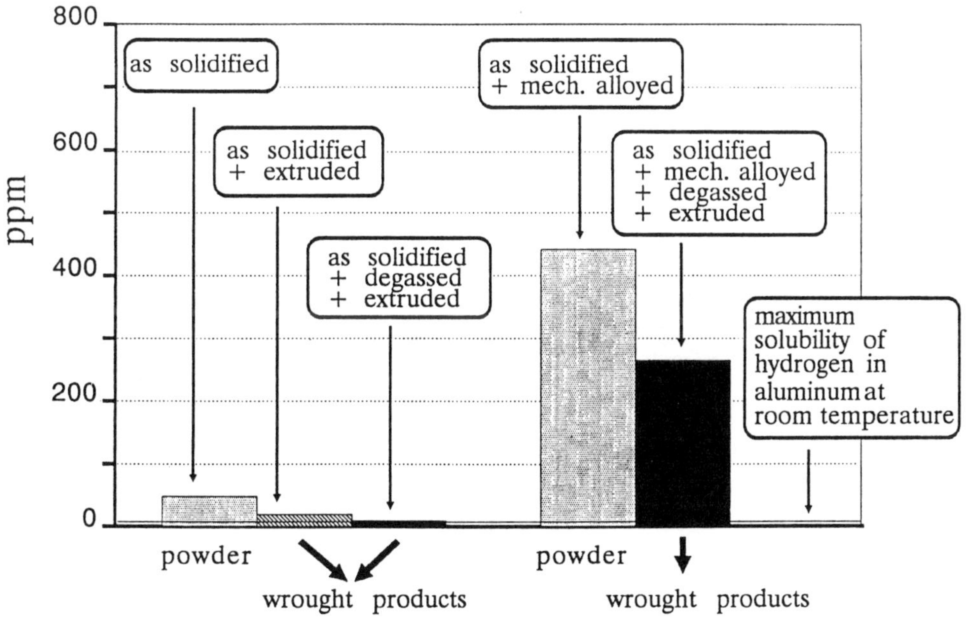

Fig. 9:
Hydrogen content of rapidly solidified and mechanically alloyed aluminum alloy.

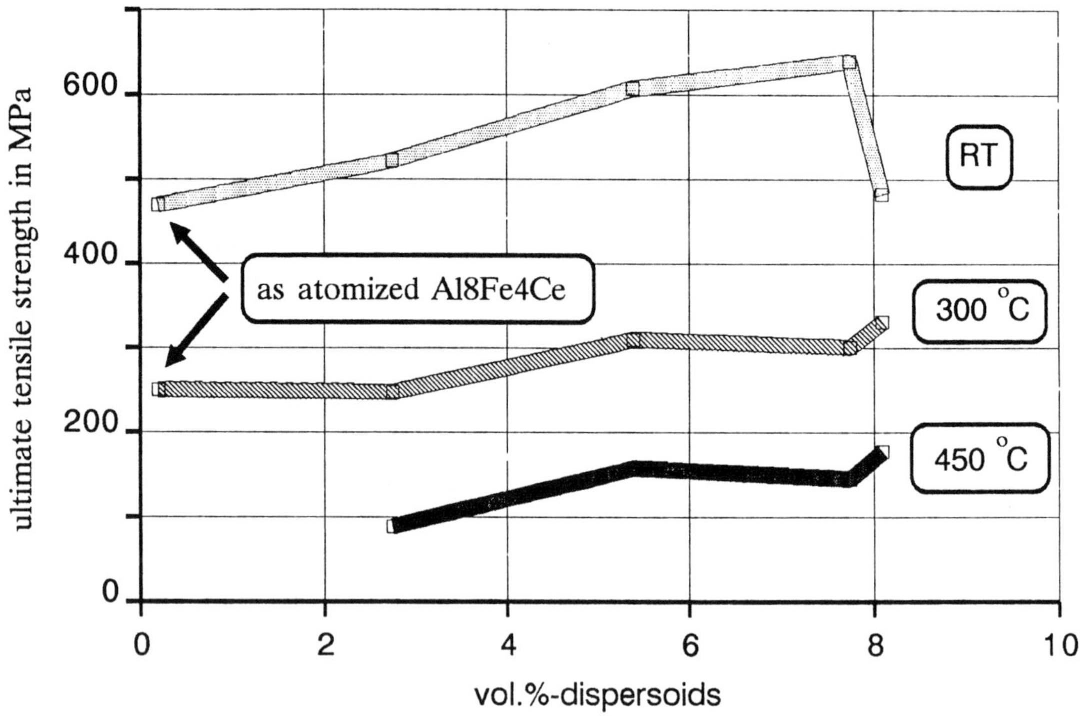

Fig. 6: Ultimate tensile strength of mechanically alloyed alloys as a function of dispersoid content.

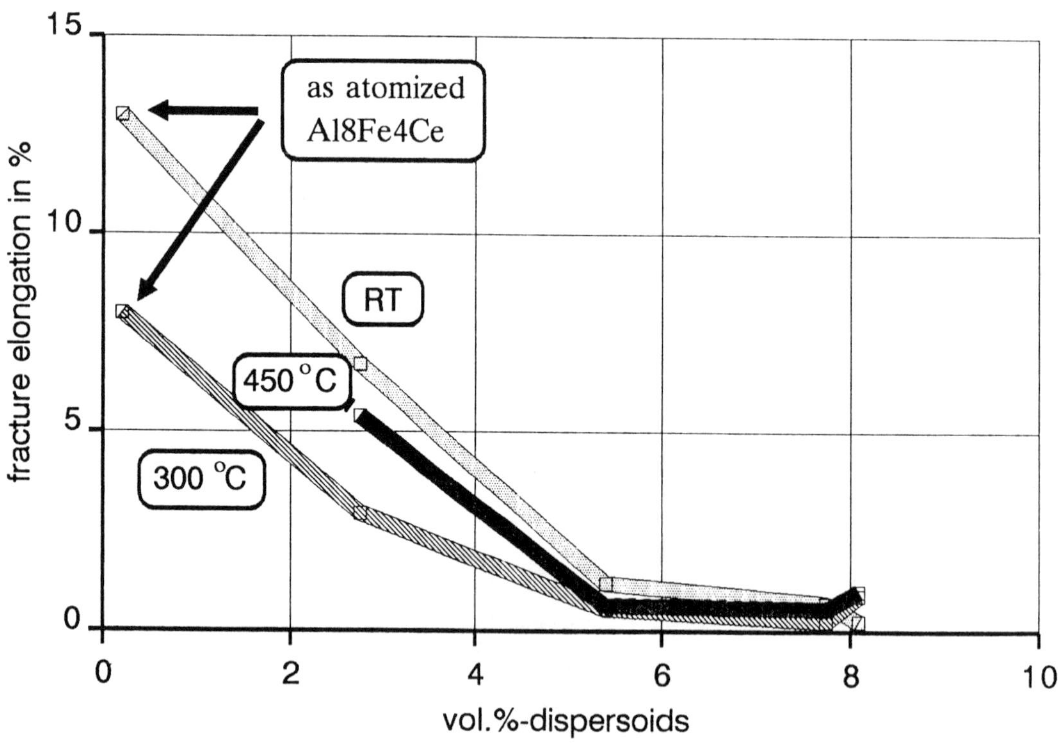

Fig.7: Fracture elongation of mechanically alloyed alloys as a function of dispersoid content.

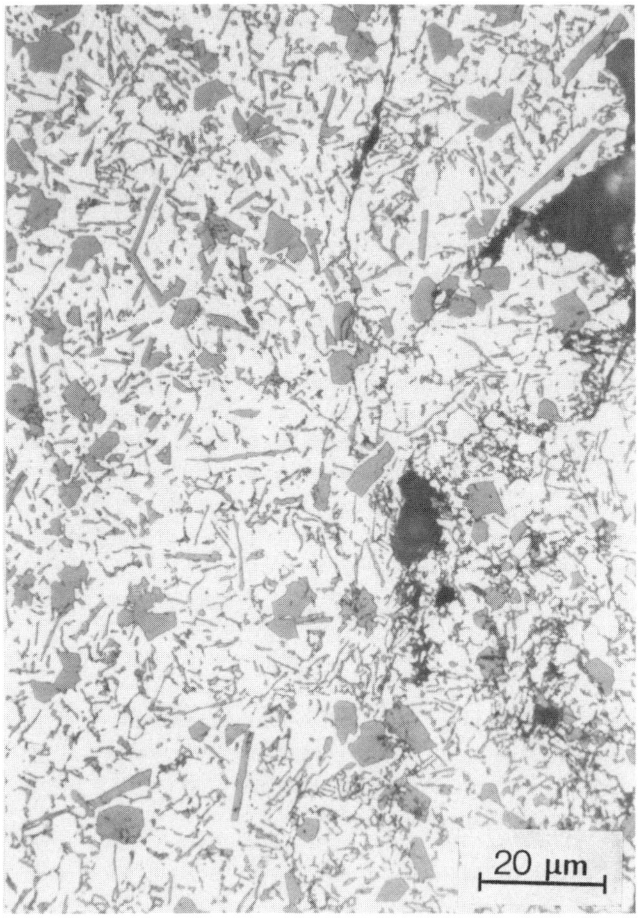

Fig. 10: Microstructure of as atomized Al20Si powder.

Fig. 11: Microstructure of Al20Si5Mn2Ti powder, mech. alloyed with 8 wt.-% Si.

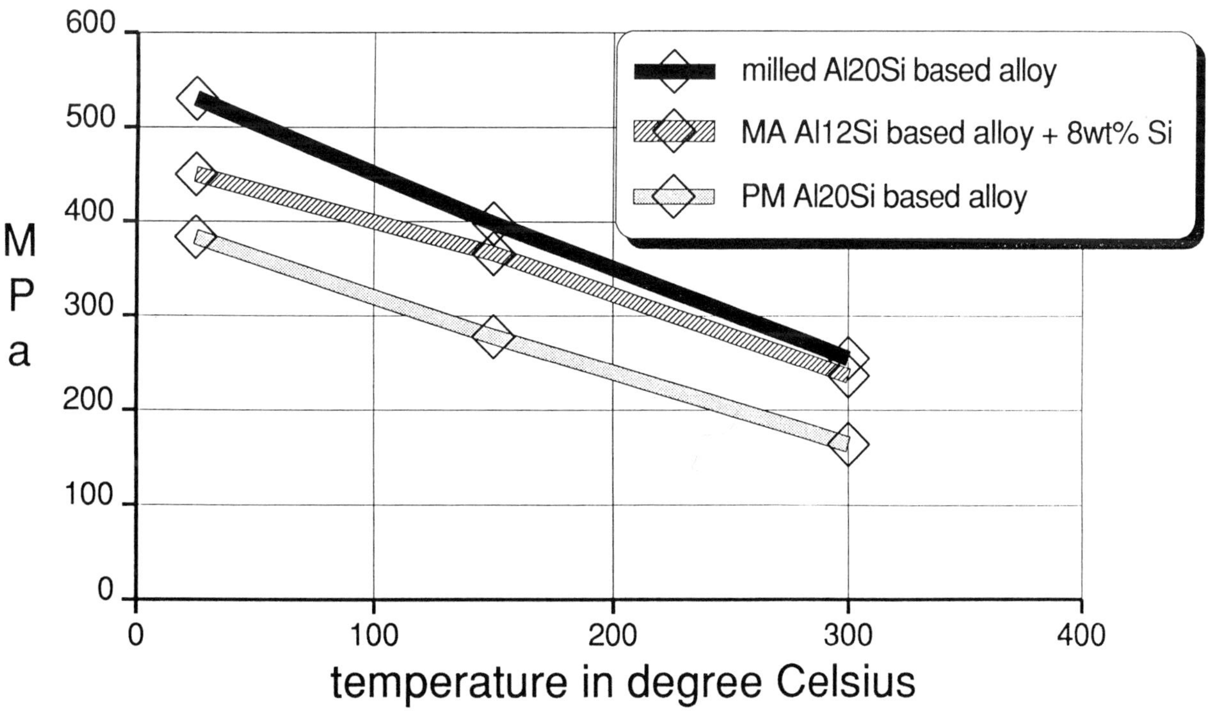

Fig. 12: Yield strength of Al20Si powder extrusions.

MECHANICAL ALLOYING OF THE Ti-Al SYSTEM

N. Burgio, W. Guo, M. Magini, F. Padella
Amorphous Materials Project
E.N.E.A.–Casaccia
Rome, Italy

S. Martelli
Materials Science Division
E.N.E.A.–Casaccia
Rome, Italy

I. Soletta
University of Sassari
Sassari, Italy

SUMMARY

Ti-Al alloys have been prepared by mechanical alloying (MA) of pure titanium and aluminium powders. In the composition range $50 \leq Ti(at\%) \leq 80$, the powder can be successfully amorphized. The milling times, required to achieve the amorphous state, strongly depend on the Al starting composition. The investigations on the early stage of MA show that the system was amorphized through two different pathes. On the Ti rich side (Ti= 75,80 at%), before the system achieves the amorphous state, the MA leads to the formation of h.c.p. αTi(Al) solid solution, whereas for the equiatomic composition (TiAl), instead of solid solution, there is the nucleation of the intermetallic Ti_3Al. However, the disorder-order transition of intermetallic superstructures constitutes the principle hindrance to the formation of the equilibrium compounds. On the Al rich side the limiting factor is represented by the nucleation of TiAl intermetallic compound.

The formation of amorphous alloys is discussed on the base of thermodynamic models. The thermodynamic properties of Ti-Al system result in the two pathes, which affects directly the reaction outcomes.

The present results confirm the existence of a metastable f.c.c. phase as final alloying stage for the $Ti_{75}Al_{25}$ composition.

THERE IS A CONSIDERABLE INTEREST in titanium aluminides because of their attractive properties for structural components operating at elevated service temperatures. Titanium aluminides, on the other hand, present low ductility and poor formability due to the intermetallic structures, which restrict considerably their industrial application[1]. A great effort has been devoted in the last years to improve the properties of Ti-Al alloys by means of chemistry modification and microstructure control .

Mechanical alloying (MA) is nowadays commercially exploited to produce aluminium titanium alloys with additional dispersoids[2]. The potentiality of supplying of large amount of amorphous powder with a relatively inexpensive process opens a wide spectrum of possible technological applications.

The feasibility of amorphising Ti-Al by MA has been already tested[3,4,5], and a thermodynamical model has been proposed which predicts the possibility to form, at room temperature, an amorphous alloy for composition Ti_xAl_{1-x} ranging from X=0.25 to X=0.8[4]. Nevertheless some disagreement has been found out in the experimental determination of the this range, especially, in the Ti-rich boundary.

Keeping in mind possible technological applications, in the present work, the Ti-Al starting powder compositions have been selected to simulate the alloys based on titanium aluminides such as α_2-alloys and Γ-alloys. Particular attention has been focussed onto the early stage of MA since the formation of intermetallic phases, may substantially alter the solid state reaction (SSR) path[6]. The knowledge about the onset of the solid state reaction is, therefore, of great interest, because it allows, in principle, to drive the process outcomes by slightly changing the chemical composition or by varing the milling conditions, as it has been demostrated for Ni-Zr[7] and Fe-Zr[8] system.

2) - EXPERIMENTAL

Pure Ti (99.0%) and Al (99.3%) (Alfa products) were used as starting powders. Mechanical alloying was carried out in a Frisch Pulverisette planetary mill with hardened steel vials and balls. The ball to powder weight ratio was approximatively 10:1. In order to minimize oxygen contamination the vials were sealed under pure argon. Four different compositions Ti_xAl_{1-x} (x= 40,50,75,80 at%) were milled in exactly the same conditions . To avoid an excessive warming up, the vials were cooled by an air flow and a milling period of 30 min was alternated with adequate rest time. At fixed times small quantities of powder ($\approx$1mg) were withdrawn from the vials for X-ray diffraction analysis.

X-ray diffraction patterns were recorded by an automatic seifert Pad IV and GSD diffractometers, both with MoK_α (= 0.07107 nm) incident radiation. To gain further informations on the intermediate state of alloying, X-ray diffraction patterns of some representative samples were taken also with a high resolution diffractometer (ITALSTRUCTURES X-RED) equiped with a 120° curved multichannel position sensitive detector and a Johanson's monochromator on the incident beam (parfocussing geometry). The used wavelength was $Co K_{\alpha 1}$ (= 0.1789 nm) and the absence of the $K_{\alpha 1}-K_{\alpha 2}$ doublet together with the high counting statistics allowed an accurate analysis of the scattering curves. The recorded diffraction patterns were corrected for the background arising from tape adhesive film (Schotch Magic) used as powder stand and the peak shape analysis was performed by a least square fitting to the experimental data with an ensemble of generalized Lorentzian curves[9] . The chemical analysis of the final powder oxygen contents was measured with a Oxygen LECO SYSTEM.

3) - RESULTS
a) - Ti_xAl_{100-x} (X= 80,75,50 at%)

In fig.1 the X-ray diffraction patterns recorded with MoK_α radiation vs increasing milling time are reported. At all compositions,the mechanical alloying process induces an amorphous state after a certain milling times, which strongly depends on the Al percentage. In fact, the $Ti_{75}Al_{25}$ composition (fig. 1-b) amorphized completely after 15 hours of milling, whereas a prolonged processing up to 22 and 27 hours has been necessary to promote the amorphization for $Ti_{80}Al_{20}$ (fig. 1-a) and $Ti_{50}Al_{50}$ (fig. 1-c) compositions. The position of broad maximum lies for all compositions, whithin the experimental uncertainty, at the same angular position (2θ =

17.7 deg.). The nearest neighbour distance can be, in first approximation, estimated by the Ehrenfest equation[10]. The resulting value of 0.288 nm agrees with the more accurate measurements obtained from the radial distribution function (RDF)[4] and corresponds to the sum of atom radii of Al and Ti.

At X=75, it is worth noticing that further MA led to the crystallization of the amorphous phase into a new phase, which could be indexed with a f.c.c. structure and a lattice parameter a= 0.42 nm in agreement with the previous reported results [4]. The suspect that the crystallization of the amorphous phase is corresponding to the formation of Ti-oxide , in the present case, can be ruled out, although the chemical analysis of the final product confirmed the presence of 1-2 wt% of oxygen. The relative peak intensities did not match the ones of the condidated Ti-oxide with same f.c.c. structure and similar lattice parameter (ASTM N=8-119.) and, morever, since the powder have been milled simultaneously in identical conditions, the $Ti_{75}Al_{25}$ and $Ti_{80}Al_{20}$ mixtures should exhibit quite similar oxidation effects, having comparable oxygen contents as measured after the same milling times. The mutual concordance of the experimental findings, obtained with different milling devices[4], supports the existence of a metastable Ti-Al cubic phase, which can not be accessed by conventional preparing techniques.

From the patterns obtained during the early stage of MA, with $Cok_{\alpha 1}$ radiation (Fig.2) some characteristic features of the SSR can be deduced. Considering the close overlapping of Ti, Al and of the main Ti-Al compounds characteristic reflexes, the analysis has been restricted to those reflections which can be easily distinguished, in order to overcome possible ambiguities. The Ti(010) reflex is well suited for the present purpose.

At all compositions, the Ti (010) peak position has already reached its maximum shift towards the higher angle after 5h of MA. Table 1 shows the peak position as calculated by the fitting procedure. Since ball milling on pure Ti only produces a line broadening, the peak shift can be consequently attribute to a shrinkage of lattice parameter due to the diffusion of Al into h.c.p. Ti. Clark et al.[11] measured the α-Ti(Al) solid solution lattice parameters as the function of Al content. According to their data, the amount of the dissolved Al can be estimated to be about 14-16at%.

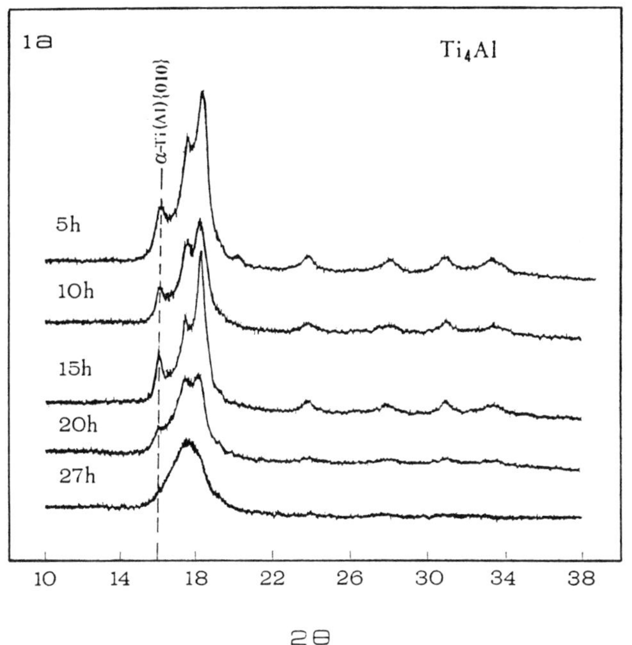

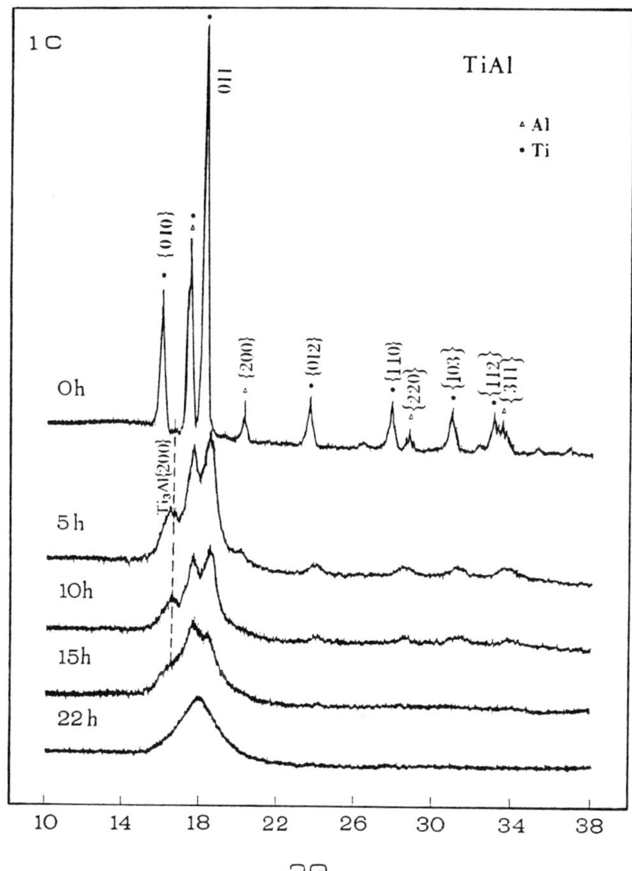

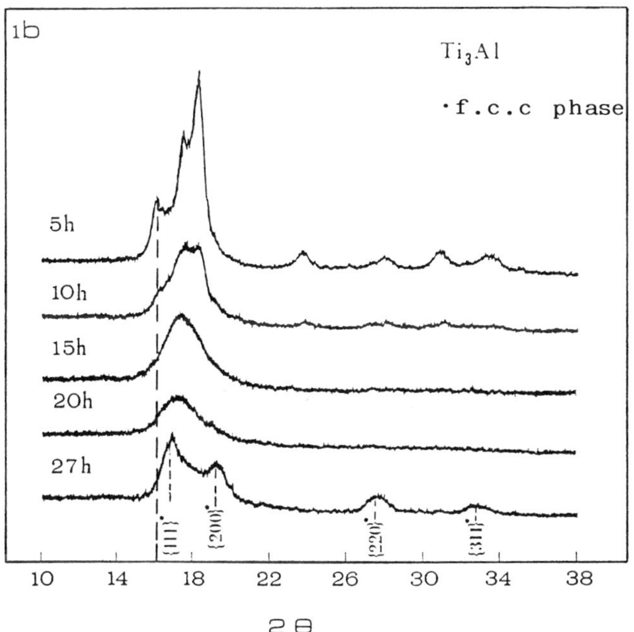

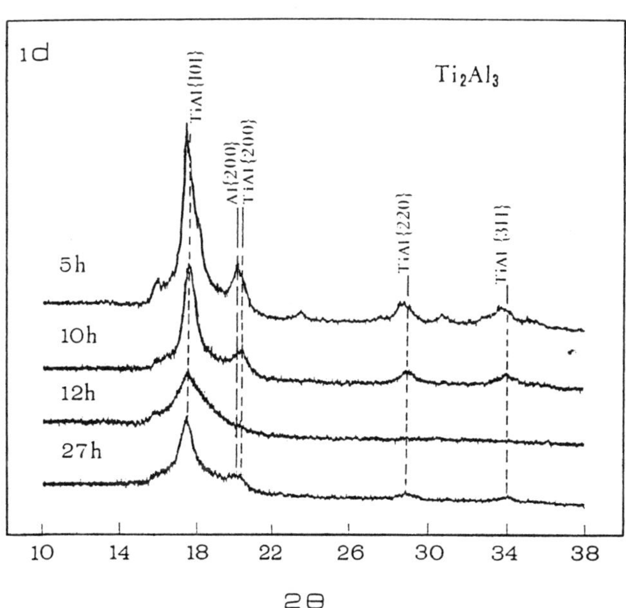

Fig.1) MoKα diffraction patterns vs. milling time of the investigated Ti-Al compositions.

The diffraction pattern of the X=50 composition (Fig.2c) shows, compared with the previous ones, that, after the first period of milling, the peaks relative to Al are almost completely disappeared. The angular positions of the most intense peaks correspond to the fundamental reflections of the DO$_{19}$ superstructure of the Ti$_3$Al intermetallic compound (ASTM N.16-867), while no superlattice reflections can be observed. The whole pattern can be in principle ascribed, either to a disordered form of the Ti$_3$Al structure, or to a supersaturated α(Ti) solid solution with an Al content $\geq$30%, having both phases the same unit cell. The asymmetry of the peak at 2$\theta \approx$ 41 deg. (fig.2d) is due, as revealed by the shape analysis, to an overlapped contribution, which can be ascribed to the Ti(Al) solid solution with an Al content of about 14%.

Morever, the profile analysis exhibits also the presence of extra reflections whose positions slightly change with Al content. For the Ti$_{80}$Al$_{20}$ and Ti$_{75}$Al$_{25}$ they could be tentativly indexed with the disorded intermetallic, while for the Ti$_{50}$Al$_{50}$ their positions are quite close to that of the f.c.c. metastable phase. These suggest thus a possibile concurrence of different phases in first stage of MA.

The patterns of the Ti$_{75}$Al$_{25}$ pow-

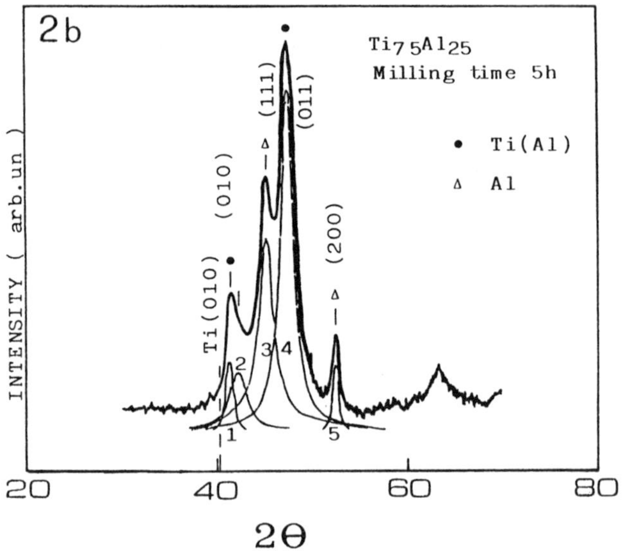

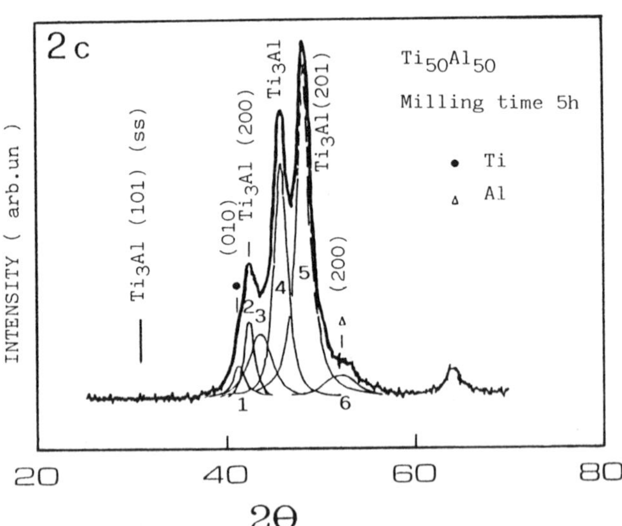

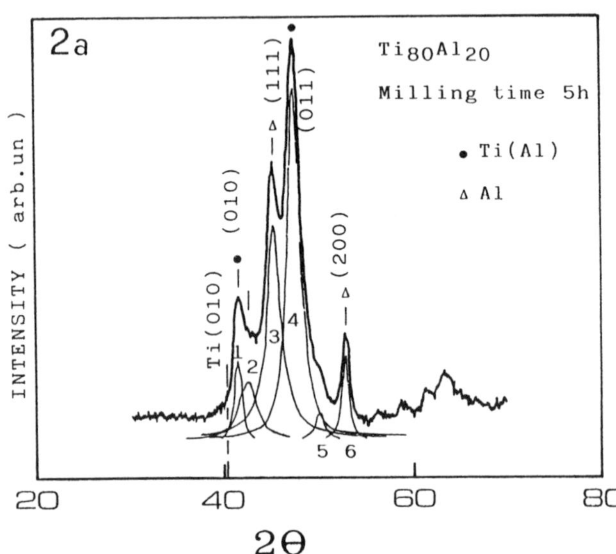

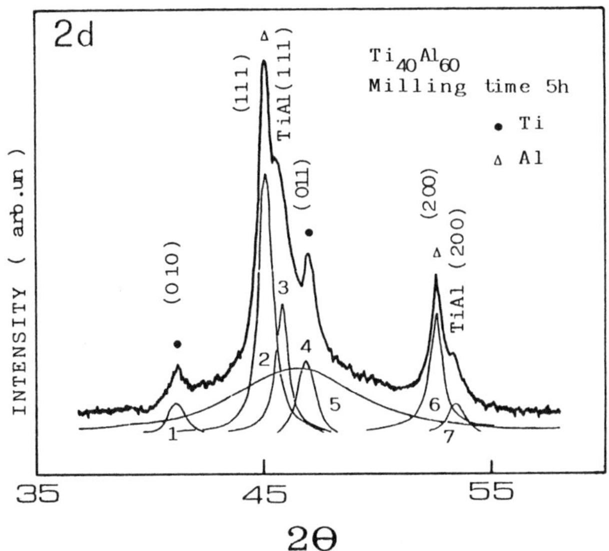

Fig.2)   CoKα$_1$ diffraction patterns re-
         corded after 5 hrs. of MA. The
         thin   lines   indicate   the
         lorentzian like profyles used
         in the analitical fitting to
         the   experimental   data.   The
         numbering refers to Tab.1.

der milled respectively 15 (amorphous)
and 27 hours (crystallized) (see fig.
1-b) and submitted to a thermal treatment up to 500 °C show the tendence of
the amorphous powder to nucleate the
corresponding intermetallic compound
(fig. 3-a), meanwhile the metastable
cubic phase treated at the same temperature exhibits only a sharpening of the
diffraction peaks, without any structural modification (fig. 3-b).

b)- Ti$_{40}$Al$_{60}$
    A distinct alloying path is
clearly recognizible for this composition. The diffraction patterns after 5
hours of MA (Fig.2d) show the nucleation onset of the TiAl intermetallic
compound (ASTM N.S-678), which is completed after additional 5 hours of MA
(Fig.1d). The only occurence of reflections, whose indices are all odd or
all even, evidences that the TiAl
intermetallic holds an A1-f.c.c structure, without forming the L1o layered
superlattice. The peak shift of the
Ti(010) is less pronounced, indicating
a preferential path to the intermetallic formation, rather than to the enrichment of the h.c.p. solid solution. An
amorphous component, whose maximum is
located at the same angular position as

for the other compositions, is already
detectable after the early stage of
alloying.The highest degree of amorphisation is achieved after 12 hours (see
fig. 1-d), however the irregular shape
of the broad maximum does not resemble
the typical Gaussian profile,indicating
that besides an amorphous component
additional contributions of the crystalline phase are still superimposed in
the global scattering curve. Upon further milling crystalline peaks grow
markedly up and, after 27 hours of MA,
the powder pattern can be again indexed
by a f.c.c. structure with a lattice
parameter a=0.41 nm.

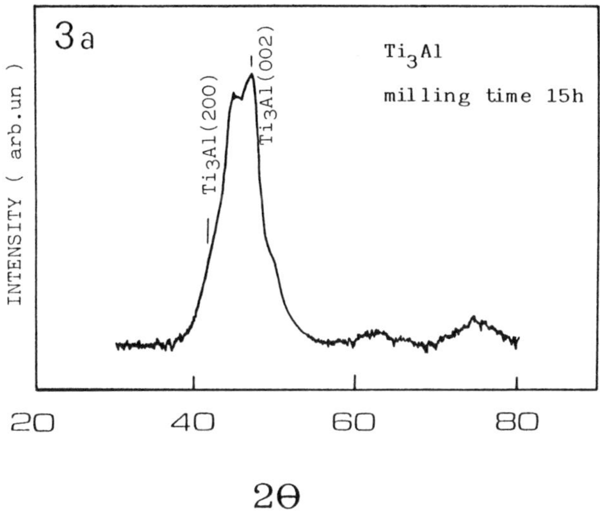

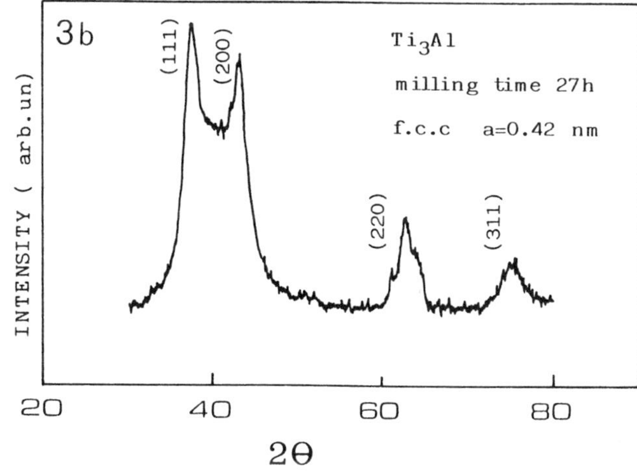

Fig.3)- CoKa1 diffraction patterns of
        the Ti$_{75}$Al$_{25}$ samples (a)- 15
        hrs of MA b)- 27 hrs. of MA)
        after thermal tratment at 500
        °C for 3 hrs. The irregular
        shape a) indicates the begin-
        ning of the Ti$_3$Al intermetallic
        nucleation (see text).

4)- DISCUSSION

a)- $Ti_xAl_{(1-x)}$ (x=80,75,50 at%.)

The analyses of the presented results show that the milling times, required to achieved the amorphous state, strongly dependend on the Al starting composition. Recalling the thermodynamic model developped by Cocco et al.[4], the main feature of the Ti-Al system, assuming a kinetical inhibition of the intermetallics compound, can be,as follow, summarized: the amorphisation window opens, at room temperature, for an Al concentrtion of about 20 at.%, and all the compositions investigated here are, in principle, able to access the amorphous state. The free energy curves[4], calculated at 0°K (fig.4), evidence that the amorphous state becomes more stable with respect to the h.c.p. Ti(Al) solid solution, starting from an Al amount of about 15% . Their free energy difference, hence the stability of the amorphous phase, increases sharply with the Al content reaching its maximum for the equiatomic composition.

Our experimental findings are in agreement with the theoretical predictions, as it concerns the amorphisation hability of the present compositions. Some discrepancies have been encountered in the amorphisation rate. In fact it has been observed that the $Ti_{75}Al_{25}$ powder amorphizes in the shortest milling time, whereas a compositional displacement either towards higher Ti or Al concentration seems to slow down the reaction velocity.In order to find a possible explanation, it should be kept in mind that, besides the chemically driven factors, an additional elastic energy term , originated by the atomic size mismatch of the solute atoms with the ones of the host lattice, may contribute to rise the free enthalpy of the solid solution and, as a consequence, to further stabilize the amorphous phase[12] . For the present case, the free enthalpy of distortion associated with the dissolution of Al atoms into the αTi matrix may be written as[13] :

$$Es(c)=6.\mu.\Omega.(1/\Omega.\delta\Omega/\delta c)^2.c \qquad (1)$$

where:

μ =shear modulus of Ti (37.92 GN/m²)
Ω= mean atomic volume
c= composition.

Supposing that the variation of

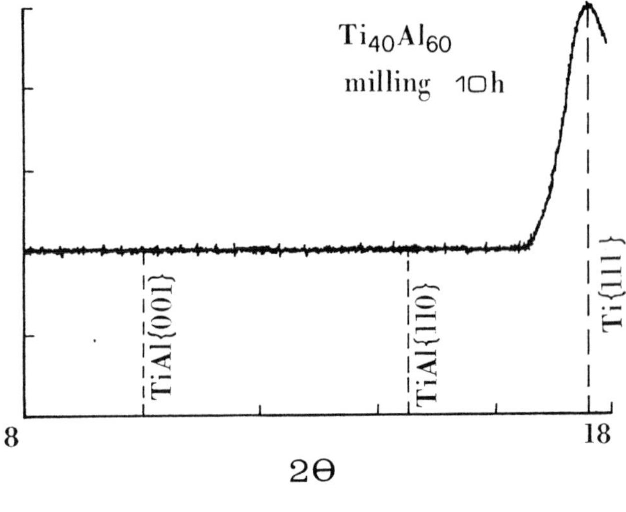

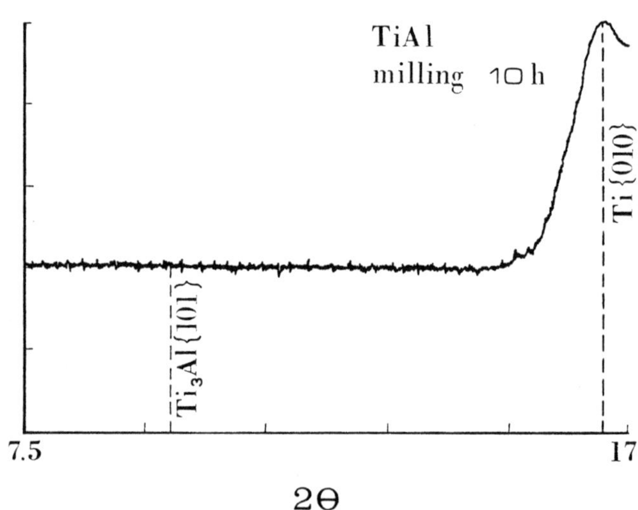

Fig.4)- MoKα low angle iffraction pattern of the $Ti_{40}Al_{60}$ sample after 10 hrs. of MA. The positions of the characteristic superstructure lines of the L1o-TiAl are indicated.

the atomic volume within the investigated composition is nearly linear[14], one may write:

$$(1/\Omega.\delta\Omega/\delta c)\approx(\Omega a-\Omega o)/\Omega o \qquad (2)$$

where:

Ωo=atomic volume Ti (1.225×10^{-29} m³)
Ωa=atomic volume Al (1.295×10^{-29} m³)

The obtained values for the additional elastic energy are ranging from 10 to 25 KJ mol^{-1} for Al concentrations between 13 and 25 at.%. Certainly they should be regarded as a first crude approximation due to the difficulty of determining the effective Al atomic volume in the αTi solid solution, but they point out the dependence of this term on the Al concentration. So that, in absence of intermetallic compounds, the acting driving force, expressed in term of negative heat of mixing and elastic energy, rises rapidly by increasing the Al content, furnishing a possible interpretation for the enhancement of the amorphisation rate, when passing from the $Ti_{80}Al_{20}$ to the $Ti_{75}Al_{25}$ composition.

It is worth noticing that, for both compositions, the Ti(010) peak shift indicates an equal upper solubility limit of Al at about 15 at.%. This limit is attained after the first milling period (5 hrs.). Upon further processing the peak intensity decreases whitout showing any additional shift. The measured concentration is in fairly good agreement with the value predicted for the amorphous phase to become more stable with respect to the h.c.p. solid solution.

On the other side, the sluggish amorphizing behaviour of the $Ti_{50}Al_{50}$ powder suggests the existence of an alloying path alternative to the collapsing of the unstabilized solid solution.

The diffraction pattern demonstrates (fig. 2c) that the average Al content has reached, already during the first stage of alloying, values (≥ 30 at.%), which are beyond the solubility limit observed for the former Ti-richer compositions.

The relative stability vs.milling time (fig. 1c) implies a more probable nucleation of a disordered form of the Ti_3Al intermetallic (unstable at low temperature), rather than a continuous enrichment of the h.c.p. solid solution.

The high negative heat of mixing combined with the local temperature rise during ball impact might be, infact, sufficient to promote the nucleation of the intermetallic compound, but, at the same time, not high enought to climb over the energy barrier associated with the h.c.p. A3->DO$_{19}$ (A_3B type) first type order transition[15].

The atomic exchange between neighbouring sublattice sites, required by the ordering transition DO$_{19}$-Ti_3Al, presupposes an elavated diffusion of both the constituent elements. The very high asymmetry of the self diffusion coefficient of Ti ($1.6 \ 10^{-14} \ cm^2 sec^{-1}$)

and of Al into h.c.p. Ti ($9.3 \ 10^{-9} \ cm^2 sec^{-1}$), measured at 900 °K, which implies a very low mobility of the Ti atoms, may account for the suppression of the ordering transition.

The nucleation of an intermetallic compound alters substantially the evolution of the alloying process, since the negative heat of mixing does not act anylonger, at least at its equilibrium composition, as concurrent driving force to the amorphisation reaction.

The transition to the glassy state, which remains kinetically favoured, as demonstrated by the experimental findings, is attained mainly through a destabilisation of the crystalline structure by introduction of lattice defects Their accumulation can rise the free enthalpy of the faulted intermetallic compound above that of the amorphous alloy[16]. Nevertheless, it should be pointed out that, owing to the composition of the starting powder, the metastable phase is driven , upon prolonged milling, far from its chemical equilibrium, as can be deduced by the homogeneity range of the ordered Ti_3Al intermetallic (22 to 39 Al at.%)[17], reinintroducing a beneficial thermodynamical contribution[18].

The step through the intermetallic phase, instead of a straightforward reaction to the amorphous phase, may, therefore, explain the longer milling time required to completely amorphize the $Ti_{50}Al_{50}$ composition.

The analysis of the diffraction data confirms the absence of the nucleation of the TiAl intermetallic compound, even if the starting powder has been selected to simulate this phase. From the Ti-Al phase diagram, one observes that the lower Al boundary of the TiAl equilibrium phase lies at about 48÷49 at.% of Al, i.e., quite close to the highest Al concentration, which is obtainable with this composition. In addition the data available in the literature[19] show that the largest negative heat of mixing of this phase is shifted towards an Al content of 60 at.%.

Assuming that the fast diffusion of Al avoids possible local Al enrichments beyond the average value, it is plausible to suppose that during the MA process the prerequisites for the formation of the TiAl intermetallic, as concerning the chemical composition, the negative heat of mixing and the energy imparted to system, are not fulfilled. The above supposition is indirectly supported by the analysis of the $Ti_{40}Al_{60}$ composition, which will be discussed in the following section.

b)- $Ti_{40}Al_{60}$

The x-ray diffraction pattern (fig. 2d) shows the formation of the A1-f.c.c. TiAl intermetallic already at the first sampling time (5 hours). The less accentuated Ti peak shift, corresponding to Al concentration in the h.c.p.-solid solution of about 7at.%, demonstrates also, that the intermetallic nucleation happens promptly during the ball impact at the expences of the parent materials.

The involved mechanism appears to be quite similar to the one observed for the $Ti_{50}Al_{50}$ composition, the higher Al initial content allows, in this case, the system to reach, during ball milling, the appropiate chemical composition for nucleating the TiAl phase. It is worth noticing that the critical temperature of the order/disorder transition lies in the near proximity of the melting point of TiAl phase ($\approx1450°C$)[1]. The total absence of the (001) and (110) lines (fig.5), which are characteristic of the L1o-TiAl superlattice, confirms the lack of long range order, supporting the hypothesis that mechanical alloying processes disclose the existence of low temperature intermediate phases[20].

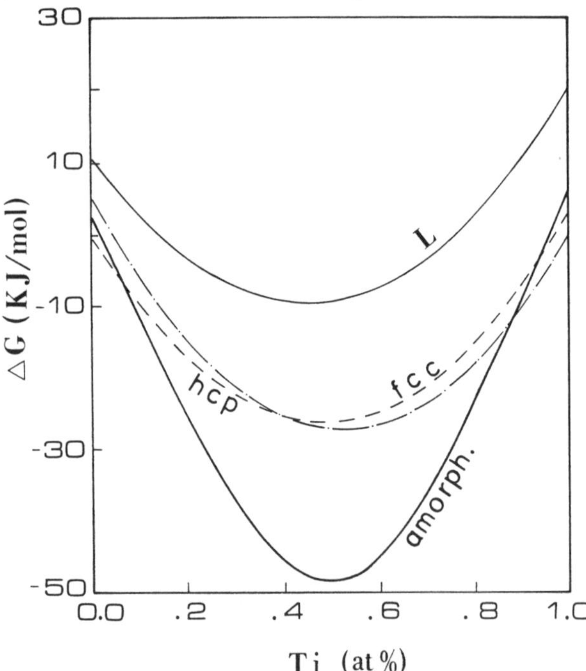

Fig.5)- Free energy curves of the b.c.c., f.c.c. solid solution calculated at 0 °K. The curve relative to the amorphous phase is computed according to the Miedema's model.

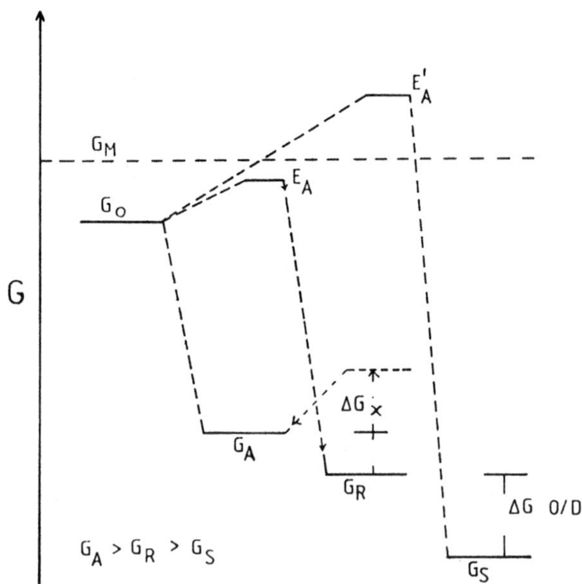

Fig.6)- Schematic energy diagram of the reaction path for $Ti_{50}Al_{50}$ and $Ti_{40}Al_{60}$ compositions. G_O- initial state, G_M- energy state during ball impact; G_A,G_R,G_S- free energy of the amorphous phase, disordered and ordered intermetallic compound; E_A,E'_A activation energy for the disordered, ordered intermetallic compound. $\Delta G_{O/D}$- free energy difference associated to the order/disorder transition; ΔG_X- free energy increase of the intermetallic compound induced by MA. $G_R + \Delta G_X > G_A$ for $Ti_{50}Al_{50}$, $G_R+\Delta G_X \approx G_A$ for $Ti_{40}Al_{60}$

The contemporaneous formation of an amorphous component suggests that the two reaction pathes, towards the amorphous phase and to the intermetallic phase, are each other competiting. The above considerations about off-stoichiometric com position do not hold for the present case, because of the wide range of homogeneity (48÷66 at.% Al) of the TiAl intermetallic phase. Furthermore, since the TiAl intermetallic already appears in its disordered form, an increase of the stored internal energy by a loss of long range order in the crystalline structure, as, for instance, observed in MA of CsCl-type intermetallic compounds[21], can be presently neglected. The kinetic constraints together with the negligible difference in the heat of formation may incline the reaction path towards the amorphous state, meanwhile the MA does not successd to rise the free energy of the TiAl intermetallic up to the necessary amount,of about $10 \div 15$ KJmol^{-1}[12], which allows to stabilize the amorphous state (fig.6).

	$Ti_{80}Al_{20}$	$Ti_{75}Al_{25}$	$Ti_{60}Al_{40}$	$Ti_{40}Al_{60}$
1)	0.2530♦ αTi(Al)(010) Al 16 at.%	0.2532 αTi(Al)(010) Al 15 at.%	0.2533 αTi(Al)(010) Al 14 at.%^	0.2543 αTi(Al)(010) Al 7 at.%
2)	0.242 unident.* +Al(111)	0.246 unident.^ +Al(111)	0.24718 Ti_3Al(020) +Al(111)	0.2331 αTi(Al)(010) +Al(111)
3)	0.2332 αTi(Al)(002) +Al(111)	0.2320 αTi(Al)(002) +Al(111)	0.2407 unident.*	0.2294 TiAl(111)
4)	0.222 αTi(Al)(011)	0.2225 αTi(Al)(011)	0.2298 Ti_3Al(002)	0.2244 αTi(Al)(011)
5)	0.213 unident.*	0.2028 Al(200)	0.21887 Ti_3Al(021)	0.2267 amorph.
6)	0.203 Al(200)		0.203 Al(200)	0.2025 Al(200)
7) ↳				0.1988 TiAl(200)

♦Ti(010): d= 0.2557

*very weak + f.c.c. phase ^ Ti_3Al(200)

Tab.1) Lattic spacings (d(nm)= $\lambda/2sin\theta$ θ-Bragg'sangle) as determined by the profile analysis.The numbers in the first column refer to the single componentes of the diff-raction patterns (see fig.2)

CONCLUSION

a) TixAl(1-x) alloys have been sucessfully amorphized by MA in the composition range $50\% \leq x \leq 80\%$. On the Ti rich side, the value experimentally determined exactly hits the theoretical prediction, whereas on the Al rich side the limiting factor is represented by the nucleation of TiAl intermetallic compound.
b) The order-disorder transition of the intermetallic superstructures constitute the principal hindrance to the nucleation of low temperature equilibrium compounds.
c) The formation of intermetallic compouds should not only regarded as a drawback, because it permits to drive, by a proper choice of the milling condition, the outcomes of the MA process.
d) The present results confirm the existence of a metastable f.c.c.-phase as final alloying stage for the $Ti_{75}Al_{25}$ composition.

REFERENCES
[1]- Y.W.KIM JOM, 41, (July 1989), 24
[2]- J.A.HAWK, K.R.LAWLESS, H.G.F. WILSDORFScripta Metallurgica, 23, (1989), 119
[3]- L.SCHULTZ Mat. Science and Engineering, 97, (1988), 15
[4]- G.COCCO, I.SOLETTA, L.BATTEZZATI, M.BARICCO, S.ENZO Phil. Mag accepted (1989).
[5]- R. SUNDARESAN, A. G. JACKSON, S. KRISHNAMURTHY ,F. H. FROHES Mat. Science and Engineering 97,(1988), 115
[6]- R. B. SCWARZ, C. C. KOCH Appl. Phys. Lett., 49,(1986),146
[7]- J.ECKERT, L.SCHULTZ, E.HELLSTERN, K.URBAN J. Appl. Phys., 64, (1988), 3224
[8]- N.BURGIO, A.IASONNA, M.MAGINI, S.MARTELLI, F.PADELLA Mat.Science and Engineering (submitted)
[9]- D.W.Marquart J.Soc.Ind.Appl. Math., 91 (1963)
[10]-A.GUINIER "Théorie et technique de la radiocristallographie" ,DUNOD (Paris) (1964)
[11]-W.B.Pearson "Handbook of lattice spacings and structures of metals" Pergamon Press (1967)
[12]-A. BLATTER, M. von ALLMEN Mat., Science and Engeneering, 97, (1988), 93
[13]-P.HAASEN " Physical Metallurgy" Cambridge University. Press (1978)
[14]-M.ELLNER J.of the Less-Common Met., 60, (1978), 15
[15]-F.REYNAUD Phys. Stat. Sol. (a), 72, (1982), 11
[16]-M.Magini,F.Padella,G.Ennas,F.Poma, M.Vittiri J.Non Cryst.Solids 110(1989), 69
[17]-T.B.MASSALSKY, "Binary Phase Diagrams" A.S.M. (1986))
[18]-A.BLATTER, M.von ALLMEN Phys. Rev.,Lett., 54, (1985)
[19]-O.KUBASCHEWSKY, W.A.DENCH, Acta Met., 3, (1955), 339 O.KUBASCHEWSKY, G.HEYMER, Trans. Faraday Soc. 56, (1960), 473
[20]-S.MARTELLI, G.MAZZONE, A.MONTONE, F.PADELLA, M.VITTORI "EUROMAT '89", Aachen (R.F.G) (1989)
[21]-E.HELLSTERN, H.J.FECHT, Z.FU, W.L.JOHNSON J. Mater. Res., 4, No.6, (1989), 1292

THE EFFECT OF DEFORMATION MECHANISM(S) ON SUPERPLASTIC ELONGATION OF MECHANICALLY ALLOYED ALUMINUM IN90211

Thomas R. Bieler
Michigan State University
Lansing, Michigan, USA

Amiya K. Mukherjee
University of California
Los Angeles, California, USA

ABSTRACT

The superplastic deformation of IN90211, a mechanically alloyed aluminum similar to 2124, has been shown to occur at unusually high strain rates. Although superplastic elongations of up to 570% have been obtained, the rate dependence of stress on strain rate, m, is low, between 0.25 and 0.3. An analysis of the threshold stresses present due to the dispersed particles indicates that the rate limiting deformation mechanism is not grain boundary sliding (which was observed), but solute atom drag of dislocations in the bulk of the grains. Deformation in the superplastic regime is without cavitation, in part due to the ability for slip to accommodate stress concentrations in the boundary. Two phenomena that reduce potential elongation are serrations in the flow curve, from solute atom breakaway, and the effects of thermal flow instabilities arising from conduction of adiabatically generated heat. Modifications to the alloy composition and/or the mechanical alloying process of the powders are proposed, that may improve the superplastic forming potential of mechanically alloyed alloys.

INTRODUCTION

Since the mechanical alloying process promotes a stable, fine grain size, which is a practical requirement for superplastic deformation, superplastic elongation of a mechanically alloyed aluminum alloy has been explored. The first superplastic elongations in a mechanically alloyed aluminum were demonstrated with IN9021 by Nieh, Gilman, and Wadsworth [1]. These were the fastest superplastic elongations ever obtained up to that time. They observed that the submicron grain size accounted for the unusually high strain rates, since superplastic deformation mechanisms are often expressed in the form of

$$\dot{\epsilon} = A\sigma^n d^{-p} \exp(-Q/RT), \tag{1}$$

where A is a geometrical constant, σ is stress, d is the grain size, p is grain size exponent, and Q is the appropriate activation energy. For a given superplastic mechanism, decreasing d will increase the strain rate. Further work has shown that superplastic elongations of 400% at a strain rate of 75/sec is possible [2] (Fig. 1).

To date, the conditions for optimum superplastic elongation (570%) are 485°C, at a decreasing true strain rate from 5/sec to 0.2/sec [3]. The decreasing strain rate is necessary because thermal conduction of adiabatically generated heat is slow compared to the deformation time, causing thermal strain instabilities that reduce elongation [3,4]. This effect results from the strong thermal dependence of stress upon temperature shown in equation 1 and Fig. 1. Reduced elongation is predicted by continuum mechanics modeling of large strain behavior where stress is temperature dependent.

There are also microstructural parameters that affect flow stability, and this paper will address these issues. The rate limiting deformation mechanisms have been analyzed using a threshold stress analysis [5]. The results of this indicate that dislocation glide is controlled by Mg atom diffusion, limiting the rate of other deformation mechanisms such as grain boundary sliding. This is an unusual result for superplastic deformation, since grain boundary sliding dependent upon grain boundary diffusion is usually the rate limiting step of deformation. In particular, the role of the ceramic dispersions that pin grain boundaries, and the breakaway stress of dislocations from solute atmospheres are of particular interest in future development of a mechanically alloyed material suitable for superplastic forming.

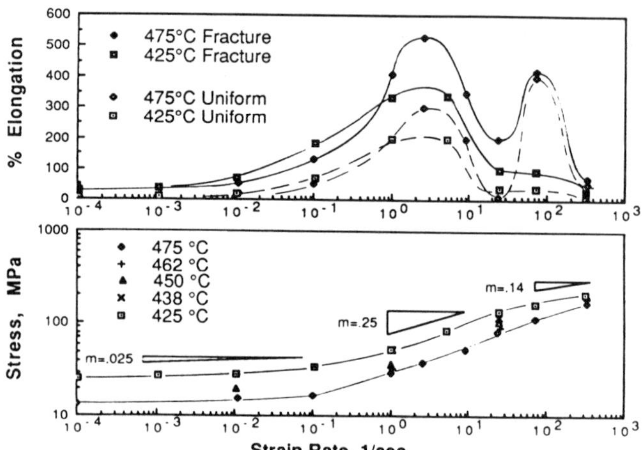

Fig. 1 - Stress and elongation as a function of strain rate and temperature.

EXPERIMENTAL PROCEDURES AND RESULTS

Tensile specimens were made from processed IN90211 sheet material produced by Novamet, Inc. (lot 85V032), and subsequently processed at the Research and Development Labs of Lockheed Missiles and Space Company [2,5]. IN90211 is similar to IN9021, which is more commonly described in the literature. From Table I, IN90211 has higher concentrations of Cu (10%) and Mg (33%). Assuming that all the O and C are in the form of Al_4C_3 and Al_2O_3, $f_v = 5.1$ vol% ceramic particles are present. Specimens were machined out of 2.5 mm thick solutionized (annealed, water quenched) sheet. The gage length was 6.35 mm and width was 2.5 mm, with 1.6 mm radius at the ends. Deformation took place at 425-485 °C in a computer controlled hydraulic testing machine. Specimens were deformed at constant true strain rate up to 1/sec, and at constant ram velocity at higher strain rates. Stress data was taken at maximum load (UTS), which occurred at a low strain, so that stress was not affected by adiabatic heating or the effect of decreasing true strain rate associated with constant ram velocity.

TABLE 1:
COMPOSITION OF IN90XX ALLOYS

wt%	Mg	Cu	C	O
IN9021:	1.5	4.0	1.1	0.8
IN90211:	2.0	4.4	1.1	0.8
IN9051:	3.9	0	.4-.8	.6-1.8
IN9052:	4.0	0	1.1	0.8

The results of the experiments were typical for superplasticity, exhibiting three deformation regimes with maximum elongations occurring where the strain rate sensitivity is highest, in the middle regime, (Fig. 1). For specimens showing large elongations, the true stress-strain curves typically exhibited an extensive stress plateau region (Fig. 2), indicating a uniform elongation prior to necking. Overall strain softening was observed after a true strain near 0.3. Flow softening increased in magnitude with decreasing temperature. A plateau was not seen in specimens deformed under constant ram velocity, since the true strain rate decreased with strain. All specimens showing superplastic elongation exhibited serrations in the flow curves. These serrations occurred at stresses between 25 and 75 MPa.

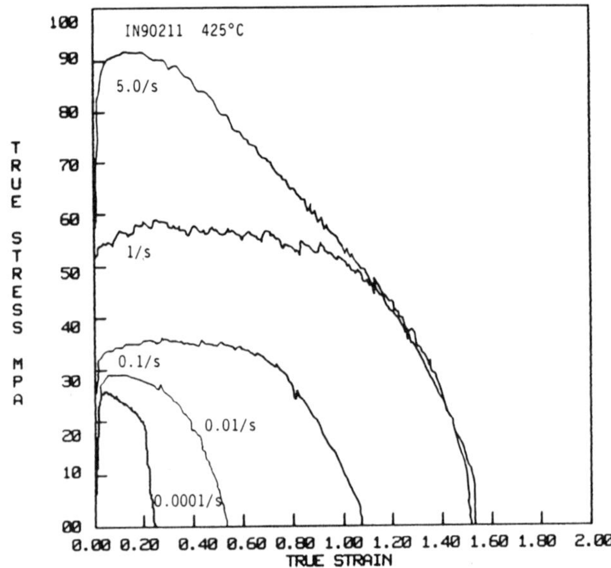

Fig. 2 - Stress-strain flow curves for several strain rates show serrated flow and slightly negative flow softening.

Flow stability theory indicates that stability is enhanced by a positive hardening exponent. The peak true stress of the specimen deformed at 2.5/sec and 475°C was 42 MPa, but the fracture stress was 56 MPa (based upon the 98.8% R.A.), which was at a much higher, but unknown temperature. Thus, the material flow hardens on a microstructural level until fracture occurs. The observed flow softening is probably due to the geometric softening resulting the adiabatic heating [3].

MICROSCOPY - TEM microscopy was used to characterize the microstructure of IN9021 specimens [6]. The particle size distribution, and the distribution of dislocation lengths between particles are summarized for 100 observations in Fig. 3. The average diameter was 13.7 μm, and the average particle spacing along dislocation lines was 46 μm. Particles were often nucleation sites for θ phase (Al_2Cu) precipitates (Fig. 4).

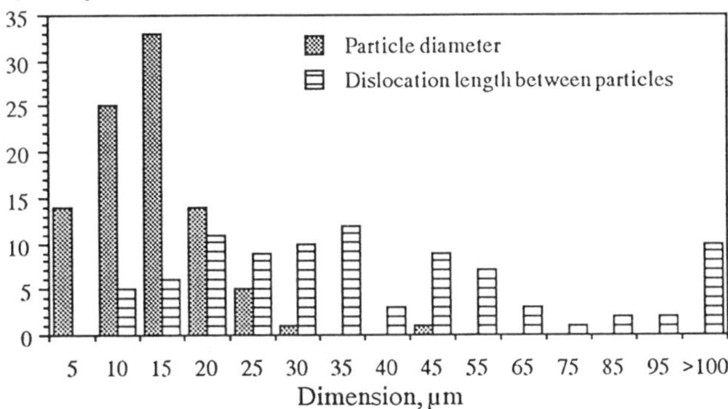

Fig. 3 - Statistical distribution of particle diameter and length of dislocation between particles.

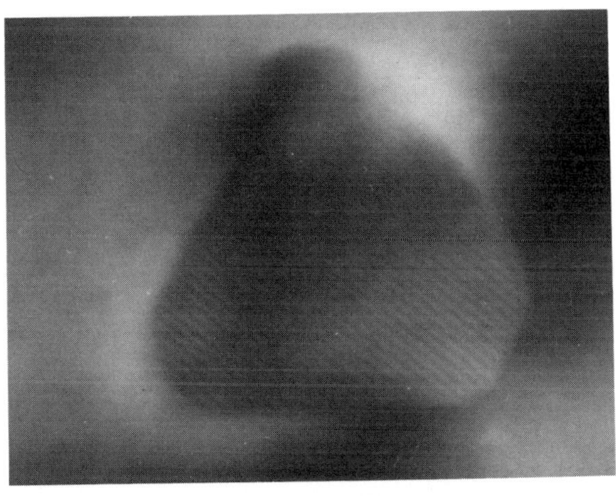

a)   20 nm

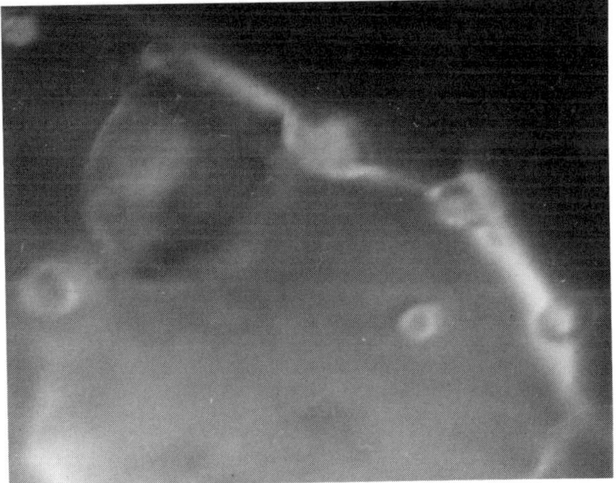

b)   20nm g_{111}

Fig. 4 - TEM Dark field: θ phase (Al_2Cu) precipitates nucleate on particles and dislocations (a,b,c).

Precipitation occurred upon air cooling after deformation. Repeated tensile deformation below the solution temperature showed an increase in flow stress due to formation of precipitates [3].

Both particles and precipitates were evident in grain interiors and boundaries, resisting both lattice and grain boundary dislocation motion (Fig. 4). Dislocations were attached to particle and precipitate interfaces. Rings of diffraction from the particles were observed in selected angle diffraction, corresponding to Al_4C_3 and Al_2O_3, similar to [7]. The typical grain morphology was of disc shaped 2-3μm grains

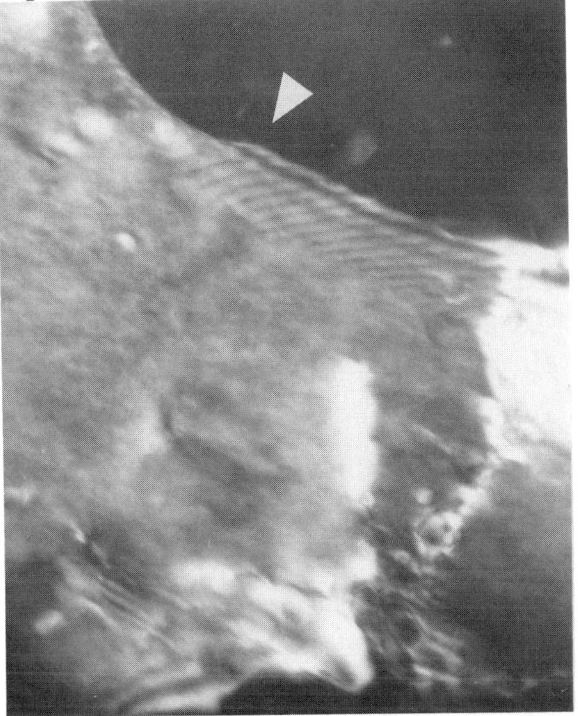

c)   50nm g_{111}

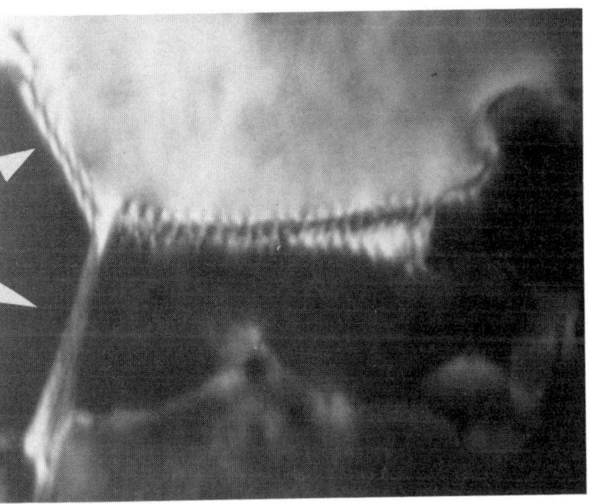

d)   50nm g_{111}

Fig. 4 - TEM Dark field: Dispersions pin both lattice and boundary dislocations (b,c,d). High angle boundaries have arrows, other boundaries are low angle.

about 0.3 μm thick (Fig. 4,5). Within each grain, a network of low angle boundaries was well established, that divided each grain into nearly equiaxed sub-micron subgrains. Subgrain boundary mismatch angles were typically about 2°, as determined from tilting experiments. The grain morphology showed no evidence of recrystallization or noticeable grain growth after elevated temperature deformation. From the grain morphology and the anisotropy in strain, the grain boundary sliding phenomena is like that of discs sliding over each other as in Fig. 6 [6]. Sliding occurs predominantly on boundaries that are high angled.

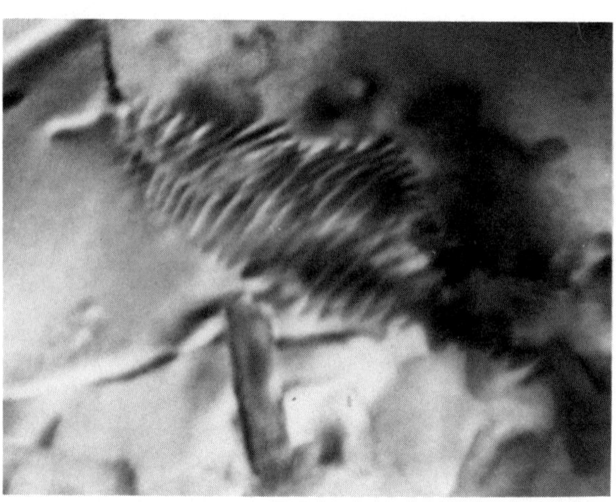

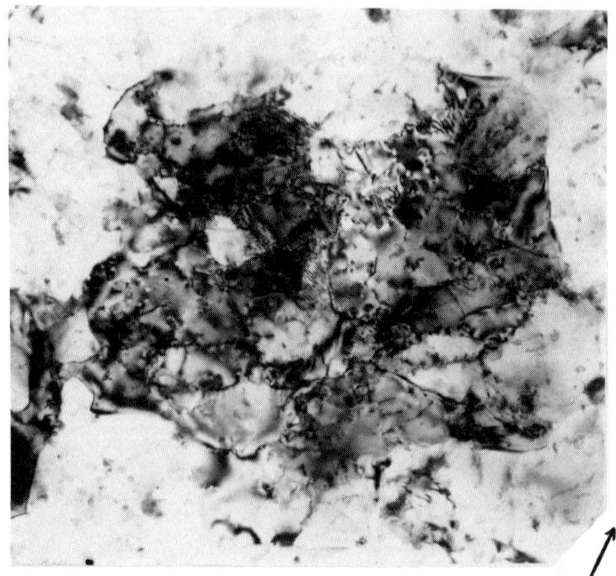

b) <u>500nm</u>   [220]
Fig. 5 - TEM Bright Field: a) Low angle grain boundary dislocations. b) Dark regions are oriented with beam parallel to 001 pole

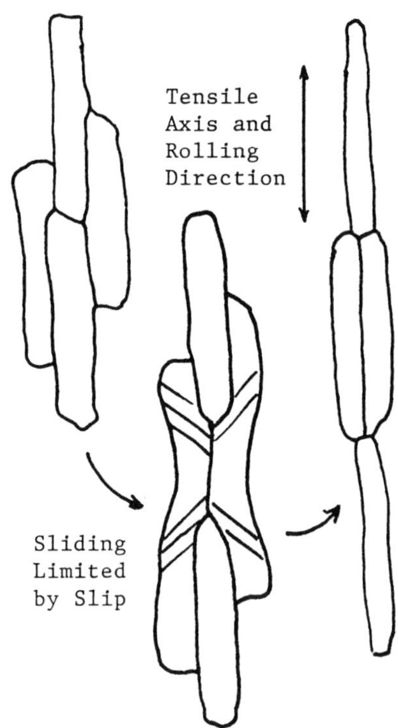

Fig. 6 - Schematic diagram of grain boundary sliding in IN90211.

ANALYSIS AND DISCUSSION

MICROSTRUCTURE - The grain size can be compared with that predicted by the Zener limit. This limit is defined when the chemical potential across a curved boundary balances the drag force for boundary motion, given by $d_{max} = 2r/3f_v \approx$ 100 nm [8]. The observed value is larger, indicating that the grain size is unlikely to grow any larger, consistent with the general observations of extremely stable grain size and grain morphology.

From the particle size information, an estimate of the Orowan looping stress can be obtained. Assuming that at high temperatures the amount of θ phase present is negligible, the Orowan stress can be estimated from

$$\tau_0 \approx 0.84bG/L_m; \qquad (2)$$

$$L_m = 2r(\pi/6f_v)^{1/3} \qquad (3)$$

where b is the Burgers vector, G is the shear modulus (G=30220-16T MPa [9]), and L_m (= 30 μm, with r=13.7μm) is the mean planar square lattice particle spacing. This value is 2/3 of the observed average spacing of particles along a dislocation line. However, many of the particles reside in grain boundaries, so the average spacing in the grain interiors is larger than the calculated value. Using the calculated value of 30 nm and the observed spacing of 46 nm, and the Von Mises criterion ($\sigma=\tau/3$), an Orowan (tensile) stress between 165 and 253 MPa is obtained. The observed break in the flow stress behavior near 180 MPa in Fig. 1 is consistent with the onset of predominantly Orowan looping behavior. Consequently, superplastic flow occurs well below the Orowan stress.

THE APPARENT STRESS EXPONENT - The data of Fig. 1 indicate a strain rate sensitivity of $m\approx0.25$, or a stress exponent of $n\approx4$. Edwards, et.al., derived a climb theory with n=4 for a zinc alloy with 0.3 and 0.6 μm alumina or tungsten dispersions [10]. There were 1 to 4 particles in each 1-3μm grain (there are 10^3 particles in each grain in IN90211). The model assumes that dislocations were generated at particle interfaces when the local stress exceeded a threshold stress. Dislocations piled up at grain boundaries are then annihilated during steady state climb, making the distance between the particle and the boundary a significant dimension in their model. Substituting parameters appropriate for IN90211, agreement was obtained at a strain rate near 1/sec, but not at higher or lower values. The Zn data also deviated from model in a similar way.

<u>Threshold Stresses in Lattice</u> - Flow below the Orowan stress occurs in creep deformation due to various climb mechanisms. Commonly, the flow stress is resisted by a temperature insensitive threshold stress of about half of the Orowan stress in particle strengthened metals where the interface is incoherent. The incoherent interfaces relax at high temperatures enough for the interface to become attractive to dislocations, allowing the line energy to be reduced. For coherent precipitates, the threshold stress is proportional to the applied stress, and the stress field of the harder precipitate repels the dislocation from the interface. These two extremes (Fig. 7) represent limiting cases for materials with small volume fractions of particles

[11-14]. The effect of these two threshold stresses on the apparent stress exponent is shown schematically in Fig. 8.

Two temperature dependent threshold stresses are operative in IN90211. One is associated with dislocation glide in the lattice, exhibiting an activation energy of 140 kJ/mol, (n=3, $\sigma\cdot\approx\sigma_0/20$), and the other with grain boundary sliding (60 kJ/mol, n=2, $\sigma\cdot\approx\sigma_0/7$) [5]. Though the n=2 values are larger than the n=3 stresses, both are smaller than half of the Orowan stress. The threshold stress analysis did not support the possibility of $n\approx4$.

The n=3 threshold stresses can be described by a mixed mode of climb resulting from partial relaxation of the interface proposed by Arzt, et. al. [12,13]. With the high homologous temperatures, lattice climb is facile, even though the particles have incoherent interfaces that attract dislocations. The presence of lattice climb, and the energy requirement to maximize the radius of

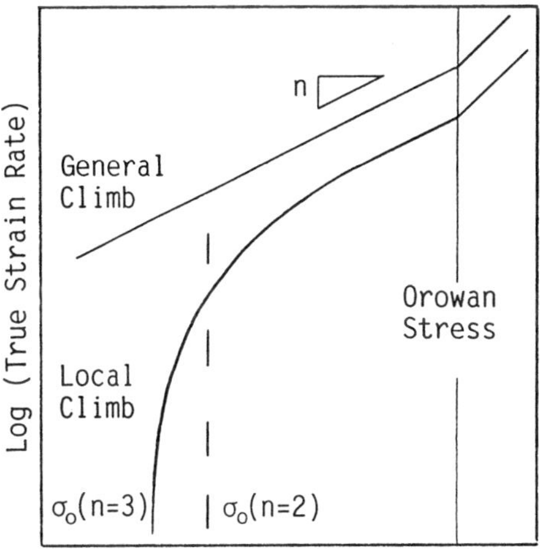

Fig. 8 - The effect on stress exponent is compared for local and general climb conditions.

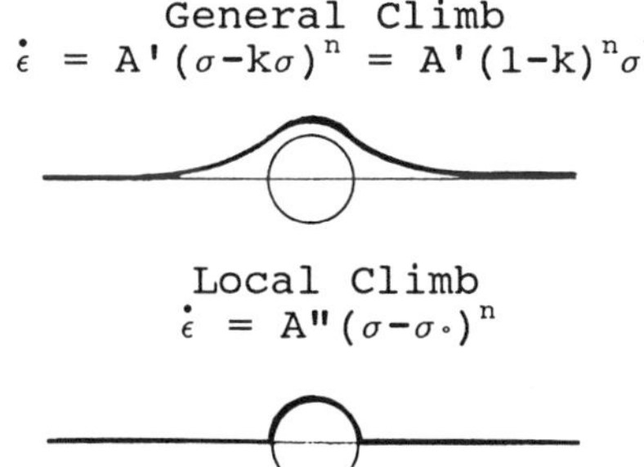

Fig. 7 - The route of dislocations passage over particles in General climb and Local climb conditions.

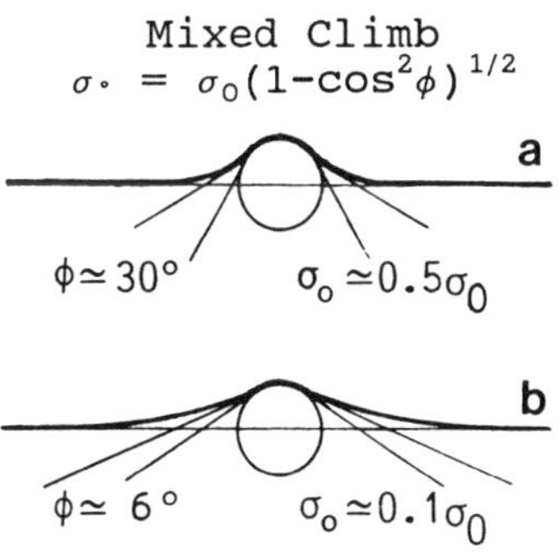

Fig. 9 - Mixed climb over particles for two different interface angles.

curvature where the dislocation separates from the particle, results in an equilibrium angle ϕ between the particle and the dislocation line. This angle then defines the magnitude of the threshold stress, as illustrated in Fig. 9.

Experimental evidence for this theory is evident in Fig. 4, and in [13]. The angle between the particle and the dislocation line is larger in Fig. 4 than predicted by the threshold stress where $\sigma_\bullet \approx \sigma_0/20$ for high temperature data, because the specimen was cooled under load slowly, and the dislocation had ample time to assume the shape predicted by a larger threshold stress resulting from the lower temperature. The threshold stress depends upon temperature in accordance with the amount of climb permitted by diffusion in the lattice and in the interface. The origin of the lattice (n=3) threshold stress is unambiguous, since the rate controlling mechanism is Mg solute atom drag of dislocation cores [5]. The particles also resist dislocation glide by means of the threshold stress.

<u>Threshold Stresses in Grain Boundaries</u> - The threshold stress for the n=2 phenomena is not as easily explained. The n=2 threshold stresses correspond with the lower limit of the superplastic mechanism, as illustrated in Figs. 1 and 8, similar to observations in mechanically alloyed MA6000 [15]. Below the n=2 threshold stresses, the grain boundary sliding process is inoperative.

Grain boundary sliding is nearly always observed in superplastic deformation [16]. Grain boundary dislocations are generally involved in theoretical treatments of superplastic deformation processes [17]. Both grain boundary sliding and grain boundary dislocations have been observed in IN90211 (Fig. 4-6, and [6]). Regardless of the source of grain boundary dislocations, they must be mobile for grain boundary sliding to occur. Boundaries on which sliding occurs must be high angled [18]. For dispersion strengthened materials to be superplastic, grain boundary dislocations must be able to overcome particles in the boundary.

Particles in the boundary resist the progress of grain boundary dislocations [18]. The geometry and chemistry of the grain boundary region is more complicated than the lattice. Models pertaining to interface phenomena have been proposed [19,20]. Threshold stresses arise in the interface in analogous ways as in the lattice, and also in some ways that are particular to the interface(s). A very complicated model derived in [20] is not applicable for these experiments, since the homologous temperatures is high enough

(0.76-0.82 T_m) for lattice diffusivity to dominate the effective diffusivity. Threshold stress predictions based strictly on geometrical considerations have been derived, [20], but since the threshold stresses vary with temperature only by the variation of the shear modulus, they are apparently irrelevant to the temperature dependent threshold stresses obtained in the analysis. Therefore only the threshold stress theories that are analogous to the local and general climb described for the lattice may be considered further.

Assuming that grain boundary dislocations overcome particles in the boundary by climbing processes, the boundary threshold stresses differ from the lattice threshold stresses in accordance with the differences in their structural properties. The Burgers vector for grain boundary dislocation depends upon whether it is low or high angled. The Burgers vector for a high angle grain boundary can be estimated as one third of the lattice Burgers vector [20]. Similarly, an estimate of the shear modulus of the boundary can be obtained, considering the difference between the thermal expansion in a polycrystal compared to a single crystal. This has been done for copper, and the expansion coefficient derived for the grain boundary is 2-5 times larger than the bulk [21]. Since expansion is a measure of the elastic bond strength, the shear modulus in the boundary can be estimated to be about one third of the bulk value. Using the above two estimates in equation 2, the boundary Orowan stress is about an order of magnitude smaller than it would be in the bulk. If local climb phenomena occur in the boundary then the local climb threshold stress can be estimated at about half the boundary Orowan stress for the IN90211 alloys, a value that is closer to the n=3 threshold stress values than the n=2 threshold stress values.

The discussion regarding the mixed local and general climb for the bulk does not immediately carry over into the boundary, for geometric reasons. It is difficult to visualize how any kind of general climb phenomena can occur if a boundary dislocation is constricted to remain in the interfaces. For this reason, boundary threshold stresses should be relatively insensitive to temperature once the temperature is high enough to permit interfacial diffusion in the particle/matrix interface. Since this result does not agree with the temperature sensitivity obtained in the above analysis, it appears that another explanation is needed. The details of the boundary threshold stress may be obscured since the grain boundary sliding process is not rate limiting.

SOLUTE ATOM DRAG - The rate limiting deformation mechanism is that of solute drag of Mg atoms in dislocation cores [5], where the Cottrell-Jaswon model predicts the deformation [9]. At high stresses within this flow regime, dislocations can break away from the solute atmospheres, producing serrations on the flow curve (the Portevin-LeChatelier effect). The stress for breakaway is given by [9]

$$\sigma_b = W^2 c/5b^3 kT, \text{ where} \tag{4}$$

$$W = -G|\delta V|(1+v)/2\pi(1-v) \tag{5}$$

where c is bulk solute concentration, δV is the difference in atomic volume between solute and solvent atoms, and v is Poisson's ratio. For the conditions of the above tests, the breakaway stress is near 20 MPa. The model for dislocation breakaway during bowing between particles in IN9051 (Table I), by Oliver and Nix [19], provides a similar value to the breakaway stress. Serrations are observed in the flow curve at stresses in excess of 20 MPa, so either model describes the emergence of the observed serrations.

In addition to the possibility of solute breakaway in the grain interiors, similar phenomena can occur in the grain boundaries. Mg segregates in Al alloys to the boundary, and at a temperatures investigated, an enrichment by a factor of 3 has been observed [22]. Estimating the breakaway stress in the boundary in a similar way as for the boundary threshold stresses, the breakaway stress is essentially the same value as that in the lattice.

The irregular dislocation motion that produces serrations causes inhomogeneous macroscopic flow, which reduces flow stability [23]. This has been demonstrated in a superplastic Al-Cu alloy that had optimum elongations near a strain rate of 10^{-3}/sec, a strain rate typical for superplasticity [24]. A dip in elongation occurred in the middle of the superplastic regime, similar to that in Fig. 1. In this case the dip was due to the stress exceeding the breakaway stress rather than adiabatic heating. Since the stress is sufficiently high for serrations to occur at all superplastic strain rates in IN90211, elongations are less than what could be obtained if no solute atom breakaway occurred.

This alloy shows optimum conditions for superplastic elongation under the following conditions: Lattice diffusivity is dominant and dislocation glide is facile. Cavitation is precluded by both of the above factors to accommodate stress concentrations in the boundaries without nucleating cavities [6]. Practical superplastic forming only requires 200-300% elongation, so the superplastic elongations are adequate.

However, since both adiabatic heating and serrated flow can reduce elongation, it may be possible to adjust the microstructure slightly to obtain greater elongations. A slightly larger grain size may be possible if milling time or the atmosphere is changed during milling. A slightly larger grain size will decrease the strain rate for optimum superplastic elongation, as well as decrease the stress needed for deformation.

Also, an adjustment in alloy composition will affect the details of serrated flow. Adding more Mg (or other solute atoms) will raise the breakaway stress. Removing the solute strengthening will remove the possibility of serrated flow, but the rate limiting process for slip may change from glide to climb. Such a change will cause the stress exponent to increase to 4 or 5, decreasing the strain rate sensitivity, and therefore, elongation. In IN9051 and IN9052 [19,25], slip is limited by climb, even though the only alloying addition is Mg. The reason why Cu promotes solute drag rather than climb in IN90211 is unclear.

If the price paid for a larger grain size is larger dispersion size (and possibly poorer ambient properties), then the details of how dislocations overcome particles in both the lattice and the boundaries will also be changed. It is unclear how much room there is for modification of this remarkably optimum microstructure and composition to optimize both superplastic and post-forming properties.

CONCLUSIONS

1. Superplastic elongations have been obtained at unusually high strain rates in mechanically alloyed aluminum.
2. The deformation mechanisms are those usually reported in superplastic deformation, however, instead of grain boundary sliding being the rate controlling mechanism, solute atom drag of lattice dislocations controls the overall deformation.
3. Grain boundary sliding occurs in the form of clusters of sub-micron subgrains in disc shaped grains. Sliding primarily occurs on the surfaces where boundaries are high angled.
4. The high rate of superplastic deformation is due to the fine grain size, but the high strain rate introduces two factors that reduce potential elongation: thermal instabilities from adiabatic heating,

and serrated flow above the breakaway stress. To reduce these two undesirable factors, a slight increase in the grain size (that would lower superplastic stress and strain rate) may be profitable.
5. A modification in alloy composition may permit solute breakaway stress to be removed, or increased, providing more homogeneous flow.

ACKNOWLEDGEMENTS

This work has been supported by the Air Force Office of Scientific Research under grant # AFOSR-86-0091, monitored by Dr. A. Rosenstein. Additional support, and the material, was provided by the Research and Development Division of Lockheed Missiles and Space Co., Inc., in collaboration with J. Wadsworth and T.G. Nieh.

REFERENCES

1. Nieh, T.G. and P.S. Gilman, J. Wadsworth, Scripta Metall., 19, 1375 (1985)

2. Bieler, T.R., T.G. Nieh, J. Wadsworth, and A.K. Mukherjee, Scripta Metall., 22, 81, (1988)

3. Bieler, T.R. and A.K. Mukherjee, "Superplasticity in Aerospace II", symposium at February TMS/AIME meeting in Anaheim, in press (1990)

4. Bieler, T.R. and A.K. Mukherjee, Scripta Metall., in press (1990)

5. Bieler, T.R. and A.K. Mukherjee, Materials Science and Engineering, in press (1990)

6. Bieler, T.R. and A.K. Mukherjee, Journal of Mat. Sci., in press (1990)

7. Singer, R.F., W.C. Oliver and W.D. Nix, Metall. Trans 11A, 1895 (1980)

8. C. Zener, private communication to C.S. Smith in Trans AIME 175, 15 (1949)

9. Mohamed, F.A. and T.G. Langdon, Acta Metall., 22, 779 (1974)

10. Edwards, G.R., T.R. McNelley, O.K. Sherby, Phil. Mag., 32, 1245 (1975)

11. Blum, W. and B. Reppich, "Creep Behaviour of Crystalline Solids", pp. 100-106, eds. B. Wilshire and R.W Evans, Pineridge Press Ltd., Swansea, U.K., (1985)

12. Arzt, E. and D.S. Wilkinson, Acta Metall., 34, 1893 (1986)

13. Arzt, E., J. Roesler and J.H. Schroeder, "Creep and Fracture of Engineering Materials and Structures", pp. 217-230, eds. B. Wilshire and R.W. Evans, The Institute of Metals, London, (1987)

14. Herrick, R.S., J.R. Weertman, R. Petkovic-Luton, and M.J. Luton, Scripta Metall., 22, 1879 (1988)

15. Gregory, J.K., J.C. Gibeling and W.D. Nix, Metall. Trans., 16A, 777 (1985)

16. Kashyap, B.P., and A.K. Mukherjee, Res Mechanica 17, 293 (1986)

17. Suery, M., and A.K. Mukherjee, "Creep Behavior of Crystalline Solids", p. 172, eds. B. Wilshire, R.W. Evans, Pineridge Press, Swansea, U.K. (1984)

18. Howell P.R., and G.L. Dunlop, "Creep and fracture of Engineering Materials an Structures", eds. B. Wilshire, D.R.J. Owen, p. 127, Pineridge Press, Swansea, U.K. (1981)

19. Oliver, W.C. and W.D. Nix, Acta Metall., 30, 1335-47 (1982)

20. Arzt, E., M.F. Ashby, and R.A. Verrall, Acta Metall., 31, 1977-1989 (1983)

21. Klam, H.J., H. Hahn, and H. Gleiter, Acta Metall., 35, 2101 (1987)

22. Lea, C., and C. Molinari, J. Mat. Sci., 19, 2336 (1984)

23. McCormick, P.G., Acta Metall., 36, 3061 (1988)

24. Chaudhury P.K., and F.A. Mohamed, Met. Trans. 18A, 2105 (1987)

25. T.G. Nieh, T.R. Bieler, unpublished research

KEYNOTE ADDRESS
MECHANICAL ALLOYING OF TITANIUM ALLOYS

C. Suryanarayana
Wright Research & Development Center WRDC/MLLS
Wright-Patterson Air Force Base, Ohio 45433-6533, USA

R. Sundaresan
Defence Metallurgical Research Laboratory
Hyderabad - 500 258, India

F. H. Froes
Institute for Materials and Advanced Processes
University of Idaho
Moscow, Idaho 83843-4140, USA

ABSTRACT

Application of mechanical alloying to titanium alloys is of recent development. A wide range of titanium-base alloys and intermetallics (both Ti_3Al and $TiAl$) have been examined and it has been shown that nanocrystalline metastable f.c.c. solid solutions can be obtained in the virtually immiscible Ti-Mg alloys. It has also been shown that the resistance to coarsening of both the grains and the dispersoids in mechanically alloyed Ti_3Al-based alloys is much higher than in comparison with rapidly solidified alloys. Formation of an amorphous phase in the Ti-Ni-Cu system and expected $TiAl$ phase formation from Al_3Ti and TiH_2 powder mixtures have also been investigated.

SEVERAL NON-EQUILIBRIUM processing techniques have been developed in recent years to modify the alloy chemistry of existing commercial alloys and also to develop entirely novel compositions to improve the material performance in a variety of demanding environments. These techniques include rapid solidification from the melt [1], mechanical alloying [2], laser processing [3], ion implantation [4] and amongst these mechanical alloying, the subject matter of the present Conference, has been receiving increased attention. Though originally developed to make high-temperature alloys combining oxide dispersion strengthening with γ' precipitation hardening in a nickel-based superalloy [5], mechanical alloying has now been shown to be beneficial in developing a number of alloy systems possessing novel combinations of strength, oxidation resistance and sulfidation resistance at temperatures up to 1100 ºC. These improvements have been found to be due to a very fine and uniform distribution of oxide dispersions in a fine-grained matrix and also due to the formation of a wide spectrum of metastable phases including supersaturated solid solutions [6], crystalline intermediate phases [7], and amorphous phases [8,9].

There is much to be gained from the mechanical alloying of titanium alloys. Since the whole process takes place entirely in the solid state, it is possible to produce alloys from virtually immiscible components. It is also possible to increase the solid solubility limits of alloying elements and form metastable crystalline and amorphous phases to improve the mechanical performance . Further improvement in the properties of these materials can also be achieved through a uniform dispersion of fine oxide or carbide particles resulting from the mechanical alloying process. The present article will deal with the structural characterization of several titanium-base alloys subjected to mechanical alloying and the results obtained will be compared with those obtained by another important non-equilibrium processing technique, viz., rapid solidification from the melt, wherever possible.

EXPERIMENTAL PROCEDURE

Mechanical alloying was carried out at room temperature in a Spex 8000 Mixer mill or in a Zegvari attritor. In either case, the grinding medium was 3/16" diameter hardened 52100 steel balls. In order to avoid/minimize oxidation during the milling operation, the premixed powder and the balls were loaded into the canister in an argon-filled glove box or a continuous flow of argon gas was maintained in the attritor. Forced air cooling during milling prevented an excessive increase of the temperature of the powder. While the powder charge was about 10 g for the Spex mill, it was about 80 g for the attritor.

The mechanically alloyed powder was taken out of the container periodically to follow the progress of alloying. X-ray diffraction, optical and transmission electron microscopy and differential scanning calorimetry techniques were employed to characterize the mechanically alloyed powder both in the as-milled condition and in the consolidated form. Consolidation was carried out either by vacuum hot pressing (VHP) at 30,000 psi or by hot isostatic pressing (HIP) at 40,000 psi at various temperatures. The hot-consolidated powder compacts were then annealed allowing the transformation behavior to be followed.

The Titanium-Magnesium System

Addition of magnesium to titanium can reduce the density and potentially produce very light alloys with a high specific strength. On the basis of theoretical considerations there should be appreciable solid solubility of magnesium in titanium. But, the solid solubility under equilibrium conditions has been reported to be small. Attempts to produce titanium alloys containing larger amounts of magnesium have met with only limited success because magnesium, with its boiling point well below the melting point of titanium, cannot be retained in the molten metal using conventional melting methods. Non-conventional methods such as rapid solidification from the melt, solid -state reaction in the form of sintering, sheath rolling, and extrusion did not improve the situation any further [10]. Hence, mechanical alloying has been tried out and this led to the production of nanocrystalline solid solutions having a non-equilibrium f.c.c. structure.

X-ray diffraction patterns of the powders mechanically alloyed even after 4 h suggested the formation of a h.c.p. solid solution of magnesium in titanium as indicated by the shift in the positions of the x-ray peaks of titanium towards lower angles and a concomitant decrease in the intensity of magnesium peaks [10,11]. Some amount of magnesium was also found to be present even after long milling times. With increased milling time, the titanium peaks broadened considerably and merged into a single peak (at 12 to 16 h of milling time in the Spex mill) suggesting the formation of either an amorphous phase or an extremely fine-grained solid solution.

Transmission electron microscopic investigations were carried out on the powders after consolidation by HIP'ing at 700 ºC and 40,000 psi pressure for 1 h. The micro-structure consists of two types of grains — one very large, about 0.4 μm in size and the other extremely fine, extending down to the nanometer level. The relative proportions of the two types of grains are not always the same in different regions of the specimens. Figure 1 shows a region where the micro-structure contains predominantly large equiaxed grains and the selected area diffraction patterns (inset) could be indexed on the basis of a h.c.p. structure with the lattice parameters corresponding to those of pure titanium. On the other hand, the fine 10 to 15 nm grains (Figure 2) give rise to continuous rings in the diffraction patterns . This diffraction pattern was most satisfactorily indexed on the basis of an f.c.c. structure with $a = 0.426$ nm. Since no f.c.c. phase has been reported under equilibrium conditions in the Ti-Mg system, this may well be a metastable phase. Energy dispersive x-ray spectra (Figure 3) indicated that while the large h.c.p. grains contained only pure titanium, the f.c.c. nanocrystalline grains contained approximately 2.5 to 3.0 wt.% magnesium and the balance titanium [12].

It has been shown that the metastable f.c.c. structure is not due to titanium hydride because the latter always forms as thin platelets and never as equiaxed grains [13]. Other contaminants from the system also are not responsible for the formation of this metastable phase, since on annealing the powder compact at sufficiently high temperature, the f.c.c. phase transformed to the two equilibrium phases, viz., elemental titanium and

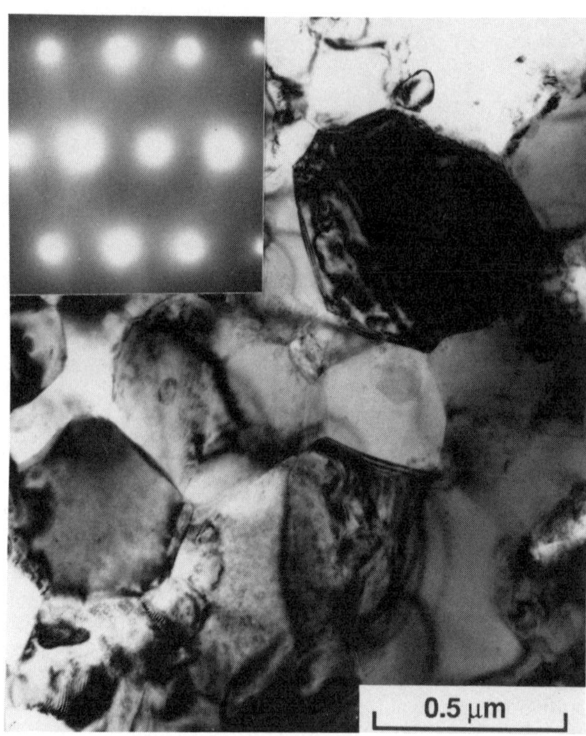

Fig. 1 - Microstructure showing large equiaxed grains. The diffraction pattern (inset) is indexed as arising from h.c.p. elemental titanium. Foil normal is [2$\bar{4}$2$\bar{3}$].

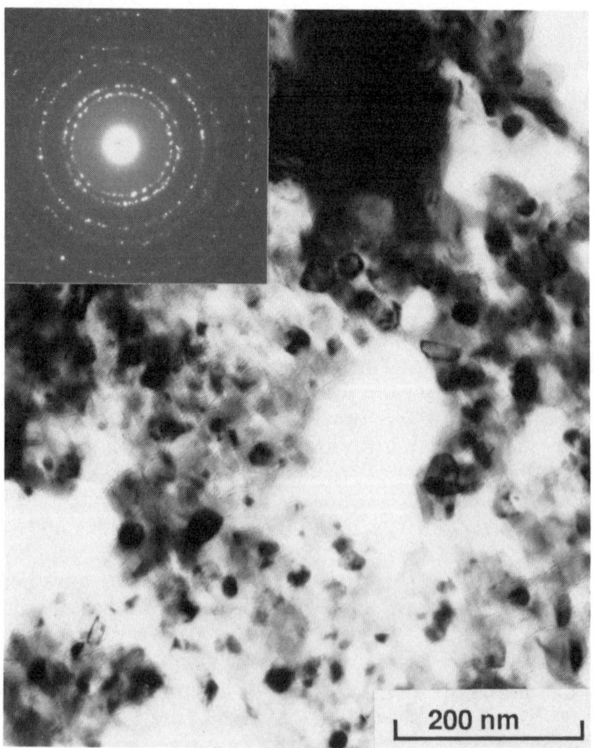

Fig. 2 - Microstructure showing an abundance of nanocrystalline grains. The diffraction pattern (inset) indicates the presence of a metastable f.c.c structure with $a = 0.426$ nm.

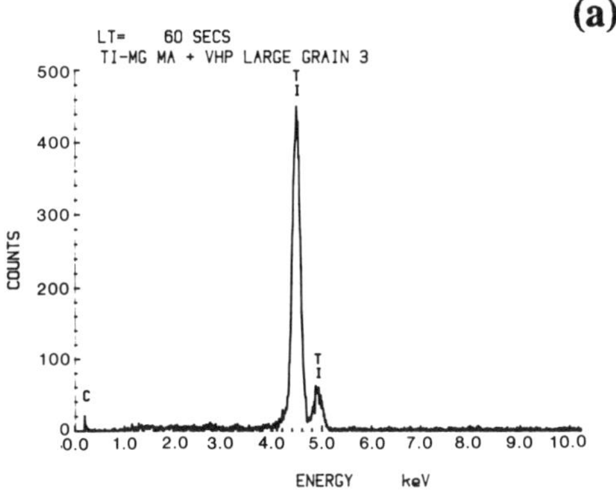

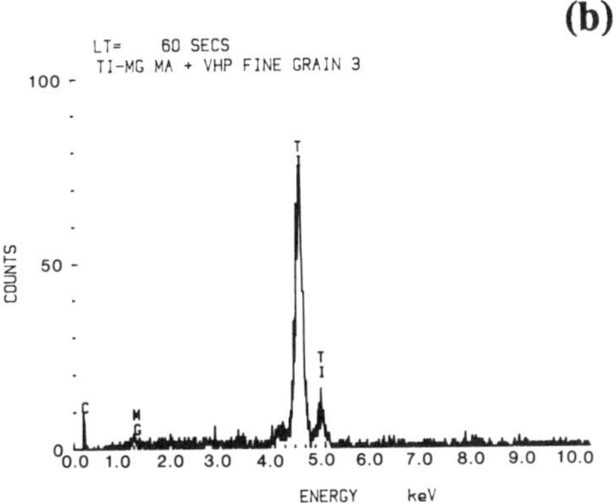

Fig.3 - X-ray spectra from (a) the large grains showing the presence of only titanium and (b) the nanometer-sized grains indicating the presence of about 3 wt.% Mg in addition to Ti.

magnesium. *In-situ* hot-stage experiments have confirmed that the fine grain size is quite stable and does not coarsen appreciably at least up to 500 °C.

The formation of the metastable f.c.c. solid solution phase can be explained in one of the following ways. Firstly, it might have formed as a result of the mechanical alloying operation *per se*. The presence of about 3 wt.% magnesium in titanium could have changed the electronic structure significantly to stabilize the f.c.c. structure. Since the X-ray diffraction patterns of the as-milled powder did not indicate the presence of an f.c.c. phase, this possibility can be discounted. Secondly, it is possible that the f.c.c. phase has formed during crystallization of an amorphous phase. However, there is no evidence for either the formation or the decomposition of an amorphous phase in the mechanically alloyed condition in the present program; this possibility may also be discounted. A third possibility is that the Ti-Mg solid solution had the h.c.p. structure as soon as forms. But, because of the heavy deformation involved, the h.c.p. structure might have transformed to the f.c.c. structure, since the introduction (or removal) of stacking faults at regular intervals can convert one close-packed structure into another. But, it should be mentioned that stacking faults could not be observed because of the extremely fine

grain size of the f.c.c. phase. However, this third possibility received additional support since the distances of closest approach of atoms in the hypothetical h.c.p. solid solution containing about 3 wt.% magnesium and the resultant f.c.c. phase are almost the same. The fact that deformation was involved was indirectly confirmed by the observation of low-angle grain boundaries (formed due to recovery) in the microstructure.

Thus, it has been shown that mechanical alloying can extend the solid solubility of magnesium in titanium to produce nanocrystalline f.c.c. solid solutions.

Amorphous Phase Formation

Amorphous phases have been known to form in suitable alloy systems through techniques such as rapid solidification from the liquid state, vapor deposition, sputtering, laser processing and ion implantation. In all these methods, there is a change in the state of matter, i.e., a solid phase is formed either from the liquid or the vapor phase, and the effective quenching rate and associated undercooling have been shown to be responsible for the amorphization process. However, in recent years there have been a number of reports on the formation of amorphous phases obtained by mechanical alloying in a variety of alloy systems, but only to a limited extent in titanium-base binary alloys [14-16]. Hence, a study was undertaken to clarify the nature of the amorphous phase formation in a mechanically alloyed Ti-Ni-Cu ternary system [17,18], a system which is susceptible to amorphous phase formation by rapid solidification because of the deep eutectic.

Both elemental blend and prealloyed Ti-Ni-Cu powders were used for mechanical alloying. In order to investigate the effect of the process control agent, stearic acid was added in some runs only. X-ray diffraction patterns after milling the powder for 16 h revealed the presence of a single diffuse peak suggesting formation of an amorphous phase. Transmission electron microscopy results indicated that a homogeneous amorphous phase formed only when the elemental blend or prealloyed powders were milled for 16 h without stearic acid, while with stearic acid both microcrystalline and amorphous phases were observed for similar milling times (Figure 4).

Table 1

Crystallization temperatures (T_X) and heats of crystallization (ΔH) for the Ti-Ni-Cu amorphous phase produced by mechanical alloying and rapid solidification

Alloy composition (wt.%)	T_X(°C)	ΔH(J/g)
Ti-18Ni-15Cu (MA/Elemental blend)	464	96.2
Ti-16Ni-15Cu (MA/Prealloyed powder)	466	62.7
Ti-16Ni-15Cu (RS Splat)	437	19.9*

* not completely amorphous

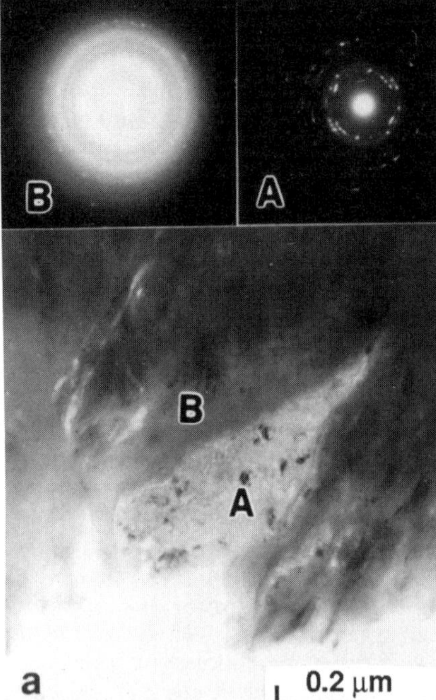

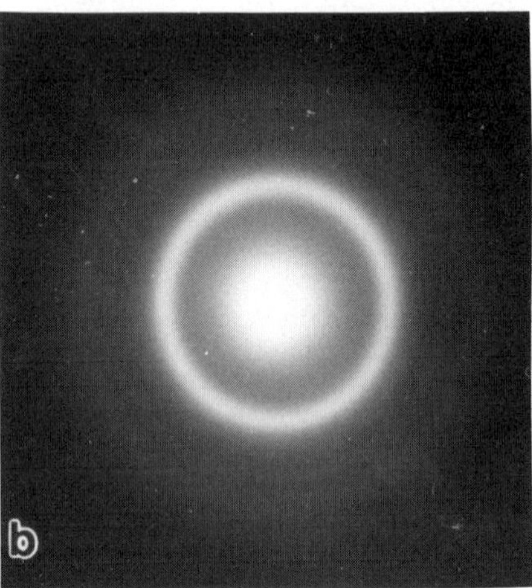

Fig. 4 - (a) Electron micrograph of the Ti-Ni-Cu elemental blend powder mechanically alloyed with stearic acid for 16 h showing the presence of both microcrystalline and amorphous regions (Marked A and B respectively). (b) electron diffraction pattern showing the presence of a homogeneous amorphous phase when milled without stearic acid.

Further confirmation of the amorphous phase formation was obtained from differential scanning calorimetry experiments by the presence of an exothermic peak corresponding to the crystallization event. The crystallization temperatures and the heats of crystallization are reported in Table 1, together with the values for the amorphous phase obtained by rapid solidification from the melt (splat).

The formation of the amorphous phase is suggested to be related to the generation of a high density of dislocations. At high enough dislocation density, diffusion through dislocation cores becomes very significant. This high atomic mobility represents a high configurational entropy and promotes the formation of an amorphous phase.

Ti₃Al (α_2)-Based Alloys

In recent years there has been an intense activity on the development of alloys based on the intermetallics Ti₃Al (α_2) and TiAl (γ) for elevated temperature applications, especially in the aerospace industry. This has been essentially due to the attractive combination of properties such as low density, high-temperature strength, modulus retention, resistance to hydrogen absorption (TiAl) and good resistance to oxidation. The inherent room temperature brittleness of the intermetallics must be overcome by combinations of phase control by alloying additions, microstructural control by processing and by introducing fine dispersions [19,20]. Although considerable work has been done to develop these intermetallics through both ingot metallurgy and rapid solidification/powder metallurgy routes, it has only yielded limited success. The present study was undertaken to evaluate the effects of mechanical alloying on Ti₃Al (α_2)-based alloy systems, both with and without rare-earth element additions. X-ray diffraction, optical and transmission electron microscopy techniques have been used extensively to characterize the distribution and thermal stability of grains and dispersoids in these alloys.

Mechanical alloying was carried out on atomized powders of Ti₃Al-based alloys in a Zegvari attritor under a flowing argon cover with chilled water cooling, and a powder to ball ratio of 1:34. Screening experiments were carried out on several alloys and based on these results the Ti-25Al-10Nb-3V-1Mo (at.%) (corresponding to Ti-14Al-20Nb-3V-2Mo (wt.%)) alloy was chosen for a detailed study. The powders milled to different extents of time were consolidated by either VHP or HIP'ing. Knoop hardness measurements were used as a primary check on room temperature mechanical properties.

Optical microscopy of the alloy consolidated by VHP at 980 ºC/1 kbar for 4 h directly from the gas atomized powder revealed the presence of a two-phase structure consisting of equiaxed α_2 with a grain size of about 4 μm and the ordered B2 phase at the intergranular regions. Consolidation of the mechanically alloyed powders led to a structure consisting of both very fine grains of about 1 μm and coarser grains of about 3 μm. Among the coarser grains in mechanically alloyed compacts, there is a much larger proportion of the lenticular grains than in the powder not subjected to mechanical alloying [21]. The addition of Gd did not lead to a fine enough dispersion (0.1 μm in size) in the mechanically alloyed powder compact to significantly enhance mechanical properties.

X-ray diffraction patterns of the as-atomized powders showed substantially the B2 structure, while the consolidated powder showed a mixture of α_2 and B2 phases. Although the mechanically alloyed and compacted powder also contained the same two phases, viz., α_2 and B2 (Figure 5), the proportion of the B2 phase retained in the mechanically alloyed consolidated

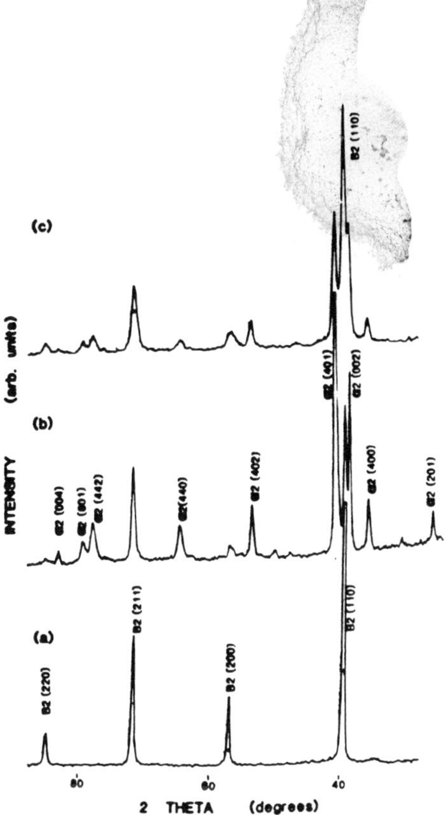

Fig. 5 - X-ray diffraction patterns from the Ti-25Al-10Nb-3V-1Mo alloy in the (a) gas atomized state showing only the B2 phase, (b) as above, but vacuum hot pressed showing both B2 and α_2 phases, and (c) mechanically alloyed Ti-25Al-10Nb-3V-1Mo + 2 wt.% Gd alloy HIP'ed and annealed at 1090 ºC.

condition was much more than in the directly consolidated condition. Further, it appears that the B2 phase retained in the mechanically alloyed HIP compact is substantially more stable, and is retained after a long exposure at 1090 ºC [21].

Table 2 shows the hardness values from the consolidated materials and it can be seen that there is a substantial increase in the room temperature hardness in the mechanically alloyed material compared to the material from the gas-atomized powder. Since the oxygen pick-up during mechanical alloying is only about 200 to 300 ppm, the hardness increase may be attributed mainly to the grain refinement.

Table 2
Knoop hardness values of consolidated Ti₃Al-based alloys with and without Gd

Condition	Fine grains	Coarse grains
Gas atomized (no Gd) + VHP	-	339
MA (no Gd) + VHP	722	434
MA (with Gd) + VHP	861	444
MA (with Gd) + HIP + 1090 ºC/6 h	711	378

Since Gd did not produce a fine enough dispersion, additions of Er were then evaluated. The starting powder was again rapidly solidified Ti-25Al-10Nb-3V-1Mo alloy powder, but produced by the plasma rotating electrode process, to which 2 wt.% Er was later added.

Detailed electron microscopic investigations were carried out on the mechanically alloyed powder compacts obtained by HIP'ing at 40,000 psi pressure and 1000 ºC for 4 h. These compacts were subsequently annealed at temperatures up to 1200 ºC to evaluate the thermal stability of both the grains and the dispersoids. Further, in order to evaluate the extent of deformation during the mechanical alloying operation, the resultant powder was separated into coarse (+140 mesh) and fine (-140 mesh) fractions. Significant differences were observed between these two sets of powders.

Figure 6 shows a typical electron micrograph of the fine powder in the as-HIP'ed condition. The microstructural features of interest are (i) the very fine grain size (about 0.2 to 0.5 μm), (ii) two different sizes dispersoids averaging 10 and 100 nm, respectively and (iii) a high density of defect structure consisting mostly of stacking faults. On the other hand, the coarse powder showed features which were different from the above. For example, Figure 7(a) shows mostly featureless equiaxed grains about 2 μm across, while Figure 7(b) shows fine 0.5 μm size grains containing a high density of dispersoids. Two different size ranges of dispersoids can again be seen in this case. Dislocations and dislocation arrays have also been observed in some regions [22]. On annealing these powder compacts at temperatures of 1100 and 1200 ºC, irrespective of the original size fraction, both the grains and dispersoids coarsened slightly. Table 3 summarizes the variation in grain and dispersoid size as a function of the annealing treatments and Figure 8 shows typical electron micrographs. Three different sizes of the dispersoids and continued presence of the deformation structure can be observed at this stage. The coarsening of both the grains and dispersoids is only marginal and that there is no preferential growth of the dispersoids at the

Fig. 6 - Electron micrograph of the fine powder from the Ti-25Al-10Nb-3V-1Mo + 2 wt.% Er alloy mechanically alloyed and HIP'ed.

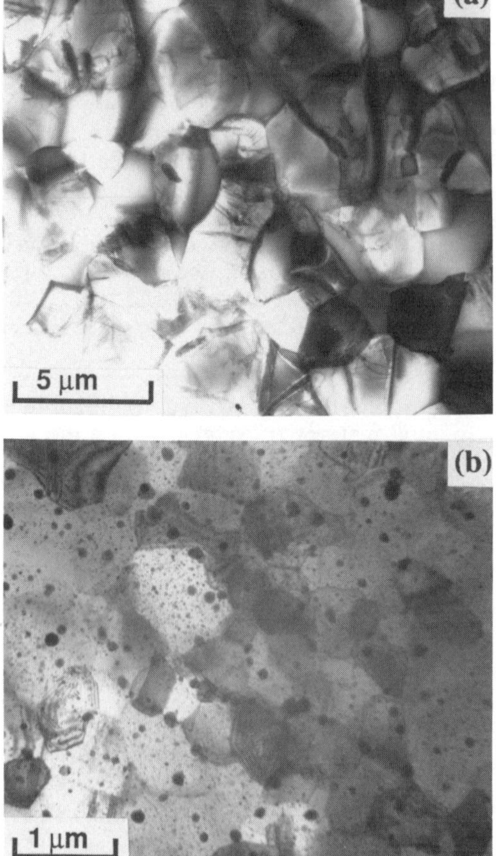

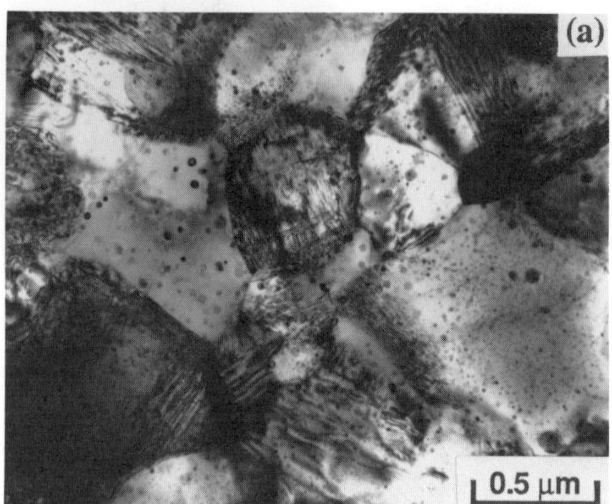

Fig. 7 - Electron micrographs of the coarse powder from the Ti-25Al-10Nb-3V-1Mo + 2 wt.% Er alloy mechanically alloyed and HIP'ed. (a) shows large, featureless equiaxed grains devoid of any dispersoid while (b) shows fine grains containing two different sizes of the dispersoids.

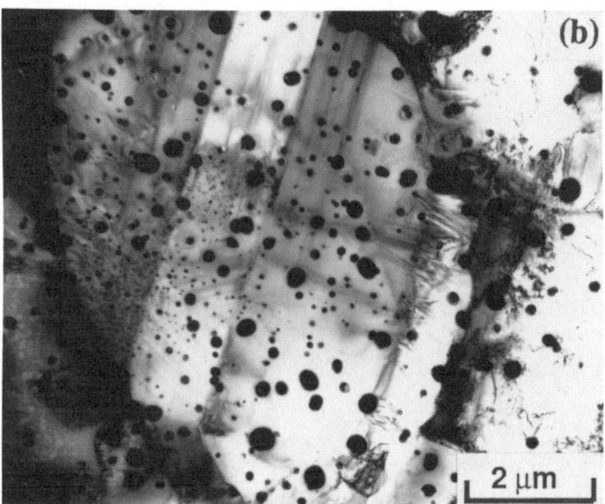

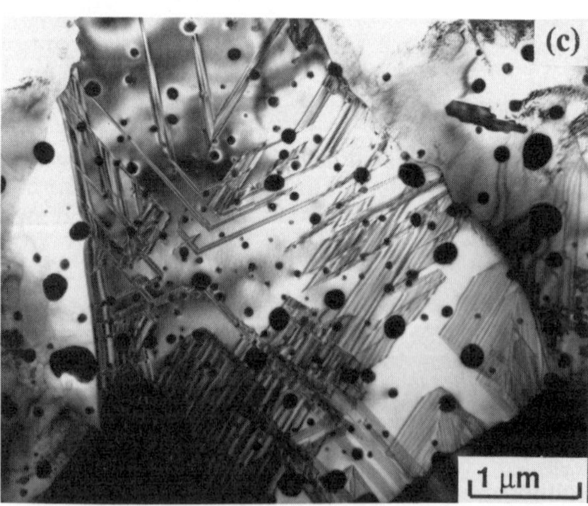

Fig. 8 - Electron micrographs of the Ti-25Al-10Nb-3V-1Mo + 2 wt.% Er alloy powder mechanically alloyed , HIP'ed and annealed at (a) 1100 ºC for 2 h, (b) and (c) 1200 ºC for 1 h. Note the continued presence of a high density of stacking faults in the powder even on annealing at 1200 ºC.

grain boundaries. Consequently, there are no dispersoid-free zones near the grain boundaries, which contrasts directly with the results obtained on rapidly solidified materials [23]. Figure 9 shows a comparison of the microstructures of HIP compacts from the rapidly solidified and mechanically alloyed products. Even after HIP'ing at a higher temperature, the mechanically alloyed material showed a fine grain size, small size of the dispersoids and the absence of dispersoid-free zones.

The observation of a high density of dislocations and stacking faults can be easily rationalized in terms of the large amount of plastic deformation experienced by the individual powder particles during the milling operation. The continued presence of the deformation structure even at 1200 ºC suggests that this structure is extremely stable and omens well for strength retention up to very high temperatures. The occurrence of three different sizes of the dispersoids in the annealed material is also easy to understand because at these high temperatures, the excess erbium from the supersaturated solid solution would have

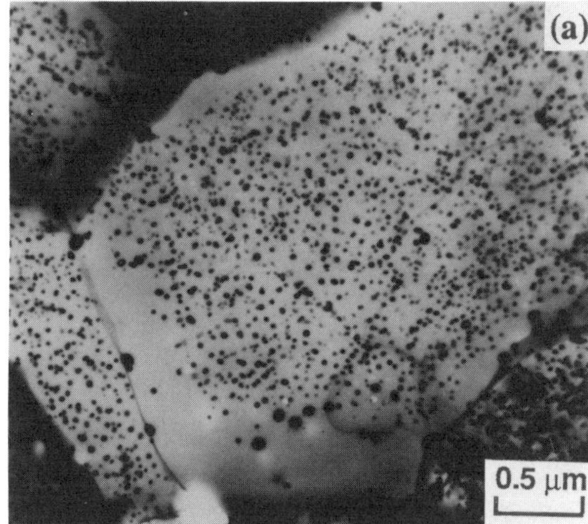

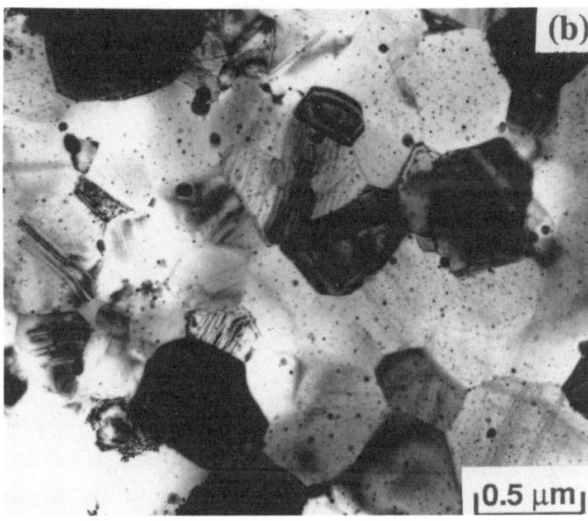

Fig. 9 - Electron micrographs showing the size and distribution of dispersoids in rapidly solidified (RS) and Mechanically alloyed (MA) condition. (a) RS Ti₃Al + 2 wt.% Er alloy HIP'ed at 850 °C and (b) MA Ti-25Al-10Nb-3V-1Mo + 2 wt.% Er alloy HIP'ed at 1000 °C. Note the fine grain size and small dispersoid size in the MA sample even after HIP'ing at a higher temperature. Note also the presence of dispersoid-free zones in the RS alloy and their absence in the MA alloy.

Table 3
Grain and dispersoid size as a function of annealing temperature in mechanically alloyed Ti-25Al-10Nb-3V-1Mo + 2 wt.% Er alloy

Treatment	Grain size (µm)	Dispersoid size (nm)		
As-HIP'ed	0.5	100	10	-
HIP'ed + 1100 °C/2h	1.0	200	40	10
HIP'ed + 1200 °C/1h	2.5	300	80	20

been rejected as very fine dispersoids. Along with this, the preexisting dispersoids in the HIP'ed compact would have coarsened, thus resulting in three different sizes.

Lastly, it should also be mentioned that the Er content in the coarse fraction of the powder was only 0.6 wt.%, while in the fine powder, it was about 2.0 wt.%. This suggests that insufficient milling time prevented the maximum dissolution of Er into solid solution in the coarse particles.

TiAl(γ) Formation

TiAl has a lower density and higher elevated temperature strength than Ti₃Al and is thus very attractive as a structural material in the aerospace industry. However, formation of the equiatomic TiAl phase from elemental powders through conventional mechanical alloying may be difficult because the extent of diffusion required is large and further, the use of process control additives is not desirable since they result in excessive pick up of detrimental interstitial impurities. Hence, a novel route involving hydrogen has been investigated.

Hydrogen can dissolve in titanium and its alloys up to 8.3 at.% (0.2 wt.%) in the α-phase and up to 60.3 at.% (3 wt.%) in the β-phase. Since the reaction of hydrogen with titanium is reversible, hydrogen can be easily removed from the alloy by vacuum annealing. Thus, when used judiciously, hydrogen can be used as a temporary alloying element leading to improved workability and beneficial effects in refining the microstructure and improving the mechanical properties of titanium alloys [24]. More specifically, addition of hydrogen reduces the β transus considerably and so diffusion can be much faster. Thus, mechanical alloying (which produces very fine-grained material) and hydrogenation can contribute to enhanced processability and final mechanical properties of these alloys.

A mixture of Al_3Ti and TiH_2 powders was mechanically alloyed in a Spex mill to form TiAl according to the equation :

$$Al_3Ti + 2\ TiH_2 \rightarrow 3\ TiAl + 2\ H_2\uparrow$$

Figure 10(a) shows the X-ray pattern of the powder mixture mechanically alloyed for 52 h. The presence of both the tetragonal TiAl and cubic $TiH_{1.924}$ phases was observed in this pattern. In contrast, the powder milled for only 28 h showed that all the three possible phases, viz., $Al_{24}Ti_8$, $TiH_{1.924}$, and TiAl were present. The relative proportions of the phases in these conditions was calculated from the intensities of the diffraction peaks and the results are presented in Table 4 [25]. Continued milling increased the volume fraction of the TiAl phase. A further increase up to 95% was achieved on HIP'ing the powder at 40,000 psi and 750 °C for 5 h (Figure 10 (b)).

With a view to determining the stability of the TiAl phase formed even after the removal of hydrogen, the mechanically alloyed powder was annealed for 1 h each at 300 and 500 °C. The X-ray pattern of the powder annealed at 500 °C (Figure 10(c)) clearly showed that the amount of the hydride phase had decreased and the TiAl phase constituted about 75%. Thus, mechanical alloying appears to be a novel way of producing the equiatomic TiAl compound from a mixture of Al_3Ti and TiH_2 powders.

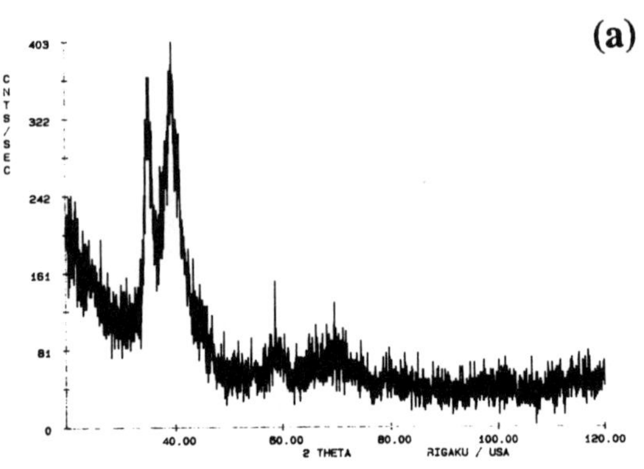

(a)

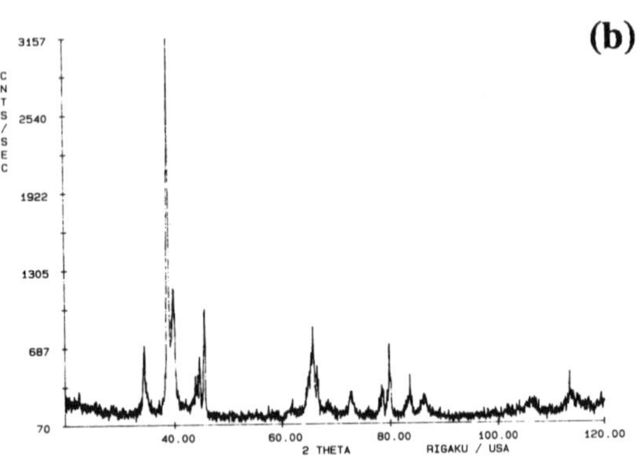

(b)

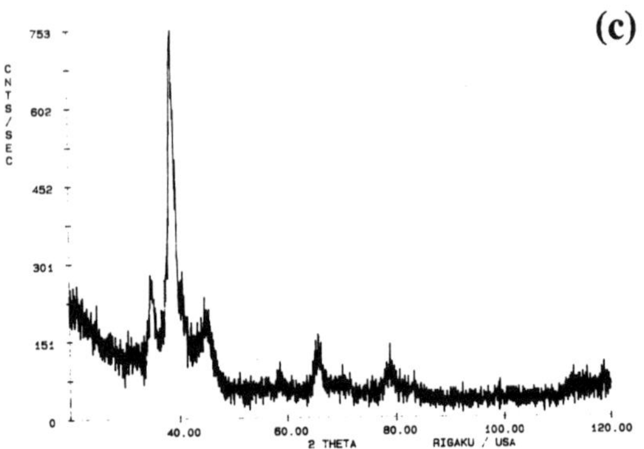

(c)

Fig. 10 - X-ray patterns of the Al₃Ti and TiH₂ powder mixture (a) mechanically alloyed for 52 h, (b) after HIP'ing for 5 h at 750 °C and 40,000 psi, and (c) after annealing for 1 h at 500 °C.

Table 4

Relative proportions (vol.%) of the phases present in the mechanically alloyed Al₃Ti and TiH₂ powder mixture

Condition	TiAl	TiH$_{1.924}$	Al$_{24}$Ti$_8$
MA/28h	33	26	41
MA/52h	55	45	-
MA+HIP'ed	95	5	-
MA+300°C/1h	53	47	-
MA+500°C/1h	75	25	-

CONCLUSIONS

From the foregoing it can be seen that mechanical alloying is a powerful non-equilibrium processing technique and can be successfully applied to titanium and its alloys. In addition to producing equiatomic TiAl and metastable f.c.c. Ti-Mg solid solutions having nanometer-sized crystallites, mechanical alloying can result in an amorphous phase in the Ti-Ni-Cu system. The fine grain size and dispersoid size in mechanically alloyed Ti₃Al-based alloys containing Er have been found to be quite resistant to coarsening even at very high temperatures.

ACKNOWLEDGEMENTS

This work was done while two of the authors (C.S. and R.S.) held National Research Council - Air Force Materials Laboratory Research Associateships.

REFERENCES

1. T.R. Anantharaman and C. Suryanarayana, "Rapidly Solidified Metals - A Technological Overview", Trans Tech Publications, Aedermannsdorf, Switzerland (1987).

2. E. Arzt and L. Schultz (eds.), "New Materials by Mechanical Alloying Techniques", Deutsche Gesellschaft fur Metallkunde, Oberursel, F.R. Germany (1989).

3. J. Mazumder, in "Interdisciplinary Issues in Materials Processing and Manufacturing", (eds.) S.K. Samanta, R. Komanduri, R. McMeeking, M.M. Chen and A. Tseng, ASME, New York, NY, 1987, pp. 599-630.

4. R.F. Hochman (ed.), "Ion Plating and Implantation - Applications to Materials", ASM, Materials Park, OH. (1986).

5. J.S. Benjamin, Metall. Trans. **1**, 2943-2951(1970).

6. L. Schultz, Mater. Sci. & Engg. **97**, 15-23 (1988).

7. E.A. Kenik, R.J. Bayuzick, M.S. Kim and C.C. Koch, Scripta Metall. **21**, 1137-1142 (1987).

8.	C.C. Koch, O.B. Cavin, C.G. McKamey and J.O. Scarbrough, Appl. Phys. Lett. **43,** 1017-1019 (1983).

9.	A.W. Weeber and H. Bakker, Physica **B153**, 93-135 (1988).

10.	R. Sundaresan and F.H. Froes, Key Engg. Mater. **29-31**, 199-206 (1989).

11.	R. Sundaresan and F.H. Froes, in "Proc. Sixth World Conference on Titanium", (eds.) P. Lacombe, R. Tricot and G. Beranger, Les Editions de Physique, Les Ulis Cedex, France, 1989, Part II, pp. 931-936.

12.	C. Suryanarayana and F.H. Froes, J. Mater. Res. (in press).

13.	I.W. Hall, Scandinavian J. Metallurgy **7**, 277-281 (1978).

14.	R.B. Schwartz, R.R.Petrich and C.K. Saw, J. Non-Cryst. Solids **76**, 281-302 (1985).

15.	C.Politis and W.L. Johnson, J. Appl. Phys. **60**, 1147-1151 (1986).

16.	J. R. Thompson and C. Politis, Europhys. Lett. **3**, 199 (1985).

17.	R. Sundaresan, A.G. Jackson, S. Krishnamurthy and F.H. Froes, Mater. Sci. & Engg. **97**, 115-119 (1988).

18.	R. Sundaresan, A.G. Jackson and F.H. Froes, in "Proc. Sixth World Conference on Titanium", (eds.) P. Lacombe, R. Tricot and G. Beranger, Les Editions de Physique, Les Ulis Cedex, France, 1989, Part II, pp. 925-930.

19.	Y-W. Kim, JOM **41**, No.7, 24-30 (1989).

20.	C. Suryanarayana, F.H. Froes and R.G. Rowe, Internat. Mater. Rev. (in press).

21.	R. Sundaresan and F. H. Froes, Metal Powder Rep. **44**, 206-208 (1989).

22.	C. Suryanarayana, R. Sundaresan and F.H. Froes, in "Advances in Powder Metallurgy", compiled by T.G. Gasbarre and W.F. Jandeska,Jr., MPIF, Princeton, NJ, 1989, Vol. 3, pp. 175-188.

23.	J.A. Sutliff and R.G. Rowe, in "Rapidly Solidified Alloys and Their Mechanical and Magnetic Properties", (eds.) B.C. Giessen, D.E. Polk and A.I. Taub, Mater. Res. Soc., Pittsburgh, PA, 1986, pp. 371-376.

24.	F.H. Froes, D. Eylon and C. Suryanarayana, JOM **42**, No.3, 26-29 (1990).

25.	C.Suryanarayana, R. Sundaresan and F.H. Froes, in "Solid State Powder Processing", (eds.) A.H. Clauer and J.J. deBarbadillo, TMS, Warrendale, PA, 1990, pp. 55-64.

QUANTITATIVE MICROSCOPY OF SiC-REINFORCED ALUMINUM COMPOSITES FABRICATED FROM MECHANICALLY ALLOYED POWDERS

M. G. McKimpson, A. N. Niemi, X. Huang
Michigan Technological University/IMP
1400 Townsend Drive
Houghton, Michigan 49931-1295, USA

Abstract
A technique has been developed for quantifying
the homogeneity of reinforcement distribution
in composites containing SiC in a matrix of
mechanically alloyed Al-4Mg. The technique
involves measurement of local reinforcement
area fractions on polished sections.
Segregation shows up as an increase in the
deviation of the measurements from the mean.
A homogeneity index is introduced which is
defined as the ratio of the average
experimental measurement deviation to the
theoretical deviation in a completely
segregated composite. In this study the index
was used to compare the dispersion quality in
particulate and whisker reinforced composite
billets and extrusions. The method should
also allow evaluation of macroscopic
variations in dispersion homogeneity within a
single sample. The advantages of this
technique over other proposed methods are ease
of measurement, wide range of application, and
adaptability to automated image analysis
equipment.

IN RECENT YEARS, mechanical alloying has been
used to produce a wide variety of aluminum-
base materials, including a number of
discontinuously reinforced metal matrix
composites. By combining mechanical alloying
technology with powder-metallurgy-based metal
matrix composite fabrication techniques, it
has been possible to produce a variety of
novel and potentially useful new engineering
materials, including reinforced aluminum, Al-
Mg, Al-Cu-Mg and Al-Li alloys[1,2,3,4].

In this study, mechanically alloyed Al-
4Mg powders were fabricated into SiC_p- and
SiC_w-reinforced composite materials using
conventional composite processing technology.
These composites were then evaluated using
quantitative stereological techniques to
describe the distribution of SiC reinforcement
in the material as a function of processing
conditions. It is generally recognized that
the uniformity of the reinforcement
distribution in a discontinuously-reinforced
composite material can have a major effect on
the mechanical properties of the material[5,6].
Materials which contain "clusters" of
reinforcement particles surrounded by
reinforcement-poor matrix material are likely
to exhibit a significant difference in
ductility and toughness from materials
containing a more uniform reinforcement
distribution.

Several approaches have been suggested
for using quantitative image analysis to
characterize the distribution of second phase
particles in a composite material such as SiC-
reinforced Al-4Mg[7,8,9,10]. Probably the most
general of these is the one discussed by Hunt
et. al.[10] which involves dividing the
microstructure of the material into Dirichelet
cells. Each cell is constructed so that it
contains a single particle and the
corresponding portion of the matrix material
which is closer to that particle than to any
other. With this approach, one can calculate
a local reinforcement area fraction for the
material surrounding each reinforcement
particle and then combine these measurements
to obtain a quantitative description of
clustering in the material. Unfortunately,

however, this Dirichelet cell technique is quite computation-intensive and not particularly well suited for routine evaluation of composite microstructures.

APPROACH

One alternative approach is to use local area fraction measurements to characterize reinforcement distribution. In solid-solid mixing operations, it is common practice to express the "mixedness" of a two-phase mixture as the statistical variance observed in the composition of small random samples extracted from the mixture[11]. By analogy, one should be able to use the variation observed in the local reinforcement content of a composite as a measure of the uniformity of reinforcement distribution. Using standard quantitative stereological techniques, one can readily determine the volume fraction reinforcement in any desired region of the composite by measuring the local area fraction reinforcement in that region. By taking local area fraction measurements at a number of random locations throughout the composite microstructure and then characterizing the "scatter" associated with these measurements, one should be able to quantitatively describe the microstructural uniformity of the material. Composites which have a very uniform distribution of reinforcement will exhibit only a small degree of scatter, whereas those having a more irregular, clustered distribution will exhibit substantially more scatter. Such an approach should be particularly useful for composite materials since the area fraction measurements should be essentially independent of the morphology of the reinforcing phase; the technique should work equally well for spherical, flake-like or even convoluted filamentary reinforcement particles.

Any useful technique for characterizing reinforcement distribution in composites based on local area fraction measurements requires that one adopt consistent conventions for determining both the size of the sample area (the "cell size") and the statistical parameter used to characterize the scatter in the local area fraction measurements. As pointed out by Danckwerts,[12] the "mixedness" (i.e., the statistical variance) experimentally observed in a two-phase system depends strongly on the size of the segregated regions in the mixture (the "scale of segregation") relative to the size of the sample being analyzed (the "scale of scrutiny"). If the size of the area used for the local area fraction measurements is substantially larger than the scale of the variations in reinforcement distribution in the composite, then all of the local measurements will simply reflect the mean value of the reinforcement content in the material. Similarly, if the size of the area used for the local area fraction measurements is substantially smaller than the scale of the variations in reinforcement distribution, the data is likely to show an unrealistically high level of scatter.

For this work, it was decided to use a "scale of scrutiny" for the SiC-reinforced Al-4Mg materials which would correspond to one "unit cell" of an idealized composite containing a uniform array of reinforcement particles. This unit cell is shown schematically in Figure 1. Note that it represents the minimum analysis area of the idealized composite for which the local area fraction reinforcement will always equal the mean area fraction reinforcement. The area, A_{cell}, of this hypothetical cell can be calculated using either of two expressions:

$$A_{cell} = 1/N_A \qquad \text{Eq. (1)}$$

where N_A represents the number of SiC particles per unit area on the appropriate plane of polish, or:

$$A_{cell} = A_{SiC}/A_f \qquad \text{Eq. (2)}$$

where A_{SiC} represents the mean cross-sectional area of the SiC particles in the reinforced Al-4Mg material and A_f represents the experimentally determined area fraction of SiC reinforcement in the material. Both of these quantities can be determined using quantitative metallography on polished sections of material. Alternately, A_f can be determined by chemical analysis.

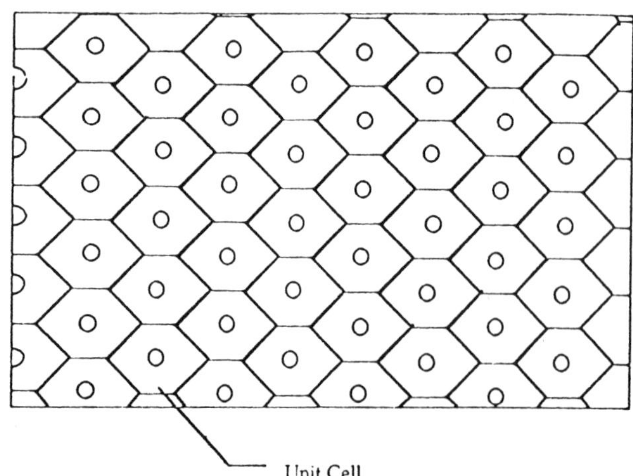

Fig. 1 - Idealized composite microstructure showing the unit cell used in determining cell size for the local area fraction measurements.

The statistical parameter most appropriate for describing the scatter in experimentally determined local area fraction measurements will depend on the character of these measurements. If the observed frequency of area fraction measurements can be described by a Gaussian distribution, then it would be appropriate to use the variance or standard deviation of these measurements, as one does in characterizing powder mixtures.[11] Many powder-metallurgy-based composites, however, will probably not exhibit a normal distribution of local area fraction measurements when analyzed using the cell size described above. The materials produced in this study, for example, contained nominally 15 vol% SiC, but exhibited local area fractions of SiC ranging from 0 to upwards of 70%. In this case, using the variance of the area measurements is probably not appropriate; the squared residual term used in calculating the variance will tend to exaggerate the contribution of high area fraction measurements, reducing the sensitivity of the parameter. Accordingly, it appears more appropriate to use a parameter related to the average deviation of the individual area fraction measurements about the mean value. This average deviation, α, is defined by the expression[13]:

$$\alpha = 1/N \; \Sigma \; |A_i - A_f| \qquad \text{Eq. (3)}$$

where N respresents the number of measurements, A_i represents the individual area fraction measurements, and A_f represents the mean area fraction SiC. As outlined in the Appendix, one can also show that for a perfectly segregated system in which all area fraction measurements consist of either pure matrix or pure reinforcement, the value of α will be given by

$$\alpha_{seg} = 2 \; A_f \; (1-A_f) \qquad \text{Eq. (4)}$$

where α_{seg} is the average deviation. One can then combine Eqns. 3 and 4 to define a homogeneity index, Q, given by the expression:

$$Q = \alpha_{composite}/\alpha_{seg}$$
$$= \alpha_{composite}/2(A_f*(1- A_f)) \qquad \text{Eq. (5)}$$

This index is very analogous to the mixing index for powder materials defined by Michaels and Puzinaukas[14]. The value of Q will range from 0.0 for a perfectly uniform dispersion of reinforcement particles to 1.0 for a perfectly segregated composite. It is Q which will be used to characterize dispersion homogeneity in this investigation.

EXPERIMENTAL PROCEDURE

The SiC-reinforced composite materials used in this study were fabricated using powder metallurgy processing technology.

Samples of an experimental mechanically alloyed Al-4Mg* powder were first screened to -200 mesh and -325 particle sizes. In both cases, the screened powders were found to consist primarily of roughly equiaxed particles having the morphology shown in Figure 2. These powders were then mixed with commercially available SiC particulate (SiC_p) and SiC whisker (SiC_w) reinforcements to prepare four composite blends:
- -200 mesh Al-4Mg powder + 15 vol% SiC_p
- -200 mesh Al-4Mg powder + 15 vol% SiC_w
- -325 mesh Al-4Mg powder + 15 vol% SiC_p
- -325 mesh Al-4Mg powder + 15 vol% SiC_w

The four blends were cold isostatically pressed into preforms, hot isostatically pressed into billets using a solid state powder consolidation practice, and extruded into round bars using an extrusion ratio of approximately 20:1. The microstructure of each material was then examined in both the as-consolidated and extruded conditions. The structures observed are shown in Figures 3 and 4.

Fig. 2 - Morphology of screened Al-4Mg mechanically alloyed powder used to produce the four experimental composite materials.

---

* Nominal composition (wt%) -- 4% Mg, 0.80% O, 1.15% C, 0.085% Fe, 0.001% Si, balance Al.

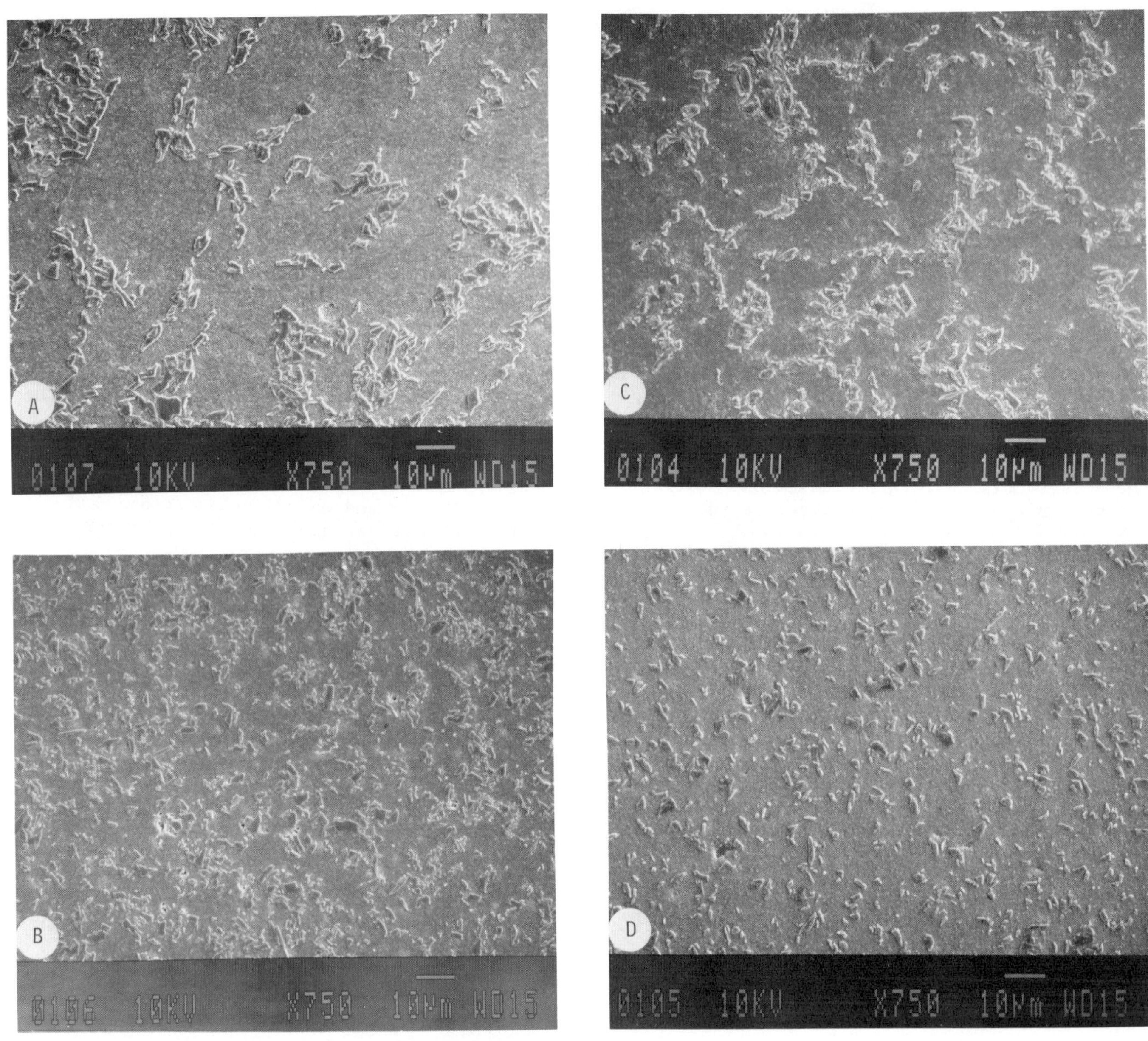

Fig. 3 - Microstructure observed on a transverse section in each of the SiC$_p$-reinforced materials. 750X. A) -200 mesh Al-4Mg+15% SiC$_p$, billet B) -200 mesh Al-4Mg+15% SiC$_p$, extrusion C) -325 mesh Al-4Mg+15% SiC$_p$, billet D) -325 mesh Al-4Mg+15% SiC$_p$, extrusion.

Quantitative image analysis measurements were then performed on each of the eight materials using a Tracor Northern Vista image analysis system interfaced with a JEOL JSM 820 scanning electron microscope. Transverse samples of each material were polished using standard metallographic techniques and then etched lightly with Graff-Sargent's reagent[15]. The samples were then placed in the SEM and analyzed in the uncoated condition. Samples coated with either carbon or gold were found to exhibit intensity variations within SiC particles which made accurate image analysis of the particles quite difficult. Even with uncoated samples, manual editing was sometimes needed to obtain accurate quantification of the binary (segmented) images.

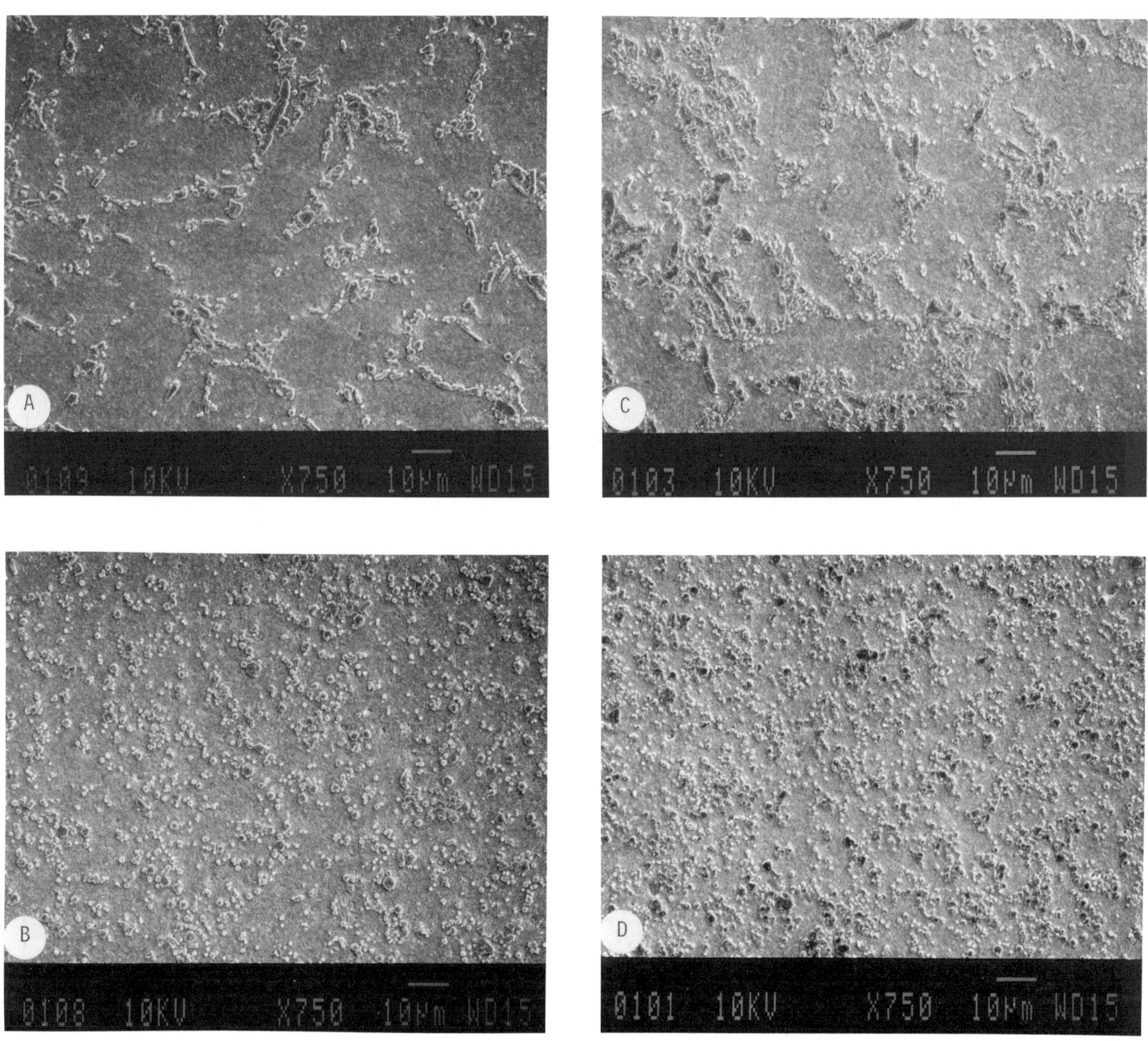

Fig. 4 - Microstructure observed on a transverse section in each of the
SiC$_w$-reinforced materials. 750X. A) -200 mesh Al-4Mg+15% SiC$_w$, billet
B) -200 mesh Al-4Mg+15% SiC$_w$, extrusion C) -325 mesh Al-4Mg+15% SiC$_w$,
billet D) -325 mesh Al-4Mg+15% SiC$_w$, extrusion.

RESULTS AND DISCUSSION

Preliminary measurements were first
obtained on each material in order to
determine an appropriate "cell size" for the
local area fraction measurements. This was
done by determining both the area fraction SiC
and number of SiC particles/unit area on five
randomly located fields at 3000X. At lower
magnifications, both parameters were found to
be a strong function of magnification due to
incomplete resolution of closely spaced
reinforcement particles. The mean area
fraction SiC values obtained on each material
are shown in Table 1. Corresponding values of
volume fraction SiC obtained on each of the
four extrusions using an acid dissolution
technique are also shown for reference. Since
both the particulate-reinforced materials were

fabricated to contain the same volume fraction
of the same SiC$_p$ reinforcement, it was decided
that the most appropriate way to specify a
cell size for these billets and extrusions was
to calculate one mean cell size based on the
average of all four materials. Based on the
N_A measurements, this cell size was determined
to be 14.4 μ^2. Similarly, since the two SiC$_w$-
reinforced alloys were fabricated to contain
the same volume fraction of the same SiC$_w$
reinforcement, one mean cell size was
calculated for these composites based on the
average of all four materials. This value was
found to be 7.8 μ^2.

Table 1 - SiC Content Measurements
Obtained on Experimental Materials[1]

	-200 Mesh Al-4Mg +15% SiC$_p$	-325 Mesh Al-4Mg +15% SiC$_p$	-200 Mesh Al-4Mg +15% SiC$_w$	-325 Mesh Al-4Mg +15% SiC$_w$
Billet	13.2	9.5	15.5	11.6
Extrusion	10.8	13.8	15.6	17.8
Chemical Analysis[2]	15.3	11.7	10.7	16.9

[1] All results expressed as volume percent
   SiC.
[2] Determined by chemical analysis of a
   bulk sample taken from the extrusion.

These cell sizes were then used to
characterize the quality of the SiC
reinforcement distribution in both the billet
and extruded forms of each composite. Local
area fraction SiC measurements were obtained
for randomly located fields of the appropriate
size on each of the eight materials. These
data were then plotted to show the cumulative
frequency of measurements with less than a
given level of SiC content. The results
obtained are shown in Figures 5 and 6. In
addition, mean SiC contents, and homogeneity
indexes, Q (as defined by Eqn. 5), were
calculated for each material based on 15 local
area fraction measurements. These results are
shown in Table 2.

Observe that the homogeneity index values
shown in Table 2 do appear to provide a way of
characterizing the various reinforcement
dispersions present in these eight materials.
As one would expect, the four extruded
materials shown in Figures 2 and 3 visually
appear to exhibit a substantially more uniform
distribution of SiC particles than the
corresponding billets. Similarly, the
composites--particularly the billets--made
using -200 mesh powder appear to be somewhat
more uniform than those made using -325 mesh
powder. In all cases, the material which
appears to exhibit the more uniform
microstructure also exhibits a smaller value
of Q. This is very encouraging.

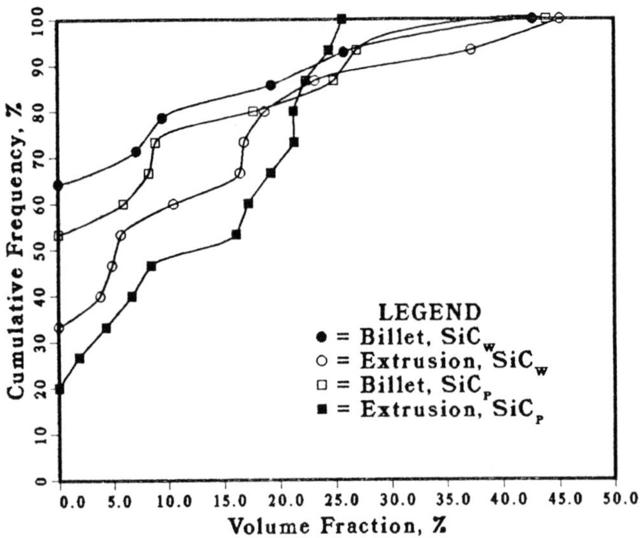

Fig. 5 - Cumulative volume fraction SiC
distribution obtained from local area fraction
measurements on each -200 mesh material.

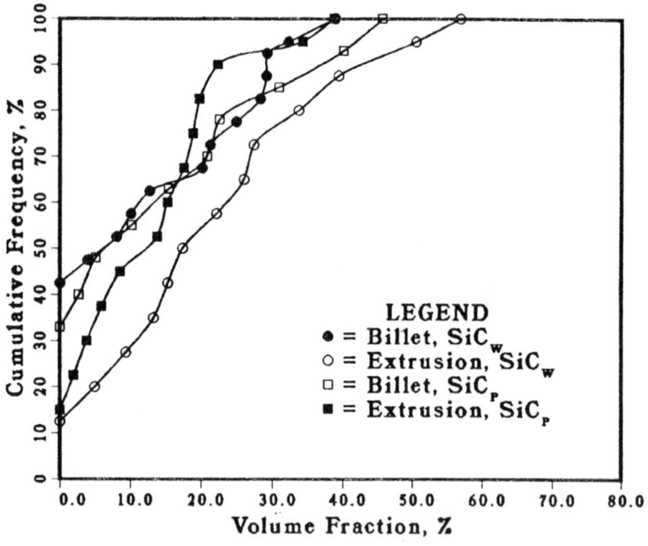

Fig. 6 - Cumulative volume fraction SiC
distribution obtained from local area fraction
measurements on each -325 mesh material.

Additional experimental work, however, is
clearly needed to more completely define the
statistical significance and the reliability
of these calculated Q values. The data
summarized in Table 2 are all based on
approximately 15 local area fraction

measurements on each material. This number of fields was chosen at the start of the program as a reasonable compromise between the (unknown) number of fields required to obtain accurate data and the costs involved in evaluating eight experimental materials.

Table 2 - Summary of Homogeneity Index Measurements on SiC-Reinforced Materials

	Mean SiC Content (vol %)	Q
-200 mesh Al-4Mg +15% SiC$_p$		
Billet	9.1	.62
Extrusion	12.6	.40
-325 mesh Al-4Mg +15% SiC$_p$		
Billet	10.4	.57
Extrusion	13.6	.45
-200 mesh Al-4Mg +15% SiC$_w$		
Billet	7.4	.70
Extrusion	12.2	.53
-325 mesh Al-4Mg +15% SiC$_w$		
Billet	13.2	.53
Extrusion	21.3	.33

To investigate how mean SiC content and calculated Q values vary as a function of the number of measurements for composite materials such as those evaluated in this program, 25 additional local area fraction measurements were obtained on the -325 mesh billet and extrusion materials. The results obtained are summarized in Figures 7 and 8. As shown, Q appears to be stabilizing after about 20-30 measurements. This suggests that one may need to analyze at least about 300 sq. microns of material in these composites in order to obtain a representative value of microscopic reinforcement distribution.

There is also a need to quantify the difference in Q values which actually correspond to significant microstructural differences in material homogeneity. In these samples, the billet and extrusion microstructures exhibit Q values which differ by about 0.2. Based on the consistency of the data, this difference appears to be significant, but additional work is clearly needed to demonstrate that this is, in fact, true. The expression used for α_{seg} when calculating Q, for example, is a reasonably strong function of mean volume fraction SiC. There is a need, therefore, to show that this

is an appropriate expression to use for normalizing out the effects of volume fraction on the value of Q. Finally, one must also verify that the material sampled for the local area fraction measurements is, in fact, representative of the the entire composite

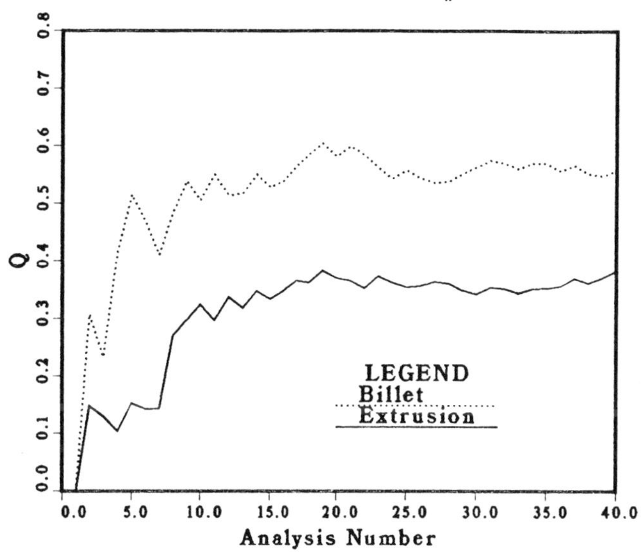

Fig. 7 - Homogeneity Index, Q, as a function of the number of local area fraction measurements made for the whisker reinforced -325 mesh Al billet and extrusion.

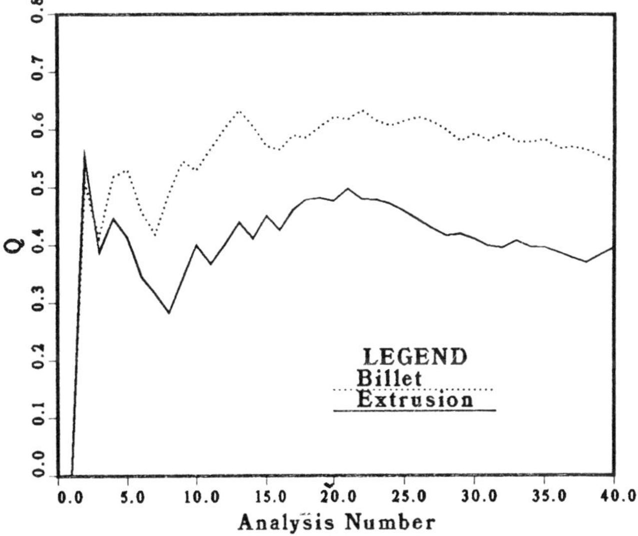

Fig. 8 - Homogeneity Index, Q, as a function of the number of local area fraction measurements made for the particulate reinforced -325 mesh Al billet and extrusion.

material. This is an age-old problem, but it is of particular concern for composites, since these materials have the potential for exhibiting substantially more variability than their unreinforced counterparts. The range in volume fraction and area fraction measurements shown in Table 1, for example, suggests that these SiC-reinforced Al-4Mg composites may have contained macroscopic variations in reinforcement content throughout their cross-section. Such inhomogeneities, if present, could have a marked effect on the character of the reinforcement distribution in the material. Even if such gradients were not present in these samples, there is no reason to assume that they could not be present in other powder-metallurgy-based composite materials.

Once these types of issues have been addressed, the homogeneity index, Q, should have significant potential as a useful parameter for characterizing microstructural uniformity in a variety of materials. The index is obtained by simply making repeated area fraction measurements at random locations in a sample. Accordingly, it should be feasible to implement Q value determinations on a wide variety of automated image analysis systems. Calculating a Q value for given microstructure is much less computer-intensive than calculating a Dirichelet tesselation grid for even one analysis field. Using local area fraction measurements with a variable cell size to assess uniformity also allows one to develop a homogeneity index which should be reasonably independent of both particle size and morphology. In addition, it allows one to characterize particles which are adjacent to-- but not quite touching--each other. Finally, the analysis for Q results in a single, somewhat intuitive parameter, rather than a distribution or histogram. This should facilitate developing correlations between reinforcement distribution and physical properties in a composite or other material. All of these characteristics should make the homogeneity index a useful tool for a variety of structure-property studies and process control investigations.

CONCLUSIONS

A technique has been developed which is useful for quantifying the homogeneity of reinforcement in composite materials. The result is a homogeneity index, Q, which is defined as the ratio of the mean deviation in local area fraction measurements to the mean theoretical deviation for a completely segregated composite. The deviation in local area fraction measurements is dependent on the analysis area, which is defined in this technique as a cell with an area equal to the inverse of the number of particles per unit area.

The homogeneity index should vary between 0.0 for an idealized distribution of particle shapes with identical areas to 1.0 for the case of complete separation of phases. Experimental determination of Q requires only standard quantitative stereological techniques for measuring area fraction and number of particles per unit area. Used in conjunction with automated image analyses, this technique offers the capability of making large numbers of measurements and the possibility of being used as a process control tool.

ACKNOWLEDGEMENTS

The authors would like to acknowledge partial funding for this work from Inco Alloys International, the assistance of Kip Paxton in preparation of samples and figures, Ruth Kramer for SEM microscopy, and Terry Anderson for typing the manuscript.

REFERENCES

1. Sundaresan, R. and F. H. Froes, Metal Powder Report, March 1989.
2. Hawkes, J. A. and H. G. F. Wilsdorf, Scripta Metal. $\underline{23}$, 125-130, (1989).
3. Fujiwara, C., J. Blucher and J. A. Cornie, U.S. Army Research Office Contract DAAG29-84-K0-174, (1988).
4. D. L. Davidson, Metall. Trans A, $\underline{18A}$, 2115-2128 (1987).
5. Lewandowski, J. J., C. Liu and W. H. Hunt, Jr., Mat. Sci and Eng., $\underline{A107}$, 241-255 (1989).
6. McKimpson, M. G. and T. E. Scott, Mat. Sci and Eng., $\underline{A107}$, 241-255 (1989).
7. Mirchandani, P., Thesis (Ph.D.) -- Michigan Tech. Univ., Houghton, MI. (1987).
8. Kelly, J. F., NBS Report NBSIR 87-3681, (1987).
9. Barryman, J. G., "Estimating Effective Moduli of Composites Using Quantitative Image Analysis," Lawrence Livermore National Laboratory, (1989).
10. J. J. Lewandowski, C. Liu and W. H. Hunt, (in) Kumar, P., K Vedula and A. Ritter, "Processing and Properties of Powder Metallurgy Composites," pp. 117-137, The Metallurgical Society, Warrendale, PA (1988).
11. Fan, L. T., S. J. Chen and C. A. Watson, Ind. and Eng. Chem., $\underline{62}$, 52-69, (1970).
12. Danckwerts, P. W., Research (London), $\underline{6}$, p. 35 (1953).
13. Bevington, P. R., "Data Reduction and Error Analysis for the Physical Sciences," p. 14, McGraw Hill Book Co., New York (1969).
14. Michaels, A. S. and V. Puzinaukas, Chem. Eng. Progr., $\underline{50}$, p. 604 (1954).
15. Graff, W. R. and D. C. Sargent, Metallography $\underline{14}$, 69-72 (1981).

APPENDIX

AVERAGE DEVIATION OF A PERFECTLY
SEGREGATED COMPOSITE - Consider an idealized,
perfectly segregated composite material
containing a mean volume fraction, V_f, of
reinforcement particles. By definition, this
perfectly segregated composite will be one in
which all possible local area fraction
measurements in the material consist of either
pure matrix material ($V_f = 0$) or pure
reinforcement ($V_f = 1$). Assuming the
equivalence of area and volume fractions in
this composite, if randomly-located area
fraction measurements are taken on this
material, a fraction of these measurements
equal to ($1-V_f$) should lie in the matrix
material, and a corresponding fraction, V_f,
should lie in the reinforcement material.
Accordingly, from Eqn. 3 of the main text, the
average deviation, α, of these measurements
should be given by:

$$\alpha = P_1 \, |A_f - A_1| + P_2 | A_f - A_2| \quad \text{Eq. (A.1)}$$

where P_1 represents the probability of
measuring phase 1($P_1 = A_f$), P_2 is the
probability of measuring phase 2 ($P_2 = 1-A_f$),
and A_1 and A_2 are the measured area fractions
($A_1 = 1$ and $A_2 = 0$ for complete segregation).
Substituting these values into Eqn. A.1:

$$\alpha_{seg} = 2A_f(1-A_f) \quad \text{Eq. (A.2)}$$

ORIGIN OF THE STRENGTH OF MECHANICALLY ALLOYED ALUMINUM ALLOYS

Tadashi Hasegawa, Tsunemasa Miura*
Department of Mechanical Engineering
Tokyo University of Agriculture and Technology
Koganei-shi, Tokyo 184, Japan

* Permanent Address: Showa Aluminum Corporation, Oyama Works, Oyama-shi 323, Japan

ABSTRACT

Stress vs strain behavior of mechanically alloyed (MA) Al alloys containing ceramic particles (SiC or Al_2O_3, a few 100 nm in size, blended in the MA process) and fine dispersoids (Al-oxide and carbide, a few 10 nm in size, formed during the MA process) was examined and analyzed by TEM and SEM. Most of dislocations in the alloys were pinned down by the fine dispersoids and their high yield strength at low temperatures ($<\sim423K$) was considered to originate from the high stress necessary for the detachment of dislocations from fine dispersoids. Further, the fine-scale dispersion of such dispersoids suppressed the annihilation of dislocations during high temperature exposure and hence the static recovery. However, the dynamic recovery under stress was not restrained so effectively as the static recovery by the dispersoids, resulting in a decrease in strength at higher temperatures ($>\sim423$ K). The increase in strength by blending ceramic particles may be caused by the same mechanism as in non-MA SAP alloys, not by the detachment mechanism. Loss of ductility at high temperatures was ascribed to the local dynamic recovery and micro-void formation due to grain boundary sliding.

THE INCREASEING NEED FOR ALUMINUM ALLOYS with high strength at room and elevated temperatures, compared to currently available ingot-metallurgy alloys, has stimulated research into powder metallurgy[1]. One of the most promising materials at present is a powder-metallurgy Al alloy produced by a mechanical alloying (MA) process[2], and the application of MA to pure Al has increased room-temperature strength up to ~400 MPa[2][3], a strength level approaching that of conventional precipitation-hardened Al alloys such as the 2000 and 7000 series.

High strength of MA alloys has been attributed to the fine-scale dispersion of thermally stable particles (usually metal-oxide) with submicron homogeneity[4]-[6]. However, the strength and its temperature dependence may not simply be explained by the Orowan mechanism, since (1) in a MA Al-Mg alloy, Orowan loops could not be observed around a particle after creep[6]; and in the present Al-base MA alloys, as will be mentioned later, (2) a reduction of yield strength with an increase in temperature occurred much faster than that expected from the mechanism and (3) a strain-softening phenomenon was observed at high temperatures. The objective of the present work is to characterize stress vs strain behavior of Al-base MA alloys at room and elevated temperatures, and to delineate operative strengthening mechanism in them.

EXPERIMENTAL PROCEDURE

Aluminum powder (~40 μm in size, 99.8 mass% purity) blended with 5 to 20 vol% SiC particles (~0.4 μm in size) or with 15 vol% Al_2O_3 particles (~0.5 μm in size) was mechanically alloyed (MA) using a high-energy ball mill (process-control agent : ethanol), cold pressed in an Al can and degassed in vacuum at 773 K. After sealing, it was hot pressed (pressure : 700 MPa) and extruded (extrusion ratio : 4.6) at 773 K. For comparison, an alloy produced only from MA Al powder was prepared by the same process. These alloys are designated MA Al-SiC alloy, MA Al-Al_2O_3 alloy and MA Al alloy, respectively, in the present paper.

Specimens for tension tests (gauge size : 12 mm length × 4 mm width × 1 mm thickness) and for compression tests (size : 3.5 mm in dia. × 4 mm height) were machined from the extruded rod, parallel to the direction of extrusion. They were deformed with an Instron-type machine in air at a nominal strain rate of 9.1×10^{-4}/s in a temperature range 293-903 K.

Thin foils for TEM were prepared from specimens before and after deformation by spark-erosion machining and electrochemical polishing, and examined with an electron microscope operated at 200 kV. Fracture surfaces were observed by SEM. The size and interparticle spacing of dispersoids were examined using an extraction replica technique.

EXPERIMENTAL RESULTS

MICROSTRUCTURE OF MA ALLOYS - An electron micrograph of an extraction replica taken from MA Al-15vol.% Al$_2$O$_3$ alloy is given in Fig. 1. Besides large ceramic particles (Al$_2$O$_3$ in Fig. 1 or SiC in MA Al-SiC alloys) blended in the MA process, one can see a uniform distribution of fine dispersoids (on the average, $\sim$35 nm in size and $\sim$88 nm in spacing) in the Al matrix. Only the fine dispersoids existed in the MA Al alloy. Their quantity could be controlled by the amount of ethanol added in the MA process, and it was 7-8 vol.% in the present alloys. These fine dispersoids may be Al-oxide and carbide[7] formed during the MA process.

Figure 2 is a TEM micrograph of the MA Al alloy. It consists of equiaxial, small grains ($\sim$0.5 μm in size), as in other MA alloys[2][6][8][9]. Further, as can be seen in a high magnification TEM micrograph (Fig.3), it contains a high density (in an order of $10^{14}/m^2$) of dis-

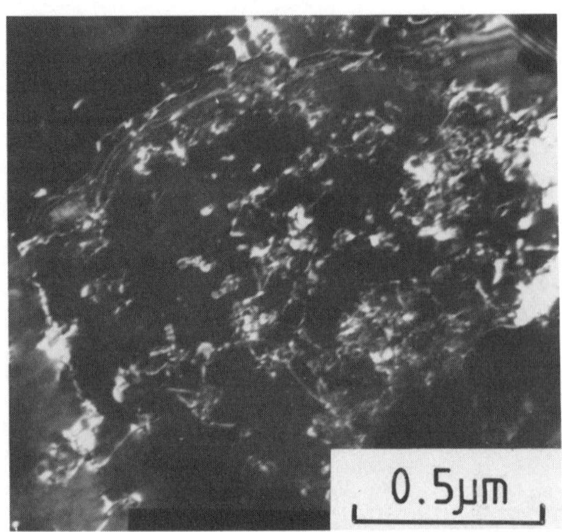

Fig. 3 - TEM micrograph of MA Al alloy before deformation (dark field).

locations most of which appear to be pinned down by the fine dispersoids that are dispersed uniformly within grains and in grain boundaries. Such a high density of dislocations indicates that the present MA alloy is in a severely work-hardened state, the dispersion of fine particles formed during the MA process being effective in storing dislocations in that process.

STRESS VS STRAIN BEHAVIOR - Nominal stress, σ_n, vs nominal plastic strain, ε_n, curves in tension for the MA Al alloy and true stress, σ_t, vs true plastic strain, ε_t, curves in compression for the MA Al-15vol.% SiC alloy are given in Figs. 4 and 5, respectively. No notable difference in stress vs strain behavior could be found in all the alloys examined, though the addition of ceramic particles in the MA process led to an increase in the flow stress and a decrease in the tensile strain to fracture. Further, the flow stress of MA Al-SiC alloy tended to be higher than that of MA Al-Al$_2$O$_3$ alloy for a given volume fraction of ceramic particles.

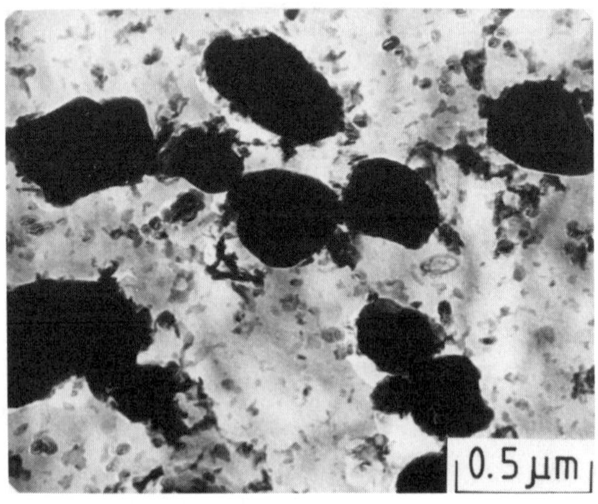

Fig. 1 - Electron micrograph of an extraction replica taken from MA Al-15vol.% Al$_2$O$_3$ alloy.

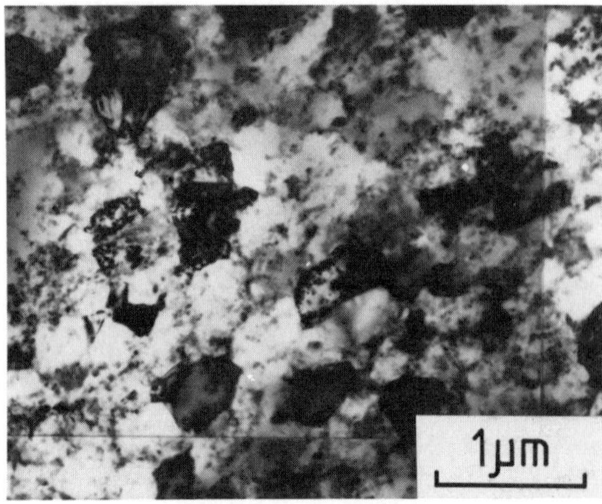

Fig. 2 - TEM micrograph of MA Al alloy before deformation (bright field).

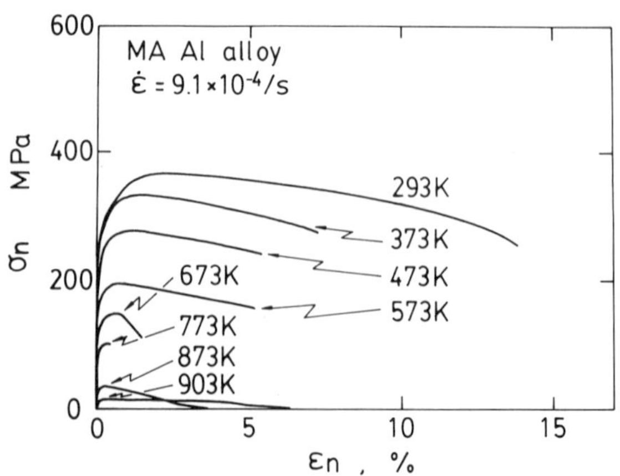

Fig. 4 - Nominal stress, σ_n, vs nominal plastic strain, ε_n, curves in tension tests of MA Al alloy.

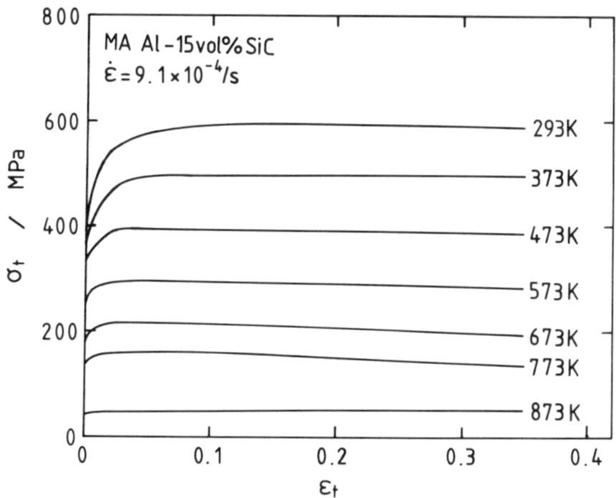

Fig. 5 - True stress, σ_t, vs true plastic strain, ε_t, curves in compression tests of MA Al-15vol.% SiC alloy

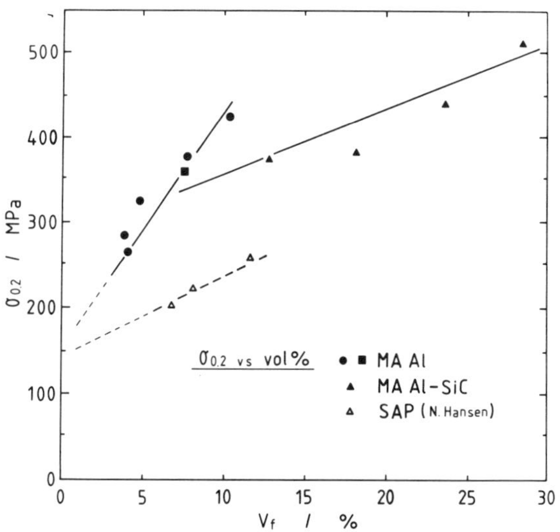

Fig. 6 - Change in 0.2% proof stress, $\sigma_{0.2}$, with volume percent of dispersoids, V_f. Circular and square marks are for MA Al alloys, and triangular ones for MA Al-SiC Alloys.

The following can be seen from the figures. (1) The yield and maximum (tensile or compressive) stresses, σ_y and σ_u, of MA Al alloy are much higher than those of non-MA powder-metallurgy aluminum (σ_y = 82 MPa, σ_u = 115 MPa). (2) Except at an early stage of deformation (< a few % strain) where strain-hardening occurs, the MA alloys deform at almost constant stresses or exhibit strain-softening over a wide range of strain. (3) Although a value of σ_u decreases almost linearly with increasing temperature, the reduction of σ_y with temperature occurs gradually up to $\sim$423 K, then rapidly at higher temperatures. (4) Strain to fracture in tension decreases with temperature up to 773 K, while specimens deform uniformly without cracking up to ε_t = 0.3 in compression.

Change in the 0.2% proof stress, $\sigma_{0.2}$, at 293 K with the volume percent of dispersoids, V_f, is shown in Fig. 6. Circular and square marks are for fine dispersoids formed during MA (circle : results obtained by modifying data of Benjamin et al.[2], square : for the MA Al alloy), and triangular marks for the MA Al-SiC alloy). Data for non-MA SAP alloys[10] are also given in the figure. It can be seen that the rate of strength-increase for V_f is much higher for the fine dispersoids formed during MA than for the ceramic particles blended in MA, indicating that strengthening mechanisms by these two kinds of dispersoids are different to each other. Further, the rate of strength-increase by SiC is approximately equal to that by Al_2O_3 in SAP.

RECOVERY BEHAVIOR - It has been known that room-temperature hardness of Al-base MA alloys does not reduce remarkably after thermal annealing at $\sim$723 K[4][11]. Effects of annealing on the room-temperature stress vs strain behavior of the MA Al alloy are shown in Fig. 7. It can be seem that high strength is retained even after annealing at 773 K, which is the final extrusion temperature of the alloy. Further, TEM revealed that the fine-scale dispersion of particles and the high density of dislocations

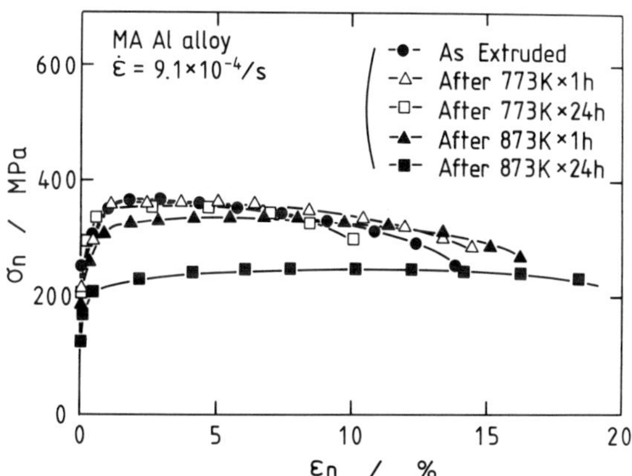

Fig. 7 - Nominal stress, σ_n, vs nominal plastic strain, ε_n, curves at 293 K in tension tests of MA Al alloy after various annealing treatments.

were almost preserved during annealing at 773 K, though notable coarsening of microstructure and a resultant decrease in the flow stress were observed after 873 K annealing (see Figs. 7 and 8). This suggests that the large resistance to the static recovery of MA alloys, at temperatures up to the final extrusion temperature, is attributed to the suppression of dislocation annihilation by the dispersion of fine particles.

DISCUSSION

STRENGTH AT LOW TEMPERATURES - As shown in Fig. 3, the present MA alloys contain high densities of dislocations. This may suggest, as a first possibility, that their high yield

Table 1 - Dislocation Densities, and Calculated and Measured Yield Stresses
in MA Al and MA Al-15vol.% SiC Alloys

Alloy	Dislocation Density (m^{-2})	Yield Stress (MPa) Cal.	Meas.
MA Al	1.0×10^{14}	240	250
MA Al-15vol.% SiC	2.8×10^{14}	392	421

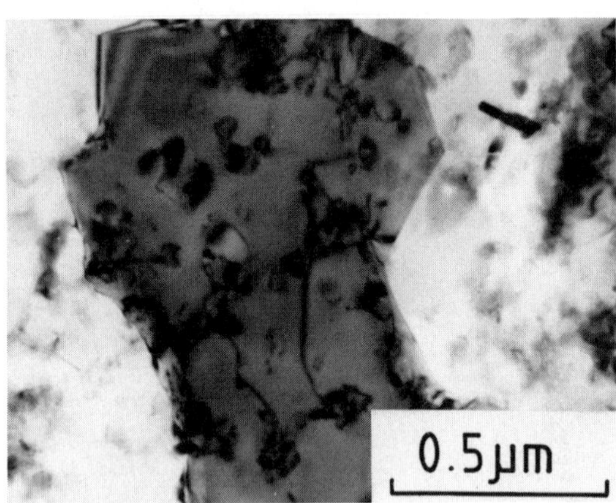

Fig. 8 - TEM micrograph of MA Al alloy after annealing at 873 K for 1 h.

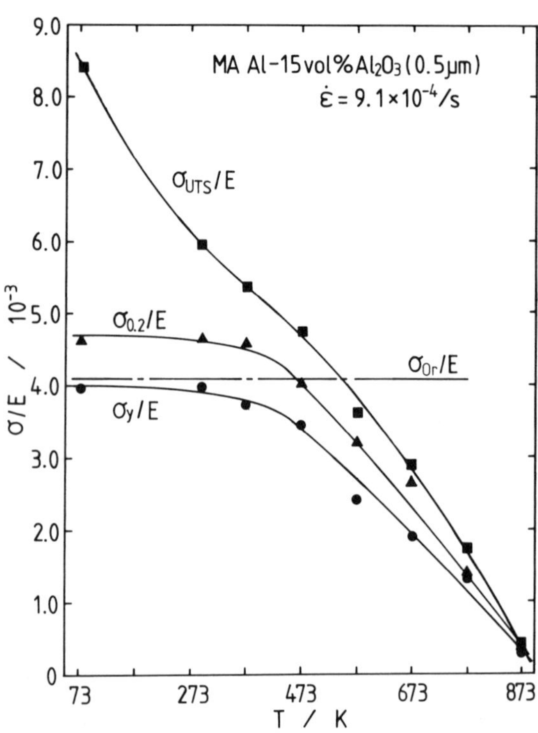

Fig. 9 - Changes in modulus-compensated yield stress, σ_y/E, 0.2% proof stress, $\sigma_{0.2}/E$, and maximum tensile stress, σ_{UTS}/E, with temperature, T, for MA Al-15vol.% Al$_2$O$_3$ alloy.

strength arises from the severe work-hardening introduced during the MA process. Let us consider here whether the room-temperature strength of MA alloys can be explained from the dislocation density.

For fcc metals, the flow stress, σ, is related to the dislocation density, ρ, by the well-known equation[12] : in the absence of a friction stress,

$$\sigma = M\alpha\mu b\sqrt{\rho}$$

where M is the Taylor factor (3.06), α a constant ($\sim$ 1)[13], μ the shear modulus (26770 MPa) and b the Burgers vector (2.86×10^{-10} m). Substituting the value of ρ into the equation, one obtains 240 and 392 MPa for MA Al and MA Al-15vol.% SiC alloys, respectively (Table 1). Taking into account a rather large error in the estimation of dislocation densities by TEM, these values are considered to approximately equal the measured yield stresses given in the table. This does not contradict the above consideration that the yield strength of the present MA alloys is determined by their work-hardened state.

One may argue that, if the yield stress is determined by the dislocation density, it will not reduce even at high temperatures (even though a rapid decrease in flow stress will occur at an early stage of plastic deformation), since the annihilation of dislocations should be accompanied with some plastic strain. This is not the case as shown in Fig. 9; the modulus-compensated yield stress remains almost constant at low temperatures ($<\sim$423 K), but decreases rapidly at higher temperatures. Further, as seen in Figs. 4 and 5, no peak of flow stress could be observed in the beginning of high temperature deformation. These facts may suggest another possible mechanism of yielding.

Srolovitz et al.[14][15] have proposed that, when the particle-matrix interface is free to slip at high temperatures, a dislocation will be attracted to the particle and be stabilized by being incorporated into the interface. The yield stress is then the stress necessary to unpin the dislocation from the particle, and was expected to be of the order of the Orowan stress [14]. In the present MA alloys which were finally extruded at 773 K, most of dislocations appeared to be pinned down at fine dispersoids by this mechanism (Fig. 3), and the value of σ_y/E at low temperatures was certainly equal to the modulus-compensated Orowan stress estimated from the distribution of fine dispersoids ($\sim$4.1 $\times$ 10^{-3}) (Fig. 9). Therefore, it may be considered, as a second possibility, that the yield stress is determined by the detachment of dislocations from fine dispersoids.

Since the average dislocation spacing ($\equiv$ 1/$\sqrt{\rho}$) is nearly equal to the interdispersoid spacing in the present MA alloys, it is diffi-

cult to decide "which (high dislocation density
or small interdispersoid spacing) the high yield
strength originates from" only from microstructural observation. At the present stage of
investigation, however, we would like to assert
that the second possibility is more probable,
because no reduction of room-temperature yield
stress could be observed even after high-
temperature deformation during which the dislocation annihilation had occurred severely (see
Figs. 10(b) and 11).

STRENGTH AT HIGHER TEMPERATURES - Most of
dislocations in the present MA alloys are pinned
down at the fine dispersoids by the mechanism
proposed by Srolovitz et al.[14][15], as in
other MA alloys[16][17]. When subjected to a
stress, dislocations will be unpinned from the
dispersoids and will glide suffering resistance
from them. If all the dislocations are mobile*,
then the dislocation glide over a distance of
interdispersoid spacing will cause a plastic
strain of ∿0.1 % which can certainly be detected
in the present experiment. Therefore, if no
coarsening of microstructure occurs, the reduc-

tion of yield stress, a stress in the beginning
of plastic flow, at high temperatures (>∿423 K,
see Fig. 9) may be ascribed to a decrease in the
stress necessary for the detachment of dislocations from fine dispersoids.

Figures 10 (a) and (b) are TEM micrographs
of MA Al alloy taken after deformation at 573 K.
No detectable coarsening of grains and dispersed
particles is observed, though a notable decrease
in dislocation density due to the dynamic recovery occurred (compare Fig. 10 (b) with Fig. 3).
Figure 11 shows effects of prestraining at 773 K
on room-temperature stress vs strain behavior in
MA Al alloy. It can be seen that the room-
temperature yield stress and flow stress are not
practically influenced by the high-temperature
prestraining, suggesting an excellent thermal
stability of microstructure of MA alloys. Therefore, the reduction of yield stress with increasing temperature should be explained from a
decrease of capability of fine dispersoids in
pinning dislocations under stress, not from the
coarsening of microstructure; the dynamic recovery is not suppressed so effectively as the
static recovery by the dispersion of fine
particles.

(a)

(b)

Fig. 10 - TEM micrographs of MA Al alloy
after compression at 573 K up to ε_t = 0.3.
(a) low magnification, bright field, (b)
high magnification, dark field.

---

* This may be realistic, because dislocations
are distributed rather uniformly in the alloys.

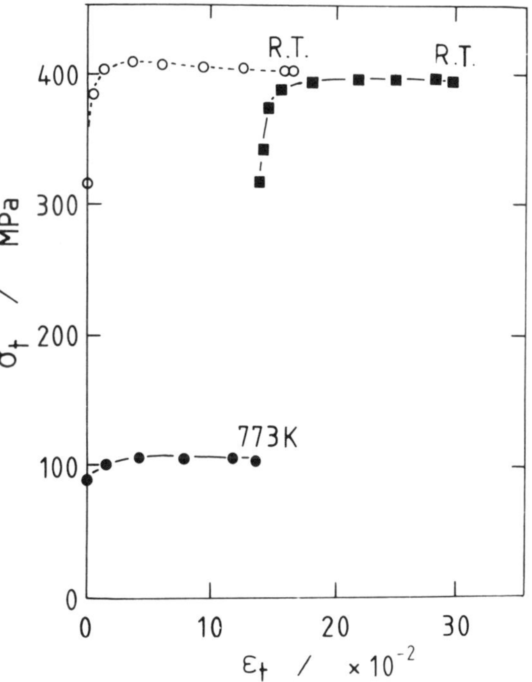

Fig. 11 - Effects of prestraining at 773 K
on room-temperature stress vs strain behavior in MA Al alloy.

FRACTURE BEHAVIOR - A decrease in the tensile strain to fracture with temperature (see
Fig. 4) is a common phenomenon in all the MA
alloys examined. Figure 12 shows SEM micrographs of fracture surface. The surface obtained at 293 K is wavy and many dimples are formed
on it, indicating that the fracture is ductile
and transgranular in nature. On the fracture
surface at 773 K, on the other hand, one can

see a pattern of individual grains, suggesting
the occurrence of intergranular fracture due to
grain boundary sliding. These results may imply
that the decrease in the fracture strain arises
from an increasing tendency to the plastic instability due to the local dynamic recovery
(that is, the local strain softening) and the
microvoid formation due to grain boundary sliding in the softened area, both of which tend to
occur more remarkably at higher temperatures.
The increase in the fracture strain above 873 K
(Fig. 4) may be due to the coarsening of microstructure.

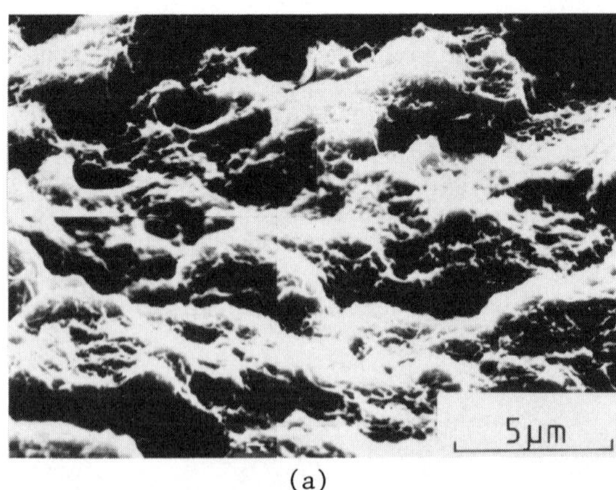

(a)

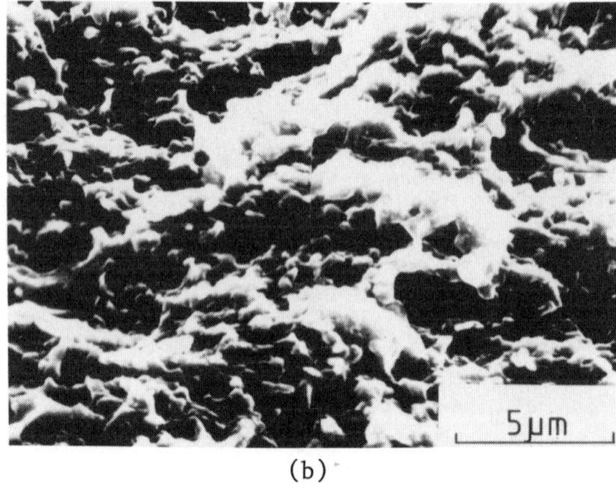

(b)

Fig. 12 — SEM micrographs of fracture surface
obtained in tension of MA Al alloy.
(a) at 293 K, (b) at 773 K.

CONCLUSIONS

1. Uniform distribution of fine dispersoids
(oxide and carbide of Al, a few 10 nm in size)
formed during the MA process is effective in
storing dislocations; the present Al-base MA alloys contain a high density (in an order of 10^{14}
$/m^2$) of dislocations and most of them have been
pinned down by these fine particles.

2. The dispersion of fine particles is stable
for high temperature exposure up to 773 K, and
suppresses the dislocation annihilation and
hence the static recovery.

3. The yield stress at low temperatures ($<\sim423$
K) is nearly equal to the Orowan stress estimated from the dispersion of fine particles, and is
considered to be determined by the stress necessary for the detachment of dislocations from
fine particles. At higher temperatures ($>\sim423$
K), however, the particles lose the capability
in pinning dislocations; the yield stress decreases with increasing temperature. The dynamic recovery is not restrained so effectively as
the static recovery by the dispersion of fine
particles.

4. The yield stress increases by blending ceramic particles (SiC or Al_2O_3, a few 100 nm in
size) in the MA process. However, the reason of
strength-increase caused by these large particles should be distinguished from that by the
fine dispersoids; the former strengthens aluminum by the same mechanism as in non-MA SAP,
while the latter by the detachment mechanism mentioned above. High strength of Al-base MA alloys essentially originates from the fine-scale
dispersion of the latter.

5. The strain to fracture in tension decreases
with increasing temperature. This may arise
from an increasing tendency to the plastic instability due to the local dynamic recovery and
the microvoid formation due to grain boundary
sliding.

ACKNOWLEDGMENTS

The authors would like to thank Dr. T. Yakou
and Mr. T. Shibuya for their help in experiments
and Prof. R. Horiuchi for discussion. This work
was supported in part by the Light Metals Educational Foundation of Japan and by Grant-in-Aid
for Scientific Research (B) from the Ministry of
Education, Science and Culture of Japan.

REFERENCES

(1) Savage, S. J. and F. H. Froes, J. of Metals
    36, 20-33(1984).
(2) Benjamin, J. S. and M. J. Bomford, Met.
    Trans. 8A, 1301-1305(1977).
(3) Hasegawa, T., T. Shibuya, T. Yakou and T.
    Miura, "Strength of Metals and Alloys",
    p.1389, Pergamon Press, Oxford(1988).
(4) Ezz, S., M. J. Koczak, A. Lawley and M. K.
    Premkumar, "High Strength Powder Metallurgy
    Aluminum Alloys II", p.287, TMS of AIME,
    Warrendale(1985).
(5) Hynnä, A., V. T. Kuokkala, T. Lepistö, T.
    Mäntylä and P. Kettunen, "Mechanical Behav-
    ior of Materials-V", p.987, Pergamon Press,
    Oxford(1987).
(6) Otsuka, M., Y. Abe and R. Horiuchi, "Creep
    and Fracture of Engineering Materials and
    Structures", p.307, The Inst. of Metals,
    London(1987).
(7) Singer, R. F., W. C. Oliver and W. D. Nix,
    Met. Trans. 11A, 1895-1901(1980).
(8) Benjamin, J. S. and R. D. Schelleng, Met.
    Trans. 12A, 1827-1832(1981).

(9) Kaneko, J., M. Sugamata and R. Horiuchi,
    Proc. 8th Intern. Light Metals Congress,
    p.776-780, Leoben-Vienna(1987).
(10)Hansen, N., Met. Trans. $\underline{1}$, 545-547(1970).
(11)Lim, S., J. Kaneko and M. Sugamata, Proc.
    of Korea-Japan Metals Symposium on Composite
    Materials, p.128-137, The Korean Inst. of
    Metals, Seoul(1988).
(12)Mughrabi, H., "Constitutive Equations in
    Plasticity", p.199, MIT Press, Cambridge
    (1975).
(13)Mecking, H. and U. F. Kocks, Acta Met. $\underline{29}$,
    1865-1875(1981).
(14)Srolovitz, D. J., R. A. Petkovic-Luton and
    M. J. Luton, Scripta Met. $\underline{18}$, 1063-1068
    (1983).
(15)Srolovitz, D. J., M. J. Luton, R. Petkovic-
    Luton, D. M. Barnett and W. D. Nix, Acta
    Met. $\underline{32}$, 1079-1088(1984).
(16)Nardone, V. C. and J. K. Tien, Scripta Met.
    $\underline{17}$, 467-470(1983).
(17)Schröder, J. H. and E. Arzt, Scripta Met.
    $\underline{19}$, 1129-1134(1985).

DEFORMATION OF MECHANICALLY ALLOYED Al-TiB$_2$ ALLOYS AT ELEVATED TEMPERATURES

M. Otsuka, T. Ishihara

Department of Metallurgical Engineering
Shibaura Institute of Technology
Sibaura 3-9
Minato-ku, Tokyo 108, Japan

M. Sugamata, J. Kaneko

Department of Mechanical Engineering
College of Industrial Technology
Nihon University
Izumicho 1-2
Narashino-shi
Chiba 275, Japan

Abstract

The mechanical behavior and microstructural stability in the temperature range between 293K and 873K have been investigated on three MA Al-TiB$_2$ alloys containing 2, 5 and 10 vol% of TiB$_2$ in order to evaluate their capability as the structural materials for high temperature use. Compression tests and microhardness tests were performed on the alloys in the as-extruded conditions. The microstructure of each alloy is stable up to about 673K. The effect of TiB$_2$ addition on the yield stress is remarkable only in the content range below 2vol% TiB$_2$. The yield stress decreases approximately in a linear manner with test temperature. The present alloys are superior in strength to a heat resistant commercial alloy (AA2218-T6). Unstable plastic flow accompanying localized shear appears at about 673K and makes the alloys very brittle. Both in the as-extruded and annealed conditions, the alloys show a remarkable yield drop at higher temperatures and strain rates which appears whenever the specimen is unloaded and subsequently reloaded. The strain rate depends much more strongly on the stress than in the case of pure metals or one phase alloys.

Introduction

CONVENTIONAL CAST AND WROUGHT ALUMINUM ALLOYS lose their strength rapidly with increasing temperature. These alloys are therefore not utilized for structural applications where the components must stand temperature beyond about 473K for long time. Thus an effort is under way to investigate processing techniques that may produce novel aluminum base alloys suitable for elevated temperature applications. Several approaches are being explored to attain this goal, including mechanical alloying(MA), powder metallurgy (P/M) and rapid solidification processing (RSP).

Mechanical alloying (hereafter referred to as MA) is a specialized technique for processing metal powders prior to consolidation into full density wrought alloys [1]. The process involves the dry, high energy milling of metal powder blends to produce particles with unique microstructures. The nature of the process permits the solid-state incorporation of large volume fraction of dispersoids within powder metal particles. Moreover, the high energy input during MA processing increases the concentration of point and line defects within the powder particles. By utilizing the structural features developed during MA, and by suitable control of post-MA processing parameters, alloys exhibiting unique structures, and often unusual combinations of properties can be produced using MA powders [2].

The purpose of this paper is to present the elevated temperature mechanical properties of four mechanically alloyed Al-TiB$_2$ alloys together with the microstructural stability against high temperature exposure.

Materials and Experimental Procedure

Materials – Atomized powder particles of pure aluminum (mean diameter 80μm) were mixed with powder particles of 0 to 10 vol% of TiB$_2$ (mean diameter 5.6μm), and mechanically alloyed in attritor in an argon atmosphere. The processing time for MA is 86.4ks(24hrs). Stearic acid was added by 1.8 mass % as a process control agent (PCA) to maintain a balance between fracturing and welding of the powder particles during processing. The PCA is initially incorporated into aluminum powder particles along with TiB$_2$ particles during the repeated welding, fracturing and rewelding of particles. The scale of PCA dispersion becomes progressively finer as milling continues. Also as the particles accumulate plastic strain, the amount of defects increases within the powder particles. With continued milling, the particles can undergo some degree of dynamic recrystallization. The structure of fully-processed powder thus consists of ultrafine grains along with hydrated oxides and carbonates generated from PCA [2].

Mechanically alloyed powder blends were cold pressed and then hot degassed in vacuum. During the latter process, the entrapped PCA was converted to stable dispersoids of oxide and carbide. These stable dispersoids together with fine dispersoids of intentionally added TiB_2 particles should serve to restrict grain growth during degassing, and the grain size in the degassed powder is thought to be only slightly greater than that in the milled powder. The stable dispersoids continue to resists to grain growth even in the subsequent thermomechanical processing (in the present case, hot extrusion) of the powder compact. Thus even though secondary recrystallization and grain growth occur during elevated temperature exposure, the present alloys should consist of very fine grains which are stable up to temperature near the melting point of aluminum.

Compression tests – Compression tests were performed mainly on as-extruded alloy and partially on an alloy which were annealed in air at 823K for 10.3ks. Compression specimens were machined out of the extruded rod of 7mm diameter. Samples were tested in an Instron universal machine at temperature ranging from 297K to 823K using strain rate ranging from 8.33 x 10^{-5} to 8.33 x 10^{-2} s^{-1}. Ultrafine graphite powder (Dag 5579, Nihon Atchison Co.) was used as lubricant in tests both at room and elevated temperatures.

Hardness test – The thermal stability of microstructure was evaluated by hardness testing on the specimens which were heated at various temperatures and then air-cooled.

Metallography – Transmission electron microscopy (TEM) was performed on a JEOL JEM-1000. JEOL T-20 scanning electron microscope was used in order to examine the surface structure of deformed samples.

Results and Discussion

Microstructure – Figure 1 shows typical microstructures observed on the sections parallel and perpendicular to the extrusion direction of Al-TiB_2 alloys. Though we can see only larger dispersoids of TiB_2 at such lower magnification level, their distribution is quite uniform and there is no strong tendency for the dispersoids to align with the direction of extrusion. Relatively large particles of several microns in diameter are considered as the TiB_2 powder particles which have not sufficiently been crushed into small sizes during MA process. The details of microstructure become more clear when the sample is observed at higher magnification

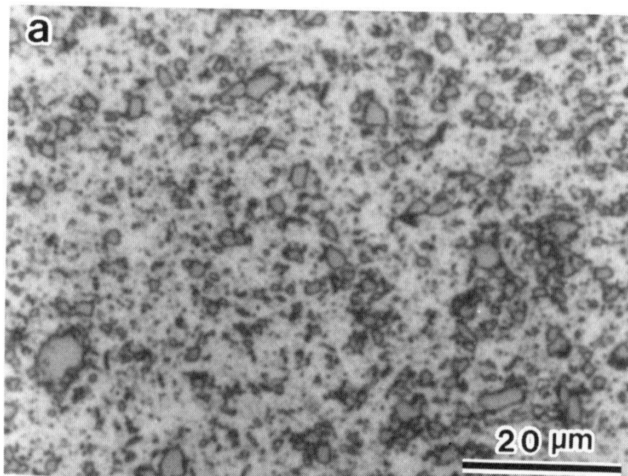
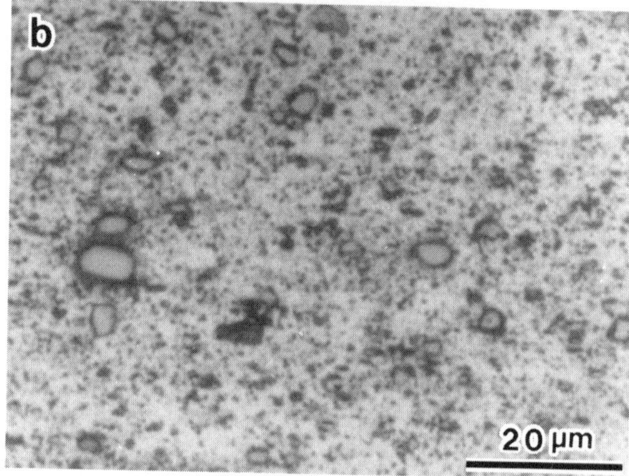

Fig. 1 Microstructure of Al-5vol%TiB_2. (a)cross section, (b)longitudinal section where extrusion direction is horizontal.

on a transmission electron microscope.

Figure 2 shows an example of TEM photograph taken on the longitudinal section of the as-extruded rod of Al-2vol%TiB_2 alloy. It can be seen that the matrix consists of very fine grains, the mean diameter of which is less than

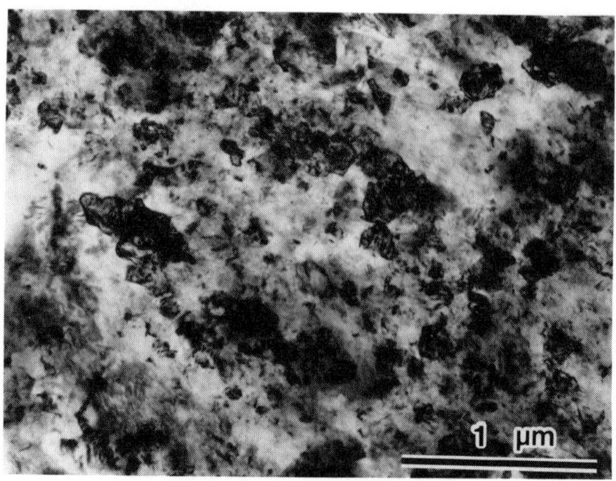

Fig. 2 TEM micrograph of Al-2vol%TiB_2 at low magnification.

$0.5\mu m$. As shown in Fig. 3, there exist many dislocations within each grain, together with a number of small dispersoids having mean size of 20 to 30nm. Though the estimation of mean spacing between dispersoids has not been made successfully due to the lack of unambiguous TEM image of dispersoids, the above mentioned high density dislocations and finely distributed dispersoids could play some essential roles in improving the mechanical properties through both work hardening and dispersion strengthening.

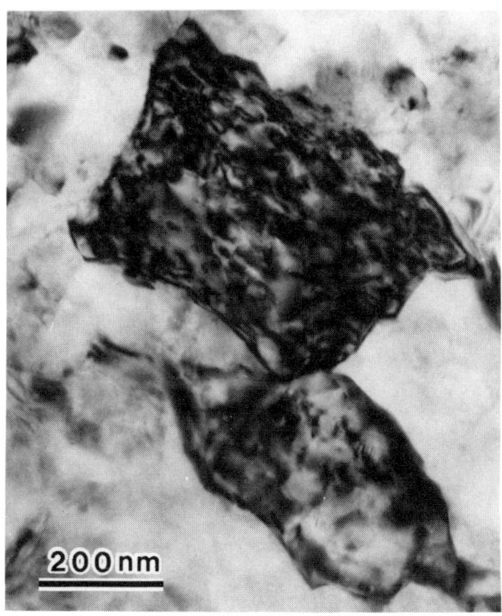

Fig. 3 TEM micrograph of Al-2vol%TiB₂ at high magnification.

The X-ray diffraction patterns of the milled powder and extruded rod for an Al-5vol%TiB₂ alloy are presented in Fig. 4. Comparison of the pattern for the extruded bar (Fig.4(b)) with that of the milled powder (Fig.4(a)) clearly shows that no remarkable phase change occurs during the hot degassing and consolidation process.

Figure 5 (a) and (b) show, respectively, the room temperature hardness as a function of duration of exposure at 673 and 873K. In the case of the exposure at 673K, all alloys investigated hardly soften even after 108ks(300hrs). On the other hand, they exhibit considerable softening after one hour exposure at 873K, although the extent of softening decreases with increase in the amount of TiB₂. It should also be noted that the effect of TiB₂ addition becomes weaker with the increase in its content both in as-extruded and annealed conditions.

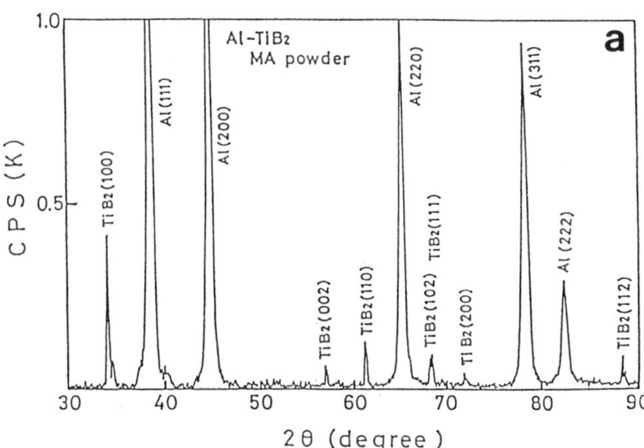

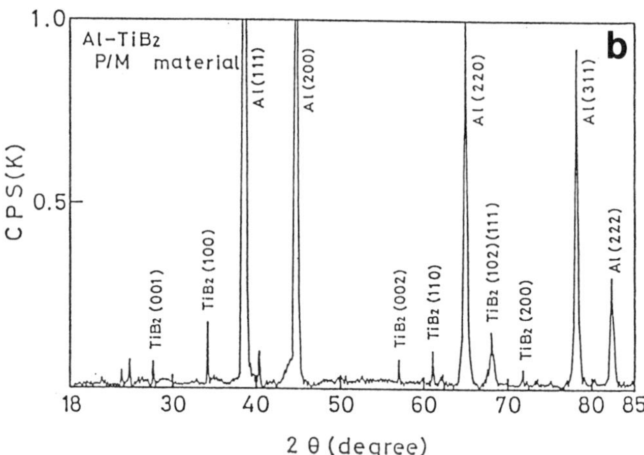

Fig. 4 X-ray diffraction patterns of (a) MA powders and (b) extruded rod. Al-5vol%TiB₂.

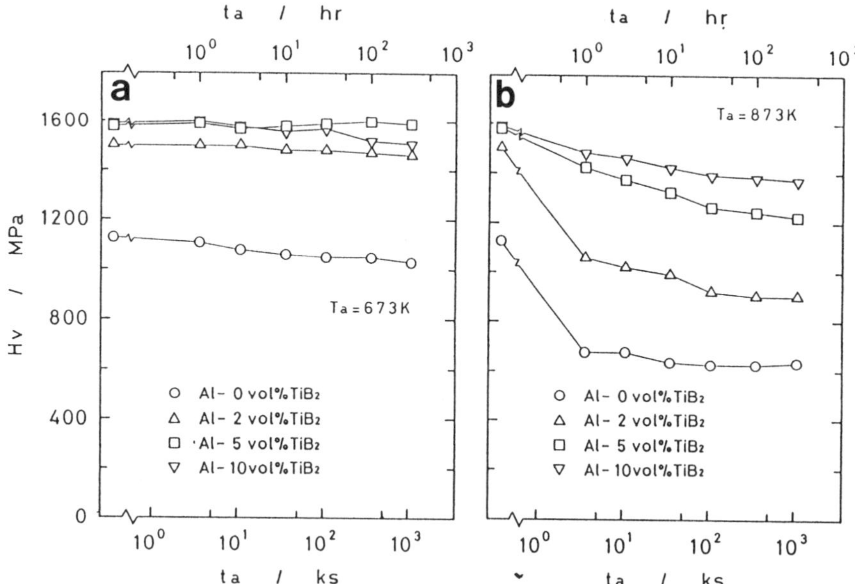

Fig. 5 Hardness vs. annealing time at (a) 673K and (b) 873K.

Figure 6 presents the room temperature hardness after the exposure at various temperatures for 3.6ks. It is clear that all alloys with or without TiB₂ begin to soften at about 673K while the amount of softening decreases with increase in the content of TiB₂.

At any rate, the above results show that the microstructures of mechanically alloyed Al-TiB₂ alloys are stable enough to withstand the exposure at 673K at least for 300hrs.

Stress-strain behavior - Figure 7(a) shows typical stress-strain diagrams of MA Al-2vol%TiB₂ alloy recorded at various temperatures at a constant strain rate of 8.33×10^{-5} s^{-1}. It is to be noted that even at room temperature the work hardening is limited to an initial stage of deformation which is followed by a prolonged work softening stage. As the flow stress decreases gradually with increase in temperature, a kind of plastic instability begins to appear around 623K and becomes most remarkable at 673K. This plastic instability accompanied with the formation of shear bands along which quite a localized material flow occurred, resulting in a rapid drop of stress. The shear bands observed

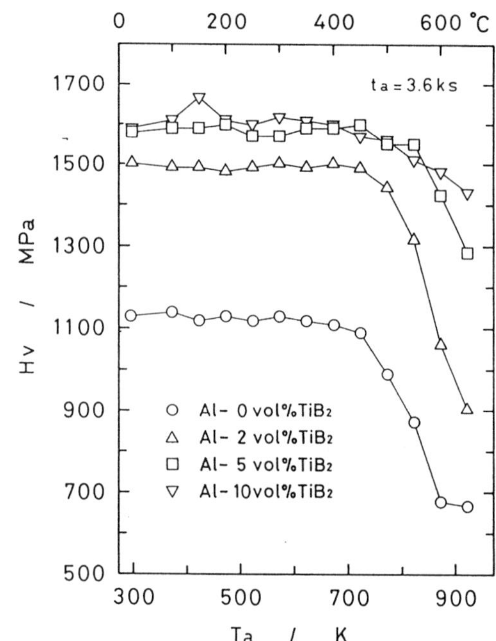

Fig. 6 Hardness vs. annealing temperature. Annealing time, 3.6ks.

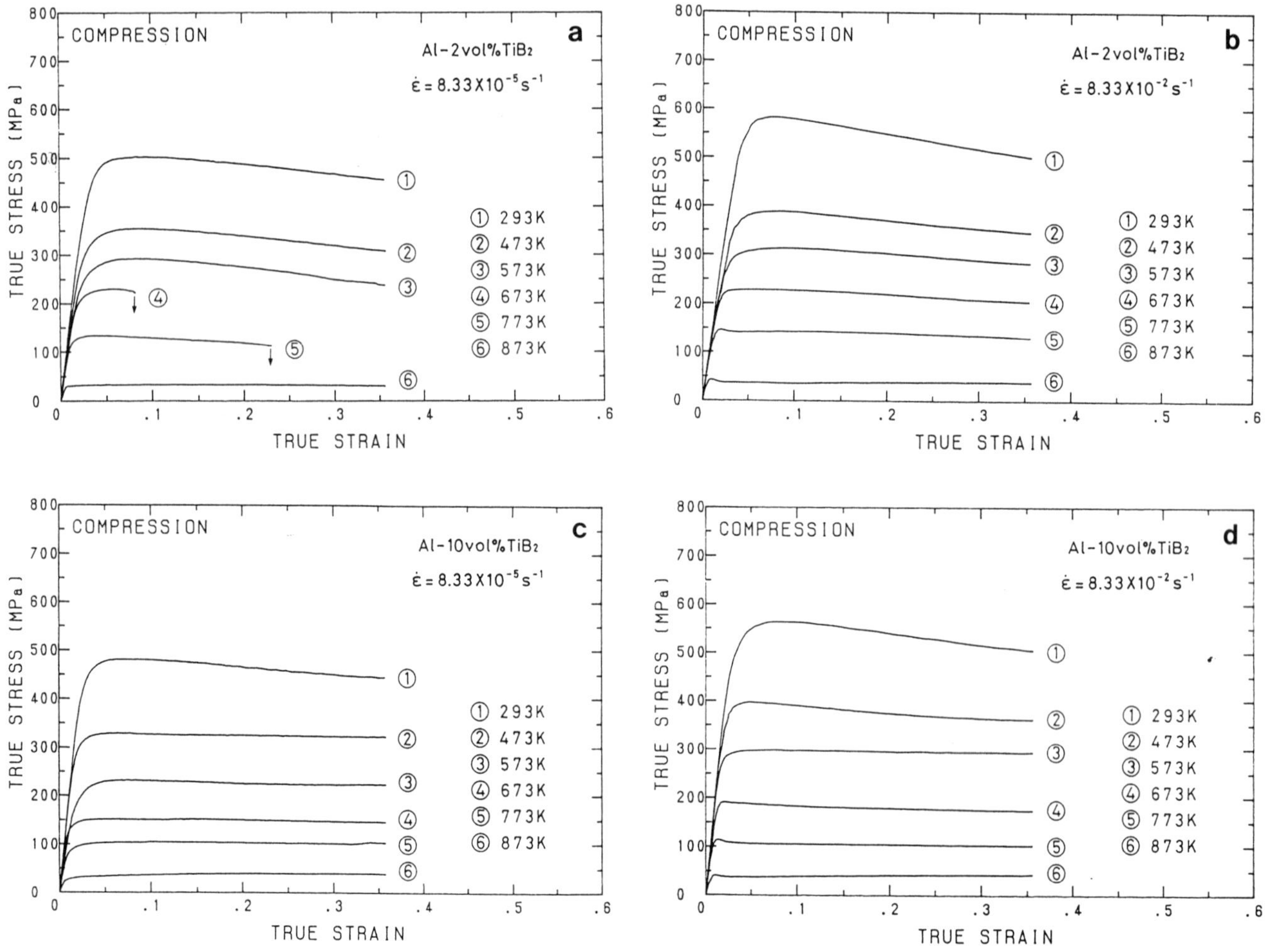

Fig. 7 Stress-strain diagrams for Al-2vol%TiB₂ (a, b) and Al-10vol%TiB₂ (c, d) at various temperatures.

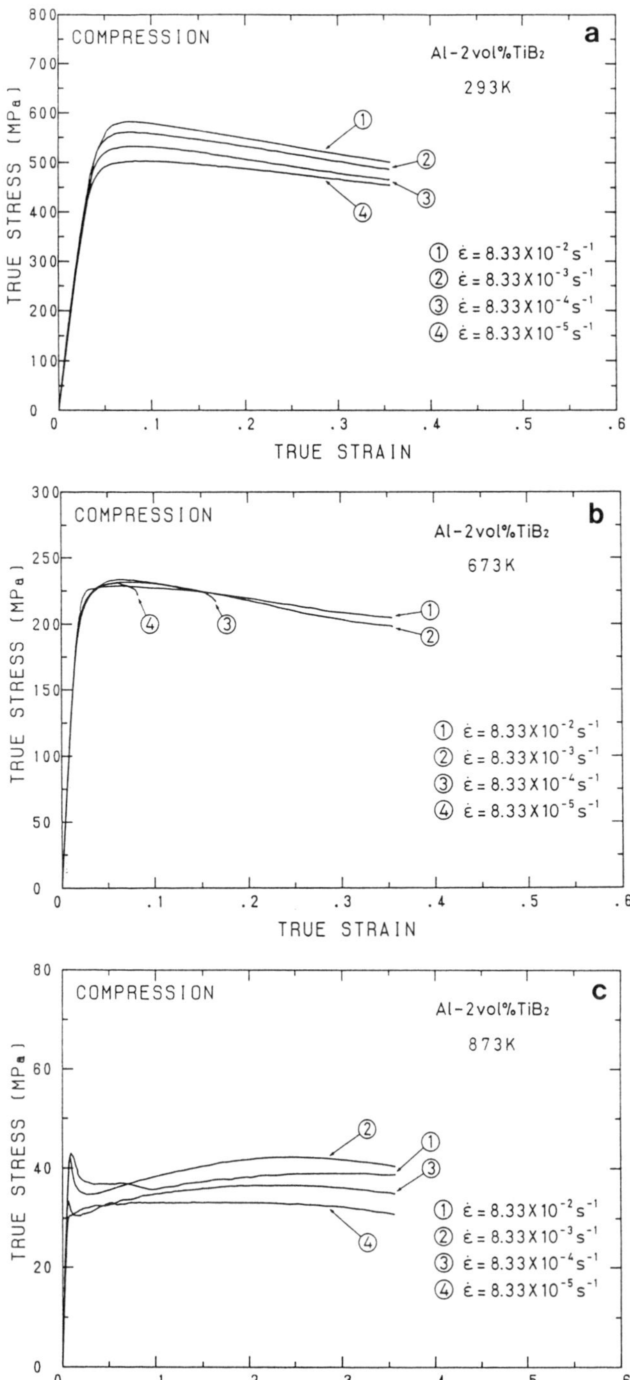

Fig. 8 Stress-strain diagrams for Al-2vol%TiB$_2$ at (a) 298K, (b) 673K and (c) 873K.

in this study are similar to those observed in MA Al-2vol%TiC alloy [3] or in MA aluminum [4].

As the strain rate increases, the plastic instability tends to appear at the later stage of deformation. For example, Fig. 7(b) shows that at a strain rate of 8.33 x 10^{-2} s^{-1} no instability occurs at any temperature.

On the other hand, such an instability is not detected at all in the compression tests of MA Al-10vol%TiB$_2$ alloy both at low and high strain rates, as shown in Fig. 7(c) and 7(d).

Figure 8 shows the influence of strain rate on the stress-strain behavior of Al-2vol%TiB$_2$ at 293K, 673K and 873K. It is found that the strain rate dependence of the flow stress becomes minimum not at room temperature but at the intermediate temperature, in contrast to the behavior of one phase alloys. Furthermore we would like to point out another very interesting phenomenon, "high temperature yield drop". The phenomenon becomes more prominent with increase in strain rate and temperature , though it does not appear at lower strain rate (see Fig. 7(a)). We can see its typical example in Fig. 8(c). Similar yield drop has been reported by Oliver and Nix [4] in the compression tests of MA aluminum and Al-Mg alloy and by the present authors in MA Al-TiC [3].

The yield drop appeared not only in the as-extruded samples but also in samples which were annealed at 823K for 3.6ks to 360ks. In addition, it appeared every time when the sample was once unloaded and soon reloaded, accompanying with the separate formation of shear bands. From these observations the yield drop is probably caused by a mechanism intimately related with the localized shear, the details of which are at present under the investigation.

Effect of temperature and strain rate on the yield stress - Figure 9 shows the yield stress defined as 0.2% proof stress ($\sigma_{0.2}$) as a function of temperature. The strain rate is 8.33 x 10^{-2} s^{-1}. $\sigma_{0.2}$ decreases almost in a linear way with increasing temperature. The magnitude and temperature dependence of $\sigma_{0.2}$ in the present alloys agree well with those observed by Hawk *et al.*[5] in Al-1.2%C-0.8%O-9.0%Ti-5.0%Y$_2$O$_3$ alloy. It can also be seen from the figure that the present alloys surpass a typical heat resistant aluminum alloy AA2218-T6 in the strength especially at temperatures above 473K.

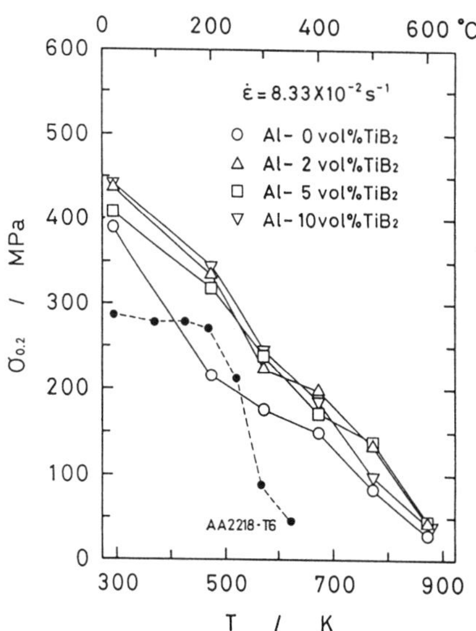

Fig. 9 Yield stress vs. temperature.

The fact that the dependence of flow stress on strain rate is negligibly small above 573K while it is relatively large below 573K is in contrast to the behavior observed in any conventional aluminum alloys.

Figure 10 shows log-log plot of temperature compensated strain rate, $\dot{\varepsilon}/D_v$, vs. modulus compensated yield stress $\sigma_{0.2}/G$, where D_v is the self diffusion coefficient of aluminum [6] and G is the shear modulus at test temperature [7]. The temperature range of the data adopted in this plot is between 473K and 873K. Two remarkable characteristics can be pointed out by comparing this plot with that for the cast and wrought alloy [8]. Firstly, at each temperature the strain rate depends quite strongly on the stress. If we dare to apply a power law, $(\dot{\varepsilon}/D_v) = A\,(\sigma_{0.2}/G)^n$, to the experimental data, the stress exponent, n, for example at 673K, is roughly evaluated as about 100.

Secondly, contrary to the expectation, the data for different temperatures are separated individually rather than fall onto a single master curve, as is often observed in metals and alloys produced by conventional ingot metallurgy process [8]. As the separation of data increases with increase in the temperature above 673K, the stress exponent n decreases to some extent. The fact that a set of data for each temperature shift towards the left hand side with increasing temperature strongly suggests that the separation could not be attributed to the activation of grain boundary sliding, because the data should shift towards right hand side if we assume that the dislocation glide and the grain boundary sliding contribute independently to the total strain. Furthermore, the fact that similar trends appear even in the specimens annealed at 873K as will be described later probably means an important role of dynamic recovery process instead of static one.

The above observation is phenomenally in agreement with the result obtained by Arzt and Rösler [9] for MA aluminum alloys containing carbide and oxide as dispersoids.

Ductility - Although tensile tests have not been carried out in the present study mainly because of the insufficient amount of materials, it could easily be expected on the analogy of the previous results on MA Al-TiC alloy prepared by just the same way as the present alloys [3] that the present alloys will exhibit very poor tensile ductility especially at elevated temperatures even if compressive flow proceeds to a considerably large strains. Taking account of the fact that the compressive strain to the initiation of shear fracture increases with increase in the content of TiB$_2$ (see Fig.7a and 7c), and that the uniformity of deformation as evaluated by the number and spacing of shear bands, is relatively higher in alloys with

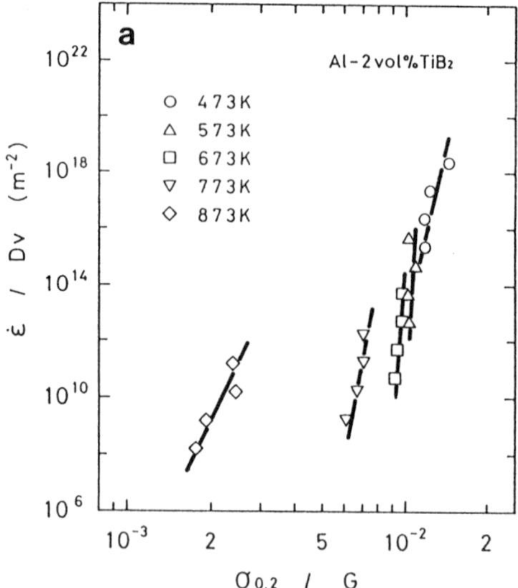

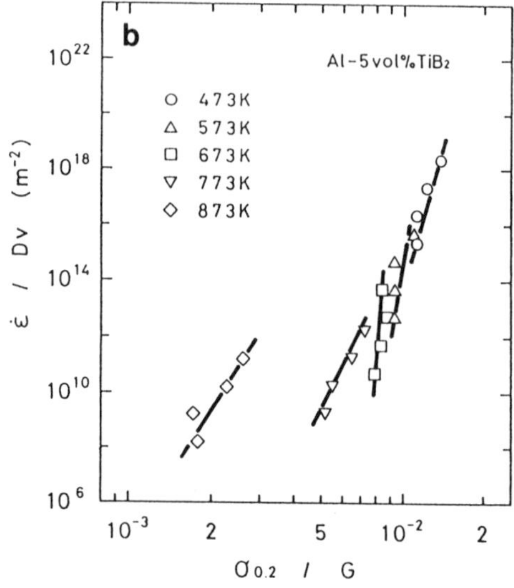

Fig. 10 The log-log plot of temperature compensated strain rate vs. normalized stress.

Fig. 11 Shear bands in Al-2vol%TiB$_2$. Temperature 773K, strain rate $8.33\times10^{-2}\,\mathrm{s}^{-1}$, strain 0.15; stress axis is vertical.

larger amount of TiB$_2$, the ductility loss can not be attributed to the existence of TiB$_2$ particles. Figure 11 shows a typical example of shear band observed in MA Al-2vol%TiB$_2$.

Effect of TiB$_2$ content on the yield stress – Figure 12 shows the yield stress at a strain rate of 8.33 x 10^{-2} s^{-1} as a function of TiB$_2$ content. Contrary to our expectations, the addition of TiB$_2$ by more than 2vol% has substantially no (or in some cases even negative) effect on the yield stress both at ambient and elevated temperatures. Similar tendency is found at each strain rate employed in this study. These results strongly suggest that under the processing conditions of this work the addition of large amount of TiB$_2$ to aluminum does not necessarily bring about a fine and uniform distribution of dispersoids enough to raise the strength of alloy product. In fact, we can hardly distinguish the difference in microstructure between MA Al-2vol%TiB$_2$ and Al-10vol%TiB$_2$.

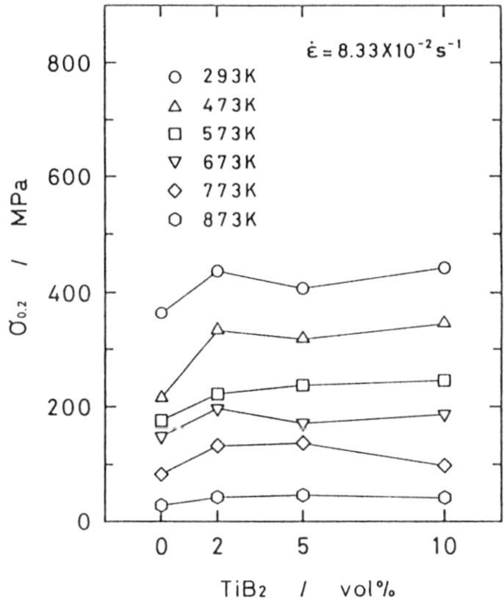

Fig.12 Yield stress vs. volume fraction of TiB$_2$.

Annealing effect – The specimens annealed at 823K for 10.8ks were also tested in compression under the conditions similar to those for the as-extruded materials. As an example, the results on Al-2vol%TiB$_2$ alloy are plotted in Fig. 13 in which the normalized strain rate vs. normalized yield stress relationship are compared between materials with and without above mentioned annealing. At each temperature between 473K and 773K, the stress dependence of strain rate of both specimens seems similar to each other, except that the stress levels of annealed specimens are slightly lower than those of as-extruded ones. Transmission electron microscopy suggests that the softening is caused mainly by the recovery due to the annihilation of dislocations and not by the recrystallization. On the other hand, annealing of Al-5vol%

TiB$_2$ and Al-10vol%TiB$_2$ alloys had often no effect on their strength, while in some cases it made them slightly stronger.

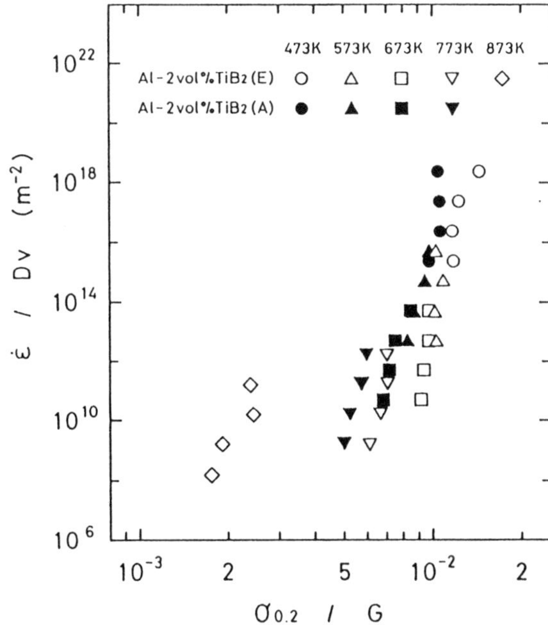

Fig. 13 Temperature compensated strain rate vs. normalized stress for as-extruded(open mark) and annealed(solid mark) alloy Al-2vol%TiB$_2$.

Comparison with the previous results on MA aluminum alloys - Most of the characteristics found in the elevated temperature deformation of three Al-TiB$_2$ alloys have previously been found in MA Al-TiC [3], Al-Al$_4$C$_3$ [10], Al-Ti [11], Al [4] [10] and so on. For example, both the shear band formation and the yield drop in the initial stage of stress-strain diagram have been reported in high temperature compression of MA aluminum [4]. Such an analogy seems quite reasonable if we remind the microstructural features common to these alloys, i.e., high density dislocations and relatively uniform distribution of very fine dispersoids.

Summary

The microstructure and elevated temperature deformation behavior of three MA Al-TiB$_2$ alloys have experimentally been investigated in the temperature range from 293K to 873K and the strain rate range from 8 x 10^{-5} s^{-1} to 8 x 10^{-2} s^{-1}. The results are summarized as follows.

(1) Mechanical alloying produces a fine grained microstructure which is stable against long term annealing at high temperatures.

(2) The yield stress decreases with increase in temperature, although it remains considerably higher than the strength of conventional heat

resistant aluminum alloy, especially above 473K.

(3) Unstable plastic flow occurs accompanying localized shear around 673K at low strain rates, resulting in very poor ductility.

(4) A yield drop phenomenon appears both in as-extruded and annealed conditions, while it tends to disappear with decrease in strain rate and temperature and with increase in TiB_2 content.

(5) The strain rate of each alloy is very sensitive to the stress, suggesting a potential ability of the present alloys as creep resistant materials.

(6) Addition of TiB_2 up to 2vol% improves compressive strength and ductility, while TiB_2 of more than 5vol% has little or no effect on them.

Acknowledgments

The authors would like to thank Dr. Y. Minonishi of Tohoku University for assistance in preparation of thin foils for transmission electron microscopy. They also wish to acknowledge Mr. M. Nakao of Shibaura Institute of Technology for his assistance in analysis of compression test data. Technical contributions by H. Tsunakawa of the University of Tokyo are greatly appreciated. This study was supported by the Ministry of Education, Science and Culture of Japan for Grant-in-Aid for Scientific Research through Contract No.01550061 and by the Light Metals Educational Foundation through Contract No.89-50.

References

[1] Benjamin, J. S., *Met. Trans.* 1, 2943-2951 (1970)

[2] Hawk, J. A., P. K. Mirchandani, R. C. Benn and H. G. F. Wilsdorf, *Dispersion Strengthened Aluminum Alloys*, (ed.) Y. -W. Kim and W. M. Griffith, p.517, The Minerals, Metals and Materials Society, Warrendale (1988)

[3] Otsuka, M., T. Ishihara, M. Sugamata and J. Kaneko, Proc. 10th RISO Int. Symp. Metallurgy and Materials Science: *Materials Architecture*, (ed.) J. B. Bilde-Sorensen *et al.*, p.491, RISO National Laboratory, Roskilde, (1989)

[4] Oliver, W. C. and W. D. Nix, *Acta Met.* 30, 1335-1347 (1982)

[5] Hawk, J. A., P. K. Mirchandani, R. C. Benn and H. G. F. Wilsdorf, *Dispersion Strengthened Aluminum Alloys*, (ed.) Y. -W. Kim and W. M. Griffith, p.551, The Minerals, Metals and Materials Society, Warrendale (1988)

[6] Lundy, T. S. and J. F. Murdock, *J. Appl. Phys.*, 33, 1671-1673 (1963)

[7] Sutton, P. M., *Phys. Rev.*, 91, 816-821(1953)

[8] Sherby, O. D. and P. M. Burke, Prog. Mater. Sci., 13, 323-390 (1966)

[9] Arzt, E. and J. Rösler, *Dispersion Strengthened Aluminum Alloys*, (ed.) Y. -W. Kim and W. M. Griffith, p.31, The Minerals, Metals and Materials Society, Warrendale (1988)

[10]Otsuka, M., M. Sugamata and J. Kaneko, unpublished work, (1987)

[11]Frazier, W. E. and M. J. Koczak, *Dispersion Strengthened Aluminum Alloys*, (ed.) Y. -W. Kim and W. M. Griffith, p.573, The Minerals, Metals and Materials Society, Warrendale (1988)

MECHANICALLY ALLOYED 2219 ALUMINUM ALLOYS CONTAINING NON-METALLIC DISPERSOIDS

Junichi Kaneko, Makoto Sugamata
Nihon University
Narashino, Chiba, Japan

Su-Gun Lim
Dong-A University
Pusan, Korea

ABSTRACT

2219 aluminum alloys strengthened by non-metallic dispersoids were fabricated by mechanical alloying and subsequent powder consolidation processes, for the purpose of obtaining P/M materials of improved mechanical properties. Powders of alumina, silicon carbide, titanium carbide and aluminum carbide were added to 2219 alloy powder and mechanically alloyed in a high energy ball mill. With the progress of ball milling, these ceramic particles became uniformly dispersed in the 2219 aluminum matrix after crushed to finer powders. P/M materials, which were fabricated by cold pressing, vacuum degassing and hot extrusion, showed high strength at room and higher temperatures in as extruded condition. However, those T6 tempered showed little increase of strength compared to I/M 2219 alloy. Age hardening was appreciably decreased in the P/M materials, and thus, expected superposition of strengthening by non-metallic dispersoids and precipitation hardening was not attainable. Nonetheless, tensile strength of as extruded P/M materials was as high as 525 MPa at room temperature, 256 MPa at 473 K and 169 MPa at 573 K.

THE MECHANICAL ALLOYING (MA) PROCESS can be used to produce alloys that are difficult or impossible to produce by conventional melting and casting techniques. It also serves as a means to produce composite metal powders with controlled and fine microstructures. It has been shown that MA is not a simple mixing process in a very fine scale. In the Al–Mg system, Mg atoms become substantially dissolved in the crystalline aluminum matrix (1). Formation of amorphous phases have been reported in several alloy systems (2-6).

MA processed aluminum contains finer and more uniform dispersion of oxide than in SAP (7). In most cases, however, the dispersion of both oxide and carbide is obtained in situ during MA process. The authors obtained more enhanced dispersion hardening in MA processed P/M aluminum materials with addition of ceramic powders (8). The present paper deals with MA of 2219 aluminum alloy with alumina, silicon carbide, titanium carbide and aluminum carbide with the purpose of improving its mechanical properties at room and higher temperatures. Alloy 2219 (Al-6.3Cu) is age hardenable, and has relatively high strength at room and elevated temperatures and high toughness at cryogenic temperatures.

Ceramic poders were added to 2219 alloy powder and MA processed in a high energy ball mill. With the progress of MA, ceramic particles became uniformly dispersed in 2219 alloy matrix after being crushed to finer particles. P/M materials were fabricated by cold pressing, vacuum degassing and hot extrusion. Microstructures and hardness were examined for both MA powders and extruded P/M materials. Age hardening behavior and mechanical properties were studied for P/M materials.

EXPERIMENTAL PROCEDURE

Machined chips of 2219 aluminum alloy were prepared from the supplied plate, the composition of which was 5.89Cu, 0.15 Fe, 0.30Mn, 0.09V, 0.13Zr and 0.08Si in wt%. The chips were blended with 2 vol% of the ceramic powder and MA processed. The characteristics of the ceramic powders used in this work are listed in

Table 1 Ceramic Powders Used in This Work

Ceramic	Purity	Density	Average particle size	Remarks
Al₂C₃	99.99%	3.9	0.6 μm	α-alumina
Al₄C₃	unspecified	2.36	3.5 μm	Al:73.24% , C :25.82%
SiC	95.8%	3.2	0.44 μm	βSiC F.C.:1.17% ,SiO₂:0.46%
TiC	unspecified	4.93	2.15 μm	C :19.69% , F.C.:0.14%

Table 1. A high energy ball mill of an
Attritor type was used for MA processing.
A powder charge of approximately 700 g
of blended powders was MA processed in
an argon atmosphere with an addition of
3 wt% of methanol with respect to the
powder charge as processing control
agent. MA was performed at 160 rpm for
36 ks.

The consolidation processes of MA
powders consisted of cold pressing,
vacuum degassing and hot extrusion, as
shown in Fig. 1 in which more detailed
processing conditions are also listed.
Finally, extruded bars of 7 mm in diameter were obtained as P/M materials and
their microstructures and mechanical properties were examined. Microstructural
observation was done by optical microscopy, TEM and SEM for both MA powders
and P/M materials. Microhardness was
measured for MA powders, and hardness
for P/M materials. Age hardening behavior
was studied for P/M materials and compared with that of I/M 2219 aluminum
alloy. Tensile tests were carried out at
room temperature, 473 and 573 K at a rate
of 0.05 mm/s.

The chemical composition of the obtained P/M materials is shown in Table 2,
along with the amount of ceramic powder
addition and the material designations
used hereafter in this paper. A significant increase in Fe content is noted for
the P/M materials with addition of ceramic particles. This is considered to be
due to abration of steel balls, impeller
blades and inside wall of the tank during

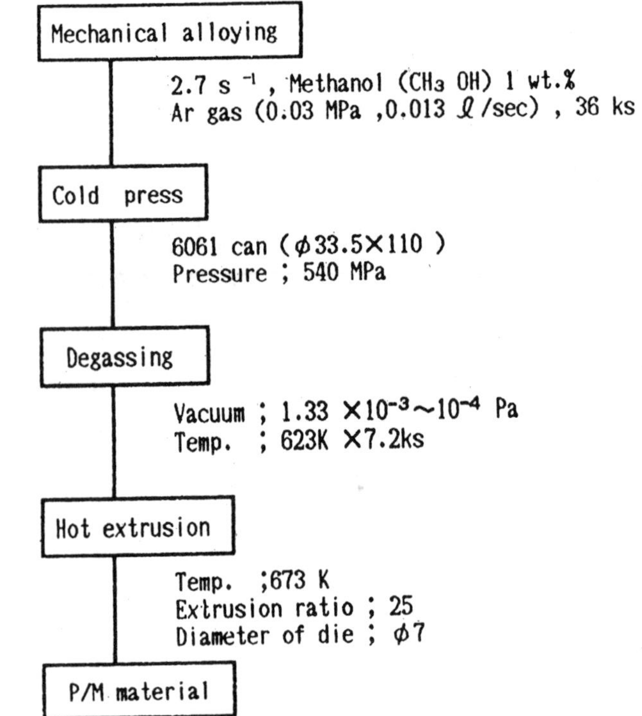

Fig. 1 Fabrication process of P/M materials

MA process. A little increase in Fe content was noted in case of no ceramic
powder addition, that is, 19MA. The contents of alumina and aluminum carbide in
19MA P/M materials listed in Table 2 are
arising from the in situ formation during
and after ball milling.

RESULTS AND DISCUSSION

MECHANICALLY ALLOYED POWDERS — With
the progress of MA processing, added
ceramic powders were crushed into fine
particles and became uniformly dispersed
in the more ductile matrix of 2219 aluminum powder. Thus, microhardness of the
powder increases with MA processing time.
The microstructures of MA powders are
shown in Fig. 2. Dispersed non-metallic
particles are much finer than their ini-

Table 2 Chemical Composition of P/M Materials

Material	vol.%	wt.%	Chemical analysis (wt.%)						Designation
			Cu	Fe	Al₂O₃	Al₄C₃	SiC	TiC	
2219	—	—	5.80	0.22	4.0	1.4	—	—	19MA
2219—Al₂O₃	2	2.69	5.73	0.26	5.7	—	—	—	19AO2
2219—Al₄C₃	2	1.66	5.65	0.57	—	2.8	—	—	19AC2
2219—SiC	2	2.23	5.76	0.50	—	—	2.3	—	19SC2
2219—TiC	2	3.40	5.53	1.04	—	—	—	3.0	19TC2

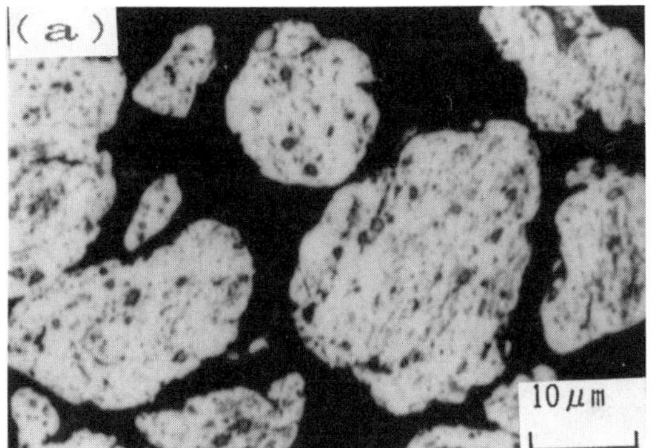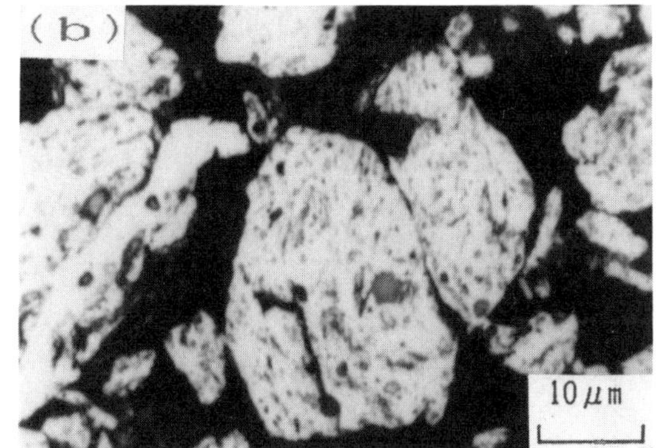

Fig. 2 Metallographs of MA powders: (a) 2vol%TiC, (b) 2vol%Al₄C₃

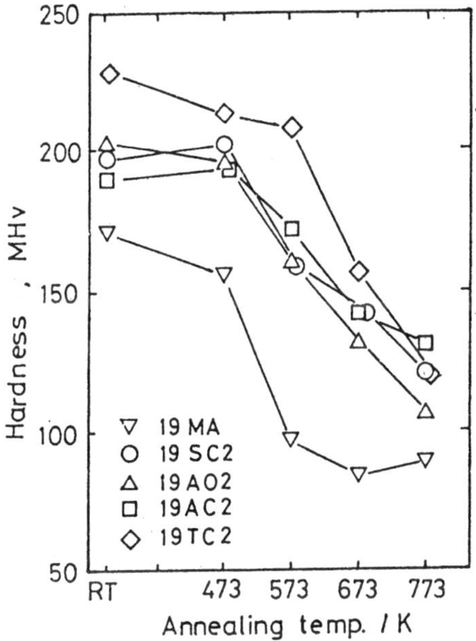

Fig. 3 Hardness changes of MA powders after isochronal annealing for 7.2 ks

to 473 K, and decreases rapidly above 573 K. The highest hardness was obtained for MA powders containing TiC particles, and those with additions of Al_2O_3, SiC and Al_4C_3 showed nearly the same hardness. The hardness of the MA powder without addition of ceramic particles (19MA) is always lower than those with ceramic particles additions. The effect of ceramic additions becomes more pronounced after annealing at above 573 K.

P/M MATERIALS — MA powders were consolidated to hot-extruded bars of 7 mm in diameter according to the processes shown in Fig. 1. The microstructures of obtained P/M materials are shown in Fig. 4, in which fine and uniform dispersion of ceramic particles is seen in 2219 aluminum matrix. The TEM micrographs of as extruded P/M materials are shown in Fig. 5. The structural features revealed are submicron subgrains pinned by a uniform and fine non-metallic dispersoids. The microstructures of P/M materials without addition of ceramic particles (19MA) are essentially the same as those with ceramic particle addition, except that coarser dispersoids are recognized in the latter.

tial size listed in Table 1. It is also noted that ceramic particles are uniformly distributed in the 2219 aluminum matrix.

The MA processed powders were isocronally annealed at various temperatures for 2 h, and the results of microhardness measurements are shown in Fig. 3. Hardness of MA powders decreases gradually with rising annealing temperature up

The changes of hardness of P/M materials on isochronal annealing are shown in Fig. 6. These materials retain fairly high hardness even after annealing up to 673 K. The P/M materials with TiC addtion

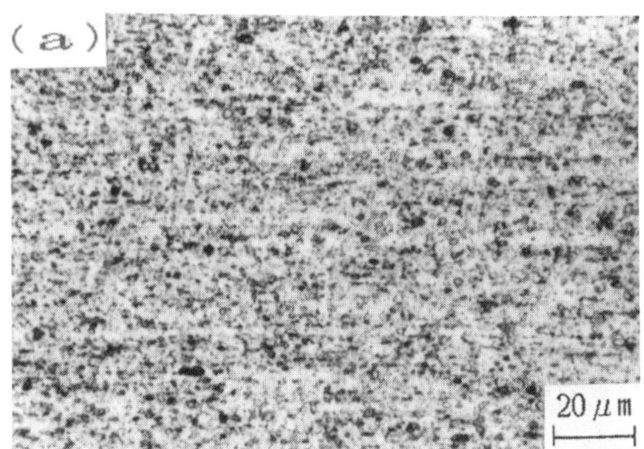

Fig. 4 Metallographs of as extruded P/M materials:
(a) 2219+2vol%TiC, (b) 2219+2vol%Al₄C₃

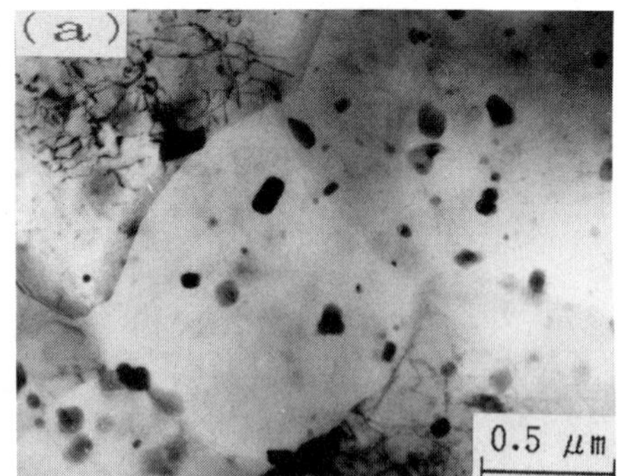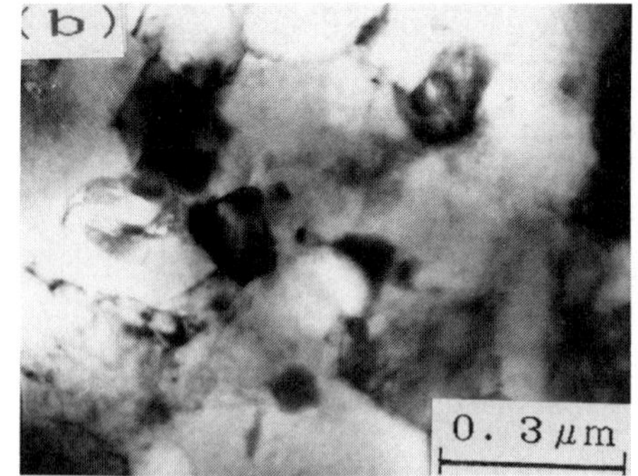

Fig. 5 TEM micrographs of as extruded P/M materials:
(a) 2219+2vol%Al₂O₃, (b) 2219+2vol%TiC

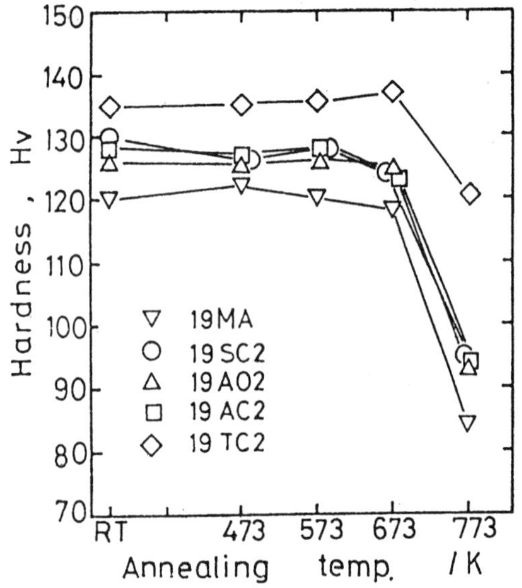

Fig. 6 Hardness changes of as extruded P/M materials

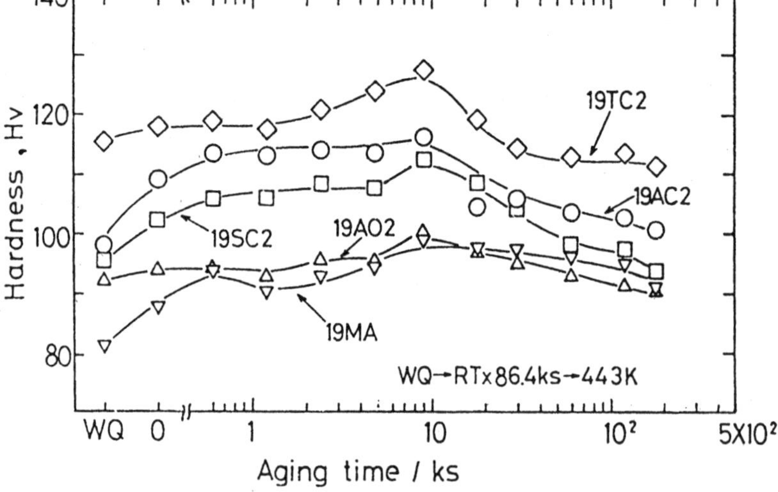

Fig. 7 Age hardening curves of P/M materials at 443 k

show the highest, and those without addition of ceramic particles the lowest hardness at various annealing temperatures.

Age hardening curves were obtained at 443 K after solutionizing at 808 K for 3.6 ks and water-quenching. The results are shown in Fig. 7. In these P/M materials, hardness in as water-quenched state is higher than that of I/M 2219 aluminum alloy. However, very low hardness increases were observed for these P/M materials during isothermal aging as shown in Fig. 7. The P/M material of 19TC shows the highest hardness among the present test materials in as water-quenched and age-hardened conditions. The hardening curves are nearly parallel each other and hardness increases attained by age hardening at 443 K are only less than 20 Hv for all the P/M materials, much less than those attained in 2219 I/M alloy.

The lattice parameter of P/M materials was measured in as water-quenched state by x-ray diffraction, and the results are listed along with the estimated solute content of Cu in Table 3. Only a

Table 3 Lattice Constant in As Water-Quenched State

Designation	Lattice constant (nm)	Cu (wt.%)
19MA	0.40434	2.59
19A02	0.40461	1.37
19AC2	0.40444	2.15
19SC2	0.40442	2.23
19TC2	0.40469	1.03

fraction of constituent Cu atoms are dissolved in aluminum matrix after water quenching, and this explains the marked decreases of age hardening in the present test P/M materials. Such decreases of age hardening in Al-Cu alloys are known to be caused by incease of Fe content (9). However, as shown in Fig. 7, age hardening curves are relatively unchanged regardless of Fe content of these P/M materials. Therefore, reduced Cu solute content in as water-quenched state was possibly caused by highly increased quench susceptibility in heavily deformed and fine

Table 4   Tensile Properties of As-Extruded P/M Materials

Material designation	σ_B (MPa) Test temp.(K)			$\sigma_{0.2}$ (MPa) Test temp.(K)			δ (%) Test temp.(K)		
	RT	473	573	RT	473	573	RT	473	573
1 9 MA	494	250	169	436	236	161	3.6	8.5	9.4
1 9 A 0 2	500	232	146	439	220	136	4.4	11.1	2.5
1 9 A C 2	525	242	159	435	229	151	4.1	13.3	6.7
1 9 S C 2	513	246	164	470	217	143	2.8	11.0	2.5
1 9 T C 2	520	247	163	455	231	152	5.1	6.0	2.2

Table 5   Tensile Properties of T6 Treated P/M Materials

Material designation	σ_B (MPa) Test temp.(K)			$\sigma_{0.2}$ (MPa) Test temp.(K)			δ (%) Test temp.(K)		
	RT	473	573	RT	473	573	RT	473	573
1 9 MA	382	228	107	361	206	99	1.6	5.1	9.4
1 9 A 0 2	378	220	106	345	198	93	2.0	4.2	6.4
1 9 A C 2	384	210	109	357	188	100	2.5	6.5	10.0
1 9 S C 2	397	223	109	352	206	101	2.2	7.5	9.7
1 9 T C 2	429	222	120	397	201	110	4.8	9.4	10.7

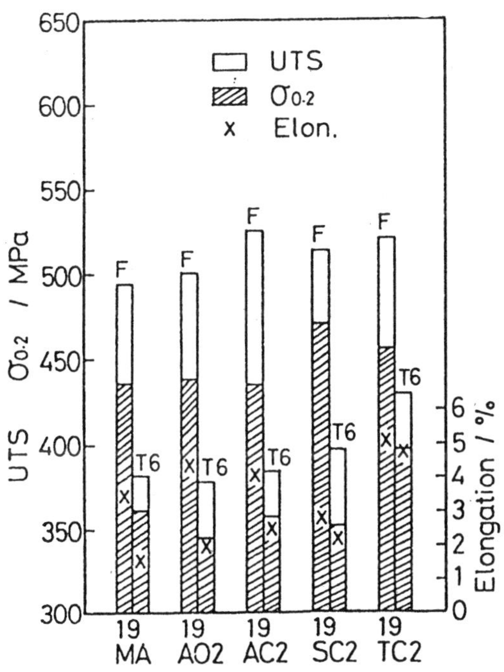

Fig. 8   Tensile properties as extruded and T6 treated P/M materials at room temperature

grained structures with high amount of non-metallic dispersoids. It was speculated that appreciable decomposition occurred during quenching of the specimens. Further experimental discussion is needed before understanding these phenomena more completely.

Tensile tests were carried out at room temperature, 473 and 573 K for as extruded and T6-treated P/M materials, and the obtained results are shown in Tables 4 and 5. Both tensile strength and proof stress are appreciably higher for as extruded materials than for T6-treated ones. Tensile strength of as extruded P/M materials is higher than 500 MPa at room temperature, approximately 250 MPa at 473 K and 160 MPa at 573 K. The strength og these P/M materials at higher temperatures is rather lower than those with pure aluminum matrix prepared in the same processes (8).

The highest tensile strength was obtained for P/M materials containing TiC in T6 temper, corresponding to the hardening curves shown in Fig. 7. However, in as extruded condition, those containing Al_4C_3 showed the highest tensile strength

233

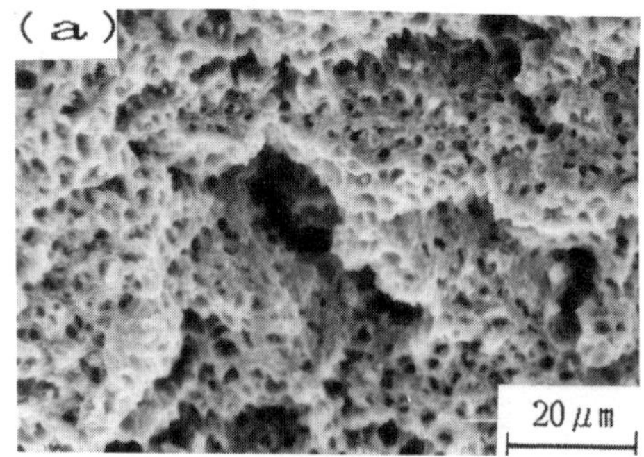

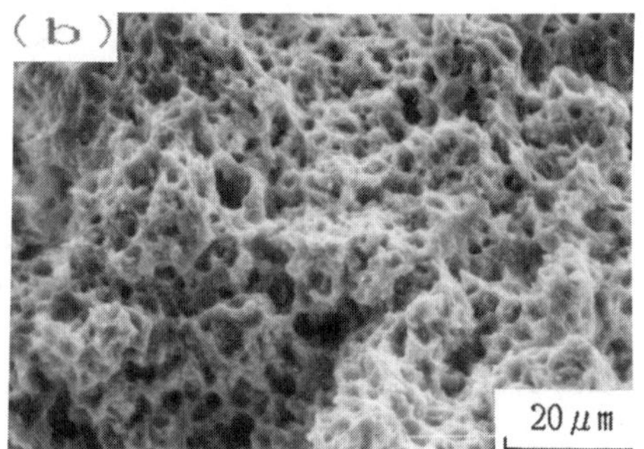

Fig. 9   Fractographs of T6 treated P/M materials without ceramic addtion:
(a) at room temperature, (b) at 573 K

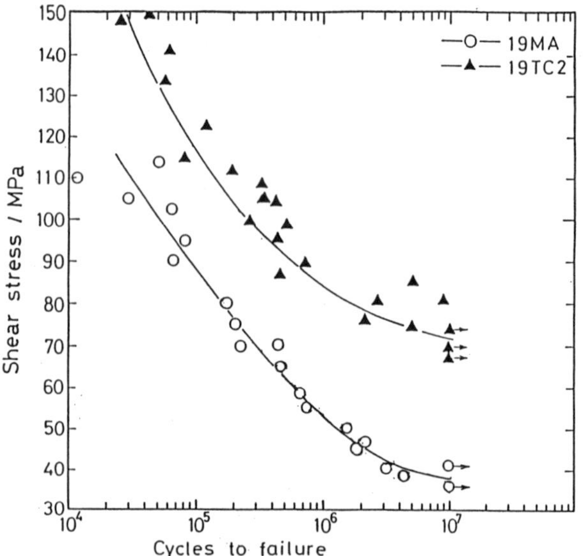

Fig. 10   S-N curves of T6 treated P/M materials

and proof stress at room temperature. Elongation is also lower for T6-treated materials than for as extruded ones at each temperature. Elongation of as extruded P/M materials is less than 5 % at room temperature, around 10 % at 473 K, and tends to decrease at 573 K. In case of T6-treated materials, elongation increases with the rise of test temperature.

Tensile properties at room temperature are shown in Fig. 8 for as extruded and T6-treated materials. Appreciable decrease of strength by T6 treatment is clearly seen. In as extruded conditions, differences in tensile properties among these P/M materials are relatively small, and P/M materials without addition of ceramic particles show lower but nearly the same strength as those with ceramic addtions. This is quite in contrast with the case of ceramic additions to pure aluminum where a significant strength increase was attained (8).

Some of the fractographs of tensile fractured surface are shown in Fig. 9.

Fractured surface at room temperature is smoother than that at higher temperature. Fracture occurs in more brittle manner at room temperature, and uniform distribution of fine dimples is observed on fractured surface at 573 K.

The results of torsional fatigue tests are shown in Fig. 10 for P/M materials of 19MA and 19TC in T6-treated condition. The fatigue strength 19TC-T6 P/M materials is appreciably higher than those without addition of ceramic particles. However, the fatigue strength of 19TC-T6 is lower when compared to those of rapidly solidified P/M 2219 alloys with addition of transition metals reported elsewhere by the authors (9).

CONCLUDING REMARKS

By applying MA process, fine and uniform dispersion of ceramic particles was attained in the matrix of 2219 aluminum alloy. The obtained P/M materials showed higher strength than conventional 2219 I/M materials in as extruded condition. This is due to fine non-metallic dispersoids and to fine subgrain structures developed in as extruded P/M materials. However, a significant decrease in age hardening was observed in these P/M materials processed by MA, and the strength of T6-treated materials was even lower than as extruded ones. The tensile strength at higher temperatures was also lower than the MA processed P/M materials with pure aluminum matrix with the addition of ceramic particles.

ACKNOWLEDGEMENTS

The authors are grateful to Professor M. Otsuka of Shibaura Institute of Technology for his valuable advice. They also would like to thank Y. Tomita and S. Nakamura for their experimental assistance. This work was supported in part by a Grant-in-Aid for Scientific Research, Ministry of Education, Science and Culture.

REFERENCES

1. J. S. Benjamin and R. D. Schelleng:
   Met. Trans., 12A(1981), 1827.
2. C. C. Koch, O. B. Cavin, C. G. McKamey
   and J. O. Scarbrough: Appl. Phys.
   Lett., 43(1983), 1017.
3. R. B. Schwarz, R. R. Petrich and C. K.
   Saw: J. Non-Cryst. Solids, 76(1985),
   281.
4. C. Politis and W. L. Johnson: J. Appl.
   Phys., 60(1986), 1174.
5. R. B. Schwarz and C. C. Koch: Appl.
   Phys. Lett., 49(1986), 146.
6. L. Schultz: Mater. Sci. Eng., 97(1988)
   , 15.
7. J. S. Benjamin and M. J. Bomford: Met.
   Trans., 8A(1977), 1301.
8. J. Kaneko, M. Sugamata and R. Horiuchi: Proc. 8th Int. Light Metals
   Congr., 1987, p776.
9. S. G. Lim, J. Kaneko and M. Sugamata:
   J. Japan Inst. Light Metals, 38(1988),
   710.

INTERACTION OF ALUMINUM WITH SILICON CARBIDE DURING MECHANICAL ALLOYING

S. I. Borovikova, E. Ja. Kaputkin, A. A. Kolesnikov
All Union Institute of Light Alloys
Moscow, USSR

Broad investigations of the Al-base composite materials manufacturing possibility are conducted for the last years. They concern both the choose of matrix, reinforcement type and their interaction.

Silicon carbide is one of the most broadly investigated and perspective reinforcement.

While developing Al-SiC composites a number of problems are solved, which include the following:

-selection of the matrix composition ensuring required level of mechanical properties,

-selection of reinforcement (shape, size),

-selection of the composite material manufacturing method ensuring uniform distribution of reinforcement, absence of the reinforcement aglomerates, optimum interaction between matrix and reinforcement.

The alloys of Al-Mg-Cu (series 2000), Al-Zn-Mg (series 6000) and other systems which have good combination of strength and ductility can be used as a composite matrix.

The particles of different shapes and sizes produced by PM technology can be used by the route of powder metallurgy.

Evaluation of quantity of reinforcement, methods of their introduction into the matrix as well as interaction of the matrix with SiC particles was conducted on the material, manufactured by mechanical mixing in ball mills and by mechanical alloying in attritors.

Powders of D16 alloy (series 2000) manufactured by atomization were used as matrix. DTA of these powders and the mixtures manufactured on their base has shown that the alloy solidus temperature shifts to the higher temperature range (by 40-45°C) as a result of rapid cooling during solidification.

Abrasive SiC powder with the mean size of $\sim 4\,\mu$m was used as reinforcement. X-ray diffraction analysis has shown that the initial powder contains two modifications: β-SiC (the main part) and α-SiC.

Two methods were used for evaluation of the uniformity of SiC particles distribution in the powder mixture manufactured by both the mechanical mixing and the mechanical alloying: apparent density variation coefficient (CVAD) determination method and SiC particles distribution (particles number) variation coefficient (CV) determination method on the unit of the compact material selection surface.

The other methods of the SiC particles homogeneity evaluation in the mixture especially manufactured by the mechanical mixing method result in the reinforcement distribution change in the investigation process.

The experiments have shown that the mechanical mixture of the powder alloy D16 with 17% SiC having the mean value of the apparent density 0,92 g/cm³ had CVAD 0,67-1,6%. This CVAD correlates with tensile strength variation coefficient (CVTS) of the rod manufactured from the mixture (fig.1).

The structure of the rod manufactured from mechanically mixed powder (fig.2) consists of the matrix material layers stretched in the direction of deformation and bordered by the SiC particles aglomerates.

This structure character indicates that in the powder mixture the reinforcement which size is smaller by the order of magnitude than that of matrix alloy

granules, dispose between separated
granules (or their aglomerates) of the
matrix. The granules are deformed under
subsequent extrusion and the SiC particles dispose over borders of the former granules. Very coarse aglomerates
of the reinforcement can be found in
some regions (fig.2a). Such heterogene-
ity of the powder mixture is characterized by the high values of CVAD and results in significant scatter of the
compact material mechanical properties
(fig.1).

Evaluation of the mixture of the
same composition manufactured by the
mechanical alloying established their
uniformity. CVAD=0,3% with the apparent density 1,3-1,4 g/cm^3 and CVTS=1,86%.
The extruded rod structure also has
more uniform distribution of the SiC
particles in the matrix as a result of
the mechanical alloying (fig.2b).

The distribution of the SiC particles in the extruded rods has been
shown as a frequency distribution of
their number in 200 equal-sized (10x10μm)
parts of the microsection (fig.3). It
has been shown that in case of manufacturing compact material from the mechanical by alloyed powder the frequency
distribution was normal that correspond-
ed to statistical particles distribution which is the optimum one. In case
of manufacturing compact material from
the powder mixture this distribution
was not achieved. Besides, it must be
taken into account that if the particles
aglomerates exist (fig.2a) the analysis
results given in fig.3 are just only
for regions which are free from the
aglomerates.

It is known that during mechanical
alloying the granules are formed whose
structure changes with increase of the
treatment time. The coarse laminar
structure (fig.4a) formed at an early
treatment stage is characterized by
more and more fine texture (fig.4b)
with increase of the treatment time
down to the layers thickness less than
0,5-1 μm which is not resolved in the
optical microscope (fig.4b). The structure lamination disappears due to tur-
bulent movement of the layers developed
during the deformation of the ductile
composite component (aluminium alloy
granules) and the SiC particles whose
hardness significantly exeeds the hardness of aluminium alloy. Presence of
the hard SiC particles promotes sufficiently rapid destruction of the aluminium granules under deformation and the
following formation of the composite
particles-granules. With that, the SiC
particles are distributed sufficiently

uniformly inside composite particle
during mechanical alloying. In view of
the fact that SiC particles are fixed in
aluminium matrix, the decomposition of
the mixture components which is observed
for instance when pouring mixture manufactured by the mixing method does not
take place. Considerably more uniform
distribution of SiC particles in the
matrix in mechanical alloying determines
the uniformity of their distribution in
the structure of extruded material too
(fig.2).

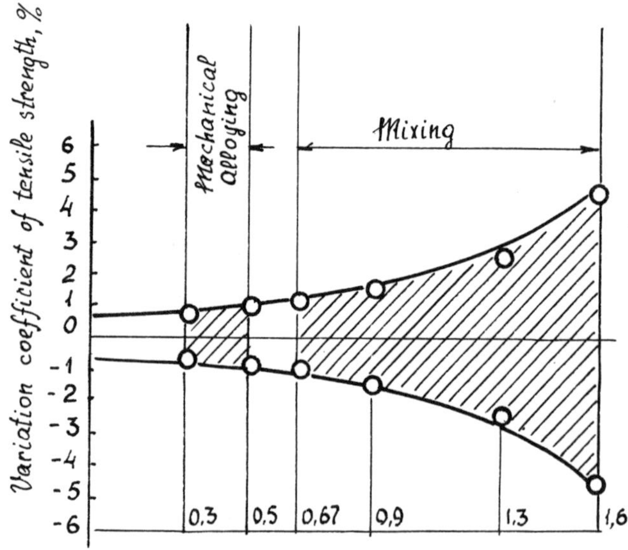

Fig.1—Correlation between variation
coefficients of the apparent density
(CVAD) and the tensile strength(CVTS).

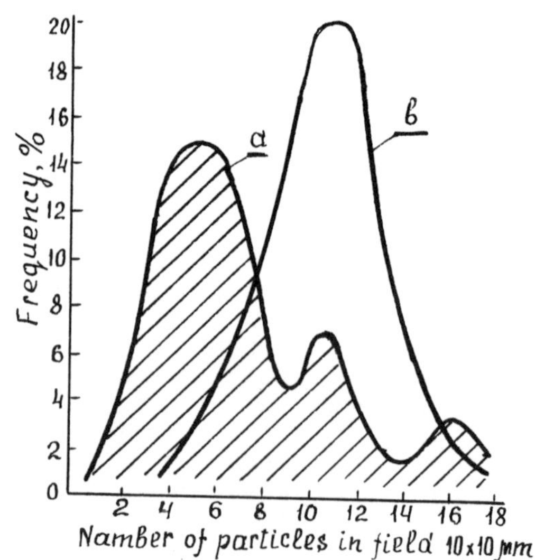

Fig.3—Distribution of the SiC particles in structure of the extruded rods
manufactured from mechanically mixed(a)
and mechanically alloyed(b) mixtures.

The phase composition and the lattice parameter of the aluminium alloy D16 and SiC initial powders, their mixtures and the rods manufactured from the mixtures were investigated by X-ray phase analysis and X-ray lines profile analysis.

The results of X-ray phase analysis show that the initial powder SiC is a mixture of two modifications:the main quantity – modification β-SiC, few part – α-SiC. These modifications remain in both the mixture and the extruded rods manufactured from them. In the rods ma-

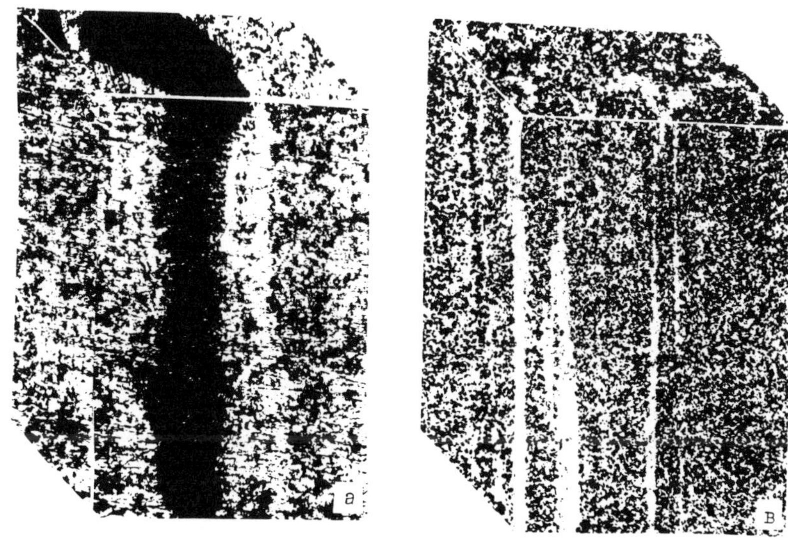

Fig.2-Microstructure (^{X}500) of the rod manufactured from the mechanical mixture (a) and the mechanically alloyed (b) powders of D16+17% SiC.

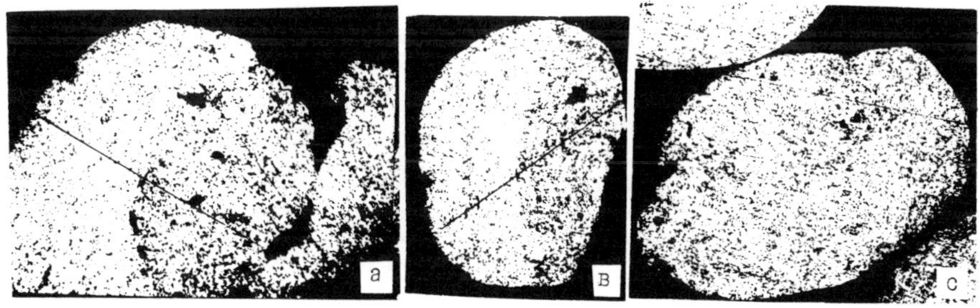

Fig.4-Structure formation (^{X}100) of the composite material Al-SiC granule during the mechanical alloying.
   a – welding of granules with simultaneous fracturing of the harder granule,
   b – laminar structure, thickness of the layer 100...200 μm,
   c – homogeneous structure.

Table 1 – Matrix Lattice Parameter Value

Composition	Condition	Lattice parameter, nm
D16	Initial powder	$0,4050_6$
SiC	Initial powder	–
D16+17%SiC	Mechanical mixing (MM)	$0,4051_0$
D16+17%SiC	Rod from MM powder	$0,4049_0$
D16+17%SiC	Mechanical alloying (MA)	$0,4046_8$
D16+17%SiC	Rod from MA powder	$0,4049_5$

nufactured from the mechanically alloyed powder beside matrix and phase SiC
(in two modifications) the insignificant
quantity of new phase appears whose iden-
tification is difficult because of its
small quantity.

The composite matrix lattice parameter analysis shows that interaction
between the matrix and the reinforcement
is absent under mixture manufacturing by
the mechanical mixing method (fig.5,
table 1). The matrix lattice period decrease is observed under mixture manufacturing by the mechanical alloying
method (fig.5, table 1) that points out
to the interaction proceeding between
the matrix and the SiC particles. The
aluminium matrix lattice parameter decreases at 0,0003 nm after mechanical
alloying as compared with the lattice
parameter before treatment. Considering
that the whole change of the matrix
lattice period has been caused by change
of the matrix lattice period has been
caused by change of the silicon content
in the solid solution as a result of interaction between the matrix and SiC, and
taking into consideration that the lattice parameter of the solid solution on
the base of aluminium decreases at
0,00017 nm with the silicon content increase at 1% (at.) it is possible to
evaluate the silicon content in the solid solution. It is about 2% (at.) that
exeeds the stable content of silicon in
aluminium under the room temperature approximately at an order of magnitude.
The matrix lattice parameter of the rod
manufactured from the mechanically alloyed powder approaches the lattice parameter of initial powder D16 (fig.5,
table 1) which indicates that the degree
of matrix alloying decreases under technological heatings.

Thus, conducted investigations of
the materials manufactured from the

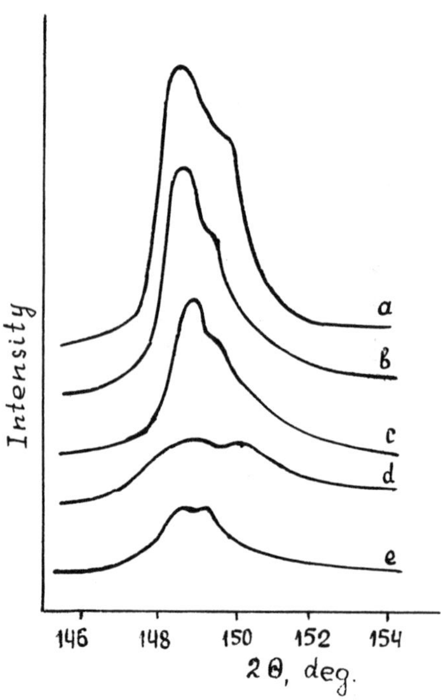

Fig.5 – Profile of X-ray line (331)$_\alpha$ of the matrix:
    a – D16, powder,
    b – D16+17% SiC, mechanical mixture (MM)
    c – D16+17% SiC, rod from MM,
    d – D16+17% SiC, granules mechanically alloyed (MA),
    e – D16+17% SiC, rod from MA.

mixed and mechanically alloyed powders
have shown that the latter had advantages in structure uniformity, particle
distribution and mechanical properties.

The anomalous supersaturated solutions manufacturing under the mechanical
alloying demands new approaches in selection of consolidation conditions, in
manufacturing of semi-finished products
and their heat treatment.

LIGHTWEIGHT AMORPHOUS Al-BASED ALLOYS
PRODUCED BY MECHANICAL ALLOYING

**Lowell E. Hazelton, Charles A. Nielsen,
Uday V. Deshmukh, Peter E. Pierini**
Dow Chemical Company
Walnut Creek Research Center
Walnut Creek, California 94598, USA

ABSTRACT

Novel lightweight amorphous Al-Mg-Ca alloys have been produced by mechanical alloying. These form a new ternary amorphous alloy system, and in addition they represent a new class of mechanically alloyed amorphous metals. They differ considerably from previously reported systems, all of which contain transition metals and have large negative free energies of mixing between the constituents. DSC results show no glass transition temperature below 235 C, indicating an unusually high reduced glass transition temperature greater than 0.70. Crystallization studies also reveal abnormally high activation energies. Large solid solubility extensions have been found, with ternary compositional ranges exceeding 50 at% and relative lattice parameter changes of over 11% for Mg_2Ca. An inverse correlation has been found between the structural complexity of a given phase and its formation stability under mechanical alloying. The density of these amorphous powders is 2.25 g/cm^3. The Vickers microhardness of the powders has been measured at 250 kg/mm^2.

THE MECHANICAL ALLOYING PROCESS, developed originally by Benjamin[1] at Inco for alloying oxide dispersion strengthened alloys, has in recent years been shown to be effective in driving a crystalline to amorphous solid state reaction in binary transition metal alloys[2-7]. The properties some have claimed are necessary for this amorphization reaction to succeed are a large negative free energy of mixing[4,5] and perhaps the rapid diffusion of one element in the other[4]. Recently, it has been found that not only early transition metal - late transition metal alloys but transition metal - metalloid alloys can be amorphized by this process[8-10]

Amorphous alloys have very high strength compared to their crystalline counterparts, and for this reason they have potential for structural uses. Such applications of amorphous alloys would require their availability in bulk form. However, their direct production in this form is difficult. Therefore consolidation of powder produced by processes such as mechanical alloying may become a key to their successful use. The present investigation has focused on lightweight alloys of potential value to the aerospace industry. We have found that a new class of alloys can be amorphized by mechanical alloying. Furthermore, while regions of glass formation by melt spinning have been reported in Al-Ca and Mg-Ca alloys above 50 at% Ca[9], to our knowledge no glass formation by melt spinning or other means has been reported in either the $Mg-Al$[10] or the ternary system.

Prior to mechanical alloying, some screening work was done by melt spinning on a series of $Mg_{73-x}Al_{27}Ca_x$ alloys, with x = 0, 8, and 47 (all compositions are in atom%). The selection of $Mg_{73}Al_{27}$ was based on the minimum in the T_o curve there[11], and on the absence of reported negative results in rapidly quenching alloys near that composition[12]. No sign of amorphous phase formation was seen in $Mg_{73}Al_{27}$ even under the most favorable quenching conditions. Introduction of 8% Ca as a glass-forming addition resulted in some amorphous phase formation, in spite of initial problems encountered with melt oxidation.

EXPERIMENTAL

Mechanical alloying was primarily done on elemental powders, though binary and ternary pre-alloyed starting powders have also been investigated. Elemental Al and Mg starting powders were normally prepared from ingots in order to reduce their oxide content. Elemental purity levels were 99.95% for Mg, 99.99% for Al, and 99.5% for Ca. In some cases commercially powdered elemental starting materials were used, with slightly inferior results. The milling was conducted in vials sealed under argon using hardened 52100 steel balls in a Spex 8000 shaker mill for periods ranging from 15 min. to 28 hours.

The milled powders were analyzed by x-ray diffraction using Cu $K\alpha$ radiation with a Norelco diffractometer for the kind and extent of crystallinity. Differential scanning calorimetry (DSC) was also conducted using a DuPont Model 2000 at heating rates ranging from 1 to 20 C/min on selected samples to determine the glass transition temperature (T_g), crystallization temperatures, and associated activation energies. DSC samples were also heated to temperatures just above each exotherm and x-rayed to determine what structural changes were occur-

ring. Optical microscopy and scanning and transmission electron microscopy (SEM and TEM) were also carried out on some of the powders to gain an understanding of the amorphization process, using a Cambridge Stereoscan 360 SEM and a JEOL JEM-100CX TEM. Elemental analysis was carried out by inductively coupled plasma. Oxygen analysis was done by neutron activation. The density of the powders was measured by the Archimedes method using toluene as the working fluid. Microhardness measurements were made using a Vickers diamond indenter at loads of 3.85, 100, and 300 g.

RESULTS AND DISCUSSION

PHASE DIAGRAM - A wide range of alloys in the Mg-Al-Ca system were mechanically alloyed, concentrating primarily on low-Ca alloys. Mechanical alloying resulted in completely or nearly amorphous powders for compositions in a region around $Mg_{38}Al_{57}Ca_5$, as shown by fig 1. Surrounding this is a multiphase region where the amorphous phase is still dominant under the mechanical alloying process, but substantial amounts of one or more crystalline phases are also present.

Amorphization of these alloys by mechanical alloying is novel in several respects. First, the free energies of mixing are relatively low. The heats of formation of Al_2Ca and Al_4Ca are moderately large in magnitude at -73 and -43 kJ/mole, respectively, but the free energy of formation of Mg_2Ca is only -10 kJ/mole and those of the Mg-Al compounds about -4 kJ/mole[13]. Since the amorphous alloys to be described contained less than 10 at% Ca, the heats of mixing are expected to be less negative than about -20 kJ/mole. Second, unlike any other alloys that have been reported to be amorphized by mechanical alloying, they contain no transition metals. In addition, atomic diffusion rates are not anomalously high in any of the three binary alloy systems involved. This shows that anomalously rapid diffusion, which appears

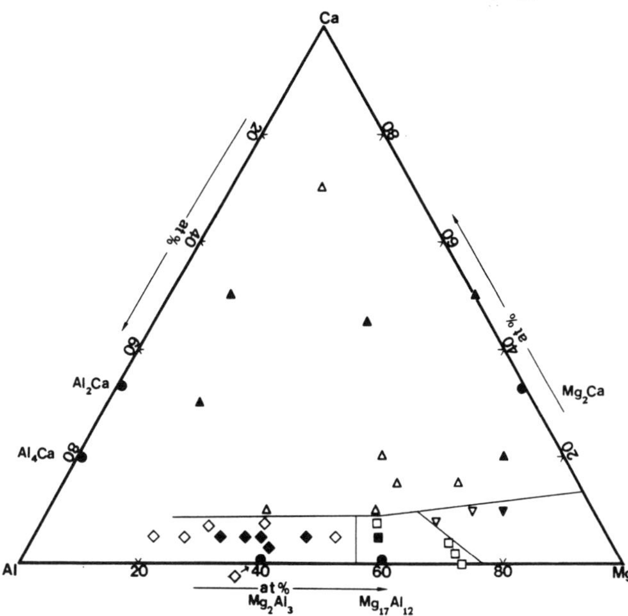

Fig. 1 - Mechanical alloying phase diagram for
the Mg-Al-Ca system. Solid symbols indicate
single phase results, open symbols indicate
which is the dominant phase in a multiphase
region. The symbols represent: ▲, Mg_2Ca;
▼, (Mg); ■, $Mg_{17}Al_{12}$; ◆, amorphous.
Solid circles (●) indicate the equilibrium
compounds. Regions over which a particular phase
is dominant are separated by lines.

to be necessary for the thermally driven solid
state amorphization reaction to succeed, probably has no general role in amorphization by
mechanical alloying and that the reaction
mechanisms for these two solid state transformation routes are substantially different.

As might be expected, under mechanical
alloying most of the elements or compounds (i.e.
Mg, Al, Mg_2Ca, $Mg_{17}Al_{12}$, and Al_4Ca) were found
to be surrounded by a substantial single-phase
region, or at least a multiphase region in which
they are the dominant phase. However, this was
not the case with Mg_2Al_3. Mechanical alloying
at this composition resulted in an aluminum
solid solution, an amorphous phase, and only a
slight amount of Mg_2Al_3 phase. The addition of
less than 3% Ca completely destabilizes the Al
solid solution, while having little or no effect

on the stability of Mg_2Al_3.

Not coincidentally, the Mg_2Al_3 fcc crystal
structure is very complex with 1832 atoms per
conventional unit cell[14], far more complex than
any of the other compounds. It is to be expected, first, that a complex phase would be more
susceptible to disruption by cold working than a
simple one. Secondly, its growth rate might be
expected to be lower under any given conditions
then that of a simple phase. This viewpoint is
supported by the DSC results, described later.
These show that nucleation and growth of
$Mg_{17}Al_{12}$, a bcc phase with 58 atoms/unit cell,
occurs at 270 C. Mg_2Al_3, however, already
present in the milled powders, does not grow
much up to 250 C and produces an exotherm,
requiring only growth, at 345 C. For both of
these reasons Mg_2Al_3 is less likely than a
simple phase to be able to grow at the rate
required to balance its rate of destruction in
the mechanical alloying process. This idea is
also borne out in fig. 2 where an average width
d of the single-phase region for each compound,
taken as the square root of the estimated
relative single-phase area, has been plotted
versus the inverse of the number of atoms in its
primitive unit cell. The good fit to a straight

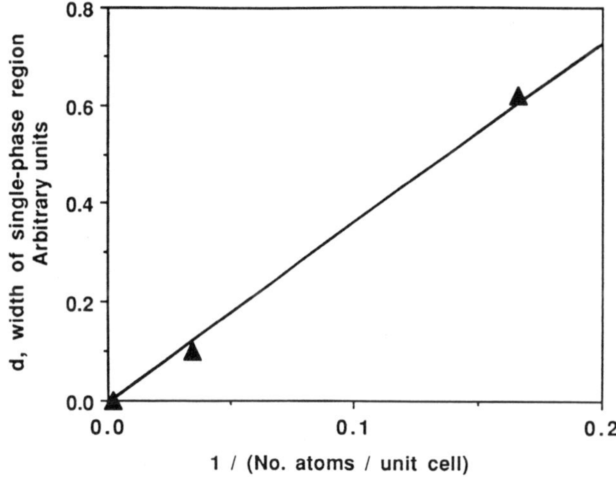

Fig. 2 - Inverse correlation between the size d
of single-phase regions and the number of atoms
per unit cell in that phase.

line is fortuitous since the estimates of d are only accurate to 20%. Furthermore, the elemental phases do not fit this correlation, being far less dominant than would be predicted on this basis. However, it appears that a phase's unit cell size may be used as at least a rough predictor of its ease of disruption by mechanical alloying.

Large changes in d-spacings were observed for some phases. By far the largest effect was seen for the Mg_2Ca phase with varying Al and Mg concentrations. The maximum lattice parameter changes seen were an 10.6% decrease in $\underline{a}$ and an 11.7% decrease in $\underline{c}$ at $Mg_{31}Al_{64}Ca_5$, giving a decrease in unit cell volume of 29.5%. This is a two phase (Mg_2Ca + amorphous) composition so the composition of the Mg_2Ca phase should be close to that at the nearest boundary of the single phase Mg_2Ca region, which lies at about 9% Ca. Using this composition a volume decrease of 29.9% is calculated from a composition-weighted sum of the pure elements' molar volumes. The excellent agreement may be fortuitous, but this tends to confirm that the composition of the Mg_2Ca phase is indeed as far from stoichiometry as it appears to be. The deviations in observed d-spacing for the other compounds, $Mg_{17}Al_{12}$, Mg_2Al_3, and Al_2Ca, rarely exceeded 1% and were never more than the 2% increase found for $Mg_{17}Al_{12}$ at $Mg_{55.5}Al_{37}Ca_{7.5}$. The largest lattice parameter changes found in terminal solid solutions were a 5% increase for (Al) at the Mg_2Al_3 composition, and a decrease of 0.4% for Mg in a $Mg_{65}Al_{27}Ca_8$ alloy.

For the amorphous compositions, broad amorphous peaks already dominate the x-ray diffraction patterns for powders milled one hour or longer, and after four hours amorphization is nearly complete (see fig. 3a). Peaks due to unreacted elemental constituents start broadening after 15 minutes, eventually saturate in width, then decrease in height with milling. The Ca lines disappear within 30 minutes, and the Mg lines within 1 hour, while the Al lines take over 4 hours. These times are inversely

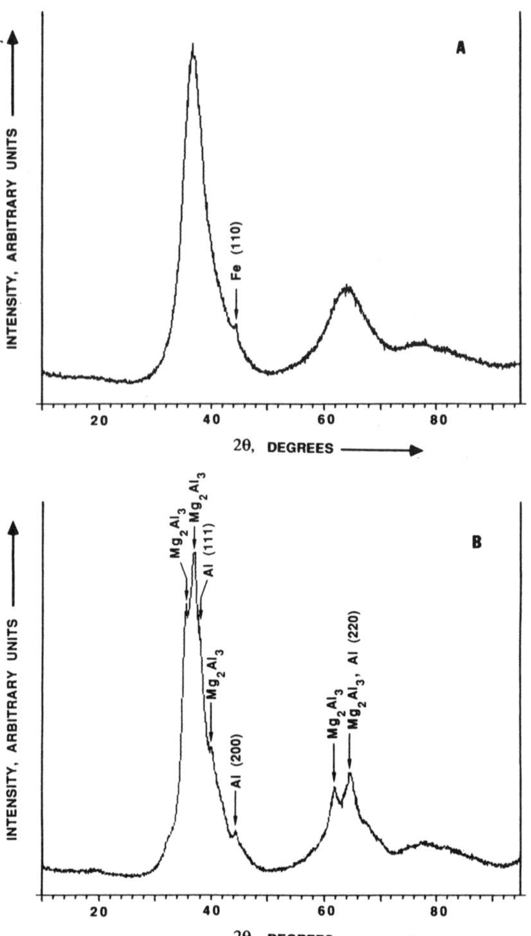

Fig. 3 - X-ray diffraction patterns for $Mg_{38}Al_{57}Ca_5$ milled (a) 4 hrs, showing complete amorphization, and (b) 1 hr., showing peaks for (Al) and Mg_2Al_3.

correlated with the distance on the phase diagram from the amorphous composition to the respective elements. This might explain these results to some extent, for the following reasons. First, a minor constituent will reach the limit of x-ray detectability when the remaining level is a larger fraction of the starting level. Second, as the characteristic early lamellar structure develops, the major constituents are the most likely to find themselves welded to another layer of the same element. Diffusional mixing at elemental interfaces will then tend to exhaust the minor constituents while leaving unreacted agglomerates of the major ones.

Other factors, however, are probably required to explain the observed elemental reaction times. The rapid reaction of Ca may be additionally explained by its heats of mixing with Mg and Al, which are much more negative than for Mg with Al. On the other hand, Al is present at a level only 50% greater than Mg and has diffusion rates in Mg similar to those of Mg in Al[17], so it should have a similar reaction time. The explanation for the difference probably lies in the fact that the two phase (Al + amorphous) region lies nearby. Furthermore, from the equilibrium Mg-Al phase diagram it appears that the free energy function for Mg in Al does not rise too steeply with increasing Mg concentration. The nearby slowly rising free energy should give a small free energy difference between Al and the amorphous phase at the compositions of interest, contributing to the less rapid conversion rate. This ignores the role of Ca, which destabilizes the Al solid solution, but evidently not too strongly.

For crystalline phases, e.g. Mg_2Al_3, calculation of a mean grain size from the full width at half maximum of its x-ray peaks after one hour of milling gives sizes in the range of 7-10 nm by use of the Scherrer formula[16]. The peak widths saturate at this level as mechanical alloying proceeds (see fig. 3b).

Under certain conditions it has also been found that early in processing $Mg_{17}Al_{12}$ appears, then disappears before one hour of milling. This is undoubtedly because Mg-Al interfaces with no Ca present will be common at early stages of processing in low-Ca alloys. $Mg_{17}Al_{12}$ is evidently the first phase to form under these conditions, which is in agreement with the earlier discussion on rates of compound formation. This is supported by the finding in $Mg_{73-x}Al_{27}Ca_x$ alloys, for x from 3 to 10%, that the Mg phase first largely disappeared then grew somewhat at the expense of $Mg_{17}Al_{12}$ after 4 hours milling time.

Examination of the powders using SEM revealed angular particles ranging in size from below 1 micron to several hundred microns. Powders sieved into different size ranges and submitted to x-ray diffraction appeared to be identical in structure. Powders were also embedded in epoxy, sectioned by microtome, and examined by TEM. Amorphous regions, identified by selected area diffraction, were relatively featureless as expected. Mg_2Al_3 diffraction patterns were not detected, probably due to their extremely fine grain size and the difficulty in discerning broad, low intensity peaks in electron diffraction patterns. Examination of incompletely amorphized powders by TEM showed grain sizes for elemental aluminum in the 5-20 nm range, in good agreement with x-ray results.

PHYSICAL AND MECHANICAL PROPERTIES - Elemental analysis yielded compositions that normally agreed with the starting composition to better than 1%. Iron contamination levels were found to be fairly low, less than 0.15 wt%. Neutron activation on both starting and milled powders revealed oxygen contents below 0.8 wt%, indicating that no significant oxidation occurred during or after milling. The thickness of the oxide surface layer measured by Auger depth profiling on milled powders was about 9 nm.

The DSC results (fig. 4) show that the present Al-rich alloys possess very high thermal

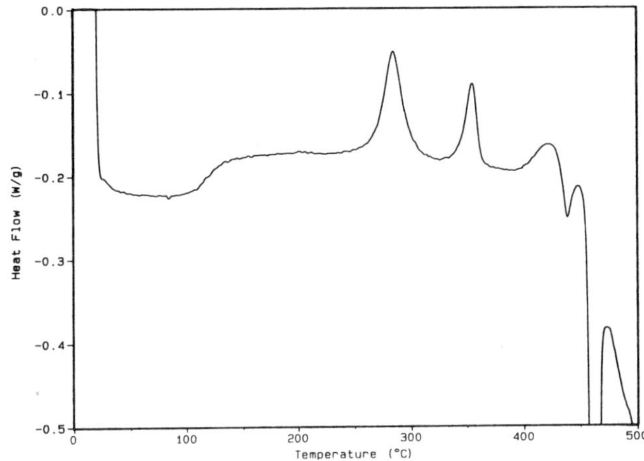

Fig. 4 - DSC run at 10 C/min on $Mg_{38}Al_{57}Ca_5$.

stability, similar to the calcium-rich Al-Ca amorphous alloys[18] For $Mg_{38}Al_{57}Ca_5$ no glass transition occurs before the onset the first crystallization peak at 245 C, so the reduced glass transition temperature $T_{rg} = T_g/T_1$ is above 0.70, where T_1 is the liquidus temperature. The highest reported T_{rg} value is 0.69 for $Ca_{65}Al_{35}$[18]. The crystallization temperature T_x measured by steepest tangent intersection is 270 C, for a reduced crystallization temperature $T_{rx} = T_x/T_1 = 0.74$, and the peak is at 285 C (see table 1).

DSC specimens heated to above each exotherm and submitted to x-ray diffraction revealed that the first exotherm at 285 C represents nucleation and growth of $Mg_{17}Al_{12}$. The second peak at 355 C results from growth of the Mg_2Al_3 phase, which for this particular specimen was already present in the milled powder at a level of about 1%. As mentioned earlier, the Mg_2Al_3 phase is very stable against coarsening, probably due to its complexity.

In order to get activation energies for the crystallization events, DSC runs were carried out at 1, 5, 10, and 20 C/min. Analysis of the peak shifts by the Kissinger method for the first two exotherms in amorphous $Mg_{38}Al_{57}Ca_5$ yielded activation energies of 243 and 266 kJ/mole. The activation energy for the strongest exothermic event in the largely crystalline Mg-rich alloy $Mg_{70}Al_{20}Ca_{10}$ was 158 kJ/mole. These values, especially for the amorphous alloy, are far above the reported activation energies for self, tracer, and chemical diffusion of Mg and Al in themselves, each other, or

their intermetallic compounds respectively, which range from 145 kJ/mole down to 57 kJ/mole[17]. Put together, the very high T_x value, the unusually high activation energies for crystallization, and the apparent resistance to coarsening by the Mg_2Al_3 phase indicate unusual thermal stability.

The density of as-milled $Mg_{38}Al_{57}Ca_5$ alloy powders was measured at 2.25 g/cm^3. The actual bulk density may be slightly higher since the particles, especially the larger ones, contain incompletely welded cracks which may not be penetrated entirely by the working fluid, lowering the measured densities. Any error is estimated to be less than 1-2%

Preliminary Vickers microhardness values have been measured on powder pellets which had been compressed at 100 kpsi for 1 minute at 220, 230, and 240 C. The measured values mostly ranged from 240 to 260 kg/mm^2. Since the powders had been held at temperature for 10-15 min, they had become crystallized to some extent. In addition, the relative densities of the pellets varied from 90 to 94%. The microhardness indentations made at 3.85 g showed cracks at the apices of some indentations, and those made at 300 g all showed cracks. This cracking could be due either to the incomplete consolidation of the pellets or to their partial crystallization, so may or may not be an indication of brittleness in the amorphous alloy. The crystalline phases are known to be extremely brittle[19]. Microhardness measurements on pellets consolidated at room temperature using higher pressures are being carried out.

Table 1 - DSC results for amorphous $Mg_{38}Al_{57}Ca_5$ alloys

<u>Exothermic Peak Temperatures, C</u>

Peak No.	Phase	Onset	Tangent[(1)]	Peak	T_{rx}	Activation Energy, kJ/mole
First	$Mg_{17}Al_{12}$	245	270	285	0.75	243
Second	Mg_2Al_3	325	345	355	0.85	266

(1) Intersection of peak's steepest tangent with baseline

Some initial work was also done on extrusion of powders to produce a fully dense amorphous metal rod. The powders were first consolidated at room temperature, using a pressure of 57 ksi, to 75% of full density. The compacts were next sealed in aluminum cans by welding, after vacuum degassing at temperatures of 190 to 200 C for 30 minutes. Then they were extruded using a 350 ton press at temperatures of 190 to 230 C. Extrusion at the higher temperatures resulted in consolidation to about 99% of theoretical density, but overheating during extrusion resulted in very substantial crystallization. Due to this the ductility of the extruded rods was low. The only mechanical property measurement made was Vickers microhardness, which was measured at 310 kg/mm^2 using a 100 g load.

CONCLUSIONS

It has been shown that the process of amorphization by mechanical alloying is not limited to transition metal alloys, or to alloys with large negative free energies of mixing. Amorphous Mg-Al-Ca alloys have been prepared by mechanical alloying in a region around $Mg_{38}Al_{37}Ca_5$. The alloy powders show unusual thermal stability, having very high reduced glass transition temperatures, very high activation energies for crystallization, and high resistance to the coarsening of the Mg_2Al_3 phase. An inverse correlation between unit cell size and phase stability under mechanical alloying has been demonstrated. Thus, Mg_2Ca was found to dominate much of the phase diagram, while the Mg_2Al_3 phase shows no dominant region at all. These alloys have low density at 2.25 g/cm^3, and very high Vickers microhardness, around 250 Kg/mm^2.

ACKNOWLEDGEMENTS

We thank R.E. Lewis, A. Joshi, A. Davinroy, and J. Herrington of Lockheed Missiles and Space Co. for providing assistance and the use of their facilities in doing the extrusions, some of the microhardness measurements, and the Auger depth profiling. We also wish to express appreciation to W.L. Johnson, D. Crooks, and especially R.E. Lewis for many helpful discussions.

REFERENCES

1. Gilman, P.S. and J.S. Benjamin, Ann. Rev. Mater. Sci. 13, 279 (1983).

2. White, R.L., Ph.D. dissertation, Stanford University, 1979.

3. Koch, C.C., O.B. Cavin, C.G. McKamey, and J.O. Scarbrough, Appl. Phys. Lett. 43, 1017 (1983).

4. Schwarz, R.B., R.R. Petrich, and C.K. Saw, J. Non-Cryst. Solids 76, 281 (1985).

5. Lee, P.Y. and C.C. Koch, Appl. Phys. Lett. 50, 1578 (1987).

6. Dolgin, B.P., M.A. Vanek, T. McGory, and D.J. Ham, J. Non-Cryst. Solids 87, 281 (1986).

7. Hellstern, E., L. Schultx, R. Bormann, and D. Lee, Appl. Phys. Lett. 53, 1399 (1988).

8. Magini, M., S. Martelli, and M. Vittori, J. Non-Cryst. Solids 101, 294 (1988).

9. St. Amand, R. and B.C. Giessen, Scripta Met. 12, 1021 (1978).

10. Hehmann, F. and H. Jones, in "Rapidly Solidified Alloys and their Mechanical and Magnetic Properties", ed. by B.C. Giessen, D.E. Polk, and A.I. Taub, p. 259, MRS, Pittsburg (1986).

11. Murray, J.L., Bull. Alloy Phase Diag. 3, 60 (1982).

12. See for example Luo, H.L., C.C. Chao, and P. Duwez, Trans. Met. Soc. AIME 230, 1488 (1964).

13. R. Hultgren, R.L. Orr, P.D. Anderson, and K.K. Kelley, "Selected Values of Thermodynamic Properties of Metals and Alloys", John Wiley, New York (1963).

14.   P. Villars and L.D. Calvert, "Pearson's
      Handbook of Crystallographic Data for
      Intermetallic Phases" vol. 2, ASM, Metals
      Park, OH (1985).

15.   E.F. Emley, "Principles of Magnesium Tech-
      nology", Pergamon, Oxford (1966).

16.   H.P. Klug and L.E. Alexander, "X-ray
      Diffraction Prodecures", John Wiley, New
      York (1974).

17.   E.A. Brandes, "Smithells Metals Reference
      Book", 6th. ed., ch. 13, Butterworths,
      London (1983).

18.   B.C. Giessen, J. Hong, L. Kabacoff, D.E.
      Polk, R. Raman, and R. St. Amand, in "Rapid-
      ly Quenched Metals III", vol. 1, p.249, The
      Metals Society, London (1978)

19.   Unpublished results, L.E. Hazelton, Dow
      Chemical Co.

KEYNOTE ADDRESS
MECHANICAL ALLOYING AS A METHOD OF PRODUCING ANOMALOUSLY SUPERSATURATED SOLID SOLUTIONS

I. S. Polkin, E. Ja. Kaputkin, A.B. Borzov
All Union Institute of Light Alloys
Moscow, USSR

Higher strengthening of various materials by alloying can be achieved by incorporating more alloying additions into the solid solution and by larger volume fraction of precipitates with optimal morphology. Unfortunately, the amount of an alloying element is, in some cases, limited either by its low solubility according to the phase diagram, or by formation of low melting equilibrium or nonequilibrium phases. The anomalous supersaturation can be achieved only by special techniques, such as rapid solidification of the melt.

Recently, more and more investigators pointed to the formation of solid solutions during mechanical alloying due to solid-phase interaction of metal powders or master alloys in high-energy ball mills (attritors). However there is practically no information about kinetics of the process of solid solution formation and about limiting concentration attained in mechanical alloying.

It was the purpose of the present investigation to study the process of structure formation, the kinitics of the process of solid solution formation and limiting concentrations achieved at solid-phase interaction during joint deformation of powders in the attritor.

Two binary alloy systems were investigated, Ni-21 at. % W-25 at. % Al or 50 at. % Cr, and Al-4 and 37 at. % Fe, with concentration of the second component exceeding the solid solution equilibrium concentration.

The investigation was performed using optical and scanning electron microscopy, as well as X-ray diffraction analysis. For examination of the layer formation mechanical alloying products on the balls, the protective copper layer preventing damage of the surface layer during the microsection preparation was applied on this layer.

The main method of the solid solution examination was the analysis of the X-ray diffraction lines profile which permitted us to evaluate the degree of solid solution alloying and its non-uniformity by change of the lattice period. When selecting the photography conditions and processing the results by a harmonic analysis method, the specific features of mechanical alloying products structure were taken into consideration, namely the influence of cold work hardening and solid solution non-uniformity on the line profile.

The formation of the mechanical alloying products structure proceeds in several stages displayed as formation of particles with laminated structure which is at the beginning displayed more clearly and then disappears with the increase of processing time. According to the present ideas which were mainly formulated by Benjamin [1], the mechanical alloying process takes place mainly in the space between the balls. They suppose simultaneous hit of several particles between the balls and the deformation of them under the balls collision, cold welding with each other, destruction and subsequent welding.

However, it must be pointed out that falling between the balls and joint deformation of a great number of powder particles simultaneously are necessary conditions for manufacturing multilayer particles observed by experiment. A probability of such a process is too small.

The structural analysis of the layer formed on the balls surfaces and investigation of the character of its fracture has shown that the morphology of particles formed during deformation is related to the processes which take place on the balls surfaces (Fig. 1).

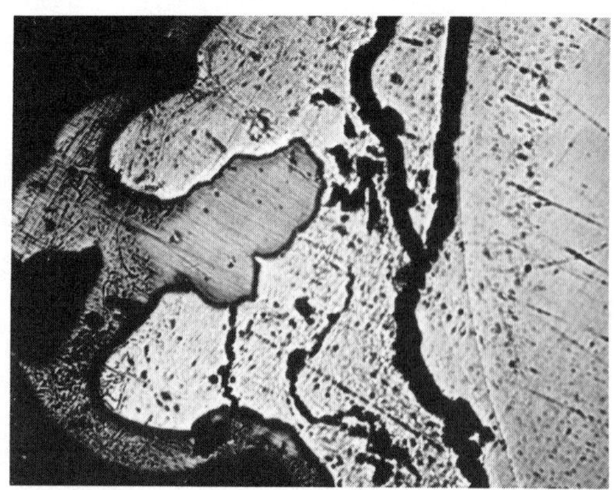

Fig. 1 Microstructure of
the layer of mecha-
nical alloying pro-
ducts on the ball
surface

Analysing the conditions required for realization of mechanical alloying, it should be pointed out that
the presence of sufficient quantity of
soft "plastic" component is of major
importance. The process seems to proceed as if by means of "dipping" hard
particles into softer component which
is located eigther on the ball surfaces, or on the surfaces of these
hard particles. If we accept this version, then the sequence of stages of
laminated structure and composite particles formation, as well as subsequent distribution of lamination can
be described in the following way
(Fig. 2).

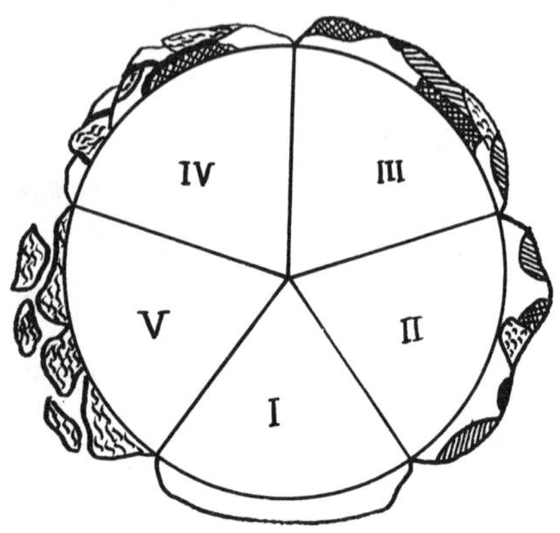

Fig. 2 Formation of composite particles during
mechanical alloying
I - adhesion of plas-
tic component
II - pressing in of
hard particles
III - formation of
laminated structure

IV - achievement of
maximum cold work
hardening, crack
initiation
V - local rupture
and separation of
composite particles

Separate particles of the plastic
component of the initial blend adhere
to the balls and to each other. Under
the impact of colliding balls or their
slipping one against another and against the particles of the blend, the
lamination is formed on the balls, in
which a hard component is pressed in a
softer one. In this case, the layer
formed on some areas of a ball becomes
strongly bonded with its surface. Particles in this layer become flat due
to impact and sliding deformation, and
this results in the formation of laminated structure. The increase in deformation time is accompanied by higher
degree of cold work hardening of the
surface layer material, as well as by
accumulation of defects in it and by
local fracture. After that, the stage
of dynamic balance between the formation of the surface layer on the balls
and its local fracture (shear deformation) starts, with composite particles being formed during this stage.
In the process of mechanical alloying,
balls and particles are exposed not
only to impact, but also to considerable shear deformation, which results in their abrasion. The deformation and shear fracture can start at
considerable lesser stresses than
under impact since ultimate shear
strength is less than ultimate compressive strength by 3-5 times [2].
Here, we should have in mind that
although during the first moments of
the mechanical alloying, "hard" particles are those of less ductile elemental powders or master alloys, after
some time cold worked particles of
blended components become "hard", and
they continue to average their composition, "dipping" again and again into
the plastic component, which may be
defined as a constant "culture medium"
(as biologists would say) and a necessary factor for the process of mechanical alloying. All the stages of the
process are proceeding simultaneously
on different parts of the ball, and
this causes the ball surface to become serrated.
The analysis of the matrix X-ray
lines profile has shown that the solid solution is formed during deformation treatment in all compositions

examinated. Thus, when examining phase
composition of Ni-25 at. % W system
and the alloying degree of the composite matrix as a function of deformation time it has been established that
just after 10 hours of the processing
unalloyed nickel, Ni-base solid solution and tungsten were present in the
mixture. Offset of the matrix X-ray
lines to lesser angles and their asym-
metrical broadening point out the formation of non-uniform solid solution
(Fig. 3). The results of the analysis
of matrix lattice period change show
that concentration of tungsten in the
solid solution of "mean" alloying is
growing from 4 at. % to 14 at. %
with increase in deformation time from
10 to 40 hours. Such concentration corresponds to the equilibrium concentration of tundsten in nickel at 800 °C,
the content of alloying element in
"lean" solid solution changing from 0
to 4-6 at. % and in "rich" solid solution from 10-12 at. % to 19-21 at. %,
and that exceeds maximum solubility of
tungsten in nickel (17.5 at. %).

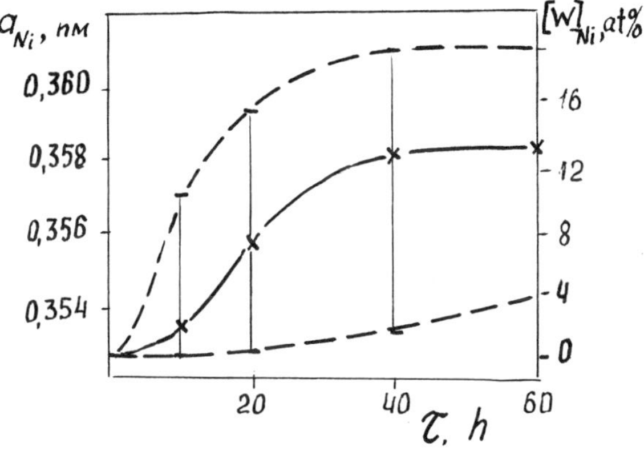

Fig. 3   Effect of processing
         time of Ni-W composi-
         tion on alloying of
         Ni-base solid solu-
         tion

Similar results have been obtained
when investigating solid solutions produced by deformation treatment, and
other compositions. So, during joint
treatment of nickel and aluminium introduced as aluminium powder or Ni-Al
master-alloys, the alloying element
content reaches about 25-27 at. % in
the solid solution of "mean" concentration, and about 30-32 at. % in solid
solution of "maximum" concentration
(Fig. 4), which sufficiently exceeds
maximum concentration of aluminium in
nickel (about 21 at. %). The increase
of deformation time results in sharp

decrease of the solid solution non-
iniformity at the expense of increase
in the degree of alloying of solid solutions of "minimum" concentration and
decrease in its quantity.

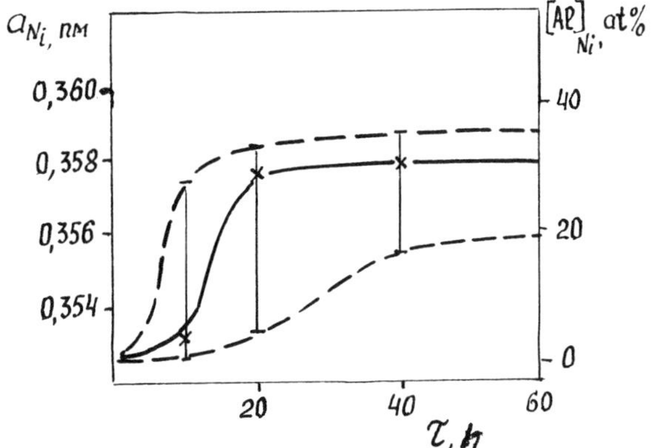

Fig. 4   Effect of processing
         time of Ni-Al compo-
         sition on alloying
         of Ni-base solid so-
         lution

The lattice period of the Ni-base
solid solution in composition Ni-50 at.
% Cr grows with the increase   in
treatment time, that of Cr-base solid
solution decreases, though in significantly lesser degree (Fig. 5). It
points out that during joint deformation two non-uniform solid solutions
are formed, alloying of the Ni-base
solid solution proceeding more intensively than that of Cr-base solid solution by nickel. This is related probably to both the different diffusion
agility of components in each other,
and significantly higher solubility
of chromium in the Ni-base solid solution, especially in temperature range
up to 600-800 °C.

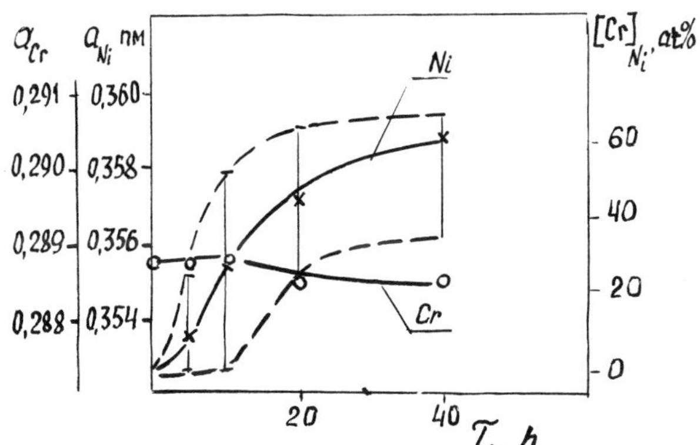

Fig. 5   Effect of processing
         time of Ni-Cr compo-
         sition on alloying of
         solid solutions

The analysis of transformations which occur during joint deformation of aluminium and iron powders (8 to 37 wt % Fe) has shown that in Al-base compositions, non-uniform solid solution is also being formed. Although chemical inhomogeneity between powder particles and inside them changes sharply over the range of processing time from 10 to 20 hours (Fig. 6), the process of solid solution formation is not completed even in 60 hours. The evaluation of iron concentration in aluminium solid solution by the analysis of profiles of matrix X-ray lines (Fig. 7) has shown that iron content in a solid solution of mean concentration is changing from 0-0.5 wt % up to 4-4.5 wt % with the increase of processing time from 10 to 60 hours, and maximum concentration mounts to 6-7 wt %. Hence, by solid-phase interaction during deformation we can achieve iron concentration in aluminium which exceeds limiting concentration for equilibrium conditions by more than two orders of magnitudes.

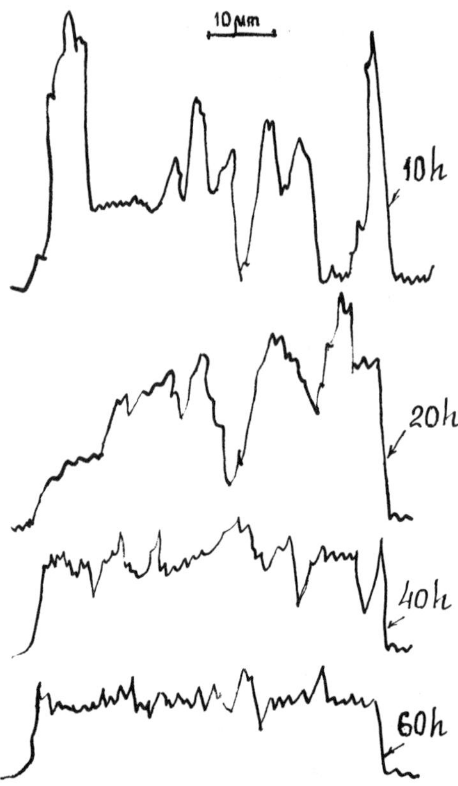

Fig. 6   Iron distribution in an Al-Fe particle depending on the time of mechanical alloying a - 10 h; b - 20 h; c - 40 h; d - 60 h.

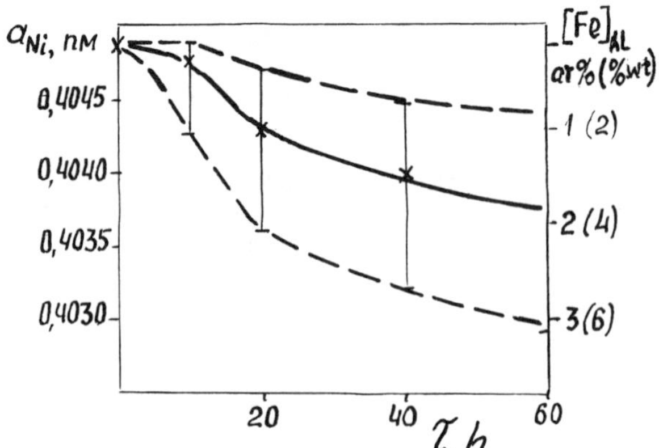

Fig. 7   Effect of processing time of Al-Fe composition on alloying of Al-base solid solution

It is interesting to note that anomalous supersaturation of solid solution in some cases causes nonequilibrium phases to appear at subsequent heating. X-ray structural analysis has shown, for example, at low temperature (<300 °C) exposure of mechanically alloyed powder of Al-8 wt % Fe alloy, the decomposition of solid solution was observed with formation first of metastable phase Al_6Fe and then of stable phase Al_3Fe. This was confirmed by calorimetric analysis (Fig. 8).

Thus, during joint deformation of opposite particles as a result of solid-phase interaction, the anomalously supersaturated solid solution is formed which is characterized by non-uniformity concentration. The realization of solid-phase interaction depends on a number of factors, the main of which being temperature, pressure and time of the process. Model examinations with special sense elements have shown that mean temperature level in the unit was about 200 °C. Consequently, local temperature "flashes" in the contact zones considerably exceed this level.

Development of the diffusion process is determined to the significant degree by the diffusion ways extent which can be characterized by the layers thickness. Flattening of the layers under deformation results in that this value does not exceed 0.1-0.2 μm. It follows from the estimation of strain rates realized under deformation treatment with the balls movement rate about 3 m/s, that maximum rate of deformation of colliding balls can reach $10^5 s^{-1}$. Similar rates are typical of high-rate impulsive loading for which the multiple masstransfer acceleration (by several

orders of magnitude) and achievement
of greater maximum concentration are
possible [3, 4].

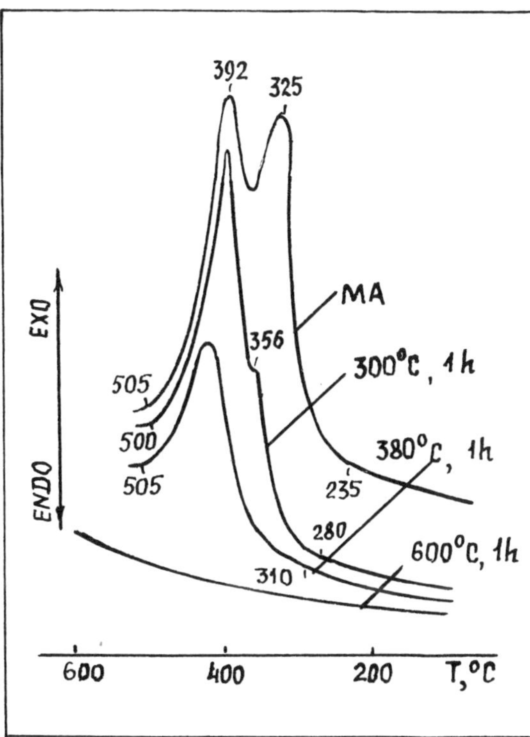

Fig. 8   DSC traces of heated
Al-Fe specimens after
mechanical alloying
(a) and various an-
nealing conditions
(b-d)

The above mentioned experimental
results allowed us to make a conclusion that by means of mechanical alloying one can achieve solid solution
supersaturations similar to those received by rapid solidification technique. The assumption that mechanical
alloying may have the same effects as
rapid solidification is not new. It
is confirmed by the facts of amorphous
phase formation in some alloys as with
rapid solidification [5]. It would be
most interesting to find a correct explanation to the mechanisms of anomalous supersaturation which are realized
at solid-phase interaction.

Before analysing possible mechanisms of solid solution supersaturation, we shall single out the factors
undoubtedly promoting this process.
These include:
- high, practically maximum density of
  lattice defects;

- high strain rates, close to explosion;
- high diffusion rates;
- low temperatures during the process.
   Taking into account these factors,
one can make the following assumptions
as to the possible mechanism of anomalous supersaturation:
- amorphization or pseudo-amorphization
  of the intermediate compound which
  facilitates deep enough diffusion of
  commonly unsolved elements;
- acceleration of the diffusion processes on account of high defect density
  and/or high strain rate;
- shifting of the phase regions on the
  equilibrium diagram under locally
  high pressures.
   The analysis of the experimental
results does not permit us to accept
any of the above versions.

The first assumption cannot be accepted because X-ray diffractions of
the alloys do not show broad halo typical of amorphous state, and DSC traces
show no crystallization peak at heating.
Usual diffusion acceleration shoul also
result in formation of metastable and
stable phases when solid solution
achieve equilibrium concentration.

Solid-phase interaction which
takes place during mechanical alloying
is a complex process. One can assume
that high density of lattice defects,
besides diffusion acceleration, results
in considerable "loosening" of the
matrix lattice. This is accompanied by
the increase in limiting solubility of
the alloying elements in the lattice.
At the same time, due to the fact that
the interaction process occurs at relatively low temperatures, there is no
time from kinetics point of view for a
new phase (phases) to form.

Thus, the alloy developed by the
above process is as if "frozen", similar to the alloy received by rapid solidification. Higher temperatures will
cause phase precipitation. This was
observed in experimental research studies when we used special high-energy
activators for mechanical alloying
where the blend temperature was much
higher than in attritors, as a result
of which intermetallics Al_3Fe precipitated in the Al-Fe system.

CONCLUSIONS

1. The mechanism of particle formation
   during mechanical alloying has been
   studied. An assumption is made that
   the whole process takes place on the
   balls surfaces under cyclic deformation and the presence of sufficient

quantity of a plastic component in
the blend is compulsory.
2. It has been shown that anomalously
supersaturated solid solutions, similar to those received by rapid
solidification technique, can be
produced by mechanical alloying(tabl).
The whole mechanism of the process
is not yet clear enough. However,
the main characteristic features of
the process are as follows:
- maximum density of lattice defects,
- anomalously high strain rates,
- relatively low temperatures of
  deformation, at which a solid
  solution is fixed in a "quasi-
  frozen" state.

REFERENCES

1. Benjamin I. S., Volin T. E., Metal. Trans., 1974, v. 5, N 8,
   p. 1929-1933.
2. Murr D., The Principles and Application of Tribonics, M.: Mir, 1978.
3. Larikov L. N., Mazanko V. F., Ne-
   moshkalenko V. V. et al., FIHOM,
   1981, N 4, p. 128-132.
4. Epshtein G. N., Structure of Metals Deformed by Explosion, M.:
   Metallurgy, 1988, p. 280.
5. Schwarz R. B. and Nach P., J. of
   Metals, 1989, January, p. 27-31.
6. Sundaresan R., Froes F.H.,
   Metal Powder Report,1989,v.44,N 3,
   p. 195-200.

Table — Anomalous Supersaturation of Solid Solution
during Mechanical Alloying

| Composition | Concentration of alloying element in solid solution, % | | | | |
| | According to phase diagram | | After mechanical alloying | | RSR |
	at room temper.	limiting solubility	average $\bar{C}$	maximum C_{max}	
Ni-W (21 at%)	~11	~17.5	12-14	19-21	32
Ni-Al(25 at%)	10	~21	25-27	30-35	
Ni-Cr(50 at%)	~32	50	55-58	60-64	
Al-Fe(8 wt%)	< 0.01	0.04	4-5	6-7	8

EXPERIMENTAL INVESTIGATIONS OF THE CRYSTAL TO AMORPHOUS PHASE TRANSITION INDUCED BY BALL-MILLING IN Si AND Si-Sn

E. Gaffett, M. Harmelin

Centre d'Etudes de Chimie Métallurgique–C.N.R.S.
15 Rue G. Urbain–F94407–Vitry/Seine Cédex–France

ABSTRACT

Based on experimental investigations, the crystal to amorphous phase transition induced by ball - milling in a pure Si powder and in a Si - Sn powder mixture - exhibiting a positive heat of mixing, have been studied : the ball - milling phase transition mechanism has to be different from the classical solid - state diffusion one.

AMORPHOUS PHASE FORMATION by mechanical alloying was first reported by A. Y. Yermakov et al. [1] and C.C. Koch et al. [2] for the Co - Y and Ni - Nb systems respectively. Subsequently such an amorphization process has been observed in many other binary alloy systems, starting from elemental crystalline powders, e.g., Fe - Zr [3], Ni - Zr [4 - 6], Ti - Pd [7], Ni - Ti, Co - Ti, Fe - Ti [8 -10], Ti - Cu [11], Co - Sn [12], or from a mixture of intermetallic compounds in the Ni - Zr system [13 - 15]. Amorphization by M.A. is usually obtained in the case of binary alloys exhibiting a large negative heat of mixing and for which one of the elements is a fast diffuser [16].

In this paper, we report on the crystal to metastable phase (i.e. amorphous phases, supersaturated solid solutions and/or non equilibrium compounds) transition induced by ball - milling in a pure element - a Si powder and in an immiscible alloy (at least, under thermal equilibrium condition, see Fig. 1) - Si/Sn. The morphological, structural and the thermal stability of the ball milled powder have been investigated by means of SEM/EDX observations, XRD pattern and DSC experiments

BALL - MILLING PROCEDURE

SI BALL - MILLING - 10 g of pure Si pieces (Hyperpure Polycrystalline Silicon from Wacker Chemitronic - Gmbh - 300 Ωcm n and 3000 Ωcm p) are introduced in a cylindrical tempered steel (12%Cr, 2%C) container of capacity 45 ml. This procedure occurs in a glove box filled with purified argon. Each container is loaded with five steel balls 1.5 cm in diameter and 14 g in mass. As the containers are sealed in the glove box with a Teflon O - ring, the milling has been proceeded in stationary argon. Ball - milling is carried out using two FRITSCH planetary high energy BM equipments (Pulverisette P7/2 and P5/2). For the last machine, two intensity settings are chosen which are mentioned later as P5/2(10) and P5/2(5). The different ball - milling conditions will be referred hereafter as Si(a), Si(b), Si(c) for the P7/2, P5/2(10), P5/2(5). The higher energetic condition corresponds to Si(a), the lower one is related to Si(c). The durations of the continuous milling processes have been 95 hours, 70 hours, 96 hours for Si(a), Si(b) and Si(c) respectively.

SI/SN BALL MILLING - Since the Sn powder higly sticks to the container walls, only the Si - rich side of the Si - Sn phase diagram has been investigated (Fig. 1). Two mixing methods have been chosen (see Table 1) :

i) - hereafter refered as Si_x - Sn_{1-x} (*), pure Si pieces and pure Sn powder (Prolabo) are introduced in a cylindrical tempered steel container of capacity 45 ml. This procedure occurs in a glove box filled with purified argon.

ii) - hereafter refered as Si_xSn_{1-x} (**), 10 g of Si are first introduced in the container and ball - milled during 72 hours in static argon.

Then, the Sn powder is introduced and the ball milling is achieved during 96 hours. All the Sn recharging sequences are performed in a glove box. Each container is loaded with five steel balls 1.5 cm in diameter and 14g in mass. As the containers are sealed in the glove box with a Teflon O - ring, the milling has been proceeded in stationary argon. Ball milling is carried out using a FRITSCH planetary high energy BM equipment (Pulverisette P5/2 - intensity setting = 10).

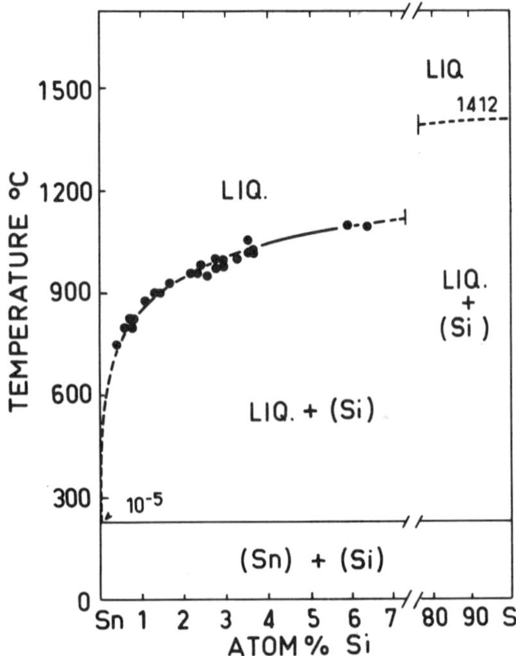

Fig. 1 - Si - Sn phase diagram
(from "The handbook of Binary Phase Diagram - W.G. Moffat - Genium Publishing Corporation - USA)

MORPHOLOGY - COMPOSITION

SEM AND EDX/SEM INVESTIGATIONS - The particle morphology has been characterized using a Digital Scanning Electron Microscope (Zeiss DSM 950 - SEI mode - Fig 2 a, b, c). The container contamination which may occur during the friction of the particles on the balls and the walls of the container has been determined by SEM/EDX analyses (Si - Li detector from TRACOR). A semi - quantitative program with internal references (SQ from TRACOR) has been used to analyze the EDX spectra. The container contamination does not exceed 3.5 and 0.5 At% in Fe and Cr respectively.

Fig. 2 : a, b) BM Si powder
c) Typical BM Si-Sn powder

END - PRODUCT STRUCTURES

X - RAY INVESTIGATIONS - The X - ray diffraction patterns of the BM powders have been obtained using a $(\theta - 2\theta)$ PHILIPS diffractometer with CoK_α radiation ($\lambda = 0.17889$ nm). A numerical method has been used in order to analyze the X - ray diffraction patterns and to obtain the position and the full-width at half height of the various peaks (For details see [17, 18]). The Bragg expression has been applied to determine the d parameter corresponding to the diffraction peak position (θ):

$$\lambda = 2\,d\,\sin\theta. \qquad (1)$$

The effective diameter of the particles (hereafter refered as Φ) has been calculated from the Scherrer expression :

$$\Phi = 0.91\,\lambda / (B\cos\theta) \qquad (2)$$

where λ is the X - ray wavelength, B the linewidth (expressed in $2\,\theta$) and θ the diffraction angle

BM Si XRD patterns - Fig 3a, 3b and 3c exhibit the various X - ray diffraction patterns which have been obtained for Si (a), Si(b), Si(c) respectively. Applying the Scherrer formula to the best fit which may be obtained by the ABFfit program (see [17,18]) leads to a mixture of microcrystallites, nanocrystallites and an amorphous phase. Such a classification appears to be schematic since, but with the ABFfit program deconvolution, it is not possible to introduce a continuous variation of the effective size of the particles. Nevertheless, it qualitatively gives the range of the effective diameters of the Si particles (see Table 2)

Si/Sn XRD patterns - The deconvolution of the XRD patterns (Fig. 4) corresponding to the Si/Sn powder lead to the detection of three phases : a Si(Sn) crystalline solid solution, Sn(Si) phase and an amorphous phase (or a new tetragonal crystalline one). The Si crystalline phase exhibits a strong increase of the diamond cubic parameter (from +0.4 to +3.9 %) as well as the Sn tetragonal crystalline lattice parameters (the relative increase ranging from +0.1 to +0.3 and from +0.2 to +0.5% for the a and c tetragonal phase parameters respectively). The c parameter exhibits a larger increase than the a one for the latter phase (see Table 3).

For the Sn richer side of the experimented composition (i.e. $Si_{70}Sn_{30}$ At. %), no more amorphous phase has been detected but a new

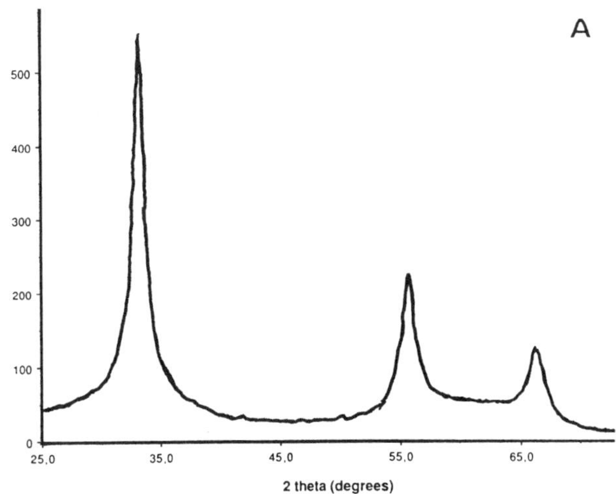

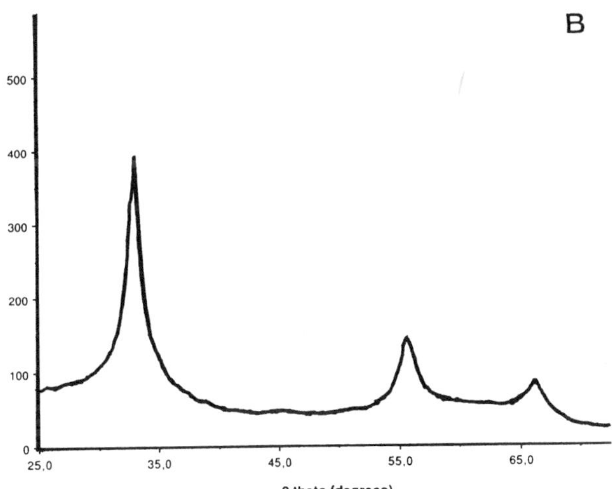

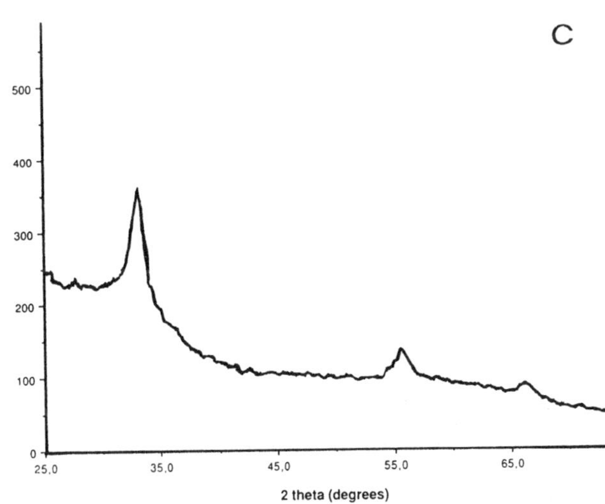

Fig. 3 - X- ray diffraction patterns of the BM Si powders corresponding to the various ball milling conditions : a) P7/2, b) P5/2(10), c) P5/2(5)

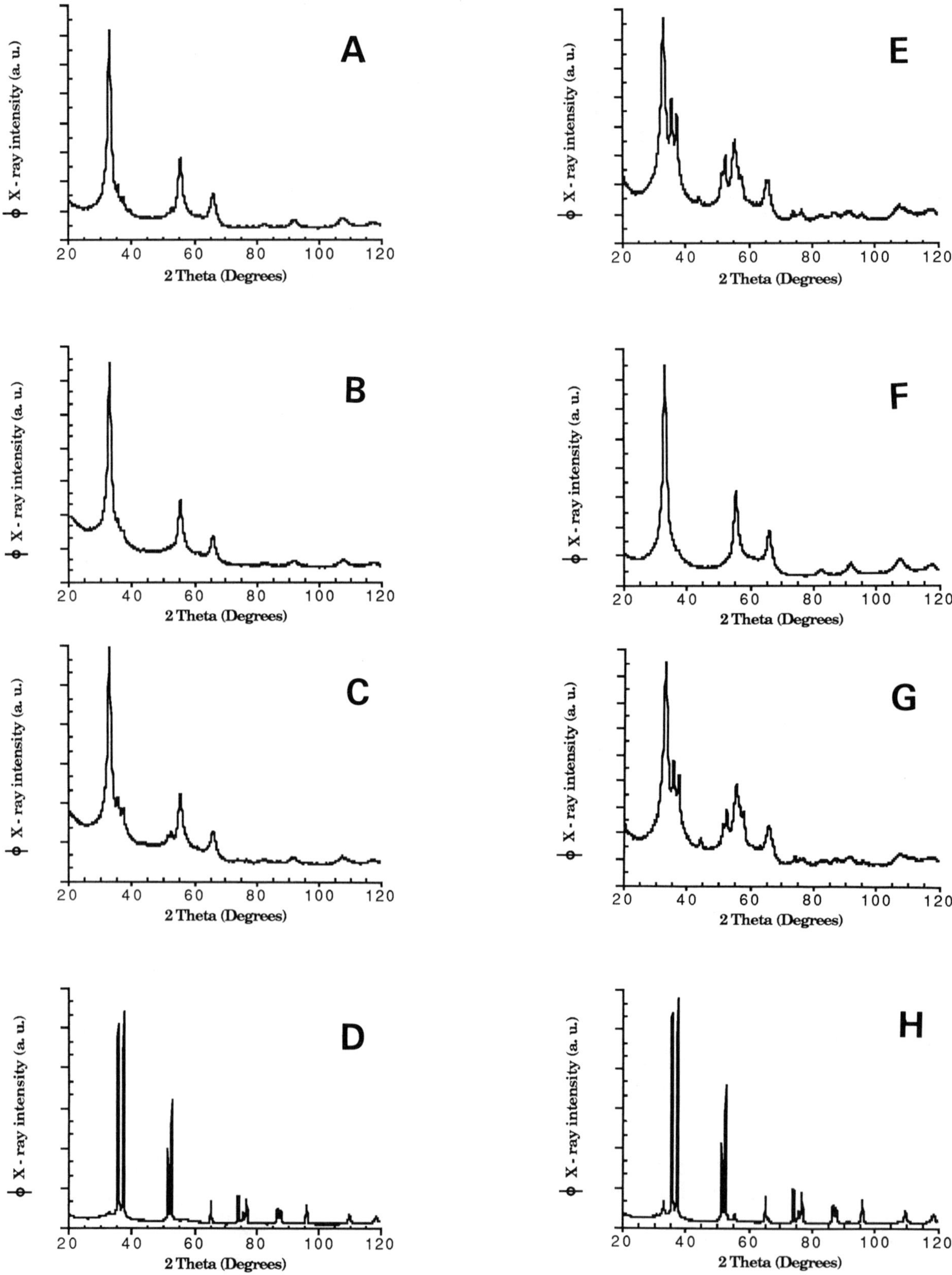

Fig. 4. - X - ray diffraction patterns of the ball - milled Si - Sn powders. - a) $Si_{95}Sn_5$[*], b) $Si_{90}Sn_{10}$[*], c) $Si_{80}Sn_{20}$[*], d) $Si_{70}Sn_{30}$[*], e) $Si_{95}Sn_5$[**], f) $Si_{90}Sn_{10}$[**], g) $Si_{80}Sn_{20}$[**], h) $Si_{70}Sn_{30}$[**]

Table 1 - List of the various experimental conditions and the end - product composition.(%At.). Si_x Sn_{1-x}(*) and Si_x Sn_{1-x}(**), see the text (n.d. $\leq$ 0.05 % At.)

Composition	End product composition
Si(a)	Fe $\leq$ 0.1, Cr $\leq$ 0.1
Si(b)	Fe $\leq$ 0.1, Cr $\leq$ 0.1
Si(c)	Fe $\leq$ 0.1, Cr n.d.
$Si_{95}Sn_5$(*)	$Si_{95.2\pm0.4}$ - $Sn_{4.8\pm0.4}$ Fe $\leq$ 0.5, Cr $\leq$ 0.1
$Si_{95}Sn_5$(**)	$Si_{93.2\pm0.4}$ - $Sn_{6.8\pm0.4}$ Fe $\leq$ 2.0, Cr $\leq$ 0.1
$Si_{90}Sn_{10}$(*)	$Si_{97.2\pm0.2}$ - $Sn_{2.8\pm0.2}$ Fe $\leq$ 0.3, Cr $\leq$ 0.1
$Si_{90}Sn_{10}$(**)	$Si_{98.1\pm0.3}$ - $Sn_{1.9\pm0.1}$ Fe $\leq$ 0.3, Cr $\leq$ 0.1
$Si_{80}Sn_{20}$(*)	$Si_{96.1\pm0.2}$ - $Sn_{3.9\pm0.2}$ Fe $\leq$ 0.7, Cr $\leq$ 0.1
$Si_{80}Sn_{20}$(**)	$Si_{94.2\pm0.3}$ - $Sn_{5.8\pm0.3}$ Fe $\leq$ 3.5, Cr $\leq$ 0.5
$Si_{70}Sn_{30}$(*)	$Si_{72.0\pm3.0}$ - $Sn_{28.0\pm3.0}$ Fe $\leq$ 3.5, Cr $\leq$ 0.5
$Si_{70}Sn_{30}$(**)	$Si_{75.3\pm3.0}$ - $Sn_{24.7\pm3.0}$ Fe $\leq$ 1.0, Cr $\leq$ 0.5

Table 2 - Effective diameter range of the BM Si crystallites which are in equilibrium with the Si amorphous phase (a_{Si} = 5.43 10^{-1} nm from [19], a_i in nm).

Si(a)	$\Phi \approx 4$ nm : a_1 = (0.5435 $\pm$0.0015) $\Phi \approx 10$ nm : a_2 = (0.5425 $\pm$0.0015) (a_1 - a_2) / a_2 $\approx$ 0.2%
Si(b)	$\Phi \approx 4$ nm : a_1 = (0.5435 $\pm$ 0.0020) $\Phi \approx 10$ nm : a_2 = (0.5430 $\pm$0.0020) (a_1 - a_2) / a_2 $\approx$ 0.1%
Si(c)	$\Phi \approx 8$ nm : a = (0.5430 $\pm$ 0.0025) (a_1 - a_{Si}) / a_{Si} $\approx$ 0 %

Table 3. - The amorphous phase parameter (d) and the Si phase crystalline parameter (a_{Si}), Sn phase crystalline parameters (a_{Sn} and c_{Sn}) as a function of the initial compositions and the ball - milling conditions for the Si/Sn system.(expressed in 10^{-1} nm)

Si_xSn_{1-x} a $Si^X(\Delta a/a_{Si})$
a($\Delta a/a_{Sn}$ - %)^X - c ($\Delta c/c_{Sn}$ - %)^X
d amorphous phase

Si a_{Si}= 5.43 [19]

BM Si   d = 3.16$\pm$0.02 [17]

Sn a_{Sn} = 5.831 - c_{Sn} = 3.182

$Si_{95}Sn_5$*a_{Si} = .48 (+0.9) - 5.45 (+ 0.4)
a_{Sn} = 5.846 (+0.3) - c_{Sn} = 3.196 (+ 0.5)
d = 3.10

$Si_{95}Sn_5$**a_{Si} = 5.50 (+ 1.3) - 5.45 (+ 0.4)
a_{Sn} = 5.846 (+ 0.3) - c_{Sn} = 3.199 (+ 0.5)
d = 3.051

$Si_{90}Sn_{10}$*a_{Si} = 5.48 (+ 0.9) - 5.45 (+ 0.4)
a_{Sn} = 5.844 (+ 0.2) - c_{Sn} = 3.192 (+ 0.3)
d = 3.127

$Si_{90}Sn_{10}$**a_{Si} = 5.64 (+ 3.9) - 5.44 (+ 0.3)
"undetermined" tetragonal phase
d = 3.148

$Si_{80}Sn_{20}$*a_{Si} = 5.48 (+ 0.9) - 5.45 (+ 0.4)
a_{Sn} = 5.839 (+0.1) - c_{Sn} = 3.195 (+ 0.4)
d = 3.100

$Si_{80}Sn_{20}$**a_{Si} = 5 .64 (+3.9) - 5.46 (+ 0.5)
a_{Sn} = 5.843 (+ 0.2) - c_{Sn} = 3.195 (+ 0.4)
d = 3.036

$Si_{70}Sn_{30}$*a_{Si} = 5.80 (+6.8) - 5.47 (+0.7)
a_{Sn} = 5.843 (+ 0.2) - c_{Sn} = 3.188 (+0.2)
a" = 5.75 (- 1.5) - c" = 2.79 (- 12.3)

$Si_{70}Sn_{30}$** a_{Si} = 5.53 (+1.8) - 5.45 (+ 0.4)
a_{Sn} = 5.844 (+ 0.2) - c_{Sn} = 3.188 (+ 0.2)
a" = 5.72 (- 1.8) - c" = 2.97 (- 6.6)

a" and c", phase parameters corresponding to a new tetragonal phase tetragonal phase has been identified with the lattice parameters which are strongly different from the pure Sn one (a relative difference ranging from -1.5 to -1.8 and from -6.6 to -12.3% for the a and c tetragonal parameters).

END - PRODUCT MICROSTRUCTURES

TEM INVESTIGATIONS - The TEM observations have been performed by means of a JEOL 2000FX TEM. The BM Si particles are directly spread on the Cu TEM grid. Therefore, the TEM study is performed on the particles which are sufficiently thin. So, no sample preparation artefact may be introduced.

Fig. 5 exhibits a typical selected area diffraction pattern on which the typical microcrystalline structure is revealed. Furthermore, in spite of the fact that the rings corresponding to the [220] and [311] indexations exhibit some clear spots, the first ring exhibits a continuous diffuse intensity. This may be explained by an overlap of different contributions. Indeed, as shown by the X - ray diffraction pattern study, the first ring corresponding to the first distance [111] in Si has several contributions : two (at least) crystalline contributions plus the first diffuse halo corresponding to the presence of the amorphous phase.

Such an overlap will lead to a continuous diffuse intensity for the first ring. Furthermore between the [220] and [311] rings, a diffuse intensity is observed (indicated by an arrow). It corresponds to the second diffuse peak (second shoulder) of the amorphous phase. Then such an electron diffraction pattern is in good agreement with the previous analyses of the X - ray diffraction patterns.

As the diameter of the selected area is about the micrometer scale, we may affirm that the both structures - crystalline (microcrystalline and/or nanocrystalline) and amorphous phase coexist at the micrometer scale.

THERMAL STABILITY

CALORIMETRIC MEASUREMENTS - The thermal analysis have been carried out using a DSC - 2C PERKIN ELMER. 20 mg sample were sealed in a copper capsule and heated from 50°C to 725°C on flowing pure argon.

Fig. 6 exhibits a typical DSC trace corresponding to the thermal response of Si(c).

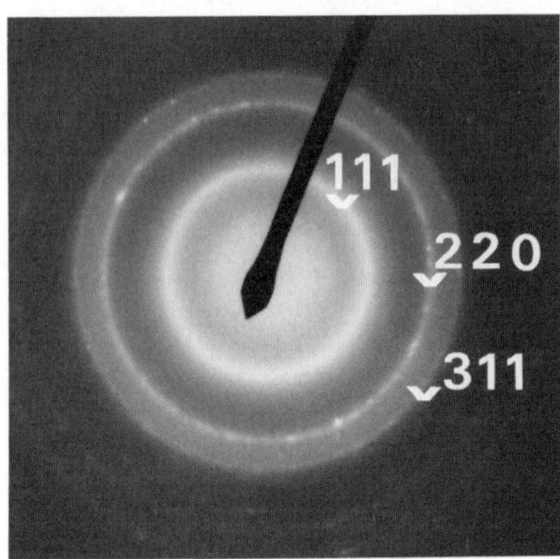

Fig. 5. - Typical SAD pattern obtained from the ball milled Si powder

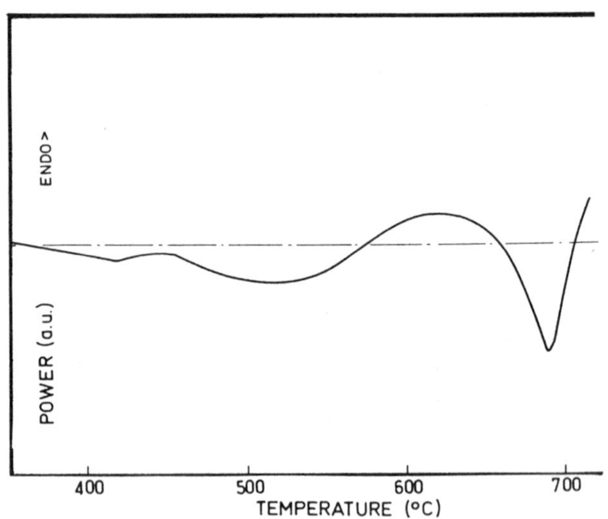

Fig. 6 - DSC traces of Si(c) at heating rate of 2.5°C min^{-1} (for details see [17])

Such a thermal response may be analyzed as a low temperature exothermic contribution, followed by an endothermic peak at intermediate temperature and a high temperature exothermic event.

The crystallization peak position is 694°C and 720°C for heating rates of 2.5 °C/min. and 10 °C/min. respectively.

The activation energy of the Si amorphous phase crystallization has been determined by the Kissinger method [20] and is equal to 323 kJmol^{-1}(or 3.35 eVat^{-1}) [17].

Further investigations have to be performed in order to compare the thermal response of the binary alloys with the BM Si powder.

DISCUSSION

BALL - MILLED SI POWDER - The experimental results lead to the existence of a dynamic equilibrium between amorphous Si - phase and polycrystalline Si - phase induced by ball - milling.

Notice that the value of the lattice parameter corresponding to the larger particle size (a_2) is in agreement with that obtained by H. Kiendl : a_{Si} = 0. 54310 nm [19], the lattice parameter corresponding to the lower particle size exhibits a relative difference of 0.2 % for Si (a), 0.1% for Si (b).

S. Veprek et al [21] studied the formation of amorphous silicon during condensation from the vaporous phase and determined a critical grain size (depending on the stress deposition conditions) below which the crystalline structure becomes unstable. According to their results, the critical particle size and the corresponding lattice expansion at which the diamond lattice becomes unstable with respect to the amorphous phase under the ball - milling conditions which have been used in our experiments, can be estimated to be :

Si(a)	Φc = 4 nm	$\Delta a_c \approx$ 0.2%
Si(b)	Φc = 4 nm	$\Delta a_c \approx$ 0.1%
Si(c)	Φc = 8 nm	$\Delta a_c \approx$ 0.0 %

Therefore, the crystal to amorphous phase transition may be explained by a refinement of the grain size of the particle which leads to a destabilization of the diamond structure, the critical size of the particle seems to be affected by the ball - milling conditions and the Δa_c value shows a strong correlation with the latter : the higher the energy input during the ball - milling, the larger the value of Δa_c.

Now, let us consider about the thermal behavior during the DSC experiments. Notice that the same kind of the various contributions, i.e. a low temperature exothermic peak, an intermediate temperature endothermic peak and a high temperature exothermic crystallization event have been also encountered for the Ni - Zr system [15].

In the just mentioned work, the origin of the various peaks have been interpreted as a relaxation of the stressed intermetallic compounds (exothermic contribution), followed by a local equilibration between the amorphous phases and the in - situ relaxed compounds leading to a change in composition of the amorphous phases (endothermic effect since the heat of mixing is negative), and at higher temperature, the crystallization of the various amorphous phases (exothermic events). Such an interpretation for the both exothermic contributions which are obtained for the ball - milled Si thermal analysis is also consistent. In the present work, considering that the (Fe - Cr) contamination is very minor, the Si powder behaves as a pure element. Thus, for explaining the intermediate temperature endothermic contribution, we propose to replace the composition parameter which may be considered as one of the components of the configuration space by another parameter - which remains to be determined - but which may be the configuration space distribution of the Si atoms - in the amorphous state - in dynamic equilibrium with the crystals.

BM SI/SN POWDERS - The XRD patterns analyses have revealed the simultaneous presence of two crystalline phase and and amorphous phase. As the crystalline phases exhibit the same structure as the pure elemental one but with an increase in the lattice parameters, we may propose that an extension of the solubility have been obtained by the ball - milling process leading to the formation of such supersaturated crystalline solid solution.

R. Birringer et al. [22] have demonstrated the influence of the crystal grain size on the solubility of some elements : the solute solubility of the nanocrystalline materials may be several orders of magnitude larger than in the case of the single crystals : the solubility of Bi (at 373K) in the nanocrystalline Cu grains (10 nm crystal size) was 4% and <10^{-4}% in the Cu single crystals. Such a similar effect may be proposed to explain the variation of the crystalline lattice parameter. A so large solute content may also explain the formation of the amorphous phase which may be due to a chemical supersaturation instability of the crystalline lattice.

CONCLUSION

Based on experimental investigations, the crystal to non equilibrium phase transition induced by ball - milling has been observed in the case of a pure element - Si - and in the case of a binary alloy which exhibits a solid - state immiscibility - Si/Sn.

The observation of such transition supports the idea that the mechanism of the non - equilibrium phase transition induced by ball - milling has to be distinct from the one leading to the amorphisation by interdiffusion of a multilayered system.

ACKNOWLEDGEMENTS :

We gratefully acknowledge Dr.A. Quivy for her help in the obtention of the X - ray patterns.

REFERENCES

[1] A.Y. Yermakov, Y.Y. Yurchikov and V.A. Barinov, Phys. Met. Metall., $\underline{52}$, 50 (1981)
[2] C.C. Koch, O.B. Cavin, C.G. McKamey and J.O. Scarbourgh, Appl. Phys. Lett., $\underline{43}$, 1017 (1983)
[3] E. Hellstern and L. Schultz, Proc. 6th Int. Conf. on Rapidly Quenched Metals, Montreal, 1987, Mat Sci. Eng., $\underline{97}$, 39 (1988)
[4] L. Schultz, E. Hellstern and A. Thoma, Europhys. Lett., $\underline{3(8)}$, 921 (1987)
[5] E. Gaffet, N. Merk, G. Martin and J. Bigot, J. Less Comm. Met., $\underline{145}$, 251 (1988)
[6] E. Gaffet, N. Merk, G. Martin and J. Bigot, DGM Conf. "New Materials by Mechanical Alloying Techniques", 95 - 100, 3 - 5 October 1988, Hirsau - West Germany, Organizers : E. Arzt and L. Schultz
[7] J. R. Thompson and C. Politis Europhys. Let., $\underline{3(2)}$, 199 (1987)
[8] B. P. Dolgin, M.A. Vanek, T. McGory and D.J. Ham, J. Non Cryst. Sol., $\underline{87}$, 281 (1986)
[9] R.B. Schwarz and C.C. Koch Appl. Phys. Lett., $\underline{49(3)}$, 146 (1986)
[10] M.S. Boldrick, D. Lee and C.N.J. Wagner, J. Less Com. Met., $\underline{106}$, 60 (1988)
[11] C. Politis and W.L. Johnson J. Appl. Phys., $\underline{60(3)}$, 1147 (1986)
[12] A. Hitaka, M.J. McKenna and C. ElBaum, Appl. Phys. Lett., $\underline{50(8)}$, 478 (1987)
[13] A. W. Weeber, H. Bakker and F.R. de Boer, Europhys. Lett., $\underline{2(6)}$, 445 (1986),
[14] P.Y. Lee and C.C. Koch Appl. Phys. Lett., $\underline{50}$, 1578 (1987)
[15] N. Merk, E. Gaffet and G. Martin, J. Less Comm. Met., $\underline{153}$, 299 - 310 (1989)
[16] L. Schultz J. Less Comm. Met., $\underline{145}$, 233 (1988)
[17] E. Gaffet and M. Harmelin J. Less Com. Met, $\underline{157}$, 201 - 222 (1990)
[18] A. Antoniadis, J. Berruyer and A. Filhol, Institute Lauë Langevin internal report (1988) 87AN22T
[19] H. Kiendl. Zeichrift für Naturfor., $\underline{22A(1)}$, 79 (1967)
[20] H.E. Kissinger J. Res. Nat. Bur. Standards, $\underline{57(4)}$, 217 (1956)
[21] S. Veprek and Z. Iqbal, F.-A. Sarott Phil. Mag. B, $\underline{45(1)}$, 1982, 137
[22] R. Birringer , H. Hahn, H. Höfler, J. Karch and H. Gleiter, Defect and Diffusion Forum, $\underline{59}$, 17 (1988)

STRENGTHENING IN COPPER ALLOYS PREPARED BY MECHANICAL ALLOYING

David G. Morris, Maria A. Morris
Institute of Structural Metallurgy
University of Neuchâtel
CH-2000 Neuchâtel, Switzerland

ABSTRACT

Dispersion strengthened copper-base alloys prepared by mechanical alloying give useful combinations of high conductivity with high strength. A wide range of dispersoid-forming materials has been examined, including bcc elements, non-metallic elements, and stable compound particles. The effectiveness of a given dispersoid is determined by the degree of refinement achieved during the milling period and by its resistance to subsequent coarsening. The degree of refinement seems to depend both on the mechanical behaviour of the dispersoid material relative to that of the matrix as well as on the chemical interaction between the components. The alloys obtained here have typically very fine dispersoid particles in a copper matrix of sub-micron grain size. The high strengths of the materials can be interpreted in terms of particle strengthening according to the Orowan model, while the fine grain size does not appear to contribute directly to the strengthening.

MECHANICAL ALLOYING IS OF PARTICULAR INTEREST for creating alloys which are dispersion strengthened by a uniform distribution of fine, stable particles. Stable particles are those of low solubility in the matrix or where diffusion of the material making up the particle is very slow. For such reasons, mechanical alloying of superalloys has concentrated on dispersing oxide particles. More recent work has examined the dispersion of stable particles in a copper matrix with the aim of achieving high strength while maintaining an unalloyed matrix of high conductivity (1-5). It has been shown that the rate of comminution of the particles in dispersion is very slow for stable compound phases and for non-metallic elements (3,5), similar to the case of oxide phases in super-alloys. For the various reasons described above, much of the work on copper alloys has concentrated on dispersions of bcc metallic elements. One of the topics presented here is a study of the factors influencing the comminution of a second phase into a fine and uniform dispersion.

High strength at room temperature may be achieved either by particle-dislocation interactions, as for example by Orowan strengthening, or by strengthening by grain boundaries since the grain size is typically very small. A recent study of strengthening in dispersion-strengthened aluminium alloys (6) suggested that both factors may contribute to the overall strength. Strengthening in dispersion-strengthened copper alloys will be examined here and the contribution of direct particle strengthening and grain boundary strengthening discussed.

MECHANICAL ALLOYING OF COPPER WITH BCC ELEMENTAL DISPERSIONS

Mechanical alloying was carried out by weighing the desired amounts of pure, elemental powders (size less than 150 μm) of copper and bcc element, which were filled into the containers of the ball mill (Fritsch, Pulverisette No 7) under inert gas, and milled for the desired period of time. Typically a ball to powder weight ratio of about 4 was used, about 5 vol % of dispersing phase was used, and milling was conducted for about 12 h. At the completion of milling, the powder was removed, again under inert gas, cold pressed to form a billet of density about 60-70 %, and then hot extruded at 700 ° C. The extrusion ratio was 10 to 1 through a die of conical shape and the product, a bar of 6 mm diameter, was quenched immediately beyond the die. For the particular tests to be described in detail here, the elements Cr, Mo, Nb, V and W were used, all at a volume concentration of 5 % in the copper matrix.

The milling time, 12 h, was chosen here because this corresponds to achieving a relatively good degree of mixing of the bcc addition in the copper matrix without severe problems of welding to the containers and balls. Fig.1 shows the structure of typical powders after mechanical alloying. A clear difference exists between the degree of mixing of each bcc element in copper, with Nb being extremely well mixed, Cr and V being fairly well mixed, but Mo and particularly W being poorly mixed. It should be noted that in the back-scattered electron images of Fig.1, particles are detected by virtue of the difference of electron number relative to copper: heavy elements (Mo, Nb and W) will appear light, while light elements (Cr and V) will appear dark. The Cu-Nb image of Fig.1 shows no light speckles, indicating that the Nb phase is extremely finely distributed, and the dark speckles seen correspond to lighter element concentrations, perhaps oxides.

The refinement of the bcc particles is confirmed in the transmission electron micrographs of Fig.2. The thin foil samples were prepared from materials extruded at 700 °C. In these micrographs it is seen that V has been rather well distributed throughout the copper leading to a fine grain size and large numbers of very small V particles, about 10 nm in size. On the other hand, the W has been poorly distributed and large particles ranging up to at least 0.5 μm are present within the area shown. In this case there are fewer particles in the size range below 50 nm and the overall grain size remains much larger than in the case of the V alloy shown.

Fig. 3 compares the mechanical properties of the materials studied, both at the stage of milled powders as well as after extrusion at 700 °C and after subsequent heat treatment at 1000 °C. This comparison is made relative to the hardness of the bcc element. It is clear that the hardness or strength after extrusion and after subsequent heat treatment is a measure both of the initial degree of refinement as well as the resistance of the microstructure to coarsening. The hardness measured on the loose powders is clearly a measure of the degree of refinement of the bcc element addition during the milling period and indeed the hardness obtained mirrors fairly well the refinement described earlier, namely good refinement for Nb, intermediate for Cr and V, and poorer for Mo and W. The hardness variation observed on the powders is similar to the strength variation of the extruded materials, indicating that little structural coarsening has occurred during extrusion. The differences in strength of the extruded materials and after heat treatment is a measure of the changes in structure occurring during the heat treatment.

The variation of strength with annealing temperature is shown in Fig. 4, which compares the as-extruded state with samples heated for 1 hour at each of the temperatures. The loss of strength for each alloy represents the increase in the scale of the microstructure, and this may be seen more explicitly in Fig. 5, which shows the size of the strengthening particles after the various heat treatments. In Fig. 4 it is seen that the Nb alloy has a substantially higher strength both after extrusion as well as after heat treatments at higher temperatures. The strengths of the Cr and V alloys fall in similar ways, particularly after heat treatments at 1000 °C. For the Mo and W alloys, which start from a relatively low strength level in the as-extruded state, there appears to be an improved retention of this strength at the highest temperature, perhaps because the larger particles in these alloys are more resistant to coarsening. Fig. 5 compares the size of the particles found in the Cr, Nb, and V alloys after extruding and after heat treatment: these values are also compared with values calculated based on the volume diffusion coarsening theory of Lifshitz, Slyozov and Wagner (7,8). Data relevant to the solubility and diffusivity of these elements in copper has been taken from various reference books (9,10,11). It may be seen that the change in size of the particles examined may be explained very well in terms of the coarsening model used: the lower resistance to coarsening of the Cr and V particles relative to the Nb particles may be understood in terms of both the higher solubility and higher diffusivity of these elements in copper.

A final point remains in understanding the mechanical alloying behaviour of the series of alloys examined, namely to explain the good breakdown and dispersion behaviour of Nb and the poor breakdown and dispersion behaviour of Mo and W. It is important to note that the degree of mixing achieved after 12 h milling on the particular ball mill used is not an absolute value, but is simply an indication of the relative rates of breakdown of the various elements during milling: it would probably be possible to achieve a fine dispersion of Mo and W but this would require much longer milling times. It is already clear from Fig. 3 that the relative rate of breakdown of the various elements cannot be explained in terms of the absolute hardness of the element (nor hardness relative to that of the copper matrix), neither is it likely to be explained by the brittleness of the bcc element. Fig. 6 shows an attempt to correlate the dgree of mixing achieved by milling with the solubility, either in the solid alloy or in liquid copper. Again it is not possible to explain the degree of mixing in this way. Finally, Fig.7 shows the degree of mixing achieved to be correlated with the chemical interaction of the copper atom and the bcc element - the thermodynamic parameter used for correlation is the heat of formation of a Cu-bcc element compound (12). This parameter, albeit an

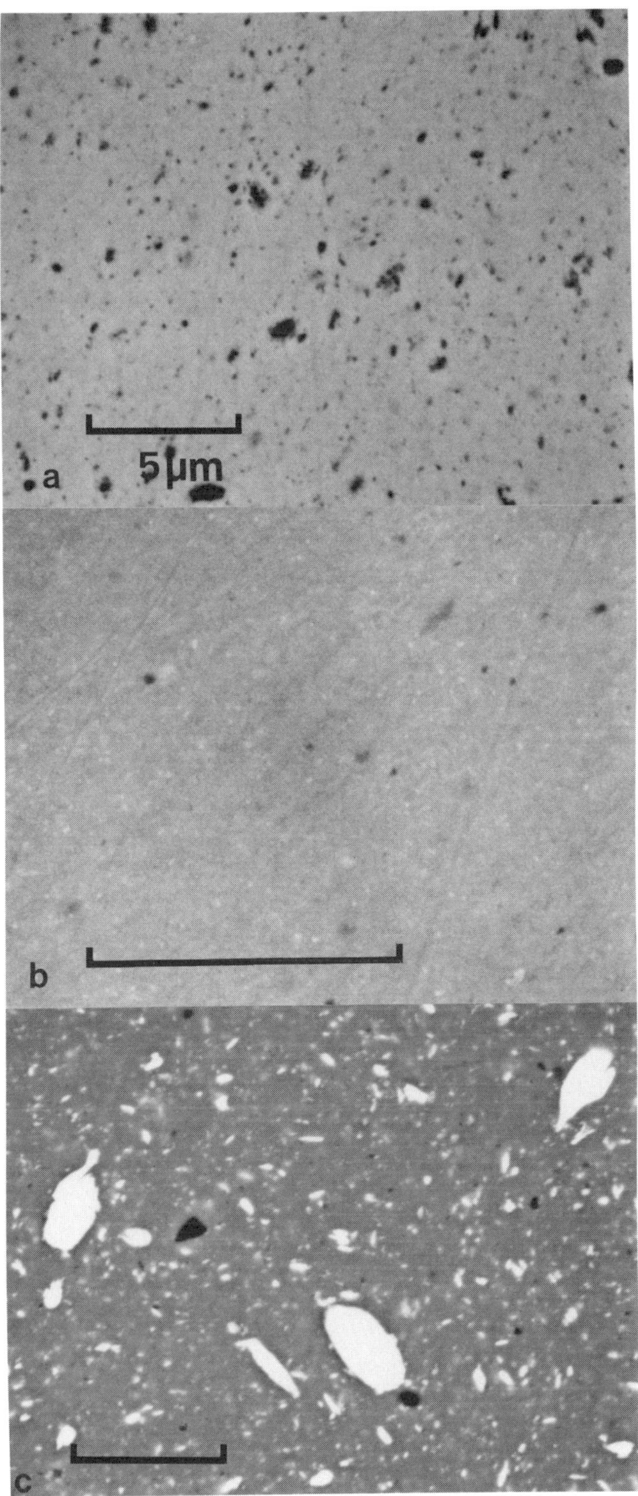

Fig.1 Back-scatter Electron images showing the refinement of (a) V, (b) Nb and (c) W in Cu after 12 h MA.

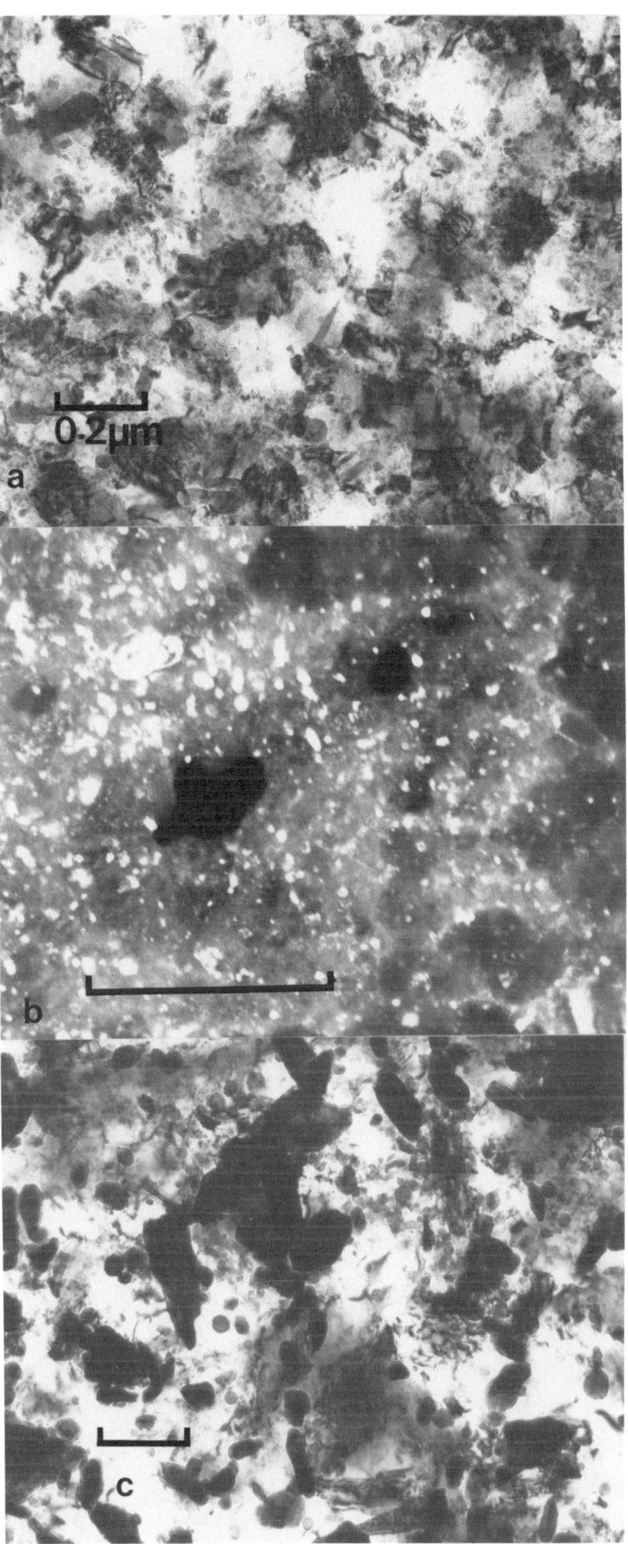

Fig.2 Transmission electron micrographs showing the fine grain size (a) and particles (b) in Cu-5V and coarser particles (c) in Cu5W: prepared by 12 h MA and extruded at 700 °C.

267

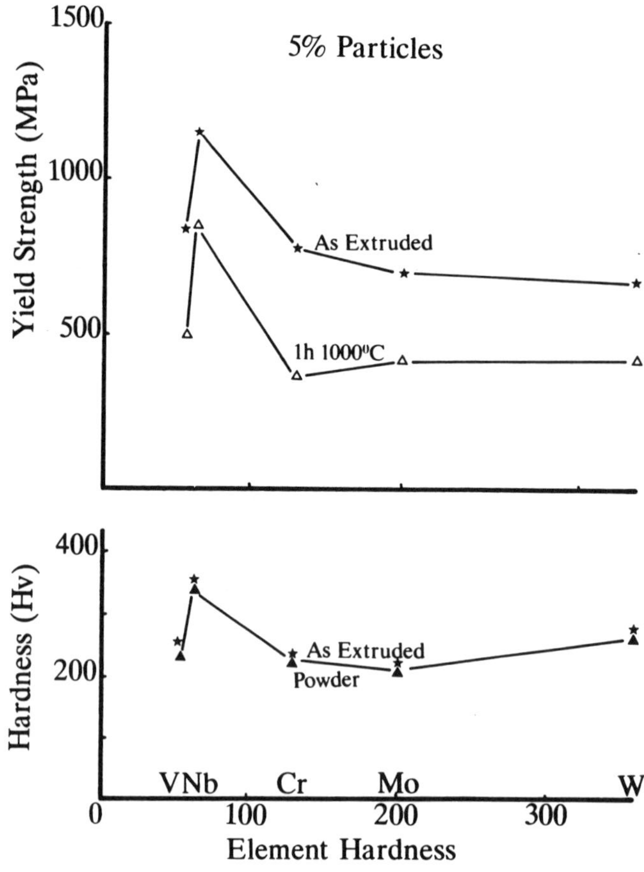

Fig.3 Mechanical properties of CuMA alloys as milled powders, after extrusion at 700 °C, and after further annealing at 1000 °C for 1 hour.

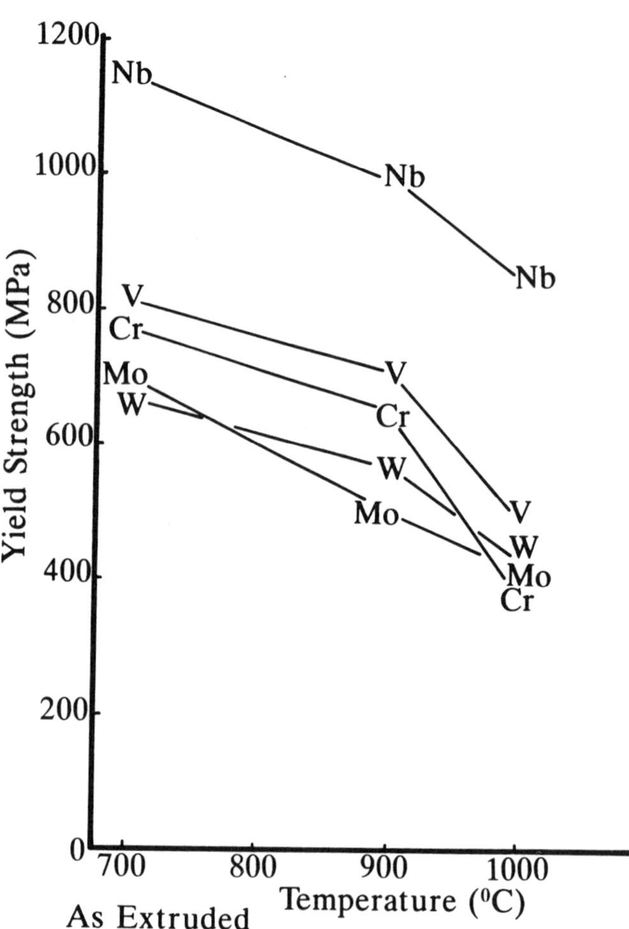

Fig. 4 Mechanical properties of CuMA alloys after extrusion at 700 °C and after heat treatments for 1 hour at higher temperatures.

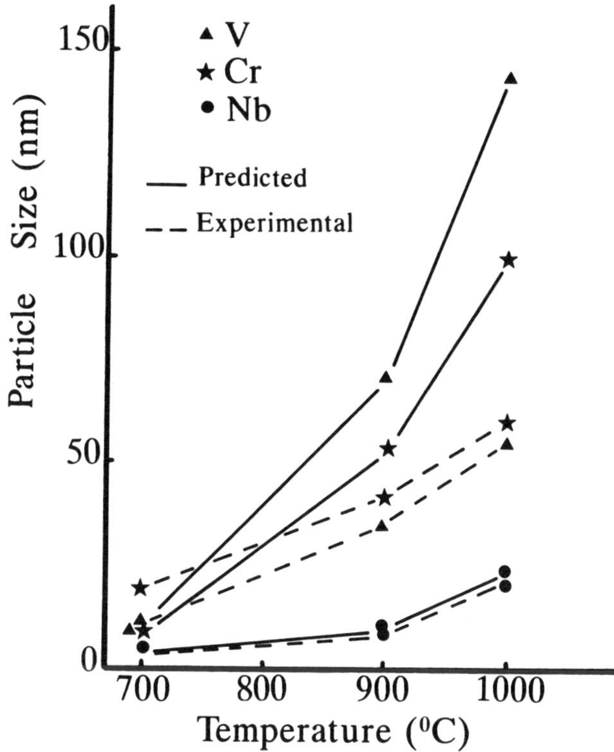

Fig.5 Coarsening of particles in copper after annealing for 1 h at high temperature. A comparison is made with bulk coarsening theories.

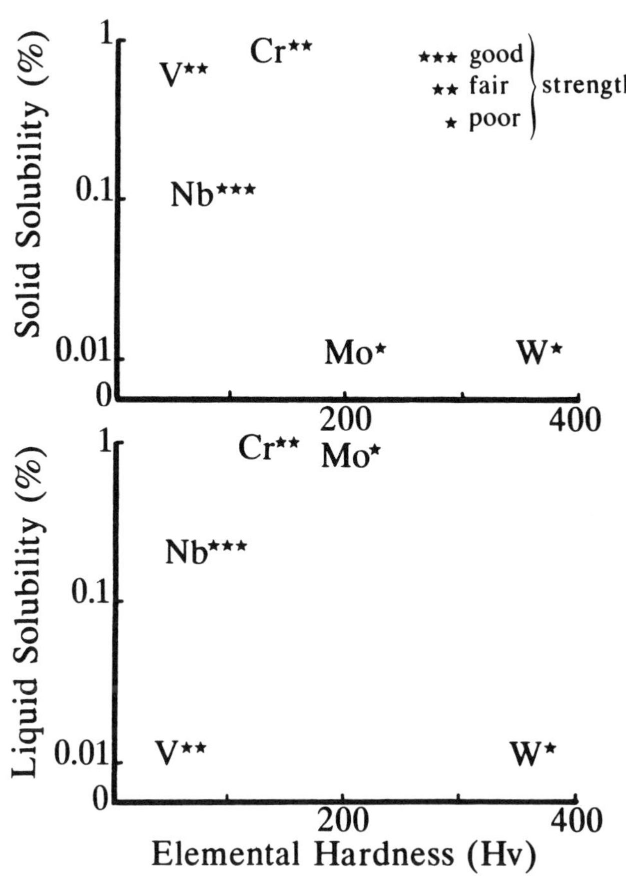

Fig.6 An attempt to correlate the degree of strengthening achieved on milling copper with bcc elements in terms of the hardness of each element and its solubility in solid and liquid copper.

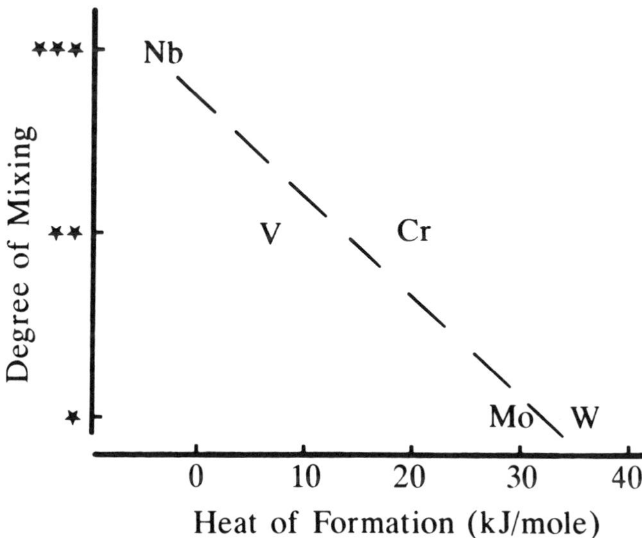

Fig.7 Correlation of the degree of mixing achieved on milling copper with a bcc element in terms of the heat of formation of a copper-bcc element compound.

imperfect one, is a measure of the heat evolved or required during the creation of Cu-bcc element bonds. It is clearly seen in Fig.7 that the degree of mixing achieved correlated very well with the tendency to have a negative heat of interaction, as in the case of Nb in Cu.

INTERPRETATION OF ROOM TEMPERATURE STRENGTH IN MA COPPER ALLOYS

Strengthening in a fine grain sized material containing large numbers of particles which will not be cut by dislocations may be caused by grain boundary strengthening and by the Orowan mechanism of dislocation strengthening. Also, as will be discussed later, there is an additional strengthening mechanism which may operate when the microstructural scale of the composite material is reduced to an extremely fine scale.

The initial approach of Orowan (13), considering dislocation bowing between the unshearable particles, leads to a material shear strength

$$\tau_{OR} = \tau_m + \mu b/\lambda \qquad 1$$

where τ_m is the matrix flow strength, given by the Peierls stress, solute strengthening and so on, μ is the shear modulus, b the Burgers vector and λ the planar centre to centre spacing of the dispersed particles. This equation was subsequently modified (14,15) to take account of a random distribution of particles, the varying line tension of a dislocation according to its screw or edge character, and the finite size of the particles on the dislocation glide plane, giving

$$\tau_{OR} = \tau_m + \frac{1}{1.18} \frac{2\mu b}{4\pi(\lambda-\emptyset)} \ln(\emptyset/2b) \qquad 2$$

In this equation the particle diameter is $\emptyset$, and the cut-off radius of the dislocation core has been taken as 2b. Substituting in suitable values for shear modulus and Burgers vector in copper we obtain

$$\tau_{OR} = \tau_m + \frac{1.55}{(\lambda-\emptyset)} \ln(\emptyset/2b) \qquad 3$$

where λ and $\emptyset$ are expressed in metres. Considering a Taylor factor of 3 for the fcc copper matrix we deduce a material strength

$$\sigma = \sigma_m + \frac{4.65}{(\lambda-\emptyset)} \ln(\emptyset/2b) \qquad 4$$

In the general case of a deforming material it is likely that both edge and screw dislocations will move and, in view of the different variation of line tension for a bowing screw dislocation, the stress required to more the screw dislocation is increased as $\tau_{screw}= \tau_{edge}/0.66$. Taking an average of the stresses required for screw and edge dislocation motion we can write

$$\sigma = \sigma_m + \frac{5.8}{(\lambda-\emptyset)} \ln(\emptyset/2b) \qquad 5$$

with an uncertainty of +/- 1.2 on the value of the numerical term 5.8. In this equation, the particle size and spacing can be related to the volume fraction (f) occupied by the particles (considering a highly flexible dislocation line) as $\lambda = 1/(N_V\emptyset)^{1/2}$, where N_V is the number of particles per volume given by $N_V\emptyset^3 = f$.

The grain boundary strengthening term can usually be described by a Hall-Petch relationship, and making the assumption that this strengthening is independent of the other strengthening terms it can be added simply (6) as

$$\sigma = \sigma_m + \frac{5.8}{(\lambda-\emptyset)} \ln(\emptyset/2b) + \frac{k}{D^{1/2}} \qquad 6$$

where k is a constant and D the grain size. The constant k is a measure of the ease of slip transfer across a grain boundary and for conventional copper alloys has a value of about 0.16 MPa m$^{1/2}$ (16).

A final point to note is the relationship between the density and size of particles and the grain size. Zener, for example, considers that particles will apply a pinning force on grain boundaries that will limit the maximum grain size (according to the boundary curvature associated with the grain size). By relating the grain size and the boundary curvature, the critical grain size, limited by particle pinning, may be written (17,18)

$$D = 2\emptyset / 3f \qquad 7$$

This relationship has been critically examined by Rios (19) who points out that the Zener equation becomes unrealistic for high volume fractions of the dispersed phase since the number of particles associated with the critically-sized grain decreases (it is for example less than 20 particles per grain at a volume fraction of dispersed phase of 10 %) and pinning must be considered on an individual boundary-particle basis rather than by an averaging calculation. An alternative approach by Haroun and Budworth (20) considers that there will be one particle per grain face (which may perhaps be located at triple points on boundaries) and predicts a grain size as

$$D = \emptyset/f^{1/2} \qquad 8$$

In both cases the stable grain size that will be achieved on allowing the grain boundaries to reach an equilibrium configuration, as for example after extrusion or heating, will be determined both by the volume fraction and by the size of the pinning particles.

The microstructural configurations and mechanical properties have been examined in detail for the copper-bcc element alloys prepared by mechanical alloying, particularly for the copper-Nb alloys (3). These results can be used to examine the validity of the grain size and strengthening models. Fig.8 shows the experimental grain size measured on the as-extruded material or after subsequent annealing treatment and compares these values with the predictions of the Zener equation for equilibrium grain size. The two sets of data are extremely close, confirming that the grain sizes achieved in the materials are very close to their stable configurations and confirming the applicability of the Zener equation. The measured flow strengths of the materials are shown in Fig.9, where they are related to $\ln(\emptyset/2b)/(\lambda-\emptyset)$. The slope of the straight line is 5.5 - 6.3 and the intercept on the stress axis is 70 - 220 MPa. This representation corresponds precisely to that of equations 5 or 6, with a matrix strength of 150 +/- 70 MPa and no grain size dependence: the experimental slope of 5.9 +/- 0.4 is in excellent agreement with the theoretical value. An alternative representation of the data in terms of a Hall-Petch relationship, that is attempting to relate strength and grain size, leads to a large negative matrix stress term which is clearly not satisfactory. Finally, we can attempt to interpret the experimental data in terms of both particle strengthening and grain boundary strengthening using equation 6 but with a numerical constant for Orowan strengthening which is less than the 5.8 deduced theoretically. For example, substituting 4.7 instead of the value 5.8 in equation 6 it is possible to interpret the experimental data with a matrix strength value of 100-150 MPa and a grain size dependency (k) of 0.067 MPa $m^{1/2}$: this is much lower than the expected value for a copper matrix, namely 0.16 MPa $m^{1/2}$. Alternatively, subsituting 3.0 instead of the value 5.8 in equation 6 allows the experimental data to be interpreted with a matrix strength of 0-30 MPa (somewhat too low) and a suitable value for grain boundary strengthening. The question to consider is whether the 5.8 term can really be lowered to 4.7 or 3.0: the value 4.7 would correspond to the situation where edge dislocations only are able to move which seems unlikely but not impossible; the value 3.0 would correspond to the case where edge dislocations only are able to move, but also using a Taylor factor of 2, which is unreasonable for a polycrystalline material of fcc crystal structure. In conclusion we can say that grain boundaries contribute essentially little or nothing to the strength of the present materials: the fine grain size is rather a consequence of the many fine particles existing and the strength is determined only by the difficulty in bypassing these particles. Fig.10 shows dislocations interacting with particles in lightly deformed materials where it is clear that the particles interfere strongly with dislocation motion. There are no signs of curved loops under the influence of the high local stresses induced.

When the grain size of a polycrystalline material approaches dimensions in the few to tens of nanometres size range, additional phenomena may occur which may change the controlling strengthening mechanism and the resulting material strength. For these materials of so-called nanoscale microstructures, dislocations may be lost at grain boundaries by image forces (21,22) or may be unable to multiply because Frank-Read sources for dislocations may not operate in the fine, multiphase materials (23). The micrographs of Fig.10 show dislocations in lightly deformed materials of grain size about 75-300 nm. In the case of the larger grain size the dislocation density is of the order of 2-3 x 10^{14}/m^2: in the case of the smaller grain size it is extremely difficult to image dislocations (in thick foils there are overlapping contrasts from grain boundaries and dislocations, in thin foils dislocations may be lost to foil surfaces) but it appears that reasonable numbers of dislocations are present - here about 1 x 10^{14}/m^2. As such it appears that the strengths of the sub-micron grain sized materials examined here may still be interpreted using classical dislocation strengthening models.

CONCLUSIONS

The work reported here has examined the creation of a uniform dispersion of fine particles in a fine-grained copper matrix by mechanical alloying. It has been shown that achieving the fine dispersion in the first place depends essentially on a suitable chemical interaction between matrix and dispersed particle phases whilst the stability of the microstructure can be explained by classical particle coarsening and grain boundary pinning theories and as such depends essentially on the solubility and diffusivity of the diffusing species. Strengthening can be explained by Orowan strengthening with no major influence of grain boundaries.

REFERENCES

1. A.N. Patel and S. Diamond, Mater.Sci.and Eng., A98, p329 (1988).
2. D.G. Morris and M.A. Morris, Mater.Sci.and Eng., A104, p201 (1988).
3. M.A. Morris and D.G. Morris, Mater.Sci.and Eng., A111, p115 (1989).
4. J.G. Schroth and V. Franetovic, J.Metals, 41.1, p37 (1989).

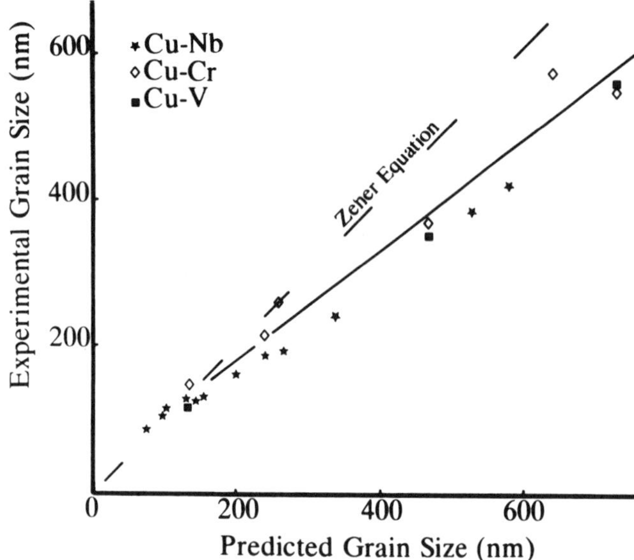

Fig.8 Comparison of experimental grain sizes and predicted values based on the Zener formula, for MA Cu alloys.

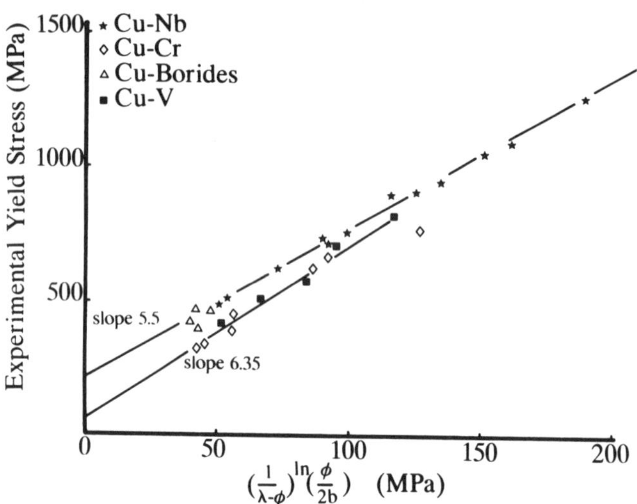

Fig.9 Correlation between experimental yield strengths of MA Cu alloys and parameters derived from the Orowan strengthening model (see text for details).

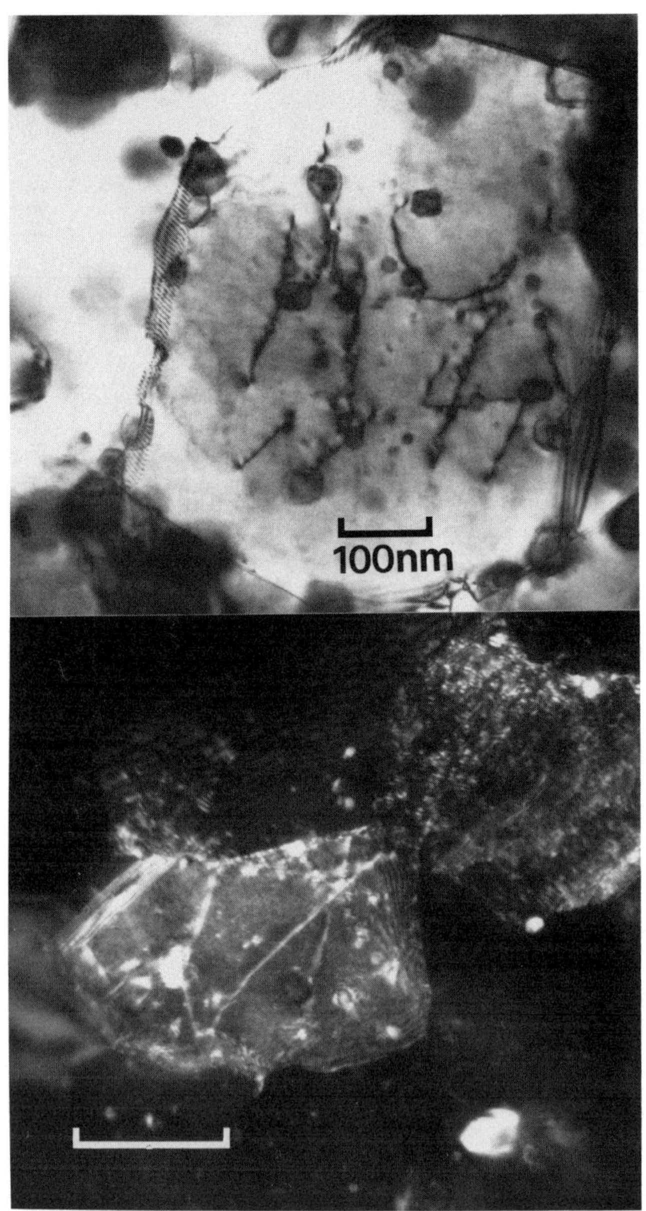

Fig.10 Dislocations in slightly-deformed CuMA alloys of medium grain size (about 500 nm) and small grain size (less than 100 nm) showing dislocations interacting with the particles.

5. M.A. Morris and D.G. Morris, Proc.7th Int.BNF Conf. "The Materials Revolution through the 90's" July 1989, Oxford, BNF, Wantage, England (1989).

6. H.G.F. Wilsdorf, Proc.Symp.on Dispersion Strengthened Aluminium Alloys, Jan.1988, TMS, Warrendale, USA (1988).

7. I.M. Lifshitz and V.V. Slyozov, J.Phys. Chem.Solids, 19, p35 (1961).

8. C. Wagner, Z.Electrochem., 65, p581 (1961).

9. Smithells Metals Reference Book, 6th Edition, Ed. E.A. Brandes, Butterworths, London (1983).

10. Binary Phase Diagrams, Ed. T.B. Massalski, ASM (1986).

11. Diffusion Rate Data and Mass Transport Phenomena for Copper Systems, INCRA Monograph V, International Copper Research Association, Ed. D.B. Butrymowics, Washington, DC (1977).

12. A.R. Miedema, Philips Technical Review, 36, p217 (1976).

13. E. Orowan, Symposium on Internal Stresses in Metals and Alloys, Discussion, Monograph and Report Series No 5, p451, Inst.of Metals, London (1948).

14. U.F. Kocks, Phil. Mag., 13, p541 (1966).

15. M.F. Ashby, in A. Kelly and R.B. Nicholson (eds), Strengthening Methods in Crystals, Elsevier, New York, p165 (1971).

16. N. Hansen, Met.Trans., A16, p2167 (1985).

17. C. Zener, private communication to C.S. Smith, Trans.Am.Inst.Min.Engs., 175, p47 (1948).

18. J.W. Martin and R.D. Doherty, Stability of Microstructures in Metallic Systems, Cambridge University Press, Cambridge, England (1976).

19. P.R. Rios, Acta Met., 35 p2805 (1987).

20. N.A. Haroun and D.W. Budworth, J.Mater. Sci., 3, p326 (1968).

21. F.H. Froes and C. Suryanarayana, J.of Metals, p12, (June 1989).

22. V.G. Gryaznov, A.M. Kaprelov and A.E. Romanov, Scripta Met., 23, p1443 (1989).

23. J.S. Koehler, Phys.Rev., B2, p547 (1970).

SYNTHESIS OF CHROMIUM SILICIDE
BY MECHANICAL ALLOYING

C. C. Bampton, C. G. Rhodes,
M. R. Mitchell, M. S. Vassiliou
Rockwell International, Science Center
Thousand Oaks, California 91360, USA

J. A. Graves
HOWMET Corporation
Whitehall, Michigan 49461, USA

Abstract

Intermetallic alloys such as Cr_3Si and $MoSi_2$ are attracting increasingly serious attention as high temperature structural materials. Mechanical alloying offers some practical advantages for synthesis of this class of materials. These issues and our preliminary experience with mechanical alloying, phase transformation of the Cr_3Si compound from elemental powders, and consolidation by conventional and explosive compactions are discussed. The mechanical alloying process does not solid solution alloy elemental Cr and Si, nor does it form an amorphous phase. A subsequent heat treatment is required to transform the mechanically mixed Cr and Si to the Cr_3Si compound.

INTERMETALLICS AS HIGH TEMPERATURE STRUCTURAL MATERIALS - The development of high temperature structural materials to date has focused primarily upon Ni-base alloys for use in gas turbine engines. Operating temperatures are now limited by these materials, with the most advanced alloys currently operating at over 80% of their melting point. The need for new, lower density, higher temperature materials has driven the investigation of titanium aluminides (Ti_3Al, α_2 and TiAl, γ) for both airframe and engine applications (e.g., NASP). These alloys have a similar operating temperature potential to the Ni-base materials but with much lower densities. Other intermetallics promise to be effective replacements for ceramics and carbon-carbon materials for even higher temperature applications.

The existence of ordered intermetallic compounds arises from there being stronger bonds between unlike nearest-neighbor atoms than between like nearest-neighbors. As a result, these compounds have melting temperatures that are significantly greater than one or both of their elemental constituents. In addition, many of these compounds have higher elastic stiffness than either of their constituents and a high elastic modulus usually implies high strength. Finally, some intermetallics exhibit an increasing strength with temperature over a significant range; a general mechanism is not currently available to explain this phenomenon but the known information has been recently reviewed.[1]

With few exceptions,[1,2] atomic ordering results in increased brittleness. Most intermetallics are brittle at less than $0.4T_m$ and many are brittle to their melting point.

The last ten years have seen the development of several aluminide based intermetallic compounds that exhibit a balance of mechanical properties that make them suitable as components for advanced propulsion and hypersonic airframe systems. However, these compounds have maximum use temperatures too low to meet all temperature requirements of future advanced structural and propulsion system applications.

Materials for use at temperatures in the range 1000°C to 1600°C (1900°F to 2900°F) will probably need to have melting temperatures in the range 1600°C to 2700°C (2900°F to 4900°F). Exceptions to this rule may be possible with intermetallic compounds that do not exhibit a ductile-brittle transition prior to melting but such materials are unlikely to meet other service requirements. For structural applications, the materials must not only possess high temperature strength but also have toughness throughout the temperature range of operation. Environmental resistance is a further necessary attribute. Finally, advanced design considerations based on minimum thickness sections dictate that materials should have the lowest possible density.

The two intermetallic compounds that have received most attention in the literature as candidates for high temperature structural materials are $MoSi_2$ and Nb_5Si_3. We have chosen to focus our work on the two metal-silicides, $MoSi_2$ and Cr_3Si. Their primary characteristics for selection as structural materials are shown in Table 1.

Although Cr_3Si has a lower melting temperature and lower oxidation resistance than $MoSi_2$, both systems are of interest since the $MoSi_2$ is a line compound, whereas Cr_3Si has a significant range of stoichiometry (see the phase diagrams in Fig. 1). The Cr_3Si phase stoichiometry range allows diffusional processes under compositional gradients and makes

Compound	Crystal Structure	Melting Temp. (C)	Density (g/cc)	Oxidation Behavior
MoSi$_2$	tI16 (tetragonal)	2032	6.2	Excellent with Ge addition
Cr$_3$	cP8 (BCC derivative)	1770	6.5	Poor > 1000°C due to volatile CrO$_3$

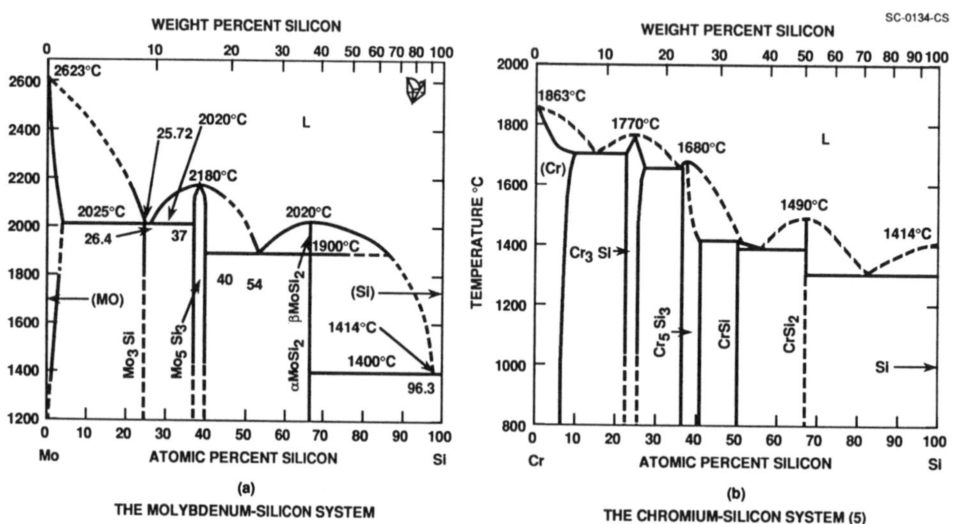

Fig. 1 (a) Equilibrium phase diagram for the molybdenium-silicon system.[3] (b) Equilibrium phase diagram for the chromium silicon system.[3]

preparation of a true single phase alloy practical, if desired. Also, the Cr$_3$Si has a BCC derivative structure, whereas the MoSi$_2$ is tetragonal. These differences may lead to fundamental differences in alloying and processing approaches and in plastic behavior.

MECHANICAL ALLOYING FOR SYNTHESIS OF INTERMETALLICS - As described above, intermetallic compounds are invariably brittle at room temperature. They require incorporation of second and possibly third phases to rend them engineeringly useful. Typically, a ductile second phase is the most effective method for improving toughness, ductility and processibility. Ceramic reinforcing phases may also be added (as particulates, whiskers or fibers) for toughening (by crack deflection), strengthening, stiffening, increasing creep resistance and lowering of density.

Very few of the intermetallic compounds of interest as structural materials have suitable ductile phases with which they are in thermodynamic equilibrium. This precludes conventional ingot metallurgy processing for these cases. Powder metallurgy processing, however, provides the opportunity for a wide range of phase blended alloys that may or may not be in thermodynamic equilibrium with the intermetallic matrix.

Figure 2 shows schematically four examples of microstructures designed to improve the balance of engineering properties of an intermetallic compound. All of these microstructures may be synthesized by various powder processing routes.

In addition to phase blending, powder metallurgy processing provides an extra degree of control over the range of microstructures attainable by allowing enhanced microstructural refinement and production of intermediate metastable phases. Metastable phases provide additional processing pathways for improved processibility and/or microstructure control.

Both rapid solidification (RS) and mechanical alloying (MA) may be utilized for synthesizing powder products which are then consolidated and processed to structural components incorporating the benefits described above. Although RS processing has received far more attention than MA processing (except for the unique case of oxide dispersion strengthened alloys), some advantages of MA merit serious consideration for intermetallic alloys. These include:

- Production of amorphous metallic alloy powders in the solid state,[4-6] which may provide additional processing pathways and microstructure control.

- Low temperature, solid state processing and avoidance of melt contamination (from the atmosphere and melt container) and segregation.

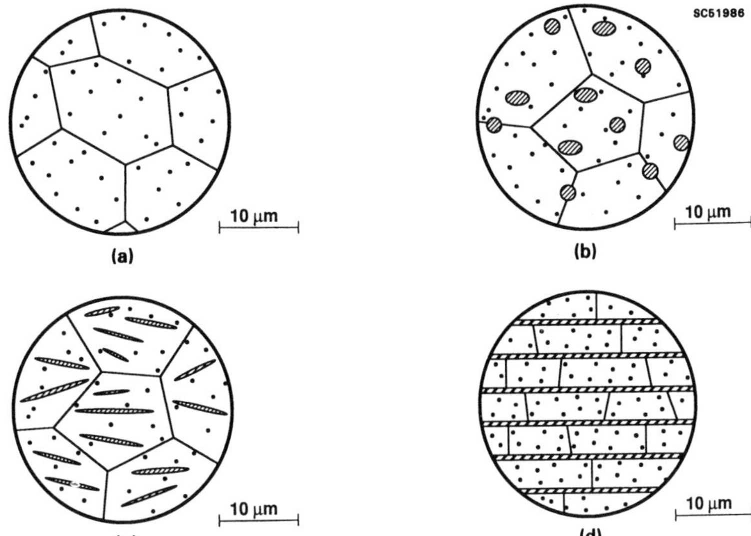

Fig. 2   Schematic of intermetallic alloy microstructural designs.   (a) Intermetallic matrix + ceramic particulates; (b) intermetallic matrix + ceramic particulates + ductile phase spherical particles; (c) intermetallic matrix + ceramic particulates + ductile phase pancake-shaped particles; and (d) intermetallic matrix + ceramic particulates layered with ductile phase.

- Production of unique, phase blended composites.
- Easy scale-up for large batches of powder.

CONSOLIDATION OF INTERMETALLIC ALLOYS POWDERS - Conventional consolidation of intermetallic alloy powders is performed by vacuum hot pressing and/or extrusion and/or hot isostatic pressing. Since all of these must be carried out at very high homologous temperatures, microstructural transformations of the mechanically alloyed powders are likely during the preheating or during the consolidation process. It may be possible to control these transformations advantageously during conventional thermomechanical processing. In some cases, however, it may be desirable to completely avoid any microstructural transformations during the consolidation stage, retaining metastable or even amorphous structures in the fully dense material. Dynamic consolidation, by methods such as explosive shock forming, is a potentially practical method for consolidation of the powder into a usable part while preserving its metastable structure. Dynamic consolidation, in which powders are consolidated via the passage of a strong shock wave, does not expose the material to high temperatures for long times, and hence has the potential to produce bulk metastable solids.

A critical practical consideration when using explosive forming is to deal with the rebounding pressure wave which would generally induce damaging tensile stresses in the compact.

GENERAL APPROACH - The work described here covers some of the preliminary studies of the synthesis of Cr_3Si by mechanical alloying from the elemental powders. The studies are focused on the microstructural evolutions during ball-milling, subsequent heat treatments, consolidation by conventional vacuum hot pressing and consolidation by explosive dynamic compaction.

EXPERIMENTAL PROCEDURE

BALL MILLING - Elemental Cr and Si powders (99.8% pure Cr at < 80 μm diameter and 99.5% pure Si at < 80 μm diameter) were pre-blended then ball-milled in an argon filled, stainless steel jar, 200 mm diameter, with tungsten carbide milling balls, 12 mm diameter, weighing 18 g, rotated at 70 RPM. Powder charges were about 1000 g. Ball/powder ratios of 6:1 and 15:1 were used with milling times from 0 to 300 h. During milling of the first two charges, small samples were extracted at 1 h intervals for analysis. This was done in a glove box with less than 4 ppm oxygen in the argon atmosphere. Two subsequent charges were milled uninterrupted. All mechanically alloyed powder samples were exposed to flowing house argon for 2 h to gradually oxidize the fresh powder surfaces prior to exposure to air.

POWDER ANAYSIS - Powders were metallographically analyzed as ball milled and after vacuum heat treatment at 900°C. Energy dispersive x-ray microanalyses were employed in addition to optical and electron microscopy. Bulk chemical analyses, including oxygen content, were made on all samples. X-ray diffraction phase analyses were also carried out on all powder samples. A limited number of powder samples were x-ray phase analyzed at multiple temperature increments in a vacuum hot stage x-ray diffraction camera, in order to gain information on phase transformation temperatures. Additional phase transformation temperature information was derived from differential thermal analysis of powder samples.

VACUUM HOT PRESS CONSOLIDATION - The ball-milled powders were consolidated in 50 mm diameter × 50 mm high Ti-6Al-4V cans by hot pressing in graphite dies. The entire die and can assemblies were sealed in evacuated stainless steel retorts. The retorts were evacuated to 1×10^{-6} Torr at 500°C. Hot pressing was carried out with a constant load of 10,000 kg at 1100°C for 30 minutes.

The consolidated ingots were examined by optical metallography and microhardness.

SHOCK-WAVE CONSOLIDATION - Cr-Si powders, filling 85% of the volume of a stainless steel container, were sealed off under vacuum, as described previously.[7][8] Consolidation was effected by a shock wave generated from the detonation of a C4 explosive charge, which propelled a stainless steel flyer plate in the sample assembly.

DISCUSSION OF EXPERIMENTAL RESULTS

BALL MILLED POWDERS - Figure 3 shows the starting powders and after 45 h of ball milling. The overall size distribution is clearly reduced by the milling, although this effect occurred during the first few hours of milling with little subsequent change in particle sizes with further milling. There appears to be a tendency for a laminated fine structure in some of the mechanically alloyed particles, although regions of more granular morphology are quite common in other particles. The lamellar structure is expected in mechanical alloying of ductile/ductile powder blends, whereas a granular morphology will result from a brittle/brittle powder blend.

Figure 4 shows two examples of x-ray diffraction peaks for powders milled 15 h and 35 h. In all cases (0-300 h) there were no Cr_3Si peaks in as-milled powder. Distinct Cr_3Si peaks were found in all powder samples after annealing 4 h at 900°C. Cr and Si peaks were found in short milling time powders with the intensities lessening with increasing milling time. Si peaks were absent in the long time, as-milled powders. Figure 5 plots ratios of certain peaks for the annealed powders, against milling time, to provide a semi-quantitative figure of merit. Lower numerical value implies greater Cr_3Si formation. Cr_3Si formation continually increases with milling time and the higher ball/powder ratio gives a slight improvement. Interestingly, in the 150 h and 300 h samples, even though Si peaks were not present in the as-milled samples, some Si phase reformed during heat treatment. Figure 5 suggests that 150 h - 15:1 ball ratio provided the greatest transformation.

Figure 6 shows lattice parameter measurements for Cr and Si in the as-ball milled powders up to 50 h milling time. There appears to be a very slight trend of the two lattice parameters converging with increasing milling time. However, this effect is not sufficiently large to indicate solid solution alloy formation.

Table 2 shows contaminant levels found in powders from 3 milling runs. In run #1, hourly interruptions were made to remove small samples for analysis. These samples, as expected, picked up more oxygen with time than the continuously milled samples. Oxygen levels were generally high in all samples. However, as milled powder suffered a significant temperature rise if opened to air, indicating that much of the powder had been unoxidized. The other, possibly significant, contaminant was Fe. This presumably came from the stainless steel container walls.

Figure 7 shows x-ray diffraction peaks for a sample of 150 h milled powder at increasing temperatures. Phase transformation, from the finely mixed Cr and Si to the Cr_3Si compound, took place between 560 and 600°C. No further changes occurred up to the highest scan, at 775°C. No Si peaks were detected in the annealed powders. These transformation temperature results were confirmed by differential thermal analysis (DTA) on the same 150 h powder. The DTA showed two exotherms. The major exotherm was at 575°C, corresponding to the $3Cr + Si \rightarrow Cr_3Si$ transformation found in the high temperature x-ray results. A smaller, unidentified exotherm occurred at 1100°C.

Scanning electron microscopy and quantitative energy dispersive x-ray microanalysis showed that increasing milling time produced composite particles with finely mixed Cr and Si in increasingly stoichiometric proportions for the Cr_3Si compound.

Figure 8 shows 150 h powders before and after annealing. The annealing process consistently appeared to perfect the Cr_3Si stoichiometry of powder particles which may have been significantly off stoichiometry as-ball milled. Figure 8(a) shows some

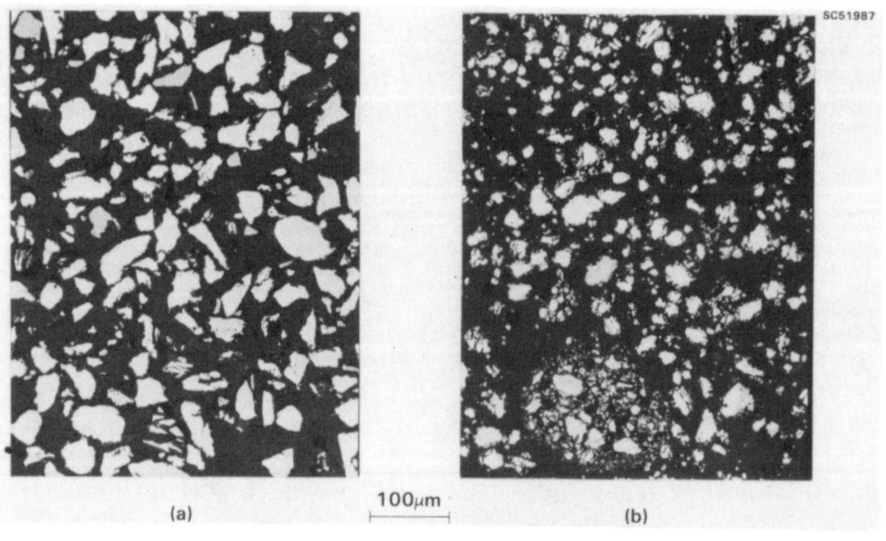

Fig. 3 Optical micrographs of (a) starting Cr + Si elemental powders, and (b) after 45 h ball milling (6:1 ball/powder).

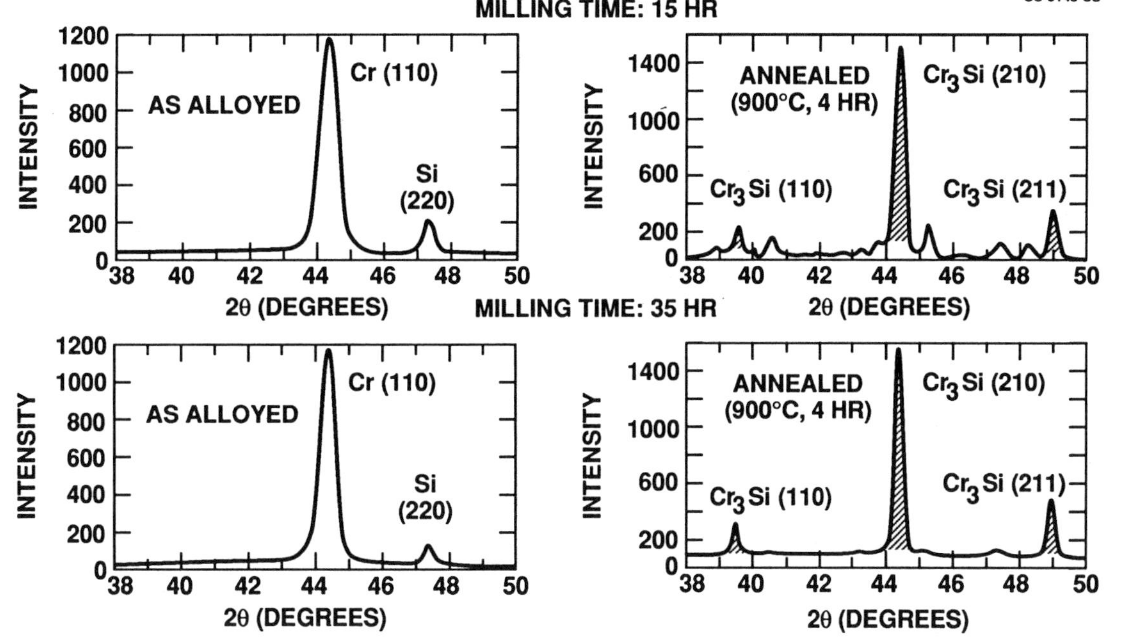

Fig. 4 X-ray diffraction peaks for 15 h and 35 h milled Cr-Si.

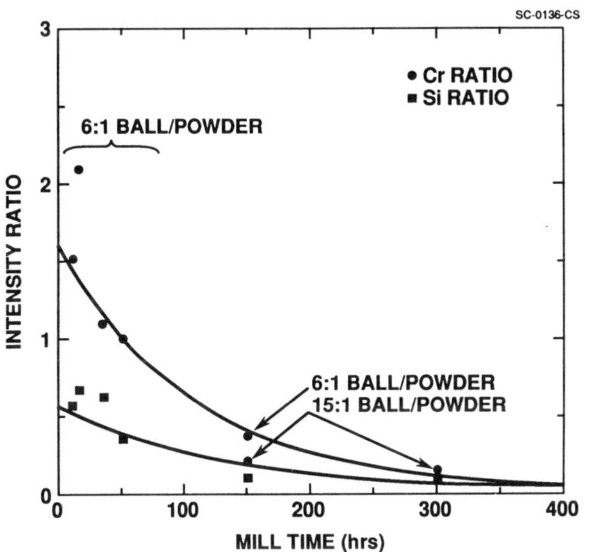

Fig. 5 X-ray intensity ratios of heat treated (4 h @ 900°C). Cr-Si samples: (Cr (211)/Cr₃Si (211); Si (111)/Cr₃Si (110).

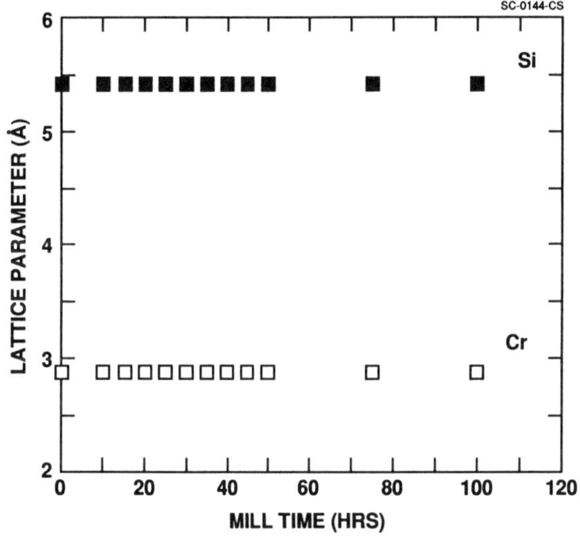

Fig. 6 Lattice parameters of ball-milled Cr and Si powders.

large granular mechanically mixed Cr + Si particles. Large Cr particles are quite common, although Si particles are difficult to find, usually being embedded in the larger composite particles. Figure 8(b) shows typical large Cr particles with surface layers or crusts of Cr₃Si. Figure 9 shows similar features in the 300 h powder, although the stoichiometry of the as-milled composite particles tend to be more nearly perfect Cr₃Si. A lamellar composite particle is also visible in Fig. 9(a).

The overall efficiency of the mechanical alloying is quite low, since about 5 volume percent of the powder remained as large Cr particles even after 300 h milling. It is believed that a significant contribution to this inefficiency was a poor ball mill design feature which allowed powder to accumulate in seams and crevices in the mill jar thus preventing continuous powder circulation and mechanical alloying.

PRESS CONSOLIDATED BILLETS - Figure 10 shows the consolidated 150 h ball milled material. The material is clearly two phase, the matrix being Cr₃Si and the large particles being Cr, as confirmed by transmission electron microscopy. Excess Si must be presumed to be present also, although this was not confirmed. Significant porosity remained in the sample, indicating insufficient consolidating pressure and/or temperature. Table 3 shows microhardness values for the Cr₃Si and Cr phases in three consolidated billets. The Cr₃Si compound had consistently very high hardness. Variations in hardness from batch to batch were most likely due to differences in contaminant level, particularly oxygen.

Table 2 - Powder Contaminants

	0 h	Run #1 10 h	50 h	Run #2 150 h	Run #3 150 h
O	0.62	1.00	1.05	0.93	0.96
N	0.032	0.097	0.087	0.19	0.17
H	0.0127	0.0049	0.0142	0.0132	0.0148
W	0.003	0.007	0.006	0.006	0.011
C	0.009	0.015	0.028	0.018	0.018
Fe	0.069	0.082	0.20	0.25	0.25
Ni	< 0.001	< 0.001	0.003	< 0.01	0.034
Co	< 0.001	< 0.001	0.004	< 0.01	< 0.001
Al	< 0.001	0.18	0.14	0.010	0.12

SC-0135-CS

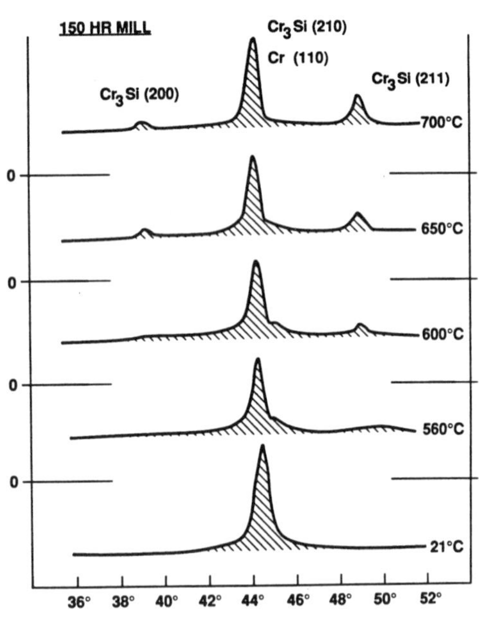

Fig. 7 Incremental temperature x-ray diffraction scans on 150 h milled Cr-Si powder.

EXPLOSIVELY CONSOLIDATED BILLETS - Powders that had been ball-milled for 150 h were dynamically consolidated by explosive shock wave compaction. The result was a densified solid consisting of bonded powder particles, Figure 11, that vary in size from 30 micrometers down to < 5 micrometers. The largest particles are essentially 100% Cr, while the smaller ones contain increasing amounts of Si. The matrix is Cr_3Si, as determined by thin-foil transmission electron microscopy. The implication is clearly that excessive heating during consolidation and/or very slow cooling allowed transformation of the finely mixed Cr and Si to the Cr_3Si compound.

There is no evidence of deformation in the extremely fine grained matrix, Figure 12, which is unusual for shock-wave consolidated powders.[9] The lack of dislocation structure may be the result of the transformation to the Cr_3Si compound at, or after, consolidation.

X-ray energy dispersive spectroscopy and selected area electron diffraction were used to identify the matrix and the large particles in the TEM. Quantitative EDS analysis in conjunction with diffraction pattern analysis confirmed that the matrix was Cr_3Si, Fig. 13.

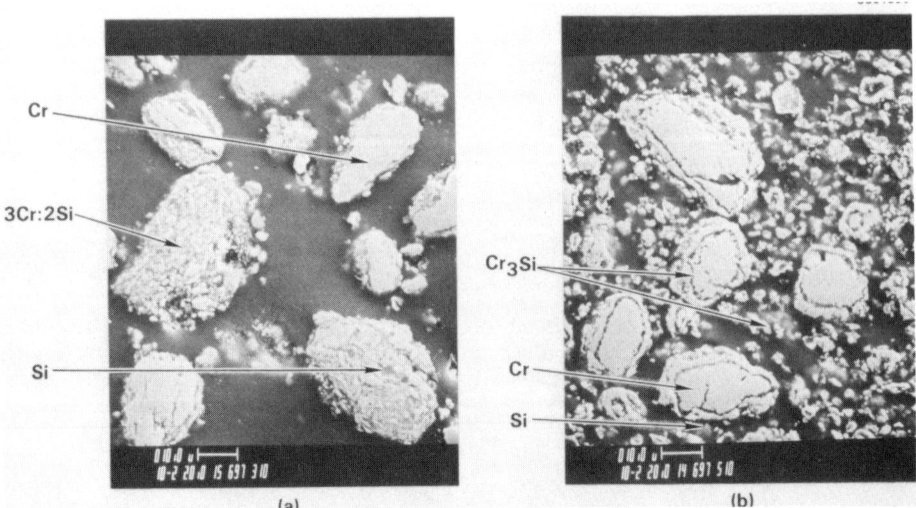

Fig. 8 Scanning electron micrographs of 150 h milled Cr-Si powders, (a) as-milled, (b) annealed 4 h at 900°C.

280

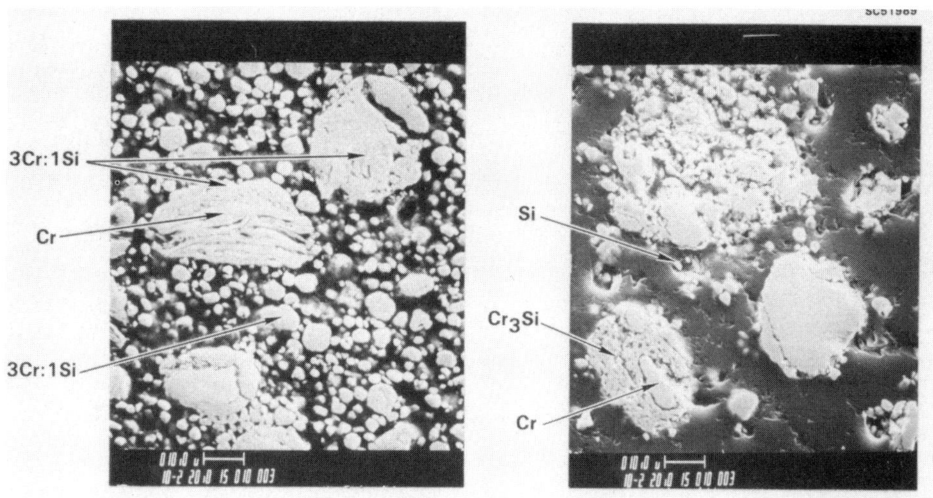

Fig. 9 Scanning electron micrographs of 300 h milled Cr-Si powders, (a) as-milled, (b) annealed 4 h at 900°C.

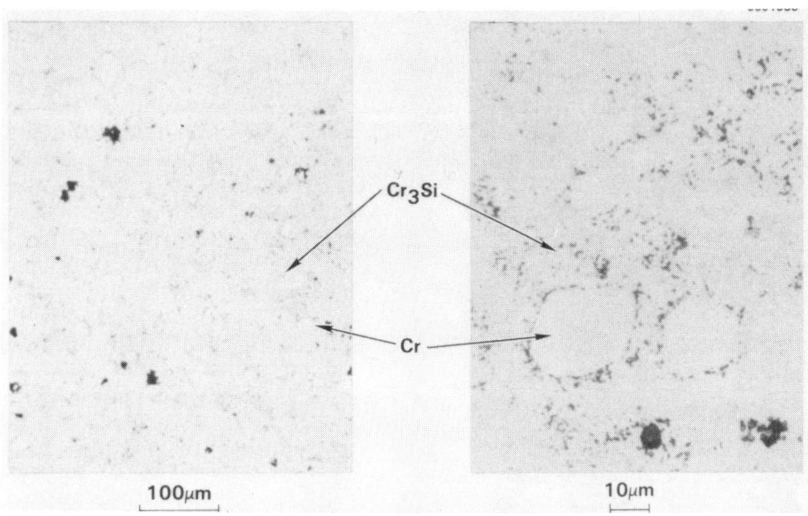

Fig. 10 Optical micrographs of vacuum hot pressed, 150 h milled Cr-Si powder.

Table 3 - Vacuum Hot Pressed Microhardnesses

Phase	Vickers hardness (HKN)		
	100 h, 6:1	150 h, 6:1	150 h, 15:1
Matrix (Cr_3Si)	1230	1187	1048
Particles (Cr)	333	317	292

Fig. 11 (a) Mechanical alloying of elemental Cr and Si powders results in powder particles that are an intimate mixture of Cr and Si; and (b) after dynamic consolidation the powder particles are retained.

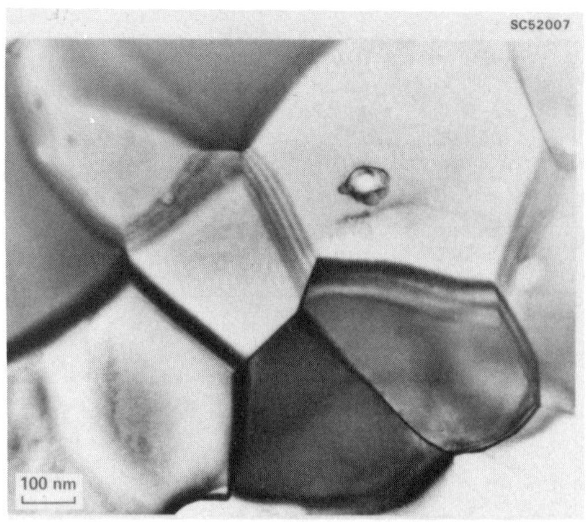

Fig. 12 TEM image of dynamically consolidated Cr-Si.

The single experiment carried out here indicates that the explosive consolidation method may not be a viable approach for retaining the as-ball-milled structure in the fully consolidated material, although the method has many variables that have significant control over the heating encountered. The method may at the very least provide an opportunity to consolidate certain nonequilibrium tough phases, pre-blended with the mechanically alloyed intermetallic powders without excessive diffusion interactions, which would otherwise result from extended high temperature exposure.

SUMMARY AND CONCLUSIONS

Elemental Cr and Si powders mechanically mix quite effectively by inert atmosphere ball milling. Solid solution alloying does not occur during ambient temperature ball milling, even for very long times (up to 300 h). Neither does amorphous phase formation occur. The elements are simply mixed increasingly finely in composite particles with insignificant interdiffusion occurring. The composite particles tend to be either granular or lamellar in morphology. Transformation to the ordered intermetallic compound Cr_3Si does occur on subsequent heating of the ball milled powder to temperatures $\geq 575°C$. The mechanical alloying efficiency in our apparatus was quite low, with a significant volume fraction (≥ 5 v%) of large, single phase Cr particles remaining after extended ball milling. We believe that redesign of the ball-mill jar would significantly improve apparatus efficiency. This redesign will also aim to reduce oxygen pickup which was higher than desired (≥ 1 wt%). Explosive dynamic compaction of the milled powders successfully produced consolidated material of the unalloyed Cr particles in a Cr_3Si matrix. This structure was similar to the conventionally hot pressed powder, except that the transformation to the Cr_3Si compound occurred during or immediately after explosive compaction; whereas the transformation occurred during the pre-heat in conventional consolidation. This provides a much lower dislocation density and generally smaller grain size in the dynamically consolidated material. Furthermore, the short time at elevated temperature in the dynamic consolidation process may provide a benefit for consolidation of phase blended toughening phases, with the mechanically alloyed Cr_3Si, if these phases are not in thermodynamic equilibrium.

REFERENCES

1. R.L. Fleischer, D.M. Dimiduk and H.A. Lipsitt: G.E. Technical Report 88CRD326, Dec. 1988.
2. N.S. Stoloff in **Strengthening Methods in Crystals**, ed. A. Kelly and R.B. Nicholson, pp 193-259, Wiley, NY. (1974).
3. Binary Alloy Phase Diagrams, ed T.B. Massalski, ASM (1986).
4. R.L. White and W.D. Nix, "New Developments and Applications in Composites, ed D. Kuhlmann-Wilsdorf and W.C. Harrigan (TMS, Warrendale, PA, 1979) p 78.
5. Koch,C.C., O.B. Cavin, C.G. McKamey and J.O. Scarborough, Appl. Phys. Lett. 43, 1017 (1983).
6. Schwarz, R.B., R.R. Petrich and C.K. Saw, J. Non-Cryst. Solids, 76, 281 (1985).
7. M.S. Vassiliou, C.G. Rhodes, M.R. Mitchell, and J.A. Graves: Scripta Met., vol. 23, 1989, pp. 1791-1794.
8. Rhodes, C.G., M.S. Vassiliou, M.R. Mitchell, and R.A. Spurling: accepted for publication in Met. Trans., 1990.
9. See, for example, L.E. Murr, A.W. Hare, and N.G. Eror: Adv. Mat. and Proc., vol. 132, 1987, pp. 36-39.

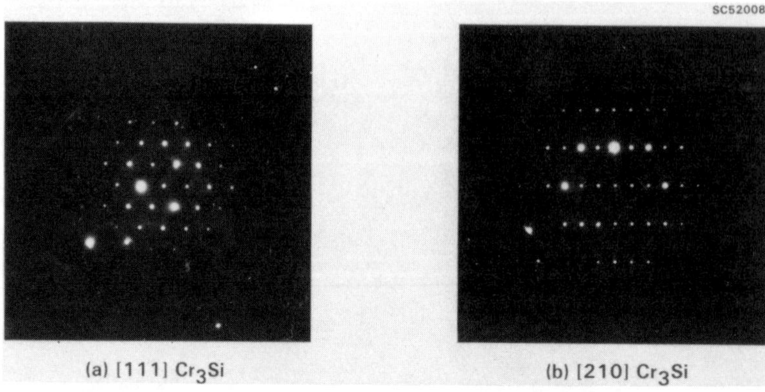

Fig. 13 SAD patterns of matrix gain shown in Fig. 12. (a) [111] Cr_3Si; (b) [210] Cr_3Si.

PRODUCTION OF "TITANIUM CARBIDE-STEEL" ALLOY BY MECHANICAL ALLOYING TECHNIQUE

O. V. Roman, A. A. Kolesnikov, V. M. Shelekhina
Byelorussian Powder Metallurgy Association
Minsk, 220600 USSR

ABSTRACT

The possibility of production of "titanium carbide-steel" alloy by mechanical alloying technique has been investigated in the present paper. The mixture of titanium and chromium-vanadium steel powders and carbon black was charged into the attritor. Then samples were compacted from the mechanically alloyed material and sintered in vacuum. The sintered samples were objected to hot second compaction under isostatic and quasi-isostatic loading.

Non-stoichiometric titanium carbide was detected during milling in the attritor. With milling time increasing carbon content in titanium carbide increases as well. Microstructural analysis of the compacted samples has shown that microstructure of the alloys produced of mechanically alloyed powders is more dispersed and homogeneous as compared to that produced of titanium carbide and steel powder mixtures. Additional hot deformation practically eliminates porosity, the size of titanium carbide particles remaining constant and less than 1 m.

Strength of the alloy, produced by means of mechanical alloying technique and hot second compaction, is much higher as compared to that of alloys, obtained from mechanically mixed powders of the same compositions.

NOWADAYS a considerable interest is expressed to sintered materials of "titanium carbide-steel" type. It is commonly known that these alloys advantageously combine the properties of both hard alloys and steels. With high wear resistance, displayed in the hardened state, they are easily machined in the annealed state. Besides, they are produced from available materials, such as powders of titanium carbide and different-grade steels. The alloys are mainly produced by dispersion of the initial components, i.e. titanium carbide and steel, compacting and liquid-phase sintering. When producing alloys from powder mixtures of carbides and metals, the degree of homogeneity in the resulted product is restricted by the size of the initial powder particles. Besides, production of alloy powders by mechanical mixing-milling technique requires preliminary obtaining of carbide and metallic constituents. Mechanical alloying is known to be a means to avoid disadvantages of mechanical mixing and milling of powders.

The paper deals with production possibilities of the mechanical alloying technique for obtaining "titanium carbide-steel" alloy. The alloy investigated was of the content 30%TiC +70% chromium vanadium steel (by weight). The mixture of titanium and chromium-vanadium steel powders and carbon black was charged into the attritor and treated in argon atmosphere under the following conditions: rotation speed of a mixer - 135 rot/min; weight ratio of the milling balls to the charge 25:1; treating period 10-40 hours. From the powders obtained samples of 10 mm in height and diameter were compacted and sintered in vacuum at residual pressure 10^{-2}Pa within temperature range of 800-1400°C during 1-2 hours.

X-ray diffraction, chemical and microstructural analyses of the powders and sintered alloys were carried out. X-ray diffraction analysis was conducted by Dron-20, Dron-3.0 diffractometers in K_α-monochromatic irradiation. Stereological characteristics of the samples were determined by the photographs, taken with Nanolab scanning microscope.

The analysis of the powders' treating in the attritor has shown the following. During treating both plastic deformation of the particles and their milling are observed. The first stage reveals rather intensive plastic deformation, then the particles become hardened and finer; the milling rate on last stage is slower, the particles become of quasispherical shape. The diffraction pattern of the powder after 10-hour treatment in the attritor displays lines of iron and titanium α -phase. The absence of other components lines is because of their low content in the mixture. In 20 hours diffraction pattern reveals the lines corresponding to TiC. With the time increasing up to 30 hours one can see more intensive lines of TiC, after 40 hours Ti lines disappear and only α -Fe and TiC lines are observed. After sintering at 800°C diffraction patterns of the powders show the lines corresponding to TiC and Fe_3C; after sintering at 1100°C only α -Fe and TiC lines are visible. Chemical analysis also revealed the presence of combined carbon during treatment of the tested powders in the attritor and the growth of its content during sintering.

With sintering temperature increasing from 800 up to 1100°C a noticeable growth of TiC unit cell dimension (lattice parameter) is observed; above 1100° C it changes slightly. For example, TiC lattice parameter for the powder treated in the attritor during 10 hours and sintered at 800°C constitutes 0.4309 HM, at 1100°C - 0.4318 HM and at 1400°C - 0.4321 HM. With the treating period increasing from 10 up to 40 hours one can observe the decrease of TiC lattice parameter in the sintered samples. It is related to oxidation of powders in the attritor and formation of titanium oxicarbides. Chemical analysis revealed significant decarburization of the sintered samples produced from the powders treated in the attritor during 30 hours.

Metallographic investigation of the alloy, obtained from the powder after 10-hour treatment in the attritor, has shown that its structure greatly differs from that of the alloy obtained from powder mixture of metallic and carbide components. The size of carbide grain in the first case was less than 1 μm, the grains being uniformly distributed in the binder, practically without conglomerates; the size of carbide grains in the latter case composed 2-2,5 μm, the structure included conglomerates of carbide grains of large sizes. The increase of the treatment time in the attritor above 10 hours is not reasonable,

for it leads to formation of large conglomerates of carbide particles (up to 200-300 μm) during sintering. The porosity of the samples obtained from the powders treated in the attritor is 2-5%; it is higher than that of the samples from dispersed mixture of steel and carbide powders (where the porosity constitutes 1%). It can be explained by both higher oxidation degree in powders after treatment in the attritor, and higher stoichiometric degree, and, thus, lower activity during sintering. When studying the relationship between hardness/bending strength in the samples and porosity (Fig.1), one can see that with porosity increasing within 0,2-5% range the decrease of hardness by 5-6 HRA and of strength by 40-45% take place.

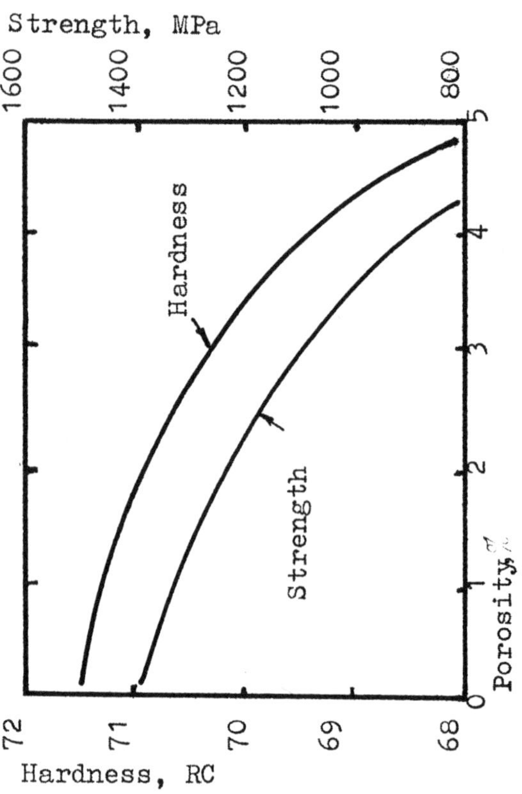

Fig.1. - Hardness and bending strength vs.porosity

The most sharp drop of properties is observed with low porosity (0-1%), that is explained by large effect of pores concentrating stresses, and by exponential relationship between the mentioned parameters and porosity. The growth of porosity within the above-mentioned range decreases failure viscosity parameter K_{1C} by 10-12% and wear resistance - by

1,5-2 times. To eliminate the residual
porosity the technique of hot second
compaction was tried out. The sintered
samples were subjected to three-axial
non-uniform compression, when loading
was transported to the sample through
a powder sheath, the powder being aluminium oxide. Another technique was also tried out, i.e. the technique of hot
second compaction at omnidirectional
uniform compression in liquid medium,
glass being selected as the latter.

The technique consists in the following: a sample in the sheath is subjected to hot deformation in the die
with a gap between the sample and the
die's walls till filling the gap with
the material. Liquid or powder sheath
allowed carrying out hot second compaction under iso- or quasi-isostatic conditions without external friction between the compact die walls and the punch
keeping required temperature when tran-
sfering the compact from the furnace to
the press and during hot compaction.
The powder sheath can deform both in
axial and radial directions. A compact
placed in such a sheath, also can be
freely deformed in the same directions.
During hot compaction of a compact in
the powder sheath shear stresses occur,
intensifying compression. They lead to
shear deformation, which depends upon
the sheath's material, its porosity,
thickness, compacting pressure and heating temperature. The described above
processes are not observed at compacting and deformation in dies and in a
liquid sheath.

Fig.2 illustrates a curve of compact cooling in alumina sheath during
hot compaction at 700 MPa pressure. As
seen from the graph, the temperature of
the compacts remains almost constant
during 5-10 sec, within the limits of
experimental accuracy. After unloading
the compacts are cooled at a high rate.

During hot compacting in a powder
sheath radial stresses approach axial
ones by values and reach the value of
0,6 of the axial. Fig.3 illustrates
porosity-to-pressure/temperature ratio
during hot second compaction . It was
established, that the increase of pressure up to 1000 MPa and of temperature
up to 1300°C during hot compaction
leads to higher density of the compacts.

Almost complete removal of residual
porosity from the compacts recompacted
is achieved at the temperatures above
1300°C, i.e. close to optimal sintering
temperature of the alloy tested (1300-
1340°C), when a major part of the binder is in the liquid state.In consequence of the above, second compaction
is mainly executed by the flow of the

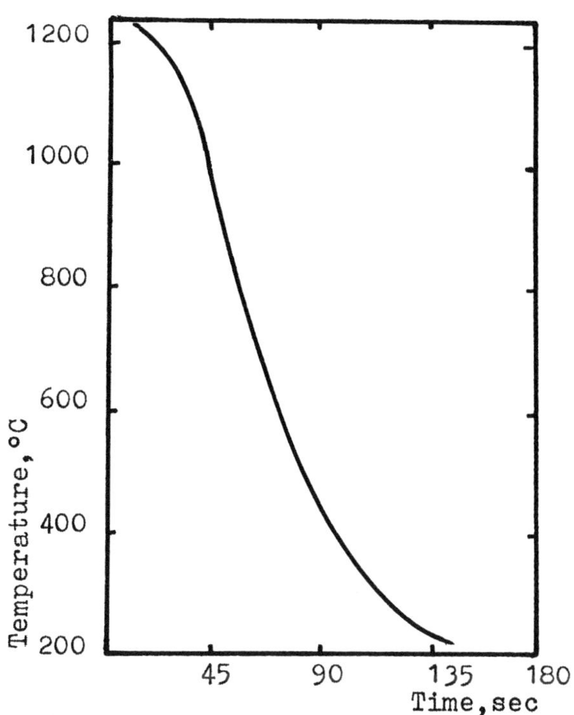

Fig.2. - Temperature-time dependence of
the compact in a powder sheath
during deformation and subsequent cooling.

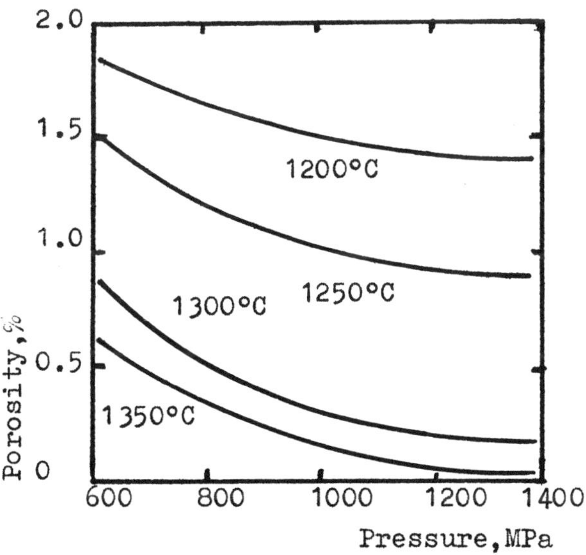

Fig.3. - Porosity of compacts vs.pres-
sure and temperature of hot
second compaction.

liquid phase and filling the pores.

A distinguishing feature of hot quasi-isostatic compaction is non-uniformity of the stressed state, which causes plastic changing of the porous body shape and dimensions. With pressure and temperature elevating axial deformation increases as well, compacts' height after second compaction under optimal conditions becomes 35-45% lower.

Elimination of the residual porosity during hot second compaction at three-axial equillibrium compression in a liquid sheath is also achieved at temperatures close to sintering, and only liquid phase accounts for the consolidation. Reducing of porosity obviously results in changing of dimensions.

Hot recompaction of the tested material both in a powder and liquid sheates leads to changing of stereological properties, destruction of certain carbide conglomerates, growth of specific area of interphase surface by 5-10%. Refining of carbide grains is not observed. The analysis of fine crystalline structure of carbide grains evidences that compaction under three-axial non-equilibrium compression results in noticeable growth of microdefects in lattices, breaking of zones with coherent dissipation, plastic deformation of the carbide phase, and deformation strengthening. Compaction with three-axial equilibrium compression is not accompanied with such essential changing in fine structure (Table 1).

Table 1 – The influence of hot compaction pressure (P) upon distorsion of lattice ($\frac{\Delta a}{a}$)

and coherent dissipation areas (D) under equilibrium and non-equilibrium compression (1,2) at t = 1300°C.

Nbs	P MPa	0	600	1000	1200
1	$\frac{\Delta a}{a}$ x10^3	2,2	3,1	3,8	4,0
2	D HM	200	160	85	80
3	$\frac{\Delta a}{a}$ x10^3	2,4	2,4	3,0	3,0
4	D HM	190	160	140	140

Hardness of alloys produced by mechanical alloying technique and subsequent recompaction is close to that of alloys obtained from mixture of steel and titanium carbide powders ($\sim$73 HRC). Bending strength of the samples, obtained under optimal conditions of recompaction with non-equilibrium three-axial compression, composes 1860 MPa, and after recompaction with equilibrium three-axial compression – 1720 MPa. Strength of the samples, obtained from powder mixture and sintering, is 1480 MPa.

CONCLUSIONS

During treating of titanium and steel powders and carbon black in the attritor plastic deformation and milling of the initial particles are observed. In 20 hours of treatment titanium carbide of non-stoichiometric content appears in the powder mixture, after 40 hours titanium fully disappears from the powder mixture. With sintering temperature increasing up to 1100°C one can observe a large increase of TiC lattice parameter. With the increase of treatment time from 10 to 40 hours TiC lattice parameter in the sintered samples decreases, due to the growth of oxidation and decarburization of the powders. The alloys obtained by mechanical alloying of the initial powders under optimal conditions are characterized by finer carbide grains (less than 1 μm) and higher porosity as compared to the alloys obtained from titanium carbide and steel powders mixture. Hot second compaction of samples under three-axial equillibrium (in a liquid sheath) and non-equillibrium (in powder sheath) compaction at the temperatures close to sintering temperatures of the tested alloy allows avoiding residual porosity and improving the alloy strength.

A MATHEMATICAL MODEL FOR THE PROCESS
OF MECHANICAL ALLOYING

B. N. Babich, V. A. Djatlenko
VIAM
Moscow, USSR

Abstract

The structural model of the process was reciewed on the basis of the system analysis. For the mathematical description of the process a system of equations, that accounts for the energy balance and inner dynamics of the apparatus, the energy and frequency of the interactions of particles of the treated materials, kinetics of the deformation and parameters of the diffusion was proposed. Numerical values of the coefficients of the equations can be determined by experiments.

THE PRESENT paper is devoted to developing the mathematical model of mechanical alloging process. The methodological basis of the paper is the principle of the system analysis, developed by V. Kafarov [1,2]. The process is considered as the generalised motion of the material system, obeying the lows of physical and chemical mechanics. Three characteristic scales or levels of the process can be singled out: I – the level of the apparatus or the system as a whole, II – the level of the apparatus local volume, III – the level of the material being treated.

The model is presented as the system of phenomenological equations for each level, linked by common parameters. Attritor mill treatment is considered as the concrete example.

I. APPARATUS LEVEL

Let's consider the attritor with periodic stock mixture charging. As this apparatus is exchanging with external systems (supersystems), mainly, as far as mechanical and thermal energy is concerned, in order to describe its behaviour towards external systems it is sufficient to have dependences of energy consumption and average temperature (or thermal flow) on regime–structural parameters.

Let's imagine the attritor with the working chamber radius R, involving the impeller with the shaft radius r, where two blades are mounted, having the thickness h, the width b and the length R, immersed into the stock mixture, consisting of the balls with the diameter d_1 and the powder with the average particle size d_2, to the depth z. The stock mixture density is assumed to be equal to ρ, angle velocity of the mixer rotation is ω. Let's define the power N, dispersed by this apparatus.

The power N, consumed while mixing the stock mixture of ρ density with the angle velocity ω in the apparatus, having radius R, can be expressed in the first approximation as

$$N = K_N \, \rho \, \omega^3 R^5 \qquad (1)$$

where K_N – power criterion.

Let's specify Eq. (1), taking into account the given structural characteristics of the apparatus. For this reason, let's consider the movement of the blade portion with the length dr_i, placed at the distance r_i from the impeller axis through the stock mixture. In the process of the blade movement the braking wedge is formed near its front edge, having the angle φ_1 at the root, which is equal to the angle of the internal friction in the stock mixture. The stock mixture elements, placed in the vicinity of the blade are exposed to the pressure P, caused by hydrostatic compression

$$P_1 = \rho\, q\, z \qquad (2)$$

where g – free-fall acceleration, centrifugal compression

$$P_2 = \tfrac{1}{2}\rho\,\omega^2 r_i^2 \qquad (3)$$

and inertial compression

$$P_3 = \tfrac{1}{2}\rho\,\omega^2 r_i^2\cos\varphi_1 \qquad (4)$$

defined by vertical acceleration of the stock mixture elements on bracking wedge running-on.

It is evident, that

$$P = P_1 + P_2 + P_3 = $$
$$= \rho\,[\,gz + \tfrac{1}{2}\omega^2 r_i^2\,(1 + \cos\varphi_1)\,] \qquad (5)$$

On the mixer rotation by the angle $\Delta\varphi$ the blade portion will move to the distance $dL = 2r_i d\varphi$, thus it will do the job dA at the expense of the stock mixture elements movement to the distance $1/2\,h$ along the sites with the cross-section $z\,dr_i$ and at the expense of the stock mixture friction against the blade along the cite with the cross-section $b\,dr_i$ with the friction angle φ_2:

$$dA = Pr_i\left(2b\,tg\,\varphi_2 + \frac{h}{2d_1}z\,tg\,\varphi_1\right)dr_i\,d\varphi \quad (6)$$

Integrating Eq. (6) by r_i, taking into account, that $\dfrac{d\varphi}{dt} = \omega$ and $\dfrac{dA}{dt} = N$, we obtain

$$N = A\omega\,(B + C\omega^2) \qquad (7)$$

where $A = \tfrac{1}{2}(R^2 - r^2)\left(\dfrac{h}{2d_1}z\,tg\,\varphi_1 + 2b\,tg\,\varphi_2\right)$

$B = \rho\,g\,z$

$C = \tfrac{1}{8}\rho\,(R^3 - r^3)\,(1 + \cos\varphi_1)$

Comparing Eq. (1) and Eq. (7) let's note, that the power criterion K_N can be represented by the coefficient only in the first approximation as it depends both on ω and R.

Knowing the power, dispersed by the process, one can determine its average temperature. Taking into account almost complete dissipativity of the process, the heat flow from the apparatus Q is equal to N. We use the known solution of thermal conductivity equation for the pipe of the outer radius R, inner radius r, length z, made of the material with the thermal conductivity λ:

$$Q = \frac{2\pi\lambda z}{\ln R/r}(T_2 - T_1) \qquad (8)$$

where T_2 and T_1 – temperatures on inner and outer walls.

Assuming the consumption Q_1, heat capacity c_1 and the cooling agent input temperature T_1 of the attritor cooling system, let's write

$$T = T_1 + N\left(\frac{\ln R/r}{2\pi z\lambda} + \frac{1}{Q_1 c_1}\right) \qquad (9)$$

Eq. (7) and Eq. (9) represent mathematical model of the first process level for the attritor. To identify the model it is sufficient to know the dimensions of the apparatus elements and to define experimentally the stock mixture friction angles and its thermal conductivity.

II. THE LEVEL OF THE APPARATUS LOCAL VOLUME

At the level of the apparatus local volume one should consider mutual movement and interaction of the stock mixture elements. For their averaged description we introduce the notion of frequency-energetic action spectrum (further on – spectrum) of the apparatus. Let's distinguish the non-working strok spectrum, showing the distribution of the mill bodies collision energies E_1 according to frequencies f_1, and the working one, related to the material being treated. Working spectrum parameters include: E_2 – the energy transmitted to the material in the single interaction (kJ/mol) and f_2 – the frequency of the interactions repetitions (Hz).

Let's determine f_1 freqency average value for the attritor as

$$f_1 = K_1\omega \qquad (10)$$

proportionaly to the impeller angle velocity, then the average energy of the mill bodies interactions will be equal to

$$E_1 = N / f_1 n_1 \qquad (11)$$

where n_1 – the number of the mill bodies.

Let's assume that on penetration of treated material particles into the zone of the mill bodies contact, the energy to the material is proportional to E_1, i.e.

$$E_2^{\text{K}} = \eta_1 E_1 \qquad (12)$$

Let's determine the average amount of the material, being involved into contact zone during treatment. Let's assume, that the contact zone boundary is the contact line of the particle with d_2 diameter of to adjacent spherical mill bodies with d_1 diameter. Then, due to geometric considerations, the contact zone radius is $r = \sqrt{1 / d_1 d_2}$ and its area $s = \frac{\pi}{2} d_1 d_z$. Let's consider, that the material being treated and mill bodies are uniformally distributed in the attritor chamber volume and that only the powder particles with cone angle at the base, equal to internal friction angle in the powder φ_3, i.e. with the cone having the base S and the height $h = r\, tg\, \varphi_3$, can be kept within the contact zone up to binding.

Calculating this cone volume and multiplying it by the average volume density of the powder in the chamber ρ_2, we cane determine the average mass m_2 of the material involved in single interaction:

$$m_2 = \frac{\pi \rho_2}{24} (d_1 d_2)^{3/2}\, tg\, \varphi_3 \qquad (13)$$

and the average frequency of the interactions repetitions

$$f_2 = f_1 n_1 \frac{m_2}{M_2} \qquad (14)$$

where M_2 – the mass of the whole stock mixture in the apparatus.

By expressing the number of the mill bodies n_1 through their common mass M_1 and the dimensions d_1, and also by transition to specific energy of intraction $E_2 = E_2^{\text{K}} / m_2$, we obtain:

$$E_2 = \frac{4\,\eta_1 N}{K_1 \omega\, M_1\, tg\, \varphi_3}\, \frac{\rho_1}{\rho_2} \left[\frac{d_1}{d_2}\right]^{3/2}$$

$$f_2 = K_1 \omega \frac{M_1}{M_2} \left[\frac{d_2}{d_1}\right]^{3/2} \frac{\rho_2}{\rho_1}\, tg\, \varphi_3 \qquad (15)$$

Let's also estimate the duration of the single effect on the material τ_1 and relative duration of the material exposure in the action zone.

Considering the balls beings absolutely rigid and assuming the material flow stress being equal to compression strength σ_1, we obtain the value of the material resistance on interaction

$$F = \sigma_1 S \qquad (16)$$

and taking into account the impulse retention

$$\tau_1 = \frac{2\sqrt{E_1 m_1}}{F} \quad , \quad \xi = \tau_1 f_2 \qquad (17)$$

or, transferring to the balls dimensions and density

$$\xi = \frac{2}{3} f_2 \frac{\rho_1 d_1^2}{\sigma_1 d_2} \sqrt{\frac{N}{f_1 M_1}} \qquad (18)$$

Eq. (15) and (18) define the parameters of frequency-energetic action spectrum on the material, i.e. they represent mathematical model of the system second level. To identify the model, it is necessary to define experimentally the values K_1, φ_3 and η_1, easily determined in model experiments.

In the first approximation one can assume $K_1 = 1$, i.e. each ball experiences one interaction with adjacent balls at every rotation of the impeller. Direct measurement of E_1 and f_1 can be carried out by placing the sample ball, made of deforming material into operating apparatus; the dimensions of indents this sample ball, resulted from mutual collision with the neiboughring balls, characterize the interactions energy E_1, and their number characterizes the frequency f_1.

In a similar way one can measure directly also the working spectrum parameters E_2 and f_2; for this reason it is necessary to use spherical monofractions of easily deforming material powders, also by placing them for a short period of time into the operating apparatus.

The deformation degree of individual particles after treatment characterizes in this case the value E_2, and their fraction in the whole amount of the powder characterizes the value f_2.

We postulate the fact, that the invariant development of mechanical alloying process, relative to the apparatus spectrum parameters, i.e. changes occurring in the material, are simply defined by the spectrum of action on this material. Thus, the description of the system third level, using the spectrum parameters doesn't depend on the apparatus used, i.e. it is of universal character.

III. THE LEVEL OF THE MATERIAL BEING TREATED

Let's write the kinetic equation of changing average values for d_2 particles dimensions in accordance with the reference [3]:

$$d_2(t) = d_2(0)\, exp\left\{ \int_0^t \kappa(t)\,[1 - K_\psi(t)]\, dt \right\} \quad (19)$$

where $d_2(0)$ – the initial dimension of the particles,

$\quad K_\psi(t)$ – equilibrium function between refining and agglomeration, introduced as $K_\psi(t) = \dfrac{\psi(t)}{\kappa(t)}$,

$\quad \kappa(t)$ – intensity of milling-fracture,

$\quad \psi(t)$ – agglomeration intensity.

Let's consider milling intensity to be proportional to the energy and frequency of interactions and inversly proportional to the energy, required for new surface formation. By expressing the surface energy through ultimate tensile strength σ_1, we obtain

$$\kappa = \eta_2\, \frac{N\rho_2}{M_2 \sigma_1} \quad (20)$$

Eq. (20) includes only the process integral power – this is true for the case, if the interaction energy E_2 is higher compared to fracture work A. Assuming the low of E_2 distribution by frequencies and natural values to gaussian and expressing the fracture work in terms of ultimate tensile strength and atomic volume Ω we write finally

$$\kappa = \eta_2 \frac{N\rho_2}{M_2 \sigma_1}\, exp\left(-\frac{\sigma_1 \Omega}{E_2}\right) \quad (21)$$

Substituting flow stress σ_2 for tensile strength in the Eq. (21) we obtain the law of structural elements deformation in the process of mechanical alloying:

$$l(t) = l_0\, exp\left(\int_0^t \kappa(t)\, dt \right) \quad (22)$$

where

$$\kappa = \eta_3\, \frac{N\rho_2}{M_2 \sigma_2}\, exp\left(-\frac{\sigma_2 \Omega}{E_2}\right),$$

l – structural element characteristic dimension (interlayer thickness in the composite particle, formed in the process of mechanical alloying).

Equilibrium function $K_\psi(t)$ between refining and agglomeration is introduced similar to equilibrium constant of direct and inverse reactions

$$K = K_0\, exp\left(-\frac{\Delta t}{RT}\right) \quad (23)$$

where activation enthalpy ΔH is replaced by fracture work A, and the temperature T is replaced by quasitemperature

$$T^* = T + \frac{E_2}{R} \quad (24)$$

then we obtain

$$K_\psi = K_0\, exp\left(-\frac{\sigma_1 \Omega}{E_2 + RT}\right) \quad (25)$$

The same approach we extend to the description of diffusion and chemical transformations in the material, assuming transformations velocities to be functions of quasitemperature T^*, included respectively into the expressions for diffusion coefficients

$$D^* = A^*\, exp\left(-\frac{Q}{RT^*}\right) \quad (26)$$

and equilibrium constants of chemical reactions

$$K^* = A_1^*\, exp\left(-\frac{\Delta H}{RT^*}\right) \quad (27)$$

As sandwich-type structures are formed as a result of mechanical alloying, one-dimensional equation is sufficient to describe diffusion transformations

$$\frac{\delta c}{\delta t} = \frac{\delta}{\delta x}\left(D^* \frac{\delta c}{\delta x}\right) \quad (28)$$

where «x» coordinate is counted in the direction of non-homogeneity l smallest dimensions, i.e. flat cellular model is considered, the solution for which relative to diffusion transformation $\alpha = C_0/C$ extent is the following function

$$\alpha = 1 - erf\left(\frac{l}{\sqrt{2D^* t}}\right) \quad (29)$$

The equation of n-order chemical reaction or, accompanying mechanical alloying process are written in a usual way

$$\frac{dc}{dt} = c^n K^* \tag{30}$$

with the following solution $\frac{c(0)}{c(t)} = exp\, K^*$.

On calculations, using Eq. (26)–(30) it should be taken into consideration, that frequency multipliers A^* and A_1^* in Arrenius equations for D^* and K^* differ considerably from multipliers, defined for transformations under normal conditions. This difference is due to pulsed-discrete character of mechanical alloying process, when the material for a certain period of time is exposed to intensive effects zone with quasitemperature T^*, and fore the rest period of time it is in the apparatus chamber with the temperature T. The criterion for the material exposure in the zone with T^* temperature is ξ parameter, introduced by the Eq. (18).

In order to use literature data in calculating D^* and K^* it is necessary to introduce the terms of Eq. (26) and (27), containing T^* parameter, taking into account ξ parameter, i.e.

$$D^* = A\{exp(-\frac{Q}{RT}) + \xi\, exp(-\frac{Q}{RT+E_2})\} \tag{31}$$

$$K = A_1\{exp(-\frac{\Delta H}{RT}) + \xi\, exp(-\frac{\Delta H}{RT+E_2})\} \tag{32}$$

or, in case if $E_2 \gg RT$ and $E_2 \cong Q$ or $E_2 \geq \Delta H$, then

$$D^* = A\xi\, exp(-\frac{Q}{E_2}) \tag{33}$$

$$K^* = A_1\xi\, exp(-\frac{\Delta H}{E_2}) \tag{34}$$

Eq. (19), (21), (28), (30) comprize mathematical model of the system third level.

Let's write the mathematical model of mechanical alloying process in the final form, for which reason we allocate to the variables the dependence on the process duration. For $d_1(t)$, $\varphi_1(t)$, $\varphi_2(t)$, $\varphi_3(t)$, $\rho(t)$, $\sigma(t)$ values such notation is formal up to establishing analytical dependences.

$$
\left\{
\begin{aligned}
N(t) &= A(t)\omega[B(t) - c(t)\omega^2]\\
A(t) &= \frac{1}{2}(R^2 - r^2)[\frac{h}{2d_1(t)}z\, tg\,\varphi_1(t) - 2b\, tg\,\varphi_2(t)]\\
B(t) &= \rho(t)gz\\
C(t) &= \frac{1}{6}\omega^2\rho(t)(R^3 - r^3)[1 + cos\,\varphi_1(t)]\\
T(t) &= T_1 + N(t)[\frac{ln\,R/r}{2\pi z\lambda} + \frac{1}{Q_1 c_1}]
\end{aligned}
\right.
$$

$$
\left\{
\begin{aligned}
E_2(t) &= \frac{4\eta_1 N(t)}{K_1\omega M_1 tg\,\varphi_3(t)}\frac{\rho_1}{\rho_2(t)}\left[\frac{d_1(t)}{d_2(t)}\right]^{3/2}\\
l_2(t) &= K_1\omega\frac{M_1}{M_2}\left[\frac{d_2(t)}{d_1(t)}\right]^{3/2}\frac{\rho_2(t)}{\rho_1}tg\,\varphi_3(t)\\
\xi(t) &= \frac{2}{3}l_2(t)\frac{\rho_1(t)d_1^2(t)}{\sigma_1(t)d_2(t)}\sqrt{\frac{N(t)}{K_1\omega M_1}}
\end{aligned}
\right.
$$

$$
\left\{
\begin{aligned}
d_2(t) &= d_2(0)\, exp\{\int_0^t \kappa(t)[1 - K_\psi(t)]\, dt\}\\
K_\psi(t) &= \frac{\psi(t)}{\kappa(t)} \tag{35}\\
\kappa(t) &= \eta_2\frac{N(t)\rho_2(t)}{M_2\sigma_1(t)}exp(-\frac{\sigma_1(t)\Omega}{E_2(t)})\\
K_\psi(t) &= K_{\psi 0}(t)\, exp(-\frac{\sigma_1(t)\Omega}{E_2(t)+RT(t)})\\
l(t) &= l_0\, exp[\int_0^t \kappa_1(t)\, dt]\\
\kappa_1(t) &= \eta_3\frac{N(t)\rho_2(t)}{M_2\sigma_2(t)}exp(-\frac{\sigma_2(t)\Omega}{E_2(t)})\\
\alpha(t) &= 1 - erf(\frac{l(t)}{\sqrt{2D^*(t)t}})\\
D^*(t) &= A\{exp(-\frac{Q}{RT(t)}) + \xi(t)\, exp(-\frac{Q}{RT(t)+E_2(t)})\}
\end{aligned}
\right.
$$

$$
\left\{
\begin{aligned}
&c(t) = c_0 \, exp\,[-K^*(t)] \\[2mm]
&K(t) = A_1 \{ exp\,(-\tfrac{\Delta H}{RT(t)}) + \\[2mm]
&\qquad\qquad + \xi(t)\, exp\,(-\tfrac{\Delta H}{RT(t)+E_2(t)}) \}
\end{aligned}
\right.
$$

where

$N(t)$ – power, consumed by the process;

ω – angle velocity of impeller rotation;

R, z – radius and leight of the attritor working chamber;

r, h, b – impeller shaft radius, the thickness and width of its blades;

$d_1(t), d_2(t)$ – current diameter of mill balls and materials particles;

$\varphi_1(t)$ – angle of friction in movable stock mixture;

$\varphi_2(t)$ – angle of friction for the stock mixture against the plane;

$\varphi_3(t)$ – angle of friction in the powder;

$\rho(t)$ – stock mixture density;

$T(t)$ – process temperature;

T_1 – cooling agent input temperature in the cooling system;

λ – stock mixture heat conductivity;

Q_1, c_1 – volume flow rate and heat capacity of the cooling agent;

$\rho_1(t), \rho_2(t)$ – bulk density of the balls and powder;

M_1, M_2 – masses of the balls and powder, charged into the attritor;

$E_2(t), f_2(t)$ – energy and frequency of the material exposure;

$\xi(t)$ – exposure time scale;

t – current time;

$\sigma_1(t), \sigma_2(t)$ – tensile strength and flow stress of the material being treated;

$l(t)$ – current dimension of inhomogeneity areas;

$\kappa(t), \psi(t)$ – intensity of milling and agglomeration;

$K_\psi(t)$ – equilibrium function between milling and agglomeration;

Ω – atomic volume of the material being treated;

α – diffusion transformation extent in the material;

A, A_1 – frequency multipliers;

D^* – diffusion coefficient on mechanical alloying;

K^* – reaction velocity constant on mechanical alloying;

$c(t)$ – reactant concentration;

ΔH – entalpy of reaction activation;

Q – activation energy;

R – universal gas constant.

Formally introduced dependences $d_1(t)$, $\varphi_1(t)$ reflect the process of scab formation on mill bodies; in this case the ball size is changed and friction angle is increased as a result of the balls roughness increase. The dependence $\varphi_3(t)$ reflects the change of the treated particles configuration, φ_3 value is decreased depending on cold work and spheridization degree of the particles. The dependences $\rho_2(t), \rho(t)$ reflect the powder densification in the process of treatment; $\sigma_1(t), \sigma_2(t)$ reflects its cold work.

Obtaining of the mentioned functions analytical dependences presents considerable difficulties, this task is not considered in the present paper. For steady-state stage of the process, the values $d_1, \varphi_1, \varphi_2, \varphi_3, \rho_2$ can be assumed as time independent constants, and functions $\sigma(t)$ can be given as aperiodic dependences with saturation

$$\sigma(t) = \sigma_{max}[1 - exp\,(-t/\tau)] \qquad (36)$$

where σ_{max} – saturation stress (hardness),

τ – process saturation time.

Conclusions

Based on the principles of system analysis mathematical model of mechanical alloying process is developed. The model includes the system of equations, taking into account the apparatus energy consumption, average energy and frequency of single effects on the material being treated, fracture kinetics, deformation and agglomeration of particles, the peculiarities of diffusion and chemical transformations.

Numerical values of coefficients in the equations can be experimentally determined.

References

1. Kafarov V.V., Dorochov I.N. System Analysis of Chemical Technology Process. Basic Strategy. Moscow, Nauka (1976) p.500
2. Kafarov V.V., Dorochov I.N., Arutunov S.U. System Analysys of Chemical Technology Processes. The Process of Milling and Mixing for Bulk Materials. Moscow, Nauka (1985) p. 440
3. Arutunov S.U. Modelling and Optimisation of Milling Process for Grain Materials. Abstract of candidate thesis. Moscow, Mendeleev Chemical Technological Institute (1982) p. 23